Beginning Algebra
Second Edition

Student Solutions Manual
and
Study Guide

Dennis Weltman
North Harris County College

Gilbert Perez
San Antonio College

Richard C. Spangler
Tacoma Community College

BROOKS/COLE

™

THOMSON LEARNING

Brooks.Cole
511 Forest Lodge Road
Pacific Grove, CA 93950
USA

For information about our products, contact us:
Thomson Learning Academic Resource Center
1-800-423-0563
http://www.brookscole.com

International Headquarters
Thomson Learning
International Division
290 Harbor Drive, 2ⁿᵈ Floor
Stamford, CT 06902-7477
USA

UK/Europe/Middle East/South Africa
Thomson Learning
Berkshire House
168-173 High Holborn
London WCIV 7AA

Asia
Thomson Learning
60 Albert Street, #15-01
Albert Complex
Singapore 189969

Canada
Nelson Thomson Learning
1120 Birchmount Road
Toronto, Ontario MIK 5G4
Canada
United Kingdom

ISBN 0-534-12445-3

The Adaptable Courseware Program consists of products and additions to existing Brooks/Cole products that are produced from camera-ready copy. Peer review, class testing, and accuracy are primarily the responsibility of the author(s).

PREFACE

This Solutions Manual and Study Guide accompanies *Beginning Algebra, Second Edition*, by Dennis Weltman and Gilbert Perez. The booklet contains worked-out solutions to odd-numbered exercises and a Study Guide for all sections in each chapter.

To use this manual:

1. For each section in the main text, read the explanations and work through the examples using pencil and paper.

2. Next, work as many exercises as assigned by your instructor.

3. Compare your answers with those at the end of your text, and rework the ones that were not correct.

4. If further help is needed, check your work against the detailed solutions provided in this manual.

5. The worked-out solutions also may be used as examples to help find solutions to other problems.

6. For additional practice try the Study Guide self-tests at the end of each chapter. Self-test solutions are also included.

CONTENTS

CHAPTER 1 THE REAL NUMBER SYSTEM AND ITS PROPERTIES

Solutions to Text Exercises

Exercises 1.1

1.
This is an intersection of sets A and B. Thus, $A \cap B = \{8\}$, since the element 8 appears in both set A and set B.

3.
This is an intersection of sets B and T. Thus, $B \cap T = \{6,9\}$, since the elements 6 and 9 appear in both set B and set T.

5.
$A \cap S = \emptyset$, since set A and set S have no element in common.

7.
$S \cap N = \{1,3,5,7, . . .\}$, since all elements in Set N appear as elements in Set S.

9.
This is a union of set A and B. Thus, $A \cup B = \{4,6,7,8,9,10,12,16,20\}$, since each element appears in either set A or set B.

11.
This is a union of set B and T. Thus, $B \cup T = \{7,8,10,0,3,6,9,12,15,...\}$, since each element appears in either set B or set T.

13.
This is a union of sets S and N. Thus, $S \cup N = \{...,-5,-3,-1,1,2,3,4,...\}$, since each element of set S appears in either set S or set N.

15.
$A \cap \emptyset = \emptyset$, since there are no elements common to both set A and the empty set.

17.
$$A \cup (B \cap T) = A \cup \{6,9\}$$
$$= \{4,6,8,9,12,16,20\}$$

First find $B \cap T$, then find the union of set A and $\{6,9\}$.

19.
$$B \cap (T \cup S) = B \cap \{...,-5,-3,-1,0,1,3,5,6,7,9,$$
$$11,12,13,...\}$$
$$= \{6,7,9\}$$

First find $T \cup S$, then find the intersection of set B and $\{...-5,-3,1,0,1,3,5,6,7,9,11,12,13,...\}$.

21.

$$\frac{21}{10} = 2.1$$

Work

```
       2.1
10)21.0
    20
    10
    10
     0
```

23.

$$\frac{4}{9} = 0.44...$$

Work

```
      0.44...
9)4.000
  3 6
   4 0
   3 6
    4 0
```

25.

$$\frac{2}{11} = 0.18...$$

Work

```
       0.1818...
11)2.00000
   1 1
    9 0
    8 8
     2 0
     1 1
      9 0
      8 8
       2 0
```

27.

$$\frac{15}{11} = 1.\overline{36}$$

Work

```
      1.3636...
  11)15.00000
     11
      4 0
      3 3
        7 0
        6 6
          4 0
          3 3
            7 0
            6 6
              4 0
```

29.

$$\frac{3}{4} = 0.75$$

Work

```
     0.75
  4)3.00
    2 8
      2 0
      2 0
        0
```

31.

$$\frac{11}{8} = 1.375$$

Work

```
      1.375
  8)11.000
     8
     3 0
     2 4
       6 0
       5 6
         4 0
         4 0
           0
```

33.

$$\frac{5}{16} = 0.3125$$

Work

```
       0.3125
  16)5.0000
      4 8
        2 0
        1 6
          4 0
          3 2
            8 0
            8 0
```

35.

$$\frac{3}{7} = 0.\overline{428571}$$

Work

```
       0.4285714...
  7)3.00000000
    2 8
      2 0
      1 4
        6 0
        5 6
          4 0
          3 5
            5 0
            4 9
              1 0
               7
               3 0
               2 8
                 2 0
```

37

$6 \in N, 6 \in W, 6 \in I, 6 \in Q$, and $6 \in R$.

39.

$\dfrac{5}{-11} \in Q$, and $\dfrac{5}{-11} \in R$.

41.

$\sqrt{3} \in H$, and $\sqrt{3} \in R$.

43.

$3.5 \in Q$, and $3.5 \in R$.

45.

$8.1535535553... \in H$ and $8.1535535553... \in R$.

47.

$\dfrac{5}{0} \notin R$

Division by 0 is not allowed.

49.

$8.0 \in I$; True.

51.

$\dfrac{6}{7} \in I$; False.

$\dfrac{6}{7} \in Q$; True.

53.
$N \cap H = \emptyset$; True.

55.
$\emptyset \subseteq R$; True.

The empty set is a subset of all sets.

57.
True

59.
False

Exercises 1.2

1. - 17.
The answers do not require worked out solutions.

19.
$|6| = 6$

21.
$|\pi| = \pi$

23.
$3 \cdot |-7| = 3(7) = 21$

25.
$|-8| + 3 = 8 + 3 = 11$

27.
$-4 > -11$
-11 is to the left of -4 on the number line

29.
$\frac{5}{2} < 3$

Note: $\frac{5}{2} = 2\frac{1}{2}$

31.
$-8 < 2$

33.
$0 > -5$

35.
$2 > \sqrt{2}$
Note: $\sqrt{2} \approx 1.414$

37.
$\frac{2}{3} > 0.6$

Note: $0.6 = \frac{3}{5}$

39.
$|-15| > |-8|$, since $15 > 8$.

41.
$|-20| > 6$, since $20 > 6$.

43.
$|-24| = |24|$, since $24 = 24$

45.
$|-12| + 5 = |1| + |16|$, since
$$12 + 5 = 1 + 16$$
$$17 = 17$$

47.
$x \geq 2$

49.
$|x| \geq 0$ or
$|x| < 0$

Exercises 1.3

1. - 23.
The answers do not require worked-out solutions.

25.
$$7 + 2\sqrt{16} \quad = 7 + 2(4)$$
$$= 7 + 8$$
$$= 15$$

27.
$42 + \sqrt{36} + 13 = 42 + 6 + 13 = 20$

29.
$$5 \cdot 2^2 - 7 \cdot 2 + 3 \quad = 5 \cdot 4 - 7 \cdot 2 + 3 \qquad \text{Powers}$$
$$= 20 - 14 + 3 \qquad \text{Rule B-1}$$
$$= 9 \qquad \text{Rule B-2}$$

31.

$2 \cdot 3^3 + 7 \cdot 3^2 - 5 \cdot 3$

$= 2 \cdot 27 + 7 \cdot 9 - 5 \cdot 3$ Powers

$= 54 + 63 - 15$ Rule B-1

$= 102$ Rule B-2

33.

$15 - 2 \cdot 4 = 15 - 8 = 7$

35.

$28 - 4 + 9 - 3 = 37 - 7 = 30$

37.

$54 + 9 + 3 = (54 + 9) + 3 = 6 + 3 = 2$

39.

$36 + (6 - 3)^2 = 36 + (3)^2 = 36 + 9 = 4$

41.

$(20 - 3) - (8 - 1) = 17 - 7 = 10$

43.

$8 - 2|\,5 - 3\,| = 8 - 2|\,2\,| = 8 - 2 \cdot 2 = 8 - 4 = 4$

45.

$5\sqrt{15 + 7^2} = 5\sqrt{15 + 49}$ Powers

$\qquad\qquad = 5\sqrt{64}$ Add

$\qquad\qquad = 5 \cdot 8 = 40$

47.

$6 \cdot 6 - 6 + 6 = 36 - 1 = 35$

49.

$6(6 - 6 + 6) = 6(6 - 1) = 6(5) = 30$

51.

$3 + 7[9 - (8 - 2)] = 3 + 7[9 - 6] = 3 + 7[3]$

$\qquad\qquad\qquad\qquad\qquad = 3 + 21 = 24$

53.

$\dfrac{24 - 8 \cdot 2}{3^2 - 5} = \dfrac{24 - 16}{9 - 5} = \dfrac{8}{4} = 2$

55.

$\dfrac{15 - (4 + 1)}{9 - 3 - 1} = \dfrac{15 - (5)}{9 - 3 - 1} = \dfrac{10}{5} = 2$

57.

$\dfrac{13 + \sqrt{13^2 - 4 \cdot 3 \cdot 12}}{2 \cdot 3} = \dfrac{13 + \sqrt{169 - 144}}{6}$

$\qquad\qquad = \dfrac{13 + \sqrt{25}}{6} = \dfrac{13 + 5}{6} = \dfrac{18}{6} = 3$

59.

$\dfrac{11 - \sqrt{11^2 - 4 \cdot 2 \cdot 14}}{2 \cdot 2} = \dfrac{11 - \sqrt{121 - 112}}{4}$

$\qquad\qquad = \dfrac{11 - \sqrt{9}}{4} = \dfrac{11 - 3}{4} = \dfrac{8}{4} = 2$

61.

$4\wedge 3 - 8 * 4 = 4^3 - 8 \cdot 4 = 64 - 32 = 32$

63.

$\dfrac{48}{2} + \dfrac{6}{3} = 24 + 2 = 26$

65.

$(7\wedge 2 - 9)/(7 - 3) = \dfrac{7^2 - 9}{7 - 3} = \dfrac{49 - 9}{4} = \dfrac{40}{4} = 10$

Exercises 1.4A

1.

$-32 + 29 = -3$ Rule 2

3.

$142 + (-69) = 73$ Rule 2

5.

$-43 + 70 = 27$ Rule 2

7.

$18 + (-18) = 0$ Rule 2

9.

$-14 + (-75) = -89$ Rule 1

11.

$-11 + (-8) + (-14)$

$= -(11 + 8 + 14) = -33$ Rule 1

13. $-16 + (-9) + 35 \quad = -25 + 35 \quad$ Rule 1
$\qquad\qquad\qquad\qquad = 10 \qquad\quad$ Rule 2

15. $19 + 32 + (-51) = 51 + (-51) \quad$ Rule 1
$\qquad\qquad\qquad\qquad = 0 \qquad\qquad$ Rule 2

17.
$-51 + 29 + (-38)$
$= -51 + (-38) + 29 \qquad$ Commutative property
$= -89 + 29 \qquad\qquad$ Rule 1
$= -60 \qquad\qquad\qquad$ Rule 2

19.
$26 + (-8) + 11 + (-24)$
$= [26 + (-8)] + [11 + (-24)]$
$= 18 + (-13) = 5 \qquad\qquad\qquad$ Rule 2

21.
$(-33) + 18 + (-9) + 62$
$= [-33 + 18] + [(-9) + 62]$
$= -15 + 53 = 38 \qquad\qquad\qquad$ Rule 2

23.
$16 + (-62) + (-17) + 33$
$= [16 + (-62)] + [-17 + 33]$
$= -46 + 16 = -30 \qquad\qquad\qquad$ Rule 2

25.
$34 + (-51) = -17 \qquad\qquad\qquad$ Rule 2

27.
$-13 + 23 + (-9) = [-13 + 23] + (-9)$
$= 10 + (-9) = 1 \qquad\qquad\qquad$ Rule 2

29.
$(-864.00) + (-59.17) + (-250.68) + 300$
$\qquad\qquad\qquad + (-15.24) = -889.09$
Suzie now owes \$889.09

Exercises 1.4B

31.
$\dfrac{10}{12} = \dfrac{\cancel{2}\cdot 5}{\cancel{2}\cdot 2\cdot 3} = \dfrac{5}{6}$

33.
$\dfrac{24}{30} = \dfrac{\cancel{2}\cdot 2\cdot 2\cdot \cancel{3}}{\cancel{2}\cdot \cancel{3}\cdot 5} = \dfrac{4}{5}$

35.
$\dfrac{15}{45} = \dfrac{\cancel{3}\cdot\cancel{5}}{3\cdot\cancel{3}\cdot\cancel{5}} = \dfrac{1}{3}$

37.
$\dfrac{75}{25} = \dfrac{3\cdot\cancel{5}\cdot\cancel{5}}{\cancel{5}\cdot\cancel{5}} = \dfrac{3}{1} = 3$

39.
$\dfrac{-70}{21} = \dfrac{-1\cdot 2\cdot 5\cdot\cancel{7}}{3\cdot\cancel{7}} = \dfrac{-10}{3}$

41.
$\dfrac{84}{63} = \dfrac{2\cdot 2\cdot 3\cdot\cancel{7}}{3\cdot 3\cdot\cancel{7}} = \dfrac{4}{3}$

43.
$\dfrac{3}{11} + \dfrac{5}{11} = \dfrac{5+3}{11} = \dfrac{8}{11}$

45.
$\dfrac{3}{20} + \dfrac{9}{20} + \dfrac{13}{20} = \dfrac{3+9+13}{20}$
$\qquad\qquad\qquad = \dfrac{25}{20} \qquad\qquad$ Rule 3
$\qquad\qquad\qquad = \dfrac{\cancel{5}\cdot 5}{\cancel{5}\cdot 4} = \dfrac{5}{4}$

47.
$-\dfrac{5}{4} + \dfrac{11}{4} = \dfrac{-5+11}{4} = \dfrac{6}{4} \text{ or } \dfrac{3}{2}$

49.
$-\dfrac{13}{45} + \left(-\dfrac{4}{45}\right)$
$= \dfrac{-13 + (-4)}{45} = -\dfrac{17}{45} \qquad$ Rules 1 and 2

51.
$\dfrac{5}{3} + \dfrac{7}{8} \qquad\qquad\qquad$ LCD = 24
$\dfrac{5}{3} + \dfrac{7}{8} = \dfrac{5}{3}\cdot\dfrac{8}{8} + \dfrac{7}{8}\cdot\dfrac{3}{3}$
$\qquad\qquad = \dfrac{40}{24} + \dfrac{21}{24}$
$\qquad\qquad = \dfrac{40+21}{24} = \dfrac{61}{24} \qquad$ Rule 3

53.

$$\frac{11}{6} + \frac{3}{4} + 2$$

Find the LCD.
$$6 = 2 \cdot 3$$
$$4 = 2 \cdot 2$$
$$LCD = 2 \cdot 2 \cdot 3 = 12$$

$$\frac{11}{6} + \frac{3}{4} + \frac{2}{1} = \frac{11}{6} \cdot \frac{2}{2} + \frac{3}{4} \cdot \frac{3}{3} + \frac{2}{1} \cdot \frac{12}{12}$$
$$= \frac{22}{12} + \frac{9}{12} + \frac{24}{12}$$
$$= \frac{22 + 9 + 24}{12} \qquad \text{Rule 3}$$
$$= \frac{55}{12}$$

55.

$$\frac{11}{12} + \frac{7}{20}$$

Find the LCD.
$$12 = 2 \cdot 2 \cdot 3$$
$$20 = 2 \cdot 2 \cdot 5$$
$$LCD = 2 \cdot 2 \cdot 3 \cdot 5 = 60$$
$$\frac{11}{12} + \frac{7}{20} = \frac{11}{12} \cdot \frac{5}{5} + \frac{7}{20} \cdot \frac{3}{3}$$
$$= \frac{55}{60} + \frac{21}{60}$$
$$= \frac{76}{60} \qquad \text{Rule 3}$$
$$= \frac{2 \cdot 2 \cdot 19}{2 \cdot 2 \cdot 3 \cdot 5} = \frac{19}{15} \qquad \text{Reduce}$$

57.

$$-\frac{7}{9} + \frac{1}{6} \qquad LCD = 18$$

$$-\frac{7}{9} + \frac{1}{6} = -\frac{7}{9} \cdot \frac{2}{2} + \frac{1}{6} \cdot \frac{3}{3}$$
$$= -\frac{14}{18} + \frac{3}{18}$$
$$= \frac{-14 + 3}{18} \qquad \text{Rule 3}$$
$$= \frac{-11}{18} \qquad \text{Rule 2}$$

59.

$$\frac{9}{22} + \left(-\frac{3}{4}\right) + \frac{20}{33}$$

Find the LCD.
$$22 = 2 \cdot 11$$
$$4 = 2 \cdot 2$$
$$33 = 3 \cdot 11$$
$$LCD = 2 \cdot 2 \cdot 3 \cdot 11 = 132$$

$$\frac{9}{22} + \left(-\frac{3}{4}\right) + \frac{20}{33} = \frac{9}{22} \cdot \frac{6}{6} + \left(-\frac{3}{4} \cdot \frac{33}{33}\right) + \frac{20}{33} \cdot \frac{4}{4}$$
$$= \frac{54}{132} + \left(\frac{-99}{132}\right) + \frac{80}{132}$$
$$= \frac{54 + (-99) + 80}{132} \qquad \text{Rule 3}$$
$$= \frac{35}{132} \qquad \text{Rules 1 and 2}$$

61.

$$-\frac{2}{3} + \frac{5}{6} + \left(-\frac{11}{18}\right) \qquad LCD = 18$$

$$= -\frac{2}{3} \cdot \frac{6}{6} + \frac{5}{6} \cdot \frac{3}{3} + \left(-\frac{11}{18}\right)$$
$$= -\frac{12}{18} + \frac{15}{18} + \left(-\frac{11}{18}\right)$$
$$= \frac{-12 + 15 + (-11)}{18} \qquad \text{Rules 1 and 2}$$
$$= \frac{-8}{18} = \frac{-2 \cdot 2 \cdot 2}{2 \cdot 9} = -\frac{4}{9} \qquad \text{Reduce}$$

63.

$$2\frac{3}{5} + 4\frac{2}{3} = \frac{13}{5} + \frac{14}{3} \qquad LCD = 15$$
$$= \frac{13}{5} \cdot \frac{3}{3} + \frac{14}{3} \cdot \frac{5}{5}$$
$$= \frac{39}{15} + \frac{70}{15}$$
$$= \frac{109}{15} \qquad \text{Rule 3}$$

65.

$$3\frac{2}{9} + 2\frac{5}{21}$$

Find the LCD.
$$9 = 3 \cdot 3$$
$$21 = 3 \cdot 7$$

#65. continued

LCD = 3 · 3 · 7 = 63

$$3\frac{2}{9} + 2\frac{5}{21} = \frac{29}{9} + \frac{47}{21}$$

$$= \frac{29}{9} \cdot \frac{7}{7} + \frac{47}{21} \cdot \frac{3}{3}$$

$$= \frac{203}{63} + \frac{141}{63}$$

$$= \frac{344}{63} \qquad \text{Rule 3}$$

67.

-5.41 + (-2.793)

Add absolute values.

$$\begin{array}{r} 5.41 \\ +\underline{2.793} \\ 8.203 \end{array}$$

The answer is -8.203 by rule 1.

69.

62.08 + (-4.93)

Subtract absolute values.

$$\begin{array}{r} 62.08 \\ -\underline{4.93} \\ 57.15 \end{array}$$

The answer is 57.15 by rule 2.

71.

-2.5 + (-20.46) + (-0.903)

Add absolute values.

$$\begin{array}{r} 2.5 \\ 20.46 \\ +\underline{0.903} \\ 23.863 \end{array}$$

The answer is -23.863 by rule 1.

73.

To find the perimeter, find the sum of the measurements of all three sides.

$$\begin{array}{r} 6.85 \text{ cm} \\ 8.13 \text{ cm} \\ +\underline{12.46} \text{ cm} \\ 27.44 \text{ cm} \end{array}$$

The perimeter is 27.44 centimeters.

75.

To find the perimeter, find the sum of the measurements of all four sides.

$$25\frac{2}{3} + 25\frac{2}{3} + 20\frac{1}{6} + 20\frac{1}{6}$$

$$= \frac{77}{3} + \frac{77}{3} + \frac{121}{6} + \frac{121}{6}$$

$$= \frac{77}{3} \cdot \frac{2}{2} + \frac{77}{3} \cdot \frac{2}{2} + \frac{121}{6} + \frac{121}{6}$$

$$= \frac{154}{6} + \frac{154}{6} + \frac{121}{6} + \frac{121}{6}$$

$$= \frac{154 + 154 + 121 + 121}{6} = \frac{550}{6} = 91\frac{2}{3}$$

The perimeter is $91\frac{2}{3}$ feet.

77.

To find the perimeter, find the sum of the measurements of all four sides.

$$3\frac{5}{6} + 3\frac{5}{6} + 3\frac{5}{6} + 3\frac{5}{6}$$

$$= \frac{23}{6} + \frac{23}{6} + \frac{23}{6} + \frac{23}{6} = \frac{92}{6} = 15\frac{1}{3}$$

The perimeter is $15\frac{1}{3}$ yards.

79.

- 3 + 7 = 4

81.

The sum is even.

83.

The sum is odd.

Exercises 1.5

1.

$$\begin{array}{ll} -18 -35 = -18 + (-35) & \text{Add opposites} \\ \qquad\quad = -53 & \text{Rule 1} \end{array}$$

3.

$$\begin{array}{ll} 26 - 48 = 26 + (-48) & \text{Add opposites} \\ \qquad\quad = -22 & \text{Rule 2} \end{array}$$

5.

$$\begin{array}{ll} - 12 - (-29) \quad = -12 + (29) & \text{Add opposites} \\ \qquad\qquad\quad = 17 & \text{Rule 2} \end{array}$$

7.

$4.39 - (-12.8) = 4.39 + (12.8)$

$$\begin{array}{r} 4.39 \\ +\ \underline{12.80} \\ 17.19 \end{array}$$

The answer is 17.19.

9.

$-94.3 - (-62.08) = -94.3 + (62.08)$

Subtract absolute values.

$$\begin{array}{r} 94.30 \\ -\ \underline{62.08} \\ 32.22 \end{array}$$

The answer is -32.22 by rule 2.

11.

$$-\frac{3}{7} - \frac{5}{14} = -\frac{3}{7} + \left(-\frac{5}{14}\right)$$

$$= -\frac{3}{7} \cdot \frac{2}{2} + \left(-\frac{5}{14}\right)$$

$$= -\frac{6}{14} + \left(\frac{-5}{14}\right)$$

$$= \frac{-6 + (-5)}{14} \qquad \text{Rule 3}$$

$$= -\frac{11}{14} \qquad \text{Rule 1}$$

13.

$$\frac{4}{25} + \left(-\frac{7}{15}\right)$$

Find the LCD.

$$\begin{aligned} 25 &= 5 \cdot 5 \\ 15 &= 3 \cdot 5 \\ \text{LCD} &= 3 \cdot 5 \cdot 5 = 75 \end{aligned}$$

$$\frac{4}{25} + \left(-\frac{7}{15}\right) = \frac{4}{25} \cdot \frac{3}{3} + \left(-\frac{7}{15} \cdot \frac{5}{5}\right)$$

$$= \frac{12}{75} + \left(\frac{-35}{75}\right)$$

$$= \frac{12 + (-35)}{75} \qquad \text{Rule 3}$$

$$= -\frac{23}{75} \qquad \text{Rule 2}$$

15.

$-6 - (-14) + (-27)$

$= -6 + 14 + (-27)$ Add opposites

---

15 continued

$= 14 + (-33)$ Rule 1

$= -19$ Rule 2

17.

$19 - (-4) + (-12) - 38$

$= 19 + 4 + (-12) + (-38)$ Add opposites

$= 23 + (-50)$ Rule 1

$= -27$ Rule 2

19.

$$\frac{7}{3} - \frac{2}{9} - \left(-\frac{5}{12}\right)$$

$$= \frac{7}{3} + \left(-\frac{2}{9}\right) + \frac{5}{12} \qquad \text{LCD = 36}$$

$$= \frac{7}{3} \cdot \frac{12}{12} + \left(\frac{-2}{9} \cdot \frac{4}{4}\right) + \frac{5}{12} \cdot \frac{3}{3}$$

$$= \frac{84}{36} + \left(\frac{-8}{36}\right) + \frac{15}{36}$$

$$= \frac{84 + (-8) + 15}{36}$$

$$= \frac{99 + (-8)}{36} \qquad \text{Rule 3}$$

$$= \frac{91}{36} \qquad \text{Rule 2}$$

21.

$$-2\frac{3}{5} - \left(1\frac{1}{2}\right) + 4\frac{2}{3}$$

$$= -2\frac{3}{5} + \left(-1\frac{1}{2}\right) + 4\frac{2}{3} \qquad \text{Add opposites}$$

$$= -\frac{13}{5} + \left(-\frac{3}{2}\right) + \frac{14}{3} \qquad \text{LCD = 30}$$

$$= \frac{-13}{5} \cdot \frac{6}{6} + \left(-\frac{3}{2} \cdot \frac{15}{15}\right) + \frac{14}{3} \cdot \frac{10}{10}$$

$$= \frac{-78}{30} + \left(\frac{-45}{30}\right) + \frac{140}{30}$$

$$= \frac{-78 + (-45) + 140}{30}$$

$$= \frac{-123 + 140}{30} \qquad \text{Rule 3}$$

$$= \frac{17}{30} \qquad \text{Rule 2}$$

23.

$$-\frac{3}{8} + 2 - \frac{5}{7} - \left(-\frac{11}{14}\right)$$

$$= -\frac{3}{8} + \frac{2}{1} + \left(-\frac{5}{7}\right) + \frac{11}{14} \qquad \text{LCD} = 56$$

$$= -\frac{3}{8} \cdot \frac{7}{7} + \frac{2}{1} \cdot \frac{56}{56} + \left(-\frac{5}{7} \cdot \frac{8}{8}\right) + \frac{11}{14} \cdot \frac{4}{4}$$

$$= \frac{-21}{56} + \frac{112}{56} + \left(\frac{-40}{56}\right) + \frac{44}{56}$$

$$= \frac{-21 + 112 + (-40) + 44}{56} \qquad \text{Rule 3}$$

$$= \frac{156 + (-61)}{56} = \frac{95}{56} \qquad \text{Rule 1 and 2}$$

25.

$$14 - (-12 - 25) = 14 - (-12 + (-25))$$
$$= 14 - (-37)$$
$$= 14 + (37) = 51$$

27.

$$-4.15 + 2.99 - (-16.8 + 9)$$

$= -4.15 + 2.99 - (-7.8) \qquad$ Rule 2

$= -4.15 + 2.99 + 7.8 \qquad$ Add opposites

$= -4.15 + 10.79 \qquad$ Rule 1

$= 6.64 \qquad$ Rule 2

29.

$$\left(\frac{2}{9} + \frac{5}{6}\right) - \left(\frac{3}{2} - \frac{11}{3}\right)$$

$$= \frac{2}{9} + \frac{5}{6} - \left[\frac{3}{2} + \left(-\frac{11}{3}\right)\right]$$

$$= \frac{2}{9} + \frac{5}{6} - \left[\frac{3}{2} \cdot \frac{3}{3} + \left(-\frac{11}{3} \cdot \frac{2}{2}\right)\right]$$

$$= \frac{2}{9} + \frac{5}{6} - \left[\frac{9}{6} + \left(-\frac{22}{6}\right)\right]$$

$$= \frac{2}{9} + \frac{5}{6} - \left(-\frac{13}{6}\right)$$

$$= \frac{2}{9} + \frac{5}{6} + \left(\frac{13}{6}\right) \qquad \text{LCD} = 18$$

$$= \frac{2}{9} \cdot \frac{2}{2} + \frac{5}{6} \cdot \frac{3}{3} + \frac{13}{6} \cdot \frac{3}{3}$$

$$= \frac{4}{18} + \frac{15}{18} + \frac{39}{18}$$

#29 continued

$$= \frac{4 + 15 + 39}{18} = \frac{58}{18} \qquad \text{Rules 1 and 3}$$

$$= \frac{29}{9} \qquad \text{Reduce}$$

31.

$$6\frac{1}{2} - \left(2\frac{5}{8} + 5\frac{1}{4}\right) - 1\frac{3}{4}$$

$$= \frac{13}{2} - \left(\frac{21}{8} + \frac{21}{4}\right) - \frac{7}{4}$$

$$= \frac{13}{2} - \left(\frac{21}{8} + \frac{21}{4} \cdot \frac{2}{2}\right) - \frac{7}{4}$$

$$= \frac{13}{2} - \left(\frac{21}{8} + \frac{42}{8}\right) - \frac{7}{4}$$

$$= \frac{13}{2} - \left(\frac{63}{8}\right) - \frac{7}{4} \qquad \text{Rule 3}$$

$$= \frac{13}{2} + \left(-\frac{63}{8}\right) + \left(-\frac{7}{4}\right) \qquad \text{LCD} = 8$$

$$= \frac{13}{2} \cdot \frac{4}{4} + \left(-\frac{63}{8}\right) + \left(-\frac{7}{4} \cdot \frac{2}{2}\right)$$

$$= \frac{52}{8} + \left(-\frac{63}{8}\right) + \left(-\frac{14}{8}\right)$$

$$= \frac{52 + (-63) + (-14)}{8} \qquad \text{Rule 3}$$

$$= \frac{52 - 77}{8} = -\frac{25}{8} \qquad \text{Rule 2}$$

33.

$$-15 - |5 - (-5 - 5)| = -15 - |5 - (-10)|$$
$$= -15 - |5 + 10|$$
$$= -15 - |15|$$
$$= -15 - 15$$
$$= -30$$

35.

$$-\frac{4}{17} - \left|\frac{1}{2} - 3\right|$$

$$= -\frac{4}{17} - \left|\frac{1}{2} - \frac{6}{2}\right|$$

$$= -\frac{4}{17} - \left|-\frac{5}{2}\right| \qquad \text{Rule 2 and 3}$$

$$= -\frac{4}{17} - \frac{5}{2}$$

$$= -\frac{4}{17} + \left(-\frac{5}{2}\right) \qquad \text{LCD} = 34$$

 # 35. continued

$$= -\frac{4}{17}\cdot\frac{2}{2} + \left(-\frac{5}{2}\cdot\frac{17}{17}\right)$$

$$= -\frac{8}{34} + \left(-\frac{85}{34}\right)$$

$$= \frac{-8 + (-85)}{34} \qquad \text{Rule 3}$$

$$= -\frac{93}{34} \qquad \text{Rule 2}$$

37.

$$-\frac{5}{13} + |\frac{5}{13} - \frac{5}{3}| + \frac{5}{3}$$

$$= -\frac{5}{13} + |\frac{5}{13}\cdot\frac{3}{3} + \left(-\frac{5}{3}\cdot\frac{13}{13}\right)| + \frac{5}{3}$$

$$= -\frac{5}{13} + |\frac{15}{39} + \left(-\frac{65}{39}\right)| + \frac{5}{3}$$

$$= -\frac{5}{13} + |\frac{15 + (-65)}{39}| + \frac{5}{3}$$

$$= -\frac{5}{13} + |\frac{-50}{39}| + \frac{5}{3}$$

$$= -\frac{5}{13} + \frac{50}{39} + \frac{5}{3}$$

$$= -\frac{5}{13}\cdot\frac{3}{3} + \frac{50}{39} + \frac{5}{3}\cdot\frac{13}{13}$$

$$= \frac{-15}{39} + \frac{50}{39} + \frac{65}{39}$$

$$= \frac{-15 + 50 + 65}{39} = \frac{100}{39} \qquad \text{Rules 1, 2 and 3}$$

39.

$$-18 - [28 - (4-19)]$$
$$= -18 - [28 - (-15)] \qquad \text{Rule 2}$$
$$= -18 - [28 + 15] \qquad \text{Add opposites}$$
$$= -18 - [43]$$
$$= -18 + (-43) = -61 \qquad \text{Rule 1}$$

41.

$$6.1 - [(3.92 - 13.5) - 8.14]$$
$$= 6.1 - [(3.92 + (-13.5)) - 8.14] \qquad \text{Rule 2}$$
$$= 6.1 - [-9.58 - 8.14] \qquad \text{Add opposites}$$
$$= 6.1 - [-9.58 + (-8.14)] \qquad \text{Rule 1}$$
$$= 6.1 - [-17.72] \qquad \text{Add opposites}$$
$$= 6.1 + 17.72$$
$$= 23.82$$

43.

$$-\frac{9}{5} + \left[\frac{3}{10} - \left(\frac{3}{2} + 2\right)\right]$$

$$= -\frac{9}{5} + \left[\frac{3}{10} - \left(\frac{3}{2} + \frac{2}{1}\right)\right]$$

$$= -\frac{9}{5} + \left[\frac{3}{10} - \left(\frac{3}{2} + \frac{2}{1}\cdot\frac{2}{2}\right)\right]$$

$$= -\frac{9}{5} + \left[\frac{3}{10} - \left(\frac{3}{2} + \frac{4}{2}\right)\right]$$

$$= -\frac{9}{5} + \left[\frac{3}{10} - \frac{7}{2}\right]$$

$$= -\frac{9}{5} + \left[\frac{3}{10} - \frac{7}{2}\cdot\frac{5}{5}\right]$$

$$= -\frac{9}{5} + \left[\frac{3}{10} + \left(-\frac{35}{10}\right)\right] \qquad \text{Add opposites}$$

$$= -\frac{9}{5} + \left[\frac{3-35}{10}\right] \qquad \text{Rule 3}$$

$$= -\frac{9}{5} + \left[-\frac{32}{10}\right]$$

$$= -\frac{9}{5} + \left[-\frac{16}{5}\right] \qquad \text{Reduce}$$

$$= \frac{-9 + (-16)}{5} = \frac{-25}{5} = -5 \qquad \text{Reduce}$$

45.

$$180° - (14.9° + 61.8°)$$
$$= 180° - (76.7°)$$
$$= 180° + (-76.7°)$$
$$= 103.3°$$

47.

$$180° - (42.18° + 108.9°)$$
$$= 180° - (151.08°)$$
$$= 180° + (-151.08°)$$
$$= 28.92°$$

49.
Begin by listing the deposits as addition and all others as subtraction.

$$- 85.31 - 30.00 + 426.50 - 75.00$$
$$- 16.38 - 295.99$$
$$= - 85.31 + (-30.00) + (-75.00) + (-16.38)$$
$$+ (-295.99) + 426.50$$
$$= (-502.6) + 426.50$$
$$= -76.18$$

Mary's checking account is over drawn by $76.18.

51.
$$- 40 - (- 23) = - 40 + 23 = - 17$$

53.
$$-\frac{3}{4} - \left(\frac{7}{12}\right) = -\frac{3}{4} + \left(-\frac{7}{12}\right) \qquad \text{LCD} = 12$$
$$= -\frac{3}{4} \cdot \frac{3}{3} + \left(-\frac{7}{12}\right)$$
$$= -\frac{9}{12} + \left(-\frac{7}{12}\right)$$
$$= -\frac{16}{12} = -\frac{4}{3} \qquad \text{Reduce}$$

55.
$$4.08 - (- 18.5) = 4.08 + 18.5 = 22.58$$

57.
$$55 - (-18) = 55 + 18 = 73$$

59.
$$-\frac{7}{15} - \left(- \frac{5}{3}\right) = -\frac{7}{15} + \frac{5}{3} \qquad \text{LCD} = 15$$
$$= -\frac{7}{15} + \frac{5}{3} \cdot \frac{5}{5}$$
$$= -\frac{7}{15} + \frac{25}{15} = \frac{18}{15} = \frac{6}{5} \qquad \text{Reduce}$$

61.
No. Example: $7 - 8 = -1$

63.
$-6 - 5 = -11$ The number is -11.

1. - 7.
The answers do not require worked out solutions.

9.
$$\left(-\frac{11}{4}\right)\left(-\frac{18}{5}\right) = \frac{-11 \cdot (-18)}{4 \cdot 5} \qquad \text{Rule 8}$$
$$= \frac{-11 \cdot (-9) \cdot \cancel{2}}{\cancel{2} \cdot 2 \cdot 5} \qquad \text{Factor}$$
$$= \frac{99}{10}$$

11.
$$\frac{1}{8} \cdot \left(-\frac{2}{13}\right)\left(\frac{26}{3}\right) = \frac{1 \cdot (-2) \cdot 26}{8 \cdot 13 \cdot 3} \qquad \text{Rule 8}$$
$$= \frac{1 \cdot (-1) \cdot \cancel{2} \cdot \cancel{2} \cdot \cancel{13}}{\cancel{2} \cdot \cancel{2} \cdot 2 \cdot \cancel{13} \cdot 3} = -\frac{1}{6}$$

13.
$$\left(-\frac{4}{7}\right)\left(-\frac{5}{6}\right)(-3) = \frac{(-4)(-5)(-3)}{7 \cdot 6} \qquad \text{Rule 8}$$
$$= \frac{-1 \cdot \cancel{2} \cdot 2 \cdot (-5)(-1) \cdot \cancel{3}}{7 \cdot \cancel{2} \cdot \cancel{3}} \qquad \text{Factor}$$
$$= -\frac{10}{7}$$

15.
$$\left(-2\frac{3}{8}\right)\left(1\frac{3}{5}\right) = \left(-\frac{19}{8}\right)\left(\frac{8}{5}\right)$$
$$= \frac{-19 \cdot 8}{8 \cdot 5}$$
$$= -\frac{19}{5}$$

Note: Only factor as much as needed for cancelling.

17.
$$\left(-4\frac{2}{3}\right)\left(-3\frac{1}{7}\right)\left(1\frac{4}{11}\right)$$
$$= \left(-\frac{14}{3}\right)\left(-\frac{22}{7}\right)\left(\frac{15}{11}\right)$$
$$= \frac{(-14)(-22) \cdot 15}{3 \cdot 7 \cdot 11} \qquad \text{Rule 8}$$

#17 continued

$$= \frac{-1 \cdot 2 \cdot \cancel{7}(-1) \cdot 2 \cdot \cancel{11} \cdot \cancel{3} \cdot 5}{\cancel{3} \cdot \cancel{7} \cdot \cancel{11}} \qquad \text{Factor}$$

$$= 20 \qquad \text{Reduce}$$

19. - 31.
The answers do not require worked out solutions.

33.
$$-6 - 11(4) + 18 \quad = -6 - 44 + 18$$
$$= -6 + (-44) + 18$$
$$= -50 + 18$$
$$= -32 \qquad \text{Rules 1 and 2}$$

35.
$$3 \cdot 8 - 8 \cdot 5 \quad = 24 - 40$$
$$= 24 + (-40) \qquad \text{Add opposites}$$
$$= -16 \qquad \text{Rule 2}$$

37.
$$\frac{2}{9} - \frac{4}{15} \cdot \frac{5}{8} = \frac{2}{9} - \left(\frac{\cancel{4}}{3 \cdot \cancel{5}} \cdot \frac{\cancel{5} \cdot 1}{2 \cdot \cancel{4}} \right)$$

$$= \frac{2}{9} - \left(\frac{1}{6} \right)$$

$$= \frac{2}{9} \cdot \frac{2}{2} - \frac{1}{6} \cdot \frac{3}{3}$$

$$= \frac{4}{18} - \frac{3}{18}$$

$$= \frac{1}{18}$$

39.
$$2\left(-\frac{11}{6} \right) + \frac{2}{21} \cdot \frac{5}{4} = \frac{\cancel{2}}{1}\left(\frac{-11}{\cancel{2} \cdot 3} \right) + \left(\frac{\cancel{2}}{21} \cdot \frac{5}{\cancel{2} \cdot 2} \right)$$

$$= \frac{-11}{3} + \frac{5}{42}$$

$$= \frac{-11}{3} \cdot \frac{14}{14} + \frac{5}{42}$$

$$= \frac{-154}{42} + \frac{5}{42}$$

$$= \frac{-154 + 5}{42} \qquad \text{Rule 3}$$

$$= \frac{-149}{42} \qquad \text{Rule 2}$$

41.
$$\frac{5}{14} + \left(\frac{-6}{17} \right)\left(\frac{4}{21} \right)\left(\frac{34}{5} \right)$$

$$= \frac{5}{14} + \frac{-1 \cdot 2 \cdot \cancel{3} \cdot 2 \cdot 2 \cdot 2 \cdot \cancel{17}}{\cancel{17} \cdot \cancel{3} \cdot 7 \cdot 5}$$

$$= \frac{5}{14} + \left(\frac{-16}{35} \right) \qquad \text{LCD} = 2 \cdot 7 \cdot 5 = 70$$

$$= \frac{5}{14} \cdot \frac{5}{5} + \left(\frac{-16}{35} \cdot \frac{2}{2} \right)$$

$$= \frac{25}{70} + \left(\frac{-32}{70} \right)$$

$$= \frac{25 + (-32)}{70} = -\frac{1}{10} \qquad \text{Reduce}$$

43.
$$4\frac{1}{2} - \left(2\frac{1}{3} \right)\left(2\frac{1}{4} \right)$$

$$= \frac{9}{2} - \left(\frac{7}{3} \right)\left(\frac{9}{4} \right)$$

$$= \frac{9}{2} - \left(\frac{7 \cdot \cancel{3} \cdot 3}{\cancel{3} \cdot 4} \right)$$

$$= \frac{9}{2} - \frac{21}{4}$$

$$= \frac{9}{2} \cdot \frac{2}{2} - \frac{21}{4}$$

$$= \frac{18}{4} - \frac{21}{4}$$

$$= \frac{18 - 21}{4} = -\frac{3}{4}$$

45.
$$3(-4)^2 - 5(-4) - 24 \quad = 3(16) + 20 - 24$$
$$= 48 + 20 - 24$$
$$= 68 - 24 = 44$$

47.
$$5(-2)^3 + 7(-2)^2 - 4(-2) + 9$$
$$= 5(-8) + 7(4) - 4(-2) + 9$$
$$= -40 + 28 + 8 + 9 = 5$$

49.
$$(-3)^2 - \left(\frac{1}{2} \right)^2 = 9 - \frac{1}{4} = 8\frac{3}{4}$$

51.

$$6\left(\frac{1}{3}\right)^2 - 4\left(-\frac{1}{3}\right) = 6\left(\frac{1}{9}\right) - 4\left(-\frac{1}{3}\right) = \frac{6}{9} + \frac{4}{3} \quad LCD = 9$$

$$= \frac{6}{9} + \frac{4}{3} \cdot \frac{3}{3} = \frac{6}{9} + \frac{12}{9} = \frac{18}{9} = 2$$

53.

$$\begin{aligned}
-3[5 - 2(4 - 11)] &= -3[5 - 2(-7)] \\
&= -3[5 + 14] \\
&= -3[19] \\
&= -57
\end{aligned}$$

55.

$$\begin{aligned}
8 - 5(-2 - 14) &= 8 - 5(-16) \\
&= 8 + (-5)(-16) \quad \text{Add opposites} \\
&= 8 + 80 = 88
\end{aligned}$$

57.

$$\begin{aligned}
(-6 - 7)(-17 + 8 \cdot 3) & \\
= (-13)(-17 + 24) & \quad \text{Rule 1} \\
= (-13)(7) = -91 & \quad \text{Rule 6}
\end{aligned}$$

59.

$$\begin{aligned}
-6.5 + 4.82(3.13 - 11.08) & \\
= -6.5 + 4.82(-7.95) & \\
= -6.5 - 38.319 & \quad \text{Multiply} \\
= -44.819 & \quad \text{Rule 2}
\end{aligned}$$

61.

$$\begin{aligned}
[1.57 - (4.6)(-2.15)](8.5) & \\
= [1.57 + 9.89](8.5) & \quad \text{Rule 7} \\
= [11.46](8.5) = 97.41 & \quad \text{Rule 3}
\end{aligned}$$

63.

$$-\frac{2}{3} + \frac{4}{5}\left(\frac{7}{6} - 2\right)$$

$$= -\frac{2}{3} + \frac{4}{5}\left(\frac{7}{6} - \frac{12}{6}\right)$$

$$= -\frac{2}{3} + \frac{4}{5}\left(-\frac{5}{6}\right) \quad \text{Rules 6 and 8}$$

$$= -\frac{2}{3} + \frac{2 \cdot 2 \cdot (-1) \cdot \cancel{5}}{\cancel{5} \cdot 2 \cdot 3}$$

$$= -\frac{2}{3} + \left(-\frac{2}{3}\right)$$

$$= \frac{-2 + (-2)}{3} = -\frac{4}{3} \quad \text{Rule 2}$$

65.

$$\frac{3}{4}\left[\frac{8}{5} + \frac{14}{9}\left(-\frac{6}{7}\right)\right]$$

$$= \frac{3}{4}\left[\frac{8}{5} + \frac{2 \cdot \cancel{7} \cdot (-1) \cdot 2 \cdot \cancel{3}}{3 \cdot 3 \cdot \cancel{7}}\right]$$

$$= \frac{3}{4}\left[\frac{8}{5} + \left(-\frac{4}{3}\right)\right] \quad \text{Rules 6 and 8}$$

$$= \frac{3}{4}\left[\frac{8 \cdot 3}{5 \cdot 3} + \left(-\frac{4 \cdot 5}{3 \cdot 5}\right)\right]$$

$$= \frac{3}{4}\left[\frac{24}{15} + \left(\frac{-20}{15}\right)\right]$$

$$= \frac{3}{4}\left[\frac{4}{15}\right]$$

$$= \frac{\cancel{3} \cdot 4}{4 \cdot \cancel{3} \cdot 5} = \frac{1}{5} \quad \text{Rules 3 and 8}$$

67.

$$\left[\left(2\frac{1}{7}\right)\left(2\frac{4}{5}\right) - 8\frac{1}{3}\right]\left(2\frac{1}{4}\right)$$

$$= \left[\left(\frac{15}{7}\cdot\frac{14}{5}\right) - \frac{25}{3}\right]\left(\frac{9}{4}\right)$$

$$= \left[\frac{3 \cdot 5 \cdot 2 \cdot \cancel{7}}{\cancel{7} \cdot 5} - \frac{25}{3}\right]\left(\frac{9}{4}\right)$$

$$= \left[\frac{6 \cdot 3}{1 \cdot 3} - \frac{25}{3}\right]\left(\frac{9}{4}\right)$$

$$= \left[\frac{18}{3} - \frac{25}{3}\right]\left(\frac{9}{4}\right)$$

$$= \left[6 - \frac{25}{3}\right]\left(\frac{9}{4}\right) \quad \text{Rule 8}$$

$$= \left[\frac{-7}{3}\left(\frac{3 \cdot 3}{4}\right)\right] \quad \text{Rules 3 and 2}$$

$$= \frac{-21}{4} \quad \text{Rules 6 and 8}$$

69.

$$\begin{aligned}
-2|-18 + 7| -7 &= -2|-11| -7 \\
&= -2(11) -7 \\
&= -22 - 7 \\
&= -29
\end{aligned}$$

71.

$$\frac{5}{6} - \frac{3}{8}\left|\frac{1}{6} - \frac{2}{9}\right|$$

$$= \frac{5}{6} - \frac{3}{8}\left|\frac{3}{18} - \frac{4}{18}\right|$$

#71 continued

$-\dfrac{5}{6} - \dfrac{3}{8}\left|\dfrac{-1}{18}\right|$

$-\dfrac{5}{6} - \dfrac{3}{8}\left(\dfrac{1}{18}\right)$

$-\dfrac{5}{6} - \dfrac{3 \cdot 1}{8 \cdot 3 \cdot 6}$

$-\dfrac{5}{6} - \dfrac{1}{48}$ Rules 7 and 8

$-\dfrac{40}{48} - \dfrac{1}{48} = \dfrac{39}{48}$ Rule 3

$-\dfrac{13}{16}$ Reduce

73.

$P = 2L + 2W$

$P = 2\left(8\dfrac{1}{3}\right) + 2\left(5\dfrac{5}{6}\right)$

$P = 2\left(\dfrac{25}{3}\right) + 2\left(\dfrac{35}{6}\right)$

$= \dfrac{50}{3} + \dfrac{70}{6}$

$= \dfrac{100}{6} + \dfrac{70}{6}$

$= \dfrac{170}{6}$ or $28\dfrac{1}{3}$

The perimeter is $28\dfrac{1}{3}$ feet.

75.

$A = \dfrac{1}{2}(b + B)h$

$A + \dfrac{1}{2}(3.8 + 6.23)\,4.5$

$A = \dfrac{1}{2}(10.03)\,4.5$

$A = \dfrac{1}{2}(45.135)$

$A = 22.5675$

The area is 22.5675 mm^2.

77.

$I = P r t$

$I = 4500\,(.0975)\left(2\dfrac{1}{2}\right)$

$I = 4500\,(.0975)\left(\dfrac{5}{2}\right)$

#77. continued

$I = 438.75\left(\dfrac{5}{2}\right)$

$I = 1096.875$ Multiply by 5, then divide by 2.

The interest is $1,096.88.

79.
$(4 \cdot 7) + 3 = 28 + 3 = 31$
The number is 31.

81.
$2(9) - 5 = 18 - 5 = 13$
The number is 13.

83.
The product is even. Example: $8 \cdot 2 = 16$

85.
The product is even. Example: $7 \cdot 4 = 28$

Exercises 1.7

1. - 13.
The answers do not require worked-out solutions.

15.

$\dfrac{4}{5} + \left(-\dfrac{12}{35}\right)$

$= \dfrac{4}{5} \cdot \left(-\dfrac{35}{12}\right)$ Rule 11

$= \dfrac{\cancel{2}\cdot\cancel{2}\cdot\cancel{5}\cdot 7 \cdot (-1)}{\cancel{5}\cdot\cancel{2}\cdot\cancel{2}\cdot 3} = -\dfrac{7}{3}$ Rule 3

17.

$\left(-\dfrac{16}{25}\right) \div \dfrac{28}{15} = \left(-\dfrac{16}{25}\right)\dfrac{15}{28}$ Rule 11

$= \dfrac{(-1)\cdot\cancel{4}\cdot 4 \cdot 3 \cdot \cancel{5}}{\cancel{5}\cdot 5 \cdot \cancel{4}\cdot 7}$

$= -\dfrac{12}{35}$ Rule 3

19.

$\left(-6\dfrac{1}{2}\right) + \left(-3\dfrac{9}{10}\right)$

$= \left(-\dfrac{13}{2}\right) + \left(-\dfrac{39}{10}\right)$

#19 continued

$$= \left(-\frac{13}{2}\right)\left(-\frac{10}{39}\right) \qquad \text{Rule 11}$$

$$= \frac{(-1)\cdot \cancel{13}\cdot(-1)\cdot 2\cdot 5}{2\cdot 3\cdot \cancel{13}} = \frac{5}{3} \qquad \text{Rules 7 and 8}$$

21.

$$-8 + 3\frac{1}{5} = -8 + \frac{16}{5}$$

$$= \frac{-8}{1}\cdot\frac{5}{16} \qquad \text{Rule 4}$$

$$= \frac{(-1)\cdot 8\cdot 5}{2\cdot 8} \qquad \text{Rule 8}$$

$$= -\frac{5}{2} \qquad \text{Rule 6}$$

23.

$$\left(-\frac{17}{9}\right) \div 4 = -\frac{17}{9}\cdot\frac{1}{4} = -\frac{17}{36}$$

25.

$$\frac{-\dfrac{18}{7}}{\dfrac{6}{-35}} = -\frac{18}{7} + \left(-\frac{6}{35}\right)$$

$$= -\frac{18}{7}\left(-\frac{35}{6}\right) \qquad \text{Rule 11}$$

$$= \frac{-11\cdot 3\cdot \cancel{6}\cdot(-1)5\cdot \cancel{7}}{\cancel{7}\cdot\cancel{6}} \qquad \text{Rule 8}$$

$$= 15 \qquad \text{Rule 7}$$

27.

$$\frac{\dfrac{12}{5}}{\dfrac{32}{15}} = \frac{12}{5} \div \frac{32}{15} = \frac{12}{5}\cdot\frac{15}{32} \qquad \text{Rule 11}$$

$$= \frac{\cancel{4}3\cdot 3\cdot \cancel{5}}{\cancel{5}\cdot 4\cdot 8} \qquad \text{Rule 8}$$

$$= \frac{9}{8}$$

29.

$$\frac{-\dfrac{10}{3}}{6} = -\frac{10}{3} \div 6$$

$$= -\frac{10}{3}\cdot\frac{1}{6} \qquad \text{Rule 11}$$

#29 continued

$$= \frac{(-1)\cancel{2}\cdot 5\cdot 1}{3\cdot\cancel{2}\cdot 3} \qquad \text{Rule 8}$$

$$= -\frac{5}{9} \qquad \text{Rule 6}$$

31.

$$\frac{-8}{\dfrac{6}{-5}} = -8 \div -\frac{6}{5}$$

$$= -\frac{8}{1}\left(-\frac{5}{6}\right) \qquad \text{Rule 11}$$

$$= \frac{(-1)\cancel{2}\cdot 4\cdot 5(-1)}{1\cdot\cancel{2}\cdot 3} \qquad \text{Rule 8}$$

$$= \frac{20}{3} \qquad \text{Rule 7}$$

33.

$$-48 \div 8 \div (-2)$$

$$= -\frac{48}{1}\cdot\frac{1}{8}\left(-\frac{1}{2}\right) \qquad \text{Rule 11}$$

$$= \frac{(-1)\cancel{2}\cdot 3\cdot\cancel{8}(-1)}{1\cdot\cancel{8}\cdot\cancel{2}} \qquad \text{Rule 8}$$

$$= 3 \qquad \text{Rule 7}$$

35.

$$2 + (12 \div 3) - 9\cdot 4$$
$$= 2 + 4 - 36$$
$$= 6 - 36$$
$$= -30 \qquad \text{Rule 2}$$

37.

$$15 + [22 + (-11)] - 8$$
$$= 15 + (-2) - 8$$
$$= 15 - 10 = 5 \qquad \text{Rule 2}$$

39.

$$75 - 16 + 4(-2)$$
$$= 75 + (-16 + 4)(-2) \qquad \text{Add opposites}$$
$$= 75 + (-4)(-2) \qquad \text{Divide}$$
$$= 75 + 8 \qquad \text{Multiply}$$
$$= 83 \qquad \text{Add}$$

41.

$$-6 + \frac{9}{2} - \frac{4}{7} + \frac{10}{21}$$

#41 continued

$$= \left(\frac{-6}{1} \cdot \frac{2}{9}\right) - \left(\frac{4}{7} \cdot \frac{21}{10}\right) \qquad \text{Rule 11}$$

$$= \left(\frac{(-1) \cdot 2 \cdot \cancel{3} \cdot 2}{\cancel{3} \cdot 3}\right) - \left(\frac{\cancel{2} \cdot 2 \cdot 3 \cdot \cancel{7}}{\cancel{7} \cdot \cancel{2} \cdot 5}\right) \qquad \text{Rule 8}$$

$$= -\frac{4}{3} - \frac{6}{5}$$

$$= -\frac{20}{15} - \frac{18}{15} = -\frac{38}{15} \qquad \text{Rules 3 and 1}$$

43.

$(4 + 16) \div (4 + 2 \cdot 2)$

$$= (20) \div \left(\frac{4}{2} \cdot 2\right)$$

$= (20) \div (2 \cdot 2)$

$= (20) \div (4)$

$= 5$

45.

$(-19 - 44) \div 7 + (-3)$

$= (-63) \div 7 + (-3) \qquad \text{Rule 1}$

$$= \frac{-63}{7} + (-3) \qquad \text{Rule 4}$$

$= -9 + (-3) \qquad \text{Rule 9}$

$$= \frac{-9}{-3} = 3 \qquad \text{Rules 11 and 9}$$

47.

$15 \cdot 14 \div 35 - 6$

$= (15 \cdot 14 \div 35) - 6 \qquad$ Multiply and divide in order

$$= \left(15 \cdot 14 \cdot \frac{1}{35}\right) - 6 \qquad \text{Rule 4}$$

$$= \left(\frac{15 \cdot 14}{35}\right) - 6$$

$$= \left(\frac{3 \cdot \cancel{5} \cdot 2 \cdot \cancel{7}}{\cancel{5} \cdot \cancel{7}}\right) - 6 \qquad \text{Rule 8}$$

$= 6 - 6 = 0 \qquad \text{Rule 2}$

49.

$$\frac{3}{4} \div \left(\frac{5}{8} - \frac{5}{2}\right)$$

$$= \frac{3}{4} \div \left(\frac{5}{8} - \frac{20}{8}\right)$$

#49 continued

$$= \frac{3}{4} \div \left(-\frac{15}{8}\right) \qquad \text{Rules 2 and 3}$$

$$= \frac{3}{4} \cdot \left(-\frac{8}{15}\right) \qquad \text{Rule 11}$$

$$= \frac{\cancel{3} \cdot (-1) \cdot \cancel{4} \cdot 2}{\cancel{4} \cdot \cancel{3} \cdot 5} = -\frac{2}{5} \qquad \text{Rules 6 and 8}$$

51.

$$\frac{6}{11} - \frac{5}{6}\left(9 \div \frac{15}{7}\right)$$

$$= \frac{6}{11} - \frac{5}{6}\left(9 \cdot \frac{7}{15}\right) \qquad \text{Rule 11}$$

$$= \frac{6}{11} - \frac{5}{6}\left(\frac{3 \cdot 3 \cdot 7}{3 \cdot 5}\right)$$

$$= \frac{6}{11} - \frac{5}{6}\left(\frac{21}{5}\right) \qquad \text{Rule 8}$$

$$= \frac{6}{11} - \frac{\cancel{5} \cdot 3 \cdot 7}{2 \cdot \cancel{3} \cdot \cancel{5}} \qquad \text{Rule 8}$$

$$= \frac{6}{11} - \frac{7}{2}$$

$$= \frac{12}{22} - \frac{77}{22} = -\frac{65}{22} \qquad \text{Rules 2 and 3}$$

53.

$$\frac{21}{10} \div \left(\frac{7}{6}\right)^2 + \frac{6}{7} = \frac{21}{10} \div \frac{49}{36} + \frac{6}{7}$$

$$= \left(\frac{21}{10} \cdot \frac{36}{49}\right) + \frac{6}{7}$$

$$= \frac{54}{35} + \frac{6}{7} \qquad \text{LCD} = 35$$

$$= \frac{54}{35} + \frac{6}{7} \cdot \frac{5}{5}$$

$$= \frac{54}{35} + \frac{30}{35} = \frac{84}{35} = \frac{12}{5} \qquad \text{Reduce}$$

55.

$-75 + (3 - 8)^2 = -75 + (-5)^2 = -75 + 25 = -3$

57.

$[4 + (12 \div 8 + 2 - 6)] \div 2$

$= [4 + (12 \div (8 + 2) - 6)] \div 2 \qquad$ Divide first

$= [4 + (12 \div 4 - 6)] \div 2$

$= [4 + (10)] \div 2 \qquad$ Rules 1 and 2

$= [14] \div 2 \qquad$ Add

$= 7 \qquad$ Divide

59.

$28 + [13 - 72 + (21 - 9)]$

$= 28 + [13 - 72 + (12)]$ Rule 2

$= 28 + [13 + (-72) + (12)]$ Add opposites

$= 28 + [13 + (-6)]$ Divide

$= 28 + [7]$ Rule

$= 4$ Divide

61.

$[6 - 2(2^3 - 5)] + 13 = [6 - 2(8 - 5)] + 13$

$= [6 - 2(3)] + 13$

$= [6 - 6] + 13 = 0 + 13 = 0$

63.

$$\frac{-4 - \sqrt{4^2 - 4(-12)}}{2} = \frac{-4 - \sqrt{16 + 48}}{2}$$

$$= \frac{-4 - \sqrt{64}}{2}$$

$$= \frac{-4 - 8}{2} = \frac{-12}{2} = -6$$

65.

$14 - |32 + (-8)| \, 5$

$= 14 - |-4| \, 5$ Divide

$= 14 - (4) \, 5$ Absolute value

$= 14 - 20$ Rule 6

$= -6$ Rule 2

67.

$$\frac{\frac{2}{5} + \frac{7}{10}}{\frac{4}{3} - 5} = \frac{\frac{4}{10} + \frac{7}{10}}{\frac{4}{3} - \frac{15}{3}}$$

$$= \frac{\frac{11}{10}}{-\frac{11}{3}}$$

$$= \frac{11}{10} + \left(-\frac{11}{3}\right)$$ Rule 11

$$= \frac{\cancel{11}}{10} \cdot \left(-\frac{3}{\cancel{11}}\right)$$

$$= -\frac{3}{10}$$ Rules 6 and 8

69.

$$\frac{\left(\frac{4}{15} \cdot \frac{10}{3}\right) - \frac{7}{6}}{\frac{9}{2} - \frac{1}{3}}$$

$$= \frac{\left(\frac{4 \cdot 2 \cdot \cancel{5}}{3 \cdot \cancel{5} \cdot 3}\right) - \frac{7}{6}}{\frac{9}{2} - \frac{1}{3}}$$ Rule 8

$$= \frac{\frac{8}{9} - \frac{7}{6}}{\frac{9}{2} - \frac{1}{3}}$$

$$= \frac{\frac{16}{18} - \frac{21}{18}}{\frac{27}{6} - \frac{2}{6}}$$

$$= \frac{-\frac{5}{18}}{\frac{25}{6}}$$ Rules 2 and 3

$$= -\frac{5}{18} + \frac{25}{6}$$

$$= -\frac{5}{18} \cdot \frac{6}{25}$$ Rule 11

$$= -\frac{(1)\cancel{5} \cdot \cancel{6}}{3 \cdot \cancel{6} \cdot \cancel{5} \cdot 5} = -\frac{1}{15}$$ Rule 8

71.

$$\frac{6\frac{3}{4} - 3\frac{5}{6}}{4\frac{1}{2} + 2\frac{3}{8}} = \frac{\frac{27}{4} - \frac{23}{6}}{\frac{9}{2} + \frac{19}{8}}$$

$$= \frac{\frac{81}{12} - \frac{46}{12}}{\frac{36}{8} + \frac{19}{8}} = \frac{\frac{35}{12}}{\frac{55}{8}}$$

$$= \frac{35}{12} + \frac{55}{8}$$

$$= \frac{35}{12} \cdot \frac{8}{55}$$ Rule 11

#71. continued

$$= \frac{5 \cdot 7 \cdot 4 \cdot 2}{3 \cdot 4 \cdot 5 \cdot 11} = \frac{14}{33}$$

73.

The formula for miles per gallon is

$$MPG = \frac{M}{G}$$

Substitute 320 for M and $12\frac{1}{2}$ for G.

$$MPG = \frac{320}{12\frac{1}{2}}$$

$$= 320 \div \frac{25}{2}$$

$$= 320 \cdot \frac{2}{25}$$

$$= 25.6 \text{ or } 25\frac{3}{5}$$

Al gets $25\frac{3}{5}$ miles per gallon.

75.

The formula for miles per hour is

$R = \frac{D}{T}$, where T represents hours and R is rate.

Substitute 240 for D and $4\frac{2}{5}$ for T.

$$R = \frac{240}{4\frac{2}{5}}$$

$$= 240 \div \frac{22}{5}$$

$$= 240 \cdot \frac{5}{22}$$

$$= \frac{240 \cdot 5}{22}$$

$$= \frac{2 \cdot 120 \cdot 5}{2 \cdot 11}$$

$$= \frac{600}{11} = 54\frac{6}{11}$$

Steve drove an average $54\frac{6}{11}$ miles per hour.

77.

To find how many minutes per mile, divide the number of minutes by the number of miles.

$$\frac{69\frac{3}{10}}{8\frac{2}{5}} = \frac{693}{10} \div \frac{42}{5}$$

$$= \frac{693}{10} \cdot \frac{5}{42}$$

$$= \frac{5 \cdot 3 \cdot 231}{2 \cdot 5 \cdot 2 \cdot 3 \cdot 7}$$

$$= \frac{231}{28} = 8\frac{1}{4}$$

Dottie ran $8\frac{1}{4}$ minutes per mile.

79.

$$-24 \div \frac{2}{3} = \frac{-24}{1} \cdot \frac{3}{2} = -12 \cdot 3 = -36$$

The number is -36.

81.

$$\frac{-17}{-85} = \frac{1}{5}$$

Exercises 1.8

1. - 53.

The answers do not require worked out solutions.

Review Exercises

1.
A ∪ B = {3,4,5,6,7,8,9,12,15,}
The union of two sets contains elements in either set A or set B.

3.
A ∩ B = {3,6,9}
The intersection of two sets contain elements common to both sets A and set B.

5.
A ∪ C = {3,4,6,7,9,10,12,13,15,16}
See exercise 1.

7.

$\frac{4}{3} = 1.33\ldots$

Work

$$\begin{array}{r} 1.33\ldots \\ 3\overline{\smash{)}4.00} \\ \underline{3} \\ 10 \\ \underline{9} \\ 10 \end{array}$$

9.

$\frac{13}{8} = 1.625$

Work

$$\begin{array}{r} 1.625 \\ 8\overline{\smash{)}13.000} \\ \underline{8} \\ 50 \\ \underline{48} \\ 20 \\ \underline{16} \\ 40 \\ \underline{40} \end{array}$$

11. - 43.

The answers do not require worked-out solutions.

45.

$25 - 6 \cdot 2 = 25 - 12 = 13$

47.

$8 \cdot 3^2 - 5 \cdot 3 + 4 = 8 \cdot 9 - 5 \cdot 3 + 4$
$= 72 - 15 + 4 = 61$

49.

$(72 + 18) \div 2 = 4 + 2 = 2$

51.

$21 - 3|8 - 4| = 21 - 3|4| = 21 - 3 \cdot 4$
$= 21 - 12 = 9$

53.

$20 - 5[16 - (3 + 9)] = 20 - 5[16 - (12)]$
$= 20 - 5[4] = 20 - 20 = 0$

55.

$\frac{38 - (1 + 7)}{15 - 6 - 3} = \frac{38 - 8}{6} = \frac{30}{6} = 5$

57.

$5\text{^}2 - 3*7 = 5^2 - 3 \cdot 7 = 25 - 21 = 4$

59.

$2\text{^}4 + 40/(23 - 3*5) = 2^4 + \frac{40}{23 - 3 \cdot 5}$

$= 16 + \frac{40}{23 - 15}$

$= 16 + \frac{40}{8} = 16 + 5 = 21$

61.

$-21 + (-14) + (-46) = -81$ Rule 1

63.

$[-42 + (-18)] + 24 = -60 + 24 = -36$

65.

$-14 + 23 + (-65) + 9 = [-14 + (-65)] + (23 + 9)$
$= -79 + 32 = -47$

67.

$\frac{48}{36} = \frac{12 \cdot 4}{12 \cdot 3} = \frac{4}{3}$

69.

$-\frac{24}{56} = -\frac{8 \cdot 3}{8 \cdot 7} = -\frac{3}{7}$

71.

$\frac{4}{13} + \frac{6}{13} = \frac{4 + 6}{13} = \frac{10}{13}$ Rule 3

73.

$\frac{3}{5} + \frac{2}{3} = \frac{9}{15} + \frac{10}{15} = \frac{19}{15}$

Note: LCD $= 5 \cdot 3 = 15$

75.

$-\frac{3}{8} + \frac{5}{9} + \frac{11}{36} = -\frac{27}{72} + \frac{40}{72} + \frac{22}{72}$

#75. Continued

$$= \frac{-27 + 40 + 22}{72}$$

$$= \frac{35}{72}$$

Note:

$$8 = 2^3$$
$$9 = 3^2$$
$$36 = 2^2 \cdot 3^2$$
$$LCD = 2^3 \cdot 3^2 = 72$$

77.

$$3\frac{4}{5} + 2\frac{3}{10} = \frac{19}{5} + \frac{23}{10}$$

$$= \frac{38}{10} + \frac{23}{10}$$

$$= \frac{61}{10}$$

79.

$$P = a + b + c$$

Let $a = 5\frac{1}{3}$, $b = 4\frac{5}{6}$, $c = 2\frac{7}{9}$

$$P = 5\frac{1}{3} + 4\frac{5}{6} + 2\frac{7}{9}$$

$$P = \frac{16}{3} + \frac{29}{6} + \frac{25}{9}$$

$$P = \frac{96}{18} + \frac{87}{18} + \frac{50}{18}$$

$$P = \frac{233}{18} = 12\frac{17}{18}$$

The perimeter is $12\frac{17}{18}$ inches.

81.

$$-22 - 14 = -22 + (-14) \qquad \text{Add opposites}$$
$$= -36 \qquad \text{Rule 1}$$

83.

$$14.79 - (-75.6) = 14.79 + 75.6 \quad \text{Add opposites}$$
$$= 90.39$$

85.

$$-\frac{4}{5} + \frac{2}{15} - \frac{9}{10} = -\frac{4}{5} + \frac{2}{15} + \left(\frac{-9}{10}\right) \text{Add opposites}$$

#85. Continued

$$= -\frac{24}{30} + \frac{4}{20} + \left(\frac{-27}{30}\right)$$

$$= \frac{-24 + 4 + (-27)}{30}$$

$$= \frac{-47}{30}$$

87.

$$-27 - (6 - 52) = -27 - (-46)$$
$$= -27 + 46$$
$$= 19$$

89.

$$-\frac{8}{7} - \left(\frac{3}{14} - \frac{1}{2}\right)$$

$$= -\frac{8}{7} - \left(\frac{3}{14} + \left(\frac{-1}{2}\right)\right)$$

$$= -\frac{8}{7} - \left(\frac{3}{14} + \left(-\frac{7}{14}\right)\right)$$

$$= -\frac{8}{7} - \left(-\frac{4}{14}\right)$$

$$= -\frac{8}{7} - \left(-\frac{2}{7}\right) \qquad \text{Reduce}$$

$$= -\frac{8}{7} + \frac{2}{7} = -\frac{6}{7}$$

91.

$$-\frac{3}{8} - \left|\frac{1}{4} - 7\right|$$

$$= -\frac{3}{8} - \left|\frac{1}{4} - \frac{28}{4}\right|$$

$$= -\frac{3}{8} - \left|\frac{-27}{4}\right|$$

$$= -\frac{3}{8} - \left(\frac{27}{4}\right)$$

$$= -\frac{3}{8} + \left(\frac{-27}{4}\right)$$

$$= -\frac{3}{8} + \left(\frac{-54}{8}\right)$$

$$= \frac{-3 + (-54)}{8}$$

$$= \frac{-57}{8}$$

93.

$$-7 - [63 - (5 - 21)] = -7 - [63 - (-16)]$$
$$= -7 - [63 + 16]$$
$$= -7 - 79 = -86$$

95.

$$-\frac{5}{2} - \frac{1}{6} = -\frac{5}{2} + \left(-\frac{1}{6}\right) = -\frac{5}{2} \cdot \frac{3}{3} + \left(-\frac{1}{6}\right) \quad LCD = 6$$
$$= -\frac{15}{6} + \left(-\frac{1}{6}\right)$$
$$= -\frac{16}{6} = -\frac{8}{3} \qquad \text{Reduce}$$

97.

$$180^{\circ} - (96.3^{\circ} + 53.8^{\circ}) = 180^{\circ} - (150.1^{\circ})$$
$$= 29.9^{\circ}$$

99.

$$(-5)(-5)(-5) = -125$$

There is an odd number of negative factors, so the product is negative.

101.

$$(-12)\left(-\frac{15}{16}\right) = \left(-\frac{12}{1}\right)\left(-\frac{15}{16}\right)$$
$$= \frac{(-12)(-15)}{16}$$
$$= \frac{(-1) \cdot 3 \cdot \cancel{4}(-1) \cdot 3 \cdot 5}{\cancel{4} \cdot 4} = \frac{45}{4}$$

103.

$$\frac{1}{9}\left(-\frac{6}{7}\right)\left(-\frac{21}{4}\right)$$
$$= \frac{1 \cdot (-6)(-21)}{9 \cdot 7 \cdot 4}$$
$$= \frac{(-1) \cdot \cancel{2} \cdot \cancel{3} \cdot (-1) \cdot \cancel{3} \cdot \cancel{7}}{\cancel{3} \cdot \cancel{3} \cdot \cancel{7} \cdot \cancel{2} \cdot 2} = \frac{1}{2}$$

105.

$$(-4)^4 = (-4)(-4)(-4)(-4) = 256$$

107.

$$(-1)^{45} = -1$$

109.

$$16 - 6 \cdot 5 - 5 = 16 - 30 - 5 = 16 - 35 = -19$$

111.

$$2\left(1\frac{3}{4}\right) + \frac{11}{8}\left(-\frac{2}{5}\right)$$
$$= 2\left(\frac{7}{4}\right) + \frac{11}{8}\left(-\frac{2}{5}\right)$$
$$= \frac{\cancel{2} \cdot 7}{2 \cdot \cancel{2}} + \frac{11 \cdot (-1) \cdot \cancel{2}}{\cancel{2} \cdot 4 \cdot 5}$$
$$= \frac{7}{2} + \left(\frac{-11}{20}\right)$$
$$= \frac{7}{2} \cdot \frac{10}{10} + \frac{-11}{20} \qquad LCD = 20$$
$$= \frac{70}{20} + \left(\frac{-11}{20}\right)$$
$$= \frac{70 + (-11)}{20} = \frac{59}{20} \qquad \text{Rules 1 and 2}$$

113.

$$8\left(-\frac{1}{4}\right)^2 - 10\left(-\frac{1}{4}\right) = 8\left(\frac{1}{16}\right) + \frac{10}{4}$$
$$= \frac{8}{16} + \frac{10}{4} \qquad LCD = 16$$
$$= \frac{8}{16} + \frac{10}{4} \cdot \frac{4}{4}$$
$$= \frac{8}{16} + \frac{40}{16} = \frac{48}{16} = 3$$

115.

$$8[3(7 - 4) + 2] = 8[3(3) + 2] = 8[9 + 2]$$
$$= 8[11] = 88$$

117.

$$-\frac{3}{5} + \frac{1}{2}\left(\frac{2}{3} - 4\right)$$
$$= -\frac{3}{5} + \frac{1}{2}\left(\frac{2}{3} - \frac{12}{3}\right) \qquad LCD = 3$$
$$= -\frac{3}{5} + \frac{1}{2}\left(-\frac{10}{3}\right) \qquad \text{Rules 2 and 3}$$
$$= -\frac{3}{5} + \frac{(-1) \cdot \cancel{2} \cdot 5}{\cancel{2} \cdot 3}$$
$$= -\frac{3}{5} + \left(\frac{-5}{3}\right) \qquad LCD = 15$$

#117 continued

$$= \frac{-9}{15} + \left(\frac{-25}{15}\right)$$

$$= \frac{-9 + (-25)}{15} = -\frac{34}{15} \qquad \text{Rules 1 and 3}$$

119.

$$\frac{7}{20} - \left[\left(\frac{11}{24} - \frac{7}{15}\right)12 + 2\right]$$

$$= \frac{7}{20} - \left[\left(\frac{55}{120} - \frac{56}{120}\right)12 + 2\right] \quad \text{LCD} = 120$$

$$= \frac{7}{20} - \left[\left(-\frac{1}{120}\right)12 + 2\right]$$

$$= \frac{7}{20} - \left[-\frac{12}{120} + 2\right] \qquad \text{Multiply by 12}$$

$$= \frac{7}{20} - \left[-\frac{1}{10} + 2\right] \qquad \text{Reduce}$$

$$= \frac{7}{20} - \left[-\frac{1}{10} + \frac{20}{10}\right] \qquad \text{LCD} = 10$$

$$= \frac{7}{20} - \left[\frac{19}{10}\right] \qquad \text{LCD} = 20$$

$$= \frac{7}{20} - \left[\frac{38}{20}\right] \qquad \text{Rule 3}$$

$$= \frac{7 - 38}{20} = \frac{-31}{20} \qquad \text{Rules 2 and 3}$$

121.

$$2\frac{1}{4} \left| \frac{11}{3} - \frac{29}{5} \right| - 3$$

$$= \frac{9}{4} \left| \frac{55}{15} - \frac{87}{15} \right| - 3 \qquad \text{LCD} = 15$$

$$= \frac{9}{4} \left| \frac{-32}{15} \right| - 3 \qquad \text{Rules 2 and 3}$$

$$= \frac{9}{4} \left(\frac{32}{15}\right) - 3 \qquad \text{Absolute value}$$

$$= \frac{3 \cdot 3 \cdot 4 \cdot 8}{4 \cdot 3 \cdot 5} - 3$$

$$= \frac{24}{5} - 3$$

$$= \frac{24}{5} - \frac{15}{5} \qquad \text{LCD} = 5$$

$$= \frac{9}{5} \qquad \text{Rules 2 and 3}$$

123.
The formula for the area of a trapezoid is
$A = \frac{1}{2}(B + b)\,h.$
Let B = 12.768, b = 6.032, and h = 4.25.

Thus,
$$A = \frac{1}{2}(12.768 + 6.032) \cdot 4.25$$

$$A = \frac{1}{2}(18.8) \cdot 4.25$$

$$A = \frac{1}{2}(79.9)$$

$$A = 39.95$$

The area is 39.95 m^2.

125.

$$-70 + (-5) = 14 \qquad \text{Rule 9}$$

127.

$$-\frac{16}{15} \div 10 = -\frac{16}{15} \cdot \frac{1}{10}$$

$$= \frac{(-1) \cdot 2 \cdot 8}{3 \cdot 5 \cdot 2 \cdot 5}$$

$$= -\frac{8}{75}$$

129.

$$\frac{-40}{\frac{5}{8}} = -40 \div \frac{5}{8}$$

$$= \frac{-40}{1} \cdot \frac{8}{5}$$

$$= \frac{(-1) \cdot 5 \cdot 8 \cdot 8}{1 \cdot 5}$$

$$= -64$$

131.

$$\frac{\frac{56}{45}}{20} = \frac{56}{45} \div 20$$

$$= \frac{56}{45} \cdot \frac{1}{20}$$

$$= \frac{7 \cdot 2 \cdot 4}{5 \cdot 9 \cdot 5 \cdot 4}$$

$$= \frac{14}{225}$$

133.

$6 \cdot 40 + 5 + 3$

$= (6 \cdot 40 + 5) + 3$ Order of operations

$= (240 + 5) + 3$

$= 48 + 3 = 51$

135.

$-32 - 6 + 99 + (-11)$

$= -32 - 6 + [99 + (-11)]$ Order of operations

$= -32 - 6 + (-9)$

$= -47$ Rule 1

137.

$4\frac{1}{2} + \left(-3\frac{3}{4}\right) + 2\frac{1}{10}$

$= \frac{9}{2} + \left(-\frac{15}{4}\right) + \frac{21}{10}$

$= \frac{9}{2}\left(-\frac{4}{15}\right)\frac{10}{21}$ Rule 11

$= \frac{\cancel{3} \cdot 3 \cdot (-1) \cdot \cancel{2} \cdot 2 \cdot 2 \cdot \cancel{5}}{\cancel{2} \cdot 3 \cdot \cancel{5} \cdot 3 \cdot 7}$

$= -\frac{4}{7}$

139.

$-27 + |6 - 5 \cdot 3|$

$= -27 + |6 - 15|$

$= -27 + |-9|$

$= -27 + 9$ Absolute value

$= -3$

141.

$\frac{3}{10} - \left(\frac{6}{5}\right)^2 + \frac{4}{15} = \frac{3}{10} - \frac{36}{25} + \frac{4}{15}$

$= \frac{3}{10} - \frac{36}{25} \cdot \frac{15}{4}$

$= \frac{3}{10} - \frac{27}{5}$ LCD = 10

$= \frac{3}{10} - \frac{54}{10} = -\frac{51}{10}$

143.

$\dfrac{-8 - \sqrt{8^2 - 4 \cdot 3 \cdot 4}}{2 \cdot 3} = \dfrac{-8 - \sqrt{64 - 48}}{6}$

$= \dfrac{-8 - \sqrt{16}}{6}$

$= \dfrac{-8 - 4}{6} = \dfrac{-12}{6} = -2$

145.

$\dfrac{2\left(\frac{4}{5}\right) + \frac{1}{2}}{\frac{5}{4} - 2} = \dfrac{\frac{8}{5} + \frac{1}{2}}{\frac{5}{4} - \frac{2}{1}} = \dfrac{\frac{16}{10} + \frac{5}{10}}{\frac{5}{4} - \frac{8}{4}}$

$= \dfrac{\frac{21}{10}}{-\frac{3}{4}} = \frac{21}{10} + \left(-\frac{3}{4}\right)$

$= \frac{21}{10}\left(-\frac{4}{3}\right) = -\frac{14}{5}$

147.

$\dfrac{84}{\frac{7}{5}} = 84 + \frac{7}{5} = \frac{84}{1} \cdot \frac{5}{7} = 60$

She averaged 60 mph.

149.

$\dfrac{6}{\frac{3}{5}} = \frac{6}{1} + \frac{3}{5} = \frac{6}{1} \cdot \frac{5}{3} = 10$

---

151. — 167.

The answers do not require worked-out solutions.

Chapter 1 Test Solutions

1. - 11.
The answers do not require solutions.

13.
$P = 2l + 2w$

$P = 2\left(6\frac{1}{4}\right) + 2\left(4\frac{1}{2}\right)$

$P = 2\left(\frac{25}{4}\right) + 2\left(\frac{9}{2}\right)$

$P = \frac{50}{4} + \frac{18}{2}$

$P = \frac{50}{4} + \frac{36}{4}$

$P = \frac{86}{4}$ or $\frac{43}{2}$ or $21\frac{1}{2}$ ft

15.

$\dfrac{40\frac{3}{10}}{6\frac{1}{5}} = 40\frac{3}{10} + 6\frac{1}{5}$

$= \frac{403}{10} + \frac{31}{5}$

$= \frac{403}{10} \cdot \frac{5}{31}$

$= \frac{13}{2} \cdot \frac{1}{1}$

$= \frac{13}{2}$ or $6\frac{1}{2}$ Answer: $6\frac{1}{2}$ min/mile

---

17.
Does not require a worked-out solution.

19.
$-63 + 17 = -46$ Rule 2

21.
$4.87 - 12.09 = 4.87 + (-12.09)$

Add opposites
$= -7.22$ Rule 2

---

23. $\dfrac{7}{10} - \dfrac{3}{4} - \dfrac{6}{5} = \dfrac{14}{20} - \dfrac{15}{20} - \dfrac{24}{20}$ LCD = 20

$= \dfrac{14 - 15 - 24}{20}$

#23 continued

$= -\dfrac{25}{20}$ or $-\dfrac{5}{4}$ Reduce

25.

$\left(-\dfrac{8}{9}\right)\left(-\dfrac{15}{22}\right) = \left(-\dfrac{4}{3}\right)\left(-\dfrac{5}{11}\right)$ Cancel

$= \dfrac{20}{33}$

27.

$52 \div (-13) = \dfrac{52}{-13} = -4$

29.

$\dfrac{5}{12} \div \left(-\dfrac{10}{27}\right) \cdot \dfrac{7}{6} \quad = \dfrac{5}{12} \cdot \left(-\dfrac{27}{10}\right) \cdot \dfrac{7}{6}$

$= \dfrac{1}{4} \cdot \left(-\dfrac{3}{2}\right) \dfrac{7}{2}$ Cancel

$= -\dfrac{21}{16}$

31.
$5 \cdot 2^3 - 6 \cdot 7 = 5 \cdot 8 - 6 \cdot 7$
$= 40 - 42 = -2$

33.
$5.4(-0.62) - 3.208 = -3.348 - 3.208$
$= -3.348 + (-3.208)$
$= -6.556$

35.

$-4 - \left(\dfrac{11}{4} - \dfrac{2}{3}\right) \quad = -4 - \left(\dfrac{33}{12} - \dfrac{8}{12}\right)$ LCD = 12

$= -4 - \left(\dfrac{25}{12}\right)$

$= -\dfrac{48}{12} - \dfrac{25}{12}$ LCD = 12

$= -\dfrac{73}{12}$

37.
$7 - 3[-12 - (-20)(4)] \quad = 7 - 3[-12 - (-80)]$
$= 7 - 3[-12 + 80]$
$= 7 - 3[68]$
$= 7 - 204 = -197$

39.

$3 + (5^2 + 3 \cdot 8) + (-7)$ $= 3 + (25 + 24) + (-7)$
$= 3 + 49 + (-7)$
$= 3 - 7 = -4$

41.

$$\frac{9 - \sqrt{(-9)^2 - 4 \cdot 20}}{2} = \frac{9 - \sqrt{81 - 80}}{2}$$
$$= \frac{9 - \sqrt{1}}{2}$$
$$= \frac{9 - 1}{2} = \frac{8}{2} = 4$$

CHAPTER 1 STUDY GUIDE

Self-Test Exercises

1. Change $\frac{29}{4}$ to a decimal.

2. Evaluate $|-9| + 9$.

3. Place the appropriate symbol , $<$, $>$, or $=$ between the pair of numbers. -8 _____ -18

4. Use the distributive property to rewrite $7 \cdot 10 + 7 \cdot 11$.

5. Reduce $\frac{42}{48}$ to lowest terms.

6. Find the area of a trapezoid with bases 15 m and 9 m with height 3m.

7. Evaluate -8^2.   8. Evaluate $-\sqrt{225}$.

Perform the indicated operations.

9. $-\frac{3}{15} + \left(-\frac{7}{6}\right)$

10. $-5\frac{3}{5} - 3\frac{9}{20}$

11. $-16 - 31 + 89$

12. $(-6)(-4)(3)(-1)$

13. $-84 \div (-28)$

14. $\frac{5}{7} - \frac{17}{54} \cdot 9$

15. $-18 - 102 \div (-6)$

16. $-17 - 7\sqrt{64}$

17. $-50 - (19 + 31)$

18. $\frac{5}{2}\left(\frac{11}{6} - \frac{4}{9} \cdot \frac{36}{14}\right)$

19. $7\sqrt{(7-2)^2 + 11}$

20. $\dfrac{\frac{3}{5} - \frac{7}{3}}{\frac{11}{6} + \frac{4}{15}}$

The worked-out solutions begin on the next page.

Self-Test Solutions

1.
7.25

Work

$$\begin{array}{r} 7.25 \\ 4\overline{)29.00} \\ \underline{28} \\ 10 \\ \underline{8} \\ 20 \\ \underline{20} \end{array}$$

2.
$|-9| + 9 = 9 + 9 = 18$

3.
$-8 > -18$

4.
$7 \cdot 10 + 7 \cdot 11 = 7(10 + 11)$

5.
$\dfrac{42}{48} = \dfrac{\cancel{6} \cdot 7}{\cancel{6} \cdot 8} = \dfrac{7}{8}$

6.
$A = \dfrac{1}{2}(b + B) \cdot h$

$A = \dfrac{1}{2}(15 + 9) \cdot 3$

$A = \dfrac{1}{2}(24) \cdot 3$

$A = 12 \cdot 3$

$A = 36$

The area is 36 sq. m

7.
$-8^2 = -1 \cdot 8 \cdot 8 = -1 \cdot 64 = -64$

8.
$-\sqrt{225} = -15$

9.
$-\dfrac{3}{15} + \left(-\dfrac{7}{6}\right) = -\dfrac{3}{15} \cdot \dfrac{2}{2} + \left(-\dfrac{7}{6} \cdot \dfrac{5}{5}\right)$

$= -\dfrac{6}{30} + \left(-\dfrac{35}{30}\right) = -\dfrac{41}{30}$

10.
$-5\dfrac{3}{5} - 3\dfrac{9}{20} = -5\dfrac{3}{5} + \left(-3\dfrac{9}{20}\right)$

Add opposites

$= -\dfrac{28}{5} + \left(-\dfrac{69}{20}\right)$ LCD = 20

$= -\dfrac{28}{5} \cdot \dfrac{4}{4} + \left(-\dfrac{69}{20}\right)$

$= -\dfrac{112}{20} + \left(-\dfrac{69}{20}\right)$

$= -\dfrac{181}{20}$ or $-9\dfrac{1}{20}$

11.
$-16 - 31 + 89 = -47 + 89 = 42$

12.
$(-6)(-4)(3)(-1) = 24(3)(-1) = 72(-1) = -72$

13.
$(-84) + (-28) = 3$

14.
$\dfrac{5}{7} - \dfrac{17}{54} \cdot 9 = \dfrac{5}{7} - \dfrac{17}{54} \cdot \dfrac{9}{1} = \dfrac{5}{7} - \dfrac{17}{6}$ Cancelling

$= \dfrac{30}{42} - \dfrac{119}{42}$ LCD = 42

$= -\dfrac{89}{42}$

15.
$-18 - 102 + (-6) = -18 + 17 = -1$

16.
$-17 - 7\sqrt{64} = -17 - 7 \cdot 8 = -17 - 56 = -73$

17.
$-50 - (19 + 31) = -50 - (50)$ $= -50 - 50$
$= -50 + (-50)$
$= -100$

18.
$\dfrac{5}{2}\left(\dfrac{11}{6} - \dfrac{4}{9} \cdot \dfrac{36}{14}\right) = \dfrac{5}{2}\left(\dfrac{11}{6} - \dfrac{8}{7}\right)$ Cancelling

$= \dfrac{5}{2}\left(\dfrac{77}{42} - \dfrac{48}{42}\right)$ LCD = 42

$= \dfrac{5}{2}\left(\dfrac{29}{42}\right) = \dfrac{145}{84}$

19.

$$7\sqrt{(7-2)^2 + 11} = 7\sqrt{5^2 + 11} = 7\sqrt{25 + 11}$$
$$= 7\sqrt{36} = 7 \cdot 6 = 42$$

20.

$$\frac{\dfrac{3}{5} - \dfrac{7}{3}}{\dfrac{11}{6} + \dfrac{4}{15}} = \frac{\dfrac{9}{15} - \dfrac{35}{15}}{\dfrac{55}{30} + \dfrac{8}{30}} = \frac{-\dfrac{26}{15}}{\dfrac{63}{30}} = -\frac{26}{15} + \frac{63}{30}$$

$$= -\frac{26}{15} \cdot \frac{30}{63} = -\frac{26}{1} \cdot \frac{2}{63} \quad \text{Cancelling}$$

$$= -\frac{52}{63}$$

CHAPTER 2 LINEAR EQUATIONS AND INEQUALITIES
IN ONE VARIABLE

Solutions to Text Exercises

15 continued

$$= 1 - \frac{35}{2}$$

$$= -\frac{33}{2}$$

Exercises 2.1

1.
$5x + 8; \quad x = 3$
$5 \cdot 3 + 8 = 15 + 8 = 23$

3.
$4 - 6x; \quad x = 4$
$4 - 6 \cdot 4 = 4 - 24 = -20$

5.
$x^2 + 7x - 4; \quad x = 2$
$(2)^2 + 7 \cdot 2 - 4 = 4 + 14 - 4 = 14$

7.
$8 - x - x^2; \quad x = -7$
$8 - (-7) - (-7)^2 = 8 + 7 - 49 = -34$

9.
$4x^2 - 9; \quad x = 3$
$4(3)^2 - 9 = 4 \cdot 9 - 9 = 36 - 9 = 27$

11.
$(x + 4)(x - 5); \quad x = -4$
$(-4 + 4)(-4 - 5) = 0(-9) = 0$

13.
$(x^2 - 4)(5x + 7); \quad x = -2$
$[(-2)^2 - 4] \, [\, 5(-2) + 7] = (4 - 4)(-10 + 7)$
$= 0(-3)$
$= 0$

15.

$2x - 5(x + 3); \quad x = \frac{1}{2}$

$2\left(\frac{1}{2}\right) - 5\left(\frac{1}{2} + 3\right) = 1 - 5\left(\frac{7}{2}\right)$

17.
$6(x + 1) - 2(x + 3); \quad x = -4$
$6(-4 + 1) - 2(-4 + 3) = 6(-3) - 2(-1)$
$= -18 + 2$
$= -16$

19.

$2(x - 4) - (2x - 1); \quad x = \frac{1}{3}$

$2\left(\frac{1}{3} - 4\right) - (2 \cdot \frac{1}{3} - 1) = 2\left(-\frac{11}{3}\right) - \left(\frac{2}{3} - 1\right)$

$= \frac{-22}{3} - \left(-\frac{1}{3}\right)$

$= \frac{-22}{3} + \frac{1}{3}$

$= \frac{-21}{3}$

$= -7$

21.
$x^2 + 2xy + y^2; \quad x = 2, y = 5$
$2^2 + 2 \cdot 2 \cdot 5 + 5^2 = 4 + 20 + 25$
$= 49$

23.
$5(x + 2y) - 4(x - 3y); \quad x = -1, y = 3$
$5(-1 + 2 \cdot 3) - 4(-1 - 3 \cdot 3) = 5(5) - 4(-10)$
$= 25 + 40$
$= 65$

25.
$5 + 3(5x + y); \quad x = -2, y = -1$
$5 + 3[5(-2) + (-1)] = 5 + 3(-10 - 1)$

#25 continued

$$= 5 + 3(-11)$$
$$= 5 - 33 = -28$$

27.

$$\frac{2x}{x+1} + \frac{x-2}{x+4}; \quad x = 5$$

$$\frac{2 \cdot 5}{5+1} + \frac{5-2}{5+4} = \frac{10}{6} + \frac{1}{3}$$

$$= \frac{6}{3} = 2$$

29.

$$\frac{1}{2}x + \frac{2}{3}y - \frac{1}{4}z; \quad x = 3, \quad y = -2, \quad z = 1$$

$$\frac{1}{2} \cdot 3 + \frac{2}{3}(-2) - \frac{1}{4} \cdot 1$$

$$= \frac{3}{2} - \frac{4}{3} - \frac{1}{4} \qquad \text{LCD} = 12$$

$$= \frac{18}{12} - \frac{16}{12} - \frac{3}{12}$$

$$= -\frac{1}{12} \qquad \text{Rule 3}$$

31.

$$\frac{2x}{y} + \frac{3y}{z} - 2; \quad x = 1, \ y = 3, \ z = 4$$

$$\frac{2 \cdot 1}{3} + \frac{3 \cdot 3}{4} - 2 = \frac{2}{3} + \frac{9}{4} - 2$$

$$= \frac{8}{12} + \frac{27}{12} - \frac{24}{12} \quad \text{LCD} = 12$$

$$= \frac{11}{12} \qquad \text{Rule 3}$$

33.

$$\frac{1}{2}(x+3) + \frac{1}{3}(y-4); \quad x = -2, \ y = -3$$

$$\frac{1}{2}(-2+3) + \frac{1}{3}(-3-4) = \frac{1}{2}(1) + \frac{1}{3}(-7)$$

$$= \frac{1}{2} - \frac{7}{3}$$

$$= \frac{3}{6} - \frac{14}{6}$$

$$= -\frac{11}{6}$$

35.

$$1.235x + 2.6(3.14x - 7.26); \quad x = 5$$
$$1.235 \cdot 5 + 2.6(3.14 \cdot 5 - 7.26)$$
$$= 6.175 + 2.6(15.7 - 7.26)$$
$$= 6.175 + 2.6(8.44)$$
$$= 6.175 + 21.944$$
$$= 28.119$$

37.

$$\frac{x+2y}{\sqrt{4z}}; \quad x = -7, \ y = 3, \ z = 4$$

$$\frac{-7 + 2 \cdot 3}{\sqrt{4 \cdot 4}} = \frac{-7 + 6}{\sqrt{16}}$$

$$= \frac{-1}{4}$$

39.

$$\sqrt{43 - x^2 + 2y^2}; \quad x = 5, \ y = -3$$

$$\sqrt{43 - 5^2 + 2(-3)^2} = \sqrt{43 - 25 + 18}$$

$$= \sqrt{36}$$

$$= 6$$

41.

$$2x + 9x = (2+9)x \quad \text{Distributive property}$$
$$= 11x$$

43.

$$5x - 8x + 7x = (5 - 8 + 7)x \quad \text{Distributive property}$$
$$= 4x$$

45.

$$-8x + 4 - 2x + 1$$
$$= -8x - 2x + 4 + 1 \qquad \text{Commutative property}$$
$$= (-8 - 2)x + 5 \qquad \text{Distributive property}$$
$$= -10x + 5$$

47.

$$6x + y - 8x - 2y$$
$$= 6x - 8x + y - 2y \qquad \text{Commutative property}$$
$$= (6 - 8)x + (1 - 2)y \quad \text{Distributive property}$$
$$= -2x - y$$

49.

$x^2 - 2x + 3z + 5x^2$

$= x^2 + 5x^2 - 2x + 3z$

$= (1 + 5)x^2 - 2x + 3z$ Distributive property

$= 6x^2 - 2x + 3z$

51.

$5x + 2y - 7$

There are no common like terms to combine.

53.

$9x^2 - 3x + 3x - 1$

$= 9x^2 + (-3 + 3)x - 1$ Distributive property

$= 9x^2 + 0x - 1$

$= 9x^2 - 1$ Since $0 \cdot x = 0$

55.

$\dfrac{5}{8}x - 1 - \dfrac{1}{4}x$

$= \dfrac{5}{8}x - \dfrac{1}{4}x - 1$ Commutative property

$= \left(\dfrac{5}{8} - \dfrac{1}{4}\right)x - 1$ Distributive property

$= \dfrac{3}{8}x - 1$

57.

$-2.839y + 1.001y - 3.86y$

$= (-2.839 + 1.001 - 3.86)y$

$= (1.001 - 6.699)y$ Distributive property

$= -5.698y$

59.

$\dfrac{1}{3}x^2 - \dfrac{1}{2}y + \dfrac{5}{6}x^2 + \dfrac{1}{4}y$

$= \dfrac{1}{3}x^2 + \dfrac{5}{6}x^2 - \dfrac{1}{2}y + \dfrac{1}{4}y$

           Commutative property

$= \left(\dfrac{1}{3} + \dfrac{5}{6}\right)x^2 + \left(-\dfrac{1}{2} + \dfrac{1}{4}\right)y$

           Distributive property

$= \left(\dfrac{2}{6} + \dfrac{5}{6}\right)x^2 + \left(-\dfrac{2}{4} + \dfrac{1}{4}\right)y$

#59 continued

$= \dfrac{7}{6}x^2 + \left(-\dfrac{1}{4}\right)y$ Rule 3

$= \dfrac{7}{6}x^2 - \dfrac{1}{4}y$

61.

$5x + 3(2x - 6)$

$= 5x + 6x - 18$ Distributive property

$= (5 + 6)x - 18$ Distributive property

$= 11x - 18$

63.

$8 - 3(x + 4)$

$= 8 - 3x - 12$ Distributive property

$= -3x + 8 - 12$ Commutative property

$= -3x - 4$

65.

$3x + 5 + 2(4x - 1)$

$= 3x + 5 + 8x - 2$ Distributive property

$= 3x + 8x + 5 - 2$ Commutative property

$= (3 + 8)x + 5 - 2$ Distributive property

$= 11x + 3$

67.

$2x - 13 - (6x - 10)$

$= 2x - 13 - 6x + 10$ Distributive property

$= 2x - 6x - 13 + 10$ Commutative property

$= (2 - 6)x - 13 + 10$ Distributive property

$= -4x - 3$

69.

$x + 15 - 2(3x + 6)$

$= x + 15 - 6x - 12$ Distributive property

$= x - 6x + 15 - 12$ Commutative property

$= (1 - 6)x + 15 - 12$ Distributive property

$= -5x + 3$

71.

$3(x + 1) + 6(x - 2)$

$= 3x + 3 + 6x - 12$ Distributive property

$= 3x + 6x + 3 - 12$ Commutative property

$= (3 + 6)x + 3 - 12$ Distributive property

$= 9x - 9$

73.
$5(2x - 1) - 3(3x + 4)$

$=10x - 5 - 9x - 12$	Distributive property
$= 10x - 9x - 5 - 12$	Commutative property
$= (10 - 9)x - 5 - 12$	Distributive property
$= x - 17$	

75.
$3(2x - 6) - (x - 2)$

$= 6x - 18 - x + 2$	Distributive property
$= 6x - x - 18 + 2$	Commutative property
$= (6 - 1)x - 18 + 2$	Distributive property
$= 5x - 16$	

77.
$5(x + y) - 3(x + y)$

$= 5x + 5y - 3x - 3y$	Distributive property
$= 5x - 3x + 5y - 3y$	CommutativeProperty
$= (5 - 3)x + (5 - 3)y$	Distributive property
$= 2x + 2y$	

79.
$2(x + y) + 3(a + b)$

$= 2x + 2y + 3a + 3b$	Distributive property

81.
$\frac{1}{3}(x + 2) - \frac{1}{2}(x - 3)$

$=\frac{1}{3}x + \frac{2}{3} - \frac{1}{2}x + \frac{3}{2}$	Distributive property
$=\frac{1}{3}x - \frac{1}{2}x + \frac{2}{3} + \frac{3}{2}$	Commutative property
$=\left(\frac{1}{3} - \frac{1}{2}\right)x + \frac{2}{3} + \frac{3}{2}$	Distributive property
$=\left(\frac{2}{6} - \frac{3}{6}\right)x + \frac{4}{6} + \frac{9}{6}$	LCD = 6
$= -\frac{1}{6}x + \frac{13}{6}$	

83.
$0.02(2x + 3) + 0.05(3x - 1)$
$= 0.04x + 0.06 + 0.15x - 0.05$
$= 0.04x + 0.15x + 0.06 - 0.05$
$= (0.04 + 0.15)x + 0.06 - 0.05$
$= 0.19x + 0.01$

85.
$5x + 3[2(x + 4) - (x + 7)]$

$= 5x + 3(2x + 8 - x - 7)$	Distributive property
$= 5x + 3(2x - x + 8 - 7)$	Commutative property
$= 5x + 3(x + 1)$	
$= 5x + 3x + 3$	Distributive property
$= 8x + 3$	

87.
$7(2x - 4y + 3) - 3(9x - 2y + 8)$
$= 14x - 28y + 21 - 27x + 6y - 24$
$= 14x - 27x - 28y + 6y + 21 - 24$
$= (14 - 27)x + (-28 + 6)y + 21 - 24$
$= -13x - 22y - 3$

89.
$25x + 2y$
Let $x = 1$ and $y = 29$
$25 \cdot 1 + 2 \cdot 29 = 25 + 58 = 83$
The monthly cost is $83.

91.
Let short side measure x.
Let medium side measure $x + 2$.
Let long side measure $x + 5$.
Then,
$P = x + (x + 2) + (x + 5)$
$P = x + x + x + 2 + 5$
$P = 3x + 7$

93.
Let short side measure x.
Let medium side measure $x + 5$.
Let long side measure $3x$.
Then,
$P = x + (x + 5) + (3x)$
$P = x + x + 3x + 5$
$P = 5x + 5$

95.
Let width measure x.
Let length measure $x + 9$.
Then,
$P = 2x + 2(x + 9)$
$P = 2x + 2x + 18$
$P = 4x + 18$
Note: $P = 2W + 2L$

97.
Let width measure x.
Let length measure 2x + 3
Then,
P = 2x + 2(2x + 3)
P = 2x + 4x + 6
P = 6x + 6
Note: P = 2W + 2L

Exercises 2.2

1.

$$x - 7 = 10$$
$$x - 7 + 7 = 10 + 7 \quad \text{Add 7 to both sides}$$
$$x = 17$$

Solution set: {17}

3.

$$4 = x - 9$$
$$4 + 9 = x - 9 + 9 \quad \text{Add 9 to both sides}$$
$$13 = x$$

Solution set: {13}

5.

$$-8 = x + 2$$
$$-8 - 2 = x + 2 - 2 \quad \text{Subtract 2 from both sides}$$
$$-10 = x$$

Solution set: {-10}

7.

$$x - 2 = -4$$
$$x + 2 + 2 = -4 + 2 \quad \text{Add 2 to both sides}$$
$$x = -2$$

Solution set: {-2}

9.

$$x - \frac{3}{4} = \frac{1}{8}$$
$$x - \frac{3}{4} + \frac{3}{4} = \frac{1}{8} + \frac{3}{4} \quad \text{Add } \frac{3}{4} \text{ to both sides}$$
$$x = \frac{1}{8} + \frac{6}{8}$$
$$x = \frac{7}{8}$$

#9 continued

Solution set: $\left\{\dfrac{7}{8}\right\}$

11.

$$x + 3 = -1$$
$$x + 3 - 3 = -1 - 3 \quad \text{Subtract 3 from both sides}$$
$$x = -4$$

Solution set: {-4}

13.

$$7.06 = x - 1.05$$
$$7.06 + 1.05 = x - 1.05 + 1.05$$
$$\text{Add 1.05 to both sides}$$
$$8.11 = x$$

Solution set: {8.11}

15.

$$3.27 = x + 7.89$$
$$3.27 - 7.89 = x + 7.89 - 7.89 \quad \text{Subtract 7.89}$$
$$-4.62 = x$$

Solution set: {-4.62}

17.

$$0 = x + 5$$
$$0 - 5 = x + 5 - 5 \quad \text{Subtract 5}$$
$$-5 = x$$

Solution set: {-5}

19.

$$x + 6 = -4$$
$$x + 6 - 6 = -4 - 6 \quad \text{Subtract 6}$$
$$x = -10$$

Solution set: {-10}

21.

$$x - 2 = 8$$
$$x - 2 + 2 = 8 + 2 \quad \text{Add 2}$$
$$x = 10$$

Solution set: {10}

23.

$$x + \frac{1}{3} = -\frac{5}{6}$$

$$x + \frac{1}{3} - \frac{1}{3} = -\frac{5}{6} - \frac{1}{3} \qquad \text{Subtract } \frac{1}{3}$$

$$x = -\frac{5}{6} - \frac{2}{6}$$

$$x = -\frac{7}{6}$$

Solution set: $\left\{ -\frac{7}{6} \right\}$

25.

$$x - 7 = -7$$

$$x - 7 + 7 = -7 + 7 \qquad \text{Add 7}$$

$$x = 0$$

Solution set: $\{0\}$

27.

$$0 = x - 11$$

$$0 + 11 = x - 11 + 11 \qquad \text{Add 11}$$

$$11 = x$$

Solution set: $\{11\}$

29.

$$x + 9 = 2$$

$$x + 9 - 9 = 2 - 9 \qquad \text{Subtract 9}$$

$$x = -7$$

Solution set: $\{-7)$

31.

$$-4 = x + 4$$

$$-4 - 4 = x + 4 - 4 \qquad \text{Subtract 4}$$

$$-8 = x$$

Solution set: $\{-8\}$

33.

$$-12 = x - 15$$

$$-12 + 15 = x - 15 + 15 \qquad \text{Add 15}$$

$$3 = x$$

Solution set: $\{3\}$

35.

$$x + 8 = 1$$

$$x + 8 - 8 = 1 - 8 \qquad \text{Subtract 8}$$

$$x = -7$$

Solution set: $\{-7\}$

37.

$$7x + 3 - 6x + 4 = 11 - 3$$

$$7x - 6x + 3 + 4 = 11 - 3$$

$$\qquad \qquad \qquad \text{Commutative property}$$

$$x + 7 = 8 \qquad \text{Simplify}$$

$$x + 7 - 7 = 8 - 7 \qquad \text{Subtract 7}$$

$$x = 1$$

Solution set: $\{1\}$

39.

$$\frac{3}{2}x - \frac{1}{2}x - 4 - 7 = -13 - 1$$

$$x - 11 = -14 \qquad \text{Simplify}$$

$$x - 11 + 11 = -14 + 11 \qquad \text{Add 11}$$

$$x = -3$$

Solution set: $\{-3\}$

41.

$$6x + 4 = 5x - 3$$

$$6x - 5x + 4 = 5x - 5x - 3 \qquad \text{Subtract 5x}$$

$$x + 4 = -3 \qquad \text{Simplify}$$

$$x + 4 - 4 = -3 - 4 \qquad \text{Subtract 4}$$

$$x = -7$$

Solution set: $\{-7\}$

43.

$$2.67x + 4.87 = 3.67x + 1.24$$

$$2.67x - 2.67x + 4.87 = 3.67x - 2.67x + 1.2$$

$$\qquad \qquad \qquad \text{Subtract } 2.67x$$

$$4.87 = x + 1.24$$

$$4.87 - 1.24 = x + 1.24 - 1.24$$

$$\qquad \qquad \qquad \text{Subtract } 1.24$$

$$3.63 = x$$

Solution set: $\{3.63\}$

Note: To eliminate an extra step, avoid solving for $-x$.

45.

$$5x - 3x + 7 = x + 7$$
$$2x + 7 = x + 7 \quad \text{Simplify}$$
$$2x + 7 - 7 = x + 7 - 7 \quad \text{Subtract 7}$$
$$2x = x$$
$$2x - x = x - x \quad \text{Subtract } x.$$
$$x = 0$$

Solution set: {0}

47.

$$13x - 3x + 5 = 11x - 4$$
$$10x + 5 = 11x - 4 \quad \text{Simplify}$$
$$10x - 10x + 5 = 11x - 10x - 4 \quad \text{Subtract 10x.}$$
$$5 = x - 4$$
$$5 + 4 = x - 4 + 4 \quad \text{Add 4}$$
$$9 = x$$

Solution set: {9}

49.

$$\frac{1}{8}x - \frac{2}{3} + \frac{1}{8}x = \frac{5}{4}x - \frac{5}{6}$$
$$\frac{1}{8}x + \frac{1}{8}x - \frac{2}{3} = \frac{5}{4}x - \frac{5}{6}$$
$$\frac{1}{4}x - \frac{2}{3} = \frac{5}{4}x - \frac{5}{6}$$
$$\frac{1}{4}x - \frac{1}{4}x - \frac{2}{3} = \frac{5}{4}x - \frac{1}{4}x - \frac{5}{6} \quad \text{Subtract } \frac{1}{4}x$$
$$-\frac{2}{3} = x - \frac{5}{6}$$
$$-\frac{2}{3} + \frac{5}{6} = x - \frac{5}{6} + \frac{5}{6} \quad \text{Add } \frac{5}{6}$$
$$-\frac{4}{6} + \frac{5}{6} = x$$
$$\frac{1}{6} = x$$

Solution set: $\left\{\frac{1}{6}\right\}$

51.

$$5.3x - 1.4 - 3.1x = 1.2x - 6.7$$
$$5.3x - 3.1x - 1.4 = 1.2x - 6.7$$
$$2.2x - 1.4 = 1.2x - 6.7$$
$$2.2x - 1.2x - 1.4 = 1.2x - 1.2x - 6.7$$
$$\text{Subtract 1.2x}$$
$$x - 1.4 = -6.7$$
$$x - 1.4 + 1.4 = -6.7 + 1.4 \quad \text{Add 1.4}$$

#51 continued

$$x = -5.3$$

Solution set: {-5.3}

53.

$$7x - 2 - 6 - 2x = 8x - 4x + 5 - 11$$
$$7x - 2x - 2 - 6 = 8x - 4x + 5 - 11$$
$$5x - 8 = 4x - 6 \quad \text{Simplify}$$
$$5x - 4x - 8 = 4x - 4x - 6 \quad \text{Subtract 4x}$$
$$x - 8 = -6$$
$$x - 8 + 8 = -6 + 8 \quad \text{Add 8.}$$
$$x = 2$$

Solution set: {2}

55.

$$7x + 8 - 9 - 7x = 2x + 5 - 3 - x$$
$$7x - 7x + 8 - 9 = 2x - x + 5 - 3$$
$$0 - 1 = x + 2$$
$$-1 - 2 = x + 2 - 2 \quad \text{Subtract 2}$$
$$-3 = x$$

Solution set: {-3}

57.

$$16x - 9 + 32x + 11 = 32x + 5 + 8x + 4$$
$$16x + 32x - 9 + 11 = 32x + 8x + 5 + 4$$
$$48x + 2 = 40x + 9 \quad \text{Simplify}$$
$$48x - 40x + 2 = 40x - 40x + 9$$
$$\text{Subtract 40x}$$
$$8x + 2 = 9$$
$$8x + 2 - 2 = 9 - 2 \quad \text{Subtract 2}$$
$$8x = 7$$
$$x = \frac{7}{8}$$

Solution set: $\left\{\frac{7}{8}\right\}$

59.

$$x + 56 - 72 + 2x = 5x + 6x - 8 + 64$$
$$x + 2x + 56 - 72 = 5x + 6x - 8 + 64$$
$$3x - 16 = 11x + 56$$
$$3x - 3x - 16 = 11x - 3x + 56$$
$$-16 = 8x + 56$$
$$-16 - 56 = 8x + 56 - 56$$
$$-72 = 8x$$
$$-9 = x$$

#59 continued

Solution set: $\{-9\}$

61.

$$\frac{1}{4}x + \frac{1}{2}x - \frac{1}{10} - \frac{2}{5} = \frac{5}{4}x + \frac{1}{2}x + \frac{9}{10} + \frac{4}{5}$$

$$\frac{3}{4}x - \frac{5}{10} = \frac{7}{4}x + \frac{17}{10}$$

$$\frac{3}{4}x - \frac{3}{4}x - \frac{5}{10} = \frac{7}{4}x - \frac{3}{4}x + \frac{17}{10}$$

$$-\frac{5}{10} = x + \frac{17}{10}$$

$$-\frac{5}{10} - \frac{17}{10} = x + \frac{17}{10} - \frac{17}{10} \quad \text{Subtract } \frac{17}{10}$$

$$-\frac{22}{10} = x$$

$$-\frac{11}{5} = x \qquad \text{Reduce}$$

Solution set: $\left\{-\dfrac{11}{5}\right\}$

63.

$$5(2x + 3) = 11x + 17$$
$$10x + 15 = 11x + 17 \quad \text{Distributive property}$$
$$10x - 10x + 15 = 11x - 10x + 17 \quad \text{Subtract } 10x$$
$$15 = x + 17$$
$$15 - 17 = x + 17 - 17$$
$$-2 = x$$

Solution set: $\{-2\}$

65.

$$2 + 3(x - 4) = 2x - 1$$
$$2 + 3x - 12 = 2x - 1 \quad \text{Distributive property}$$
$$3x - 10 = 2x - 1 \quad \text{Simplify}$$
$$3x - 2x - 10 = 2x - 2x - 1 \quad \text{Subtract } 2x$$
$$x - 10 = -1$$
$$x - 10 + 10 = -1 + 10 \quad \text{Add } 10$$
$$x = 9$$

Solution set: $\{9\}$

67.

$$6(2x + 1) + 3x = 8(2x - 3) - 1$$
$$12x + 6 + 3x = 16x - 24 - 1$$
$$\text{Distributive property}$$

#67 continued

$$12x + 3x + 6 = 16x - 24 - 1$$
$$\text{Commutative property}$$
$$15x + 6 = 16x - 25$$
$$15x - 15x + 6 = 16x - 15x - 25 \quad \text{Subtract } 15x$$
$$6 = x - 25$$
$$6 + 25 = x - 25 + 25 \quad \text{Add } 25$$
$$31 = x$$

Solution set: $\{31\}$

69.

$$\frac{1}{2}(8x + 3) - 2 = \frac{1}{3}(9x - 5) + 2$$

$$4x + \frac{3}{2} - 2 = 3x - \frac{5}{3} + 2$$

$$\text{Distributive property}$$

$$4x + \left(-\frac{1}{2}\right) = 3x + \frac{1}{3}$$

$$4x - 3x + \left(-\frac{1}{2}\right) = 3x - 3x + \frac{1}{3} \quad \text{Subtract } 3x$$

$$x + \left(-\frac{1}{2}\right) = \frac{1}{3}$$

$$x + \left(-\frac{1}{2}\right) + \frac{1}{2} = \frac{1}{3} + \frac{1}{2} \qquad \text{Add } \frac{1}{4}$$

$$x = \frac{5}{6}$$

Solution set: $\left\{\dfrac{5}{6}\right\}$

71.

$$5(2x + 7) - 3(3x + 4) = 7$$
$$10x + 35 - 9x - 12 = 7$$
$$\text{Distributive property}$$
$$10x - 9x + 35 - 12 = 7$$
$$\text{Commutative property}$$
$$x + 23 = 7 \quad \text{Simplify}$$
$$x + 23 - 23 = 7 - 23 \quad \text{Subtract } 23$$
$$x = -16$$

Solution set: $\{-16\}$

73.

$$4(3 - x) + 2(x + 3) = 25 - 3x$$
$$12 - 4x + 2x + 6 = 25 - 3x$$
Distributive property
$$-4x + 2x + 12 + 6 = 25 - 3x$$
Commutative property
$$-2x + 18 = 25 - 3x$$
$$-2x + 3x + 18 = 25 - 3x + 3x \quad \text{Add } 3x$$
$$x + 18 = 25$$
$$x + 18 - 18 = 25 - 18 \quad \text{Subtract } 18$$
$$x = 7$$

Solution set: {7}

75.

$$3(2x + 1) + 4(x - 2) = 3(3x + 1)$$
$$6x + 3 + 4x - 8 = 9x + 3$$
Distributive property
$$6x + 4x + 3 - 8 = 9x + 3$$
Commutative property
$$10x - 5 = 9x + 3$$
$$10x - 9x - 5 = 9x - 9x + 3 \quad \text{Subtract } 9x$$
$$x - 5 = 3$$
$$x - 5 + 5 = 3 + 5 \quad \text{Add } 5$$
$$x = 8$$

Solution set: {8}

77.

$$8(2x - 1) - 2(x + 3) = 3(5x - 6)$$
$$16x - 8 - 2x - 6 = 15x - 18$$
Distributive property
$$16x - 2x - 8 - 6 = 15x - 18$$
Commutative property
$$14x - 14 = 15x - 18$$
$$14x - 14x - 14 = 15x - 14x - 18$$
Subtract $14x$
$$-14 = x - 18$$
$$-14 + 18 = x - 18 + 18 \quad \text{Add } 18$$
$$4 = x$$

Solution set: {4}

79.

$$8(x + 3) + 2(x - 4) = 6(x - 3) + 3(x + 2)$$
$$8x + 24 + 2x - 8 = 6x - 18 + 3x + 6$$
Distributive property
$$8x + 2x + 24 - 8 = 6x + 3x - 18 + 6$$
$$10x + 16 = 9x - 12$$

#79 continued

$$10x - 9x + 16 = 9x - 9x - 12 \quad \text{Subtract } 9x$$
$$x + 16 = -12$$
$$x + 16 - 16 = -12 - 16 \quad \text{Subtract } 16$$
$$x = -28$$

Solution set: {-28}

81.

$$3(2x - 1) - 4(x + 2) = 5(2x + 3) - 7(x - 2)$$
$$6x - 3 - 4x - 8 = 10x + 15 - 7x + 14$$
$$6x - 4x - 3 - 8 = 10x - 7x + 15 + 14$$
$$2x - 11 = 3x + 29$$
$$2x - 2x - 11 = 3x - 2x + 29$$
$$-11 = x + 29$$
$$-11 - 29 = x + 29 - 29$$
$$-40 = x$$

Solution set: {-40}

83.

$$2[4(2x - 1) - (5x - 3)] = 7x + 2$$
$$2[8x - 4 - 5x + 3] = 7x + 2$$
$$2[3x - 1] = 7x + 2$$
$$6x - 2 = 7x + 2$$
$$6x - 6x - 2 = 7x - 6x + 2$$
$$-2 = x + 2$$
$$-2 - 2 = x + 2 - 2$$
$$-4 = x$$

Solution set: {-4}

85.

$$5[3(2x + 1) - 4(x + 1)] = 3(3x + 1) - 5$$
$$5[6x + 3 - 4x - 4] = 9x + 3 - 5$$
$$5[2x - 1] = 9x - 2$$
$$10x - 5 = 9x - 2$$
$$10x - 9x - 5 = 9x - 9x - 2$$
$$x - 5 = -2$$
$$x - 5 + 5 = -2 + 5$$
$$x = 3$$

Solution set: {3}

87.

$$3x + 2(2x - 9) = 3[4(x - 1) - 2(x + 3)]$$
$$3x + 4x - 18 = 3[4x - 4 - 2x - 6]$$
$$7x - 18 = 3[2x - 10]$$
$$7x - 18 = 6x - 30$$

#87 continued

$$7x - 6x - 18 = 6x - 6x - 30$$
$$x - 18 = -30$$
$$x - 18 + 18 = -30 + 18$$
$$x = -12$$

Solution set: {-12}

89.
$$3[5(x - 3) + 2(x + 9)] = 5[2(x + 3) + 2(x - 7)]$$
$$3[5x - 15 + 2x + 18] = 5[2x + 6 + 2x - 14]$$
$$3[7x + 3] = 5[4x - 8]$$
$$21x + 9 = 20x - 40$$
$$21x - 20x + 9 = 20x - 20x - 40$$
$$x + 9 = -40$$
$$x + 9 - 9 = -40 - 9$$
$$x = -49$$

Solution set: {-49}

Exercises 2.3

1.
$$4x = 32$$
$$\frac{4x}{4} = \frac{32}{4} \qquad \text{Divide by 4}$$
$$x = 8$$

Solution set: {8}

3.
$$\frac{3}{5}x = 9$$
$$\frac{5}{3}\left(\frac{3}{5}x\right) = \frac{5}{3} \cdot 9$$

$$\text{Multiply by } \frac{5}{3} \text{ the reciprocal of } \frac{3}{5}$$

$$\left(\frac{5}{3} \cdot \frac{3}{5}\right)x = \frac{5}{3} \cdot 9 \qquad \text{Associative property}$$
$$1 \cdot x = 15 \qquad \text{Simplify}$$
$$x = 15$$

Solution set: {15}

5.
$$-x = 3$$
$$-1 \cdot x = 3 \qquad\qquad -x = -1 \cdot x$$
$$-1(-1 \cdot x) = -1 \cdot 3 \qquad \text{Multiply by -1}$$
$$x = -3$$

Solution set: {-3}

7.
$$24 = -2x$$
$$\frac{24}{-2} = \frac{-2x}{-2} \qquad \text{Divide by -2}$$
$$-12 = x$$

Solution set: {-12}

9.
$$-6 = \frac{x}{3}$$
$$3(-6) = 3\left(\frac{x}{3}\right) \qquad \text{Multiply by 3}$$
$$3(-6) = \frac{3x}{3}$$
$$-18 = x$$

Solution set: {-18}

11.
$$7x = 0$$
$$\frac{7x}{7} = \frac{0}{7} \qquad \text{Divide by 7}$$
$$x = 0$$

Solution set: {0}

13.
$$3x = 1$$
$$\frac{3x}{3} = \frac{1}{3} \qquad \text{Divide by 3}$$
$$x = \frac{1}{3}$$

Solution set: $\left\{\frac{1}{3}\right\}$

15.

$$\frac{3}{5}x = \frac{9}{10}$$

$$\frac{5}{3}\left(\frac{3}{5}x\right) = \frac{5}{3} \cdot \frac{9}{10}$$

Multiply by $\frac{5}{3}$, the reciprocal of $\frac{3}{5}$

$$1 \cdot x = \frac{3}{2}$$ Simplify

$$x = \frac{3}{2}$$

Solution set: $\left\{\frac{3}{2}\right\}$

17.

$$10 = -56x$$

$$\frac{10}{-56} = \frac{-56x}{-56}$$ Divide by -56

$$\frac{10}{-56} = x$$ Simplify

$$-\frac{5}{28} = x$$ Reduce

Solution set: $\left\{-\frac{5}{28}\right\}$

19.

$$15x = 40$$

$$\frac{15x}{15} = \frac{40}{15}$$ Divide by 15

$$x = \frac{40}{15}$$ Simplify

$$x = \frac{8}{3}$$ Reduce

Solution set: $\left\{\frac{8}{3}\right\}$

21.

$$\frac{5}{8} = -x$$

$$\frac{5}{8} = -1 \cdot x$$

$$-1 \cdot \frac{5}{8} = -1(-1 \cdot x)$$

#21 continued

$$-\frac{5}{8} = x$$

Solution set: $\left\{-\frac{5}{8}\right\}$

23.

$$-\frac{3}{4} = 7x$$

$$\frac{1}{7}\left(-\frac{3}{4}\right) = \frac{1}{7}(7x)$$ Multiply by $\frac{1}{7}$

$$\frac{1}{7}\left(-\frac{3}{4}\right) = \frac{7x}{7}$$

$$-\frac{3}{28} = 1 \cdot x$$ Simplify

$$-\frac{3}{28} = x$$

Solution set: $\left\{-\frac{3}{28}\right\}$

25.

$$-x = -4$$

$$-1(-x) = -1(-4)$$ Multiply by -1

$$x = 4$$

Solution set: $\{4\}$

27.

$$6x = \frac{2}{3}$$

$$\frac{1}{6}(6x) = \frac{1}{6}\left(\frac{2}{3}\right)$$ Multiply by $\frac{1}{6}$

$$\frac{6x}{6} = \frac{1}{6}\left(\frac{2}{3}\right)$$

$$1 \cdot x = \frac{1}{9}$$

$$x = \frac{1}{9}$$

Solution set: $\left\{\frac{1}{9}\right\}$

29.

$$-\frac{6}{7}x = -\frac{6}{7}$$

$$-\frac{7}{6}\left(-\frac{6}{7}x\right) = -\frac{7}{6}\left(-\frac{6}{7}\right) \quad \text{Multiply by } -\frac{7}{6}$$

$$1 \cdot x = 1 \quad \text{Simplify}$$

$$x = 1$$

Solution set: {1}

31.

$$-420x = 40$$

$$\frac{-420x}{-420} = \frac{40}{-420} \quad \text{Divide by -420}$$

$$x = \frac{40}{-420} \quad \text{Simplify}$$

$$x = -\frac{2}{21} \quad \text{Reduce}$$

Solution set: $\left\{-\frac{2}{21}\right\}$

33.

$$\frac{x}{5} = -\frac{1}{9}$$

$$5\left(\frac{x}{5}\right) = 5\left(-\frac{1}{9}\right) \quad \text{Multiply by 5}$$

$$\frac{5x}{5} = 5\left(-\frac{1}{9}\right)$$

$$x = -\frac{5}{9}$$

Solution set: $\left\{-\frac{5}{9}\right\}$

35.

$$-7x = -49$$

$$\frac{-7x}{-7} = \frac{-49}{-7} \quad \text{Divide by -7}$$

$$x = 7$$

Solution set: {7}

37.

$$0 = -\frac{5}{7}x$$

$$-\frac{7}{5}(0) = -\frac{7}{5}\left(-\frac{5}{7}x\right) \quad \text{Multiply by } -\frac{7}{5}$$

$$0 = 1 \cdot x \quad \text{Simplify}$$

$$0 = x$$

Solution set: {0}

39.

$$8x = -5$$

$$\frac{8x}{8} = -\frac{5}{8} \quad \text{Divide by 8}$$

$$x = -\frac{5}{8}$$

Solution set: $\left\{-\frac{5}{8}\right\}$

41.

$$-\frac{9}{10} = -x$$

$$-1\left(-\frac{9}{10}\right) = -1(-1 \cdot x) \quad \text{Multiply by -1}$$

$$\frac{9}{10} = 1 \cdot x \quad \text{Simplify}$$

$$\frac{9}{10} = x$$

Solution set: $\left\{\frac{9}{10}\right\}$

43.

$$\frac{x}{8} = \frac{3}{4}$$

$$8\left(\frac{x}{8}\right) = 8\left(\frac{3}{4}\right) \quad \text{Multiply by 8}$$

$$\frac{8x}{8} = 8\left(\frac{3}{4}\right)$$

$$x = 6$$

Solution set: {6}

45.

$$1.6x = 4.8$$
$$10(1.6x) = 10(4.8) \quad \text{Multiply by 10}$$
$$16x = 48$$
$$\frac{16x}{16} = \frac{48}{16} \quad \text{Divide by 16}$$
$$x = 3$$

Solution set: {3}

47.

$$5.2x = -20.8$$
$$10(5.2x) = 10(-20.8)$$
$$52x = -208$$
$$\frac{52x}{52} = \frac{-208}{52} \quad \text{Divide by 52}$$
$$x = -4$$

Solution set: {-4}

49.

$$3x + 5 = 14$$
$$3x + 5 - 5 = 14 - 5 \quad \text{Subtract 5}$$
$$3x = 9$$
$$\frac{3x}{3} = \frac{9}{3} \quad \text{Divide by 3}$$
$$x = 3$$

Solution set: {3}

51.

$$7x - 8 = 13$$
$$7x - 8 + 8 = 13 + 8 \quad \text{Add 8}$$
$$7x = 21$$
$$\frac{7x}{7} = \frac{21}{7} \quad \text{Divide by 7}$$
$$x = 3$$

Solution set: {3}

53.

$$3 - 4x = 11$$
$$-3 + 3 - 4x = -3 + 11 \quad \text{Add -3}$$
$$-4x = 8$$
$$\frac{-4x}{-4} = \frac{8}{-4} \quad \text{Divide by -4}$$
$$x = -2$$

#53 continued

Solution set: {-2}

55.

$$-13 = 4x - 9$$
$$-13 + 9 = 4x - 9 + 9 \quad \text{Add 9}$$
$$-4 = 4x$$
$$\frac{-4}{4} = \frac{4x}{4} \quad \text{Divide by 4}$$
$$-1 = x$$

Solution set: {-1}

57.

$$23 = 6x + 5$$
$$23 - 5 = 6x + 5 - 5 \quad \text{Subtract 5}$$
$$18 = 6x$$
$$\frac{18}{6} = \frac{6x}{6} \quad \text{Divide by 6}$$
$$3 = x$$

Solution set: {3}

59.

$$10 = 7 - 2x$$
$$-7 + 10 = -7 + 7 - 2x \quad \text{Add -7}$$
$$3 = -2x$$
$$\frac{3}{-2} = \frac{-2x}{-2} \quad \text{Divide by -2}$$
$$-\frac{3}{2} = x$$

Solution set: $\left\{-\frac{3}{2}\right\}$

61.

$$-x - 2 = 1$$
$$-x - 2 + 2 = 1 + 2 \quad \text{Add 2}$$
$$-x = 3$$
$$-1(-x) = -1 \cdot 3 \quad \text{Multiply by -1}$$
$$x = -3$$

Solution set: {-3}

63.

$$\frac{2}{3}x + 1 = -3$$

$$\frac{2}{3}x + 1 - 1 = -3 - 1 \qquad \text{Subtract 1}$$

$$\frac{2}{3}x = -4$$

$$\frac{3}{2}\left(\frac{2}{3}x\right) = \frac{3}{2}(-4) \qquad \text{Multiply by } \frac{3}{2}$$

$$\left(\frac{3}{2} \cdot \frac{2}{3}\right)x = \frac{3}{2}(-4)$$

$$1 \cdot x = -6$$

$$x = -6$$

Solution set: {-6}

65.

$$\frac{1}{2}x - \frac{2}{3} = \frac{1}{6}$$

$$6\left(\frac{1}{2}x - \frac{2}{3}\right) = 6\left(\frac{1}{6}\right) \qquad \text{Multiply by 6, the LCD}$$

$$6 \cdot \frac{1}{2}x - 6 \cdot \frac{2}{3} = 6 \cdot \frac{1}{6} \qquad \text{Distributive property}$$

$$3x - 4 = 1 \qquad \text{Simplify}$$

$$3x - 4 + 4 = 1 + 4 \qquad \text{Add 4}$$

$$3x = 5$$

$$\frac{3x}{3} = \frac{5}{3} \qquad \text{Divide by 3}$$

$$x = \frac{5}{3}$$

Solution set: $\left\{\frac{5}{3}\right\}$

67.

$$2.6x + 1.4 = 9.2$$

$$10(2.6x + 1.4) = 10(9.2) \qquad \text{Multiply by 10}$$

$$10 \cdot 2.6x + 10 \cdot 1.4 = 10 \cdot 9.2$$

$$\text{Distributive property}$$

$$26x + 14 = 92$$

$$26x + 14 - 14 = 92 - 14 \qquad \text{Subtract 14}$$

$$26x = 78$$

$$\frac{26x}{26} = \frac{78}{26} \qquad \text{Divide by 26}$$

$$x = 3$$

#67 continued

Solution set: {3}

69.

$$3.7x - 1.9 = -9.3$$

$$10(3.7x - 1.9) = 10(-9.3) \qquad \text{Multiply by 10}$$

$$10 \cdot 3.7x - 10 \cdot 1.9 = 10(-9.3)$$

$$\text{Distributive property}$$

$$37x - 19 = -93$$

$$37x - 19 + 19 = -93 + 19 \qquad \text{Add 19}$$

$$37x = -74$$

$$\frac{37x}{37} = \frac{-74}{37} \qquad \text{Divide by 37}$$

$$x = -2$$

Solution set: {-2}

71.

$$10x = 8x - 4$$

$$-8x + 10x = -8x + 8x - 4 \qquad \text{Add } - 8x$$

$$2x = -4$$

$$\frac{2x}{2} = \frac{-4}{2} \qquad \text{Divide by 2}$$

$$x = -2$$

Solution set: {-2}

73.

$$\frac{1}{2}x + 2 = \frac{1}{3}x$$

$$6\left(\frac{1}{2}x + 2\right) = 6\left(\frac{1}{3}x\right) \qquad \text{Multiply by 6, the LCD}$$

$$6 \cdot \frac{1}{2}x + 6 \cdot 2 = 6 \cdot \frac{1}{3}x$$

$$\text{Distributive property}$$

$$3x + 12 = 2x$$

$$3x - 2x + 12 - 12 = 2x - 2x - 12$$

$$\text{Subtract 2x and 12}$$

$$x = -12$$

Solution set: {-12}

75.

$$2x + 7 = 6x - 5$$

$$2x - 2x + 7 = 6x - 2x - 5 \qquad \text{Subtract 2x}$$

$$7 = 4x - 5$$

$$7 + 5 = 4x - 5 + 5 \qquad \text{Add 5}$$

$$12 = 4x$$

#75 continued

$$\frac{12}{4} = \frac{4x}{4}$$

$$3 = x$$

Solution set: {3}

77.

$$5x - 2 = 3x - 5$$

$$5x - 3x - 2 = 3x - 3x - 5 \qquad \text{Subtract } 3x$$

$$2x - 2 = -5$$

$$2x - 2 + 2 = -5 + 2 \qquad \text{Add } 2$$

$$2x = -3$$

$$\frac{2x}{2} = \frac{-3}{2} \qquad \text{Divide by } 2$$

$$x = -\frac{3}{2}$$

Solution set: $\left\{-\frac{3}{2}\right\}$

79.

$$3x - 8 = 4x - 1$$

$$3x - 3x - 8 = 4x - 3x - 1 \qquad \text{Subtract } 3x$$

$$-8 = x - 1$$

$$-8 + 1 = x - 1 + 1 \qquad \text{Add } 1$$

$$-7 = x$$

Solution set: {-7}

81.

$$2x + 5 = 9x - 1$$

$$2x - 2x + 5 = 9x - 2x - 1 \qquad \text{Subtract } 2x$$

$$5 = 7x - 1$$

$$5 + 1 = 7x - 1 + 1 \qquad \text{Add } 1$$

$$6 = 7x$$

$$\frac{6}{7} = \frac{7x}{7} \qquad \text{Divide by } 7$$

$$\frac{6}{7} = x$$

Solution set: $\left\{\frac{6}{7}\right\}$

83.

$$6x + 12 = 8x + 60$$

$$6x - 6x + 12 = 8x - 6x + 60 \qquad \text{Subtract } 6x$$

#83 continued

$$12 = 2x + 60 \qquad \text{Simplify}$$

$$12 - 60 = 2x + 60 - 60 \qquad \text{Subtract } 60$$

$$-48 = 2x \qquad \text{Simplify}$$

$$\frac{-48}{2} = \frac{2x}{2} \qquad \text{Divide by } 2$$

$$-24 = x \qquad \text{Simplify}$$

Solution set: {-24}

85.

$$\frac{3}{4}x + \frac{1}{2} = \frac{5}{2}x - \frac{1}{4}$$

$$4\left(\frac{3}{4}x + \frac{1}{2}\right) = 4\left(\frac{5}{2}x - \frac{1}{4}\right)$$

Multiply by 4, the LCD

$$4\left(\frac{3}{4}x\right) + 4\left(\frac{1}{2}\right) = 4\left(\frac{5}{2}x\right) - 4\left(\frac{1}{4}\right)$$

$$3x + 2 = 10x - 1$$

$$3x - 3x + 2 = 10x - 3x - 1 \qquad \text{Subtract } 3x$$

$$2 = 7x - 1$$

$$2 + 1 = 7x - 1 + 1 \qquad \text{Add } 1$$

$$3 = 7x$$

$$\frac{3}{7} = \frac{7x}{7} \qquad \text{Divide by } 7$$

$$\frac{3}{7} = x$$

Solution set: $\left\{\frac{3}{7}\right\}$

87.

$$4x - 3 + 2x = 8x - 9$$

$$6x - 3 = 8x - 9$$

$$6x - 6x - 3 = 8x - 6x - 9 \qquad \text{Subtract } 6x$$

$$-3 = 2x - 9$$

$$-3 + 9 = 2x - 9 + 9 \qquad \text{Add } 9$$

$$6 = 2x$$

$$\frac{6}{2} = \frac{2x}{2} \qquad \text{Divide by } 2$$

$$3 = x$$

Solution set: {3}

89.

$$7 - 2x - 4x = 7x + 72$$

$$7 - 6x = 7x + 72$$

#89 continued

$$7 - 6x + 6x = 7x + 6x + 72 \quad \text{Add } 6x$$
$$7 = 13x + 72$$
$$7 - 72 = 13x + 72 - 72 \quad \text{Subtract } 72$$
$$-65 = 13x$$
$$\frac{-65}{13} = \frac{13x}{13} \quad \text{Divide by } 13$$
$$-5 = x$$

Solution set: $\{-5\}$

91.

$$3x + 7 - 9x = 2x + 4$$
$$-6x + 7 = 2x + 4$$
$$-6x + 6x + 7 = 2x + 6x + 4 \quad \text{Add } 6x$$
$$7 = 8x + 4$$
$$7 - 4 = 8x + 4 - 4 \quad \text{Subtract } 4$$
$$3 = 8x$$
$$\frac{3}{8} = \frac{8x}{8} \quad \text{Divide by } 8$$
$$\frac{3}{8} = x$$

Solution set: $\left\{\dfrac{3}{8}\right\}$

93.

$$2x + x - 13 = 12x + 5$$
$$3x - 13 = 12x + 5$$
$$3x - 3x - 13 = 12x - 3x + 5 \quad \text{Subtract } 3x$$
$$-13 = 9x + 5$$
$$-13 - 5 = 9x + 5 - 5 \quad \text{Subtract } 5$$
$$-18 = 9x$$
$$\frac{-18}{9} = \frac{9x}{9}$$
$$-2 = x$$

Solution set: $\{-2\}$

95.

$$4x - 4 = 18x - 16 - 6x$$
$$4x - 4 = 12x - 16 \quad \text{Simplify}$$
$$4x - 4x - 4 = 12x - 4x - 16 \quad \text{Subtract } 4x$$
$$-4 = 8x - 16 \quad \text{Simplify}$$
$$-4 + 16 = 8x - 16 + 16 \quad \text{Add } 16$$
$$12 = 8x \quad \text{Simplify}$$
$$\frac{12}{8} = \frac{8x}{8} \quad \text{Divide by } 8$$

#95 continued

$$\frac{3}{2} = x \quad \text{Reduce}$$

Solution set: $\left\{\dfrac{3}{2}\right\}$

97.

$$\frac{5}{6}x - \frac{1}{8}x + \frac{3}{4} = \frac{2}{3}x - \frac{1}{12}$$
$$24\left(\frac{5}{6}x - \frac{1}{8}x + \frac{3}{4}\right) = 24\left(\frac{2}{3}x - \frac{1}{12}\right)$$
$$\text{Multiply by } 24, \text{ the LCD}$$
$$20x - 3x + 18 = 16x - 2$$
$$17x + 18 = 16x - 2$$
$$17x - 16x + 18 = 16x - 16x - 2$$
$$x + 18 = -2$$
$$x + 18 - 18 = -2 - 18$$
$$x = -20$$

Solution set: $\{-20\}$

99.
Let x be the distance between home and Los Angeles.

Let 2x be the distance between home and San Francisco.

Since the distance between the two cities is 327 miles, the equation is:

$$x + 2x = 327$$

$$3x = 327 \quad \text{Collect common terms}$$
$$\frac{3x}{3} = \frac{327}{3} \quad \text{Divide by } 3$$
$$x = 109$$

They live 109 miles from Los Angeles.

101.
Let x be the M.S.R.P.

Since the selling price is $\frac{7}{8}$ of the M.S.R.P., the equation is

$$\frac{7}{8}x = 10,871$$

#101 continued

$$\frac{8}{7}\left(\frac{7}{8}x\right) = \frac{8}{7} \cdot 10{,}871$$

$$x = \frac{8 \cdot 10{,}871}{7}$$

$$x = \frac{8 \cdot \not{7} \cdot 1553}{\not{7}}$$

$$x = 12{,}424$$

The M.S.R.P. is $12,424.

| Exercises 2.4 |

1.

$$5(2x + 4) = 8x - 12$$
$$10x + 20 = 8x - 12 \qquad \text{Step 1}$$
$$10x - 8x + 20 = 8x - 8x - 12 \qquad \text{Step 2}$$
$$2x + 20 = -12$$
$$2x + 20 - 20 = -12 - 20 \qquad \text{Step 3}$$
$$2x = -32$$
$$\frac{2x}{2} = \frac{-32}{2} \qquad \text{Step 4}$$
$$x = -16$$

Solution set: {-16}

3.

$$4 + 2(3x - 2) = 16$$
$$4 + 6x - 4 = 16 \qquad \text{Step 1}$$
$$6x = 16$$
$$\frac{6x}{6} = \frac{16}{6} \qquad \text{Step 4}$$
$$x = \frac{8}{3} \qquad \text{Reduce}$$

Solution set: $\left\{\frac{8}{3}\right\}$

5.

$$3x - 4(2x - 1) = 5x$$
$$3x - 8x + 4 = 5x \qquad \text{Step 1}$$
$$-5x + 4 = 5x$$
$$-5x + 5x + 4 = 5x + 5x \qquad \text{Step 2}$$
$$4 = 10x$$
$$\frac{4}{10} = \frac{10x}{10}$$

#5 continued

$$\frac{2}{5} = x$$

Solution set: $\left\{\frac{2}{5}\right\}$

7.

$$5 - (2x + 6) = -4$$
$$5 - 2x - 6 = -4 \qquad \text{Step 1}$$
$$-2x - 1 = -4 \qquad \text{Collect like terms}$$
$$-2x - 1 + 1 = -4 + 1 \qquad \text{Step 3}$$
$$-2x = -3$$
$$\frac{-2x}{-2} = \frac{-3}{-2}$$
$$x = \frac{3}{2}$$

Solution set: $\left\{\frac{3}{2}\right\}$

9.

$$2 + 3(3x - 7) = 9x - 19$$
$$2 + 9x - 21 = 9x - 19 \qquad \text{Step 1}$$
$$9x - 19 = 9x - 19$$
$$0 = 0$$

Solution set: {all real numbers}

11.

$$\frac{1}{4} + \frac{1}{3}(5x + 2) = -2$$

$$12\left[\frac{1}{4} + \frac{1}{3}(5x + 2)\right] = 12(-2)$$

Multiply by 12 to change fractions to integers

$$3 + 4(5x + 2) = -24 \quad \text{Distributive property}$$
$$3 + 20x + 8 = -24 \qquad \text{Step 1}$$
$$20x + 11 = -24$$
$$20x + 11 - 11 = -24 - 11 \qquad \text{Step 3}$$
$$20x = -35$$
$$\frac{20x}{20} = \frac{-35}{20} \qquad \text{Step 4}$$
$$x = -\frac{7}{4} \qquad \text{Reduce}$$

Solution set: $\left\{-\frac{7}{4}\right\}$

13.

$$\frac{2}{3}x + \frac{1}{6}(3x - 2) = -\frac{1}{3}$$

$$6\left[\frac{2}{3}x + \frac{1}{6}(3x - 2)\right] = 6\left(-\frac{1}{3}\right)$$

Multiply by 6 to change fractions to integers

$$4x + 1 \cdot (3x - 2) = -2$$
$$4x + 3x - 2 = -2$$
$$7x - 2 = -2$$

$$7x - 2 + 2 = -2 + 2 \qquad \text{Step 3}$$
$$7x = 0$$

$$\frac{7x}{7} = \frac{0}{7} \qquad \text{Step 4}$$
$$x = 0$$

Solution set: {0}

15.

$$3(5x - 7) = 4(2x - 7)$$
$$15x - 21 = 8x - 28 \qquad \text{Step 1}$$
$$15x - 8x - 21 = 8x - 8x - 28 \qquad \text{Step 2}$$
$$7x - 21 = -28$$
$$7x - 21 + 21 = -28 + 21 \qquad \text{Step 3}$$
$$7x = -7$$
$$\frac{7x}{7} = \frac{-7}{7} \qquad \text{Step 4}$$
$$x = -1$$

Solution set: {-1}

17.

$$4(6 - 2x) - 2 = 9(3 - 2x) + 6$$
$$24 - 8x - 2 = 27 - 18x + 6 \qquad \text{Step 1}$$
$$-8x + 22 = -18x + 33$$
$$-8x + 18x + 22 = -18x + 18x + 33 \qquad \text{Step 2}$$
$$10x + 22 = 33$$
$$10x + 22 - 22 = 33 - 22 \qquad \text{Step 3}$$
$$10x = 11$$
$$\frac{10x}{10} = \frac{11}{10} \qquad \text{Step 4}$$
$$x = \frac{11}{10}$$

Solution set: $\left\{\dfrac{11}{10}\right\}$

19.

$$3 + 2(8x - 1) = 5 + 4(4x - 6)$$
$$3 + 16x - 2 = 5 + 16x - 24 \qquad \text{Step 1}$$
$$16x + 1 = 16x - 19$$
$$16x - 16x + 1 = 16x - 16x - 19 \qquad \text{Step 2}$$
$$1 = -19 \qquad \text{Not True}$$

No solution

21.

$$4(3x - 5) = 3(5x - 9)$$
$$12x - 20 = 15x - 27 \qquad \text{Step 1}$$
$$12x - 12x - 20 = 15x - 12x - 27 \qquad \text{Step 2}$$
$$-20 = 3x - 27$$
$$-20 + 27 = 3x - 27 + 27 \qquad \text{Step 3}$$
$$7 = 3x$$
$$\frac{7}{3} = \frac{3x}{3} \qquad \text{Step 4}$$
$$\frac{7}{3} = x$$

Solution set: $\left\{\dfrac{7}{3}\right\}$

23.

$$\frac{1}{3}(2x + 5) = \frac{3}{4}(x - 1) + \frac{5}{4}$$

$$12\left[\frac{1}{3}(2x + 5)\right] = 12\left[\frac{3}{4}(x - 1) + \frac{5}{4}\right]$$

$$4(2x + 5) = 9(x - 1) + 15$$
$$8x + 20 = 9x - 9 + 15 \qquad \text{Step 1}$$
$$8x + 20 = 9x + 6$$
$$8x - 8x + 20 = 9x - 8x + 6 \qquad \text{Step 2}$$
$$20 = x + 6$$
$$20 - 6 = x + 6 - 6 \qquad \text{Step 3}$$
$$14 = x$$

Solution set: {14}

25.

$$1.2(2.3x + 7.4) = 3.9(4.1x + 1.5) - 41.952$$
$$2.76x + 8.88 = 15.99x + 5.85 - 41.952$$
$$2.76x + 8.88 = 15.99x - 36.012$$
$$2.76x - 2.76x + 8.88 = 15.99x - 2.76x - 36.102$$
$$8.88 = 13.23x - 36.102$$
$$8.88 + 36.102 = 13.23x - 36.102 + 36.102$$
$$44.982 = 13.23x$$
$$3.4 = x$$

#25 continued

Solution set: {3.4}

27.
$5(2x + 1) + 4(3x - 6) = -85$
$10x + 5 + 12x - 24 = -85$
$22x - 19 = -85$
$22x - 19 + 19 = -85 + 19$
$22x = -66$
$$\frac{22x}{22} = \frac{-66}{22}$$
$x = -3$

Solution set: {-3}

29.
$6(x - 5) - (3x + 4) = -19$
$6x - 30 - 3x - 4 = -19$ Step 1
$3x - 34 = -19$
$3x - 34 + 34 = -19 + 34$ Step 3
$3x = 15$
$$\frac{3x}{3} = \frac{15}{3}$$ Step 4
$x = 5$

Solution set: {5}

31.
$2(3x + 4) - 5(x + 6) = 4x - 23$
$6x + 8 - 5x - 30 = 4x - 23$ Step 1
$x - 22 = 4x - 23$
$x - x - 22 = 4x - x - 23$ Step 2
$-22 = 3x - 23$
$-22 + 23 = 3x - 23 + 23$ Step 3
$1 = 3x$
$$\frac{1}{3} = \frac{3x}{3}$$ Step 4
$$\frac{1}{3} = x$$

Solution set: $\left\{\dfrac{1}{3}\right\}$

33.
$3(8x + 1) - 5(4x + 1) = 4x - 2$
$24x + 3 - 20x - 5 = 4x - 2$ Step 1
$4x - 2 = 4x - 2$
$0 = 0$

#33 continued

Solution set: {all real numbers}

35.
$4(2x - 1) + 2(x - 2) = 8$
$8x - 4 + 2x - 4 = 8$ Step 1
$10x - 8 = 8$
$10x - 8 + 8 = 8 + 8$ Step 3
$10x = 16$
$$\frac{10x}{10} = \frac{16}{10}$$ Step 4
$$x = \frac{16}{10} \text{ or } \frac{8}{5}$$

Solution set: $\left\{\dfrac{8}{5}\right\}$

37.
$$\frac{3}{4}(4x + 1) - \frac{1}{2}(x - 2) = \frac{5}{8}x + \frac{1}{2}$$
$$8\left[\frac{3}{4}(4x + 1) - \frac{1}{2}(x - 2)\right] = 8\left(\frac{5}{8}x + \frac{1}{2}\right)$$
$6(4x + 1) - 4(x - 2) = 5x + 4$
$24x + 6 - 4x + 8 = 5x + 4$ Step 1
$20x + 14 = 5x + 4$
$20x - 5x + 14 = 5x - 5x + 4$ Step 2
$15x + 14 = 4$
$15x + 14 - 14 = 4 - 14$ Step 3
$15x = -10$
$$\frac{15x}{15} = \frac{-10}{15}$$ Step 4
$$x = -\frac{2}{3}$$ Reduce

Solution set: $\left\{-\dfrac{2}{3}\right\}$

39.
$3.2(x - 5) + 4.2(3x + 1) = -108.18$
$3.2x - 16.0 + 12.6x + 4.2 = -108.18$
$15.8x - 11.8 = -108.18$
$15.8x - 11.8 + 11.8 = -108.18 + 11.8$
$15.8x = -96.38$
$x = -6.1$

Solution set: {-6.1}

41.

$5(x - 2) - 3(x + 2) = 2(2x - 7)$

$5x - 10 - 3x - 6 = 4x - 14$	Step 1
$2x - 16 = 4x - 14$	
$2x - 2x - 16 = 4x - 2x - 14$	Step 2
$-16 = 2x - 14$	
$-16 + 14 = 2x - 14 + 14$	Step 3
$-2 = 2x$	
$\dfrac{-2}{2} = \dfrac{2x}{2}$	Step 4
$-1 = x$	

Solution set: {-1}

43.

$3(2x + 1) + 4(x - 6) = 5(2x + 3)$

$6x + 3 + 4x - 24 = 10x + 15$	Step 1
$10x - 21 = 10x + 15$	
$10x - 10x - 21 = 10x - 10x + 15$	
$-21 = 15$	Not true

No solution

45.

$2(2x + 1) + 5(x - 7) = 4(2x + 1) - 23$

$4x + 2 + 5x - 35 = 8x + 4 - 23$

$9x - 33 = 8x - 19$

$9x - 8x - 33 = 8x - 8x - 19$

$x - 33 = -19$

$x - 33 + 33 = -19 + 33$

$x = 14$

Solution set: {14}

47.

$5(2x - 1) - 2(4x + 3) = 2(5x - 1) + 7$

$10x - 5 - 8x - 6 = 10x - 2 + 7$	
$2x - 11 = 10x + 5$	
$2x - 2x - 11 = 10x - 2x + 5$	
$-11 = 8x + 5$	
$-11 - 5 = 8x + 5 - 5$	Step 3
$-16 = 8x$	
$\dfrac{-16}{8} = \dfrac{8x}{8}$	Step 4
$-2 = x$	

Solution set: {-2}

49.

$9(3x + 1) + 12(4x - 7) = 15x + 15$

$27x + 9 + 48x - 84 = 15x + 15$

$75x - 75 = 15x + 15$

$75x - 15x - 75 = 15x - 15x + 15$

$60x - 75 = 15$

$60x - 75 + 75 = 15 + 75$

$60x = 90$

$\dfrac{60x}{60} = \dfrac{90}{60}$

$x = \dfrac{3}{2}$

Solution set: $\left\{\dfrac{3}{2}\right\}$

51.

$\dfrac{3}{4}(5x + 2) - \dfrac{1}{2}(3x - 7) = \dfrac{3}{8}x + \dfrac{5}{4}$

$8\left[\dfrac{3}{4}(5x + 2) - \dfrac{1}{2}(3x - 7)\right] = 8\left(\dfrac{3}{8}x + \dfrac{5}{4}\right)$

$6(5x + 2) - 4(3x - 7) = 3x + 10$

$30x + 12 - 12x + 28 = 3x + 10$

$18x + 40 = 3x + 10$

$18x - 3x + 40 = 3x - 3x + 10$

$15x + 40 = 10$

$15x + 40 - 40 = 10 - 40$

$15x = -30$

$x = -2$

Solution set: {-2}

53.

$4.2(x - 3) + 1.7(2x + 5) = 6.4x - 6.62$

$4.2x - 12.6 + 3.4x + 8.5 = 6.4x - 6.62$

$7.6x - 4.1 = 6.4x - 6.62$

$7.6x - 6.4x - 4.1 = 6.4x - 6.4x - 6.62$

$1.2x - 4.1 = -6.62$

$1.2x - 4.1 + 4.1 = -6.62 + 4.1$

$1.2x = -2.52$

$x = -2.1$

Solution set: {-2.1}

55.

$5[7 + 2(3x-4)] = 24x - 9$ Step 1

$5(7 + 6x - 8) = 24x - 9$

$5(6x - 1) = 24x - 9$

$30x - 5 = 24x - 9$ Step 1

$30x - 24x - 5 = 24x - 24x - 9$ Step 2

$6x - 5 = -9$

$6x - 5 + 5 = -9 + 5$ Step 3

$6x = -4$

$\dfrac{6x}{6} = \dfrac{-4}{6}$ Step 4

$x = -\dfrac{2}{3}$

Solution set: $\left\{-\dfrac{2}{3}\right\}$

57.

$5[8-2(3x + 4)] = -60$

$5(8 - 6x - 8) = -60$

$5(-6x) = -60$

$-30x = -60$

$\dfrac{-30x}{-30} = \dfrac{-60}{-30}$

$x = 2$

Solution set: $\{2\}$

59.

$3[2(4x + 1) -3(2x + 6)] = 5x - 52$

$3(8x + 2 - 6x - 18) = 5x - 52$

$3(2x - 16) = 5x - 52$

$6x - 48 = 5x - 52$

$6x - 5x - 48 = 5x - 5x - 52$

$x - 48 = -52$

$x - 48 + 48 = -52 + 48$

$x = -4$

Solution set: $\{-4\}$

61.

$2[4(2x + 3) -3(x + 3)] = 10x + 15$

$2(8x + 12 - 3x - 9) = 10x + 15$ Step 1

$2(5x + 3) = 10x + 15$

$10x + 6 = 10x + 15$ Step 1

$6 = 15$ Not true

No solution

63.

$3[4 + 2(2x + 1)] = 5(3x + 2)$

$3(4 + 4x + 2) = 15x + 10$ Step 1

$3(4x + 6) = 15x + 10$

$12x + 18 = 15x + 10$ Step 1

$12x - 12x + 18 = 15x - 12x + 10$ Step 2

$18 = 3x + 10$

$18 - 10 = 3x + 10 - 10$ Step 3

$8 = 3x$

$\dfrac{8}{3} = \dfrac{3x}{3}$ Step 4

$\dfrac{8}{3} = x$

Solution set: $\left\{\dfrac{8}{3}\right\}$

65.

$2[2(x + 1) + 3(x - 1)] = 4(2x - 3) + 20$

$2(2x + 2 + 3x - 3) = 8x -12 + 20$ Step 1

$2(5x - 1) = 8x + 8$

$10x - 2 = 8x + 8$ Step 1

$10x - 8x -2 = 8x - 8x + 8$ Step 2

$2x - 2 = 8$

$2x - 2 + 2 = 8 + 2$ Step 3

$2x = 10$

$\dfrac{2x}{2} = \dfrac{10}{2}$ Step 4

$x = 5$

Solution set: $\{5\}$

67.

$6[3(2x + 8) - 5(x + 4)] = 3(x - 4)$

$6(6x + 24 - 5x - 20) = 3x - 12$

$6(x + 4) = 3x - 12$

$6x + 24 = 3x - 12$

$6x - 3x + 24 = 3x -3x - 12$

$3x + 24 = -12$

$3x + 24 - 24 = -12 - 24$

$3x = -36$

$\dfrac{3x}{3} = \dfrac{-36}{3}$

$x = -12$

Solution set: $\{-12\}$

Exercises 2.5

1. - 9.
The answers do not require worked out solutions.

11.
$|3x| = 15$

Case I   Case II

$3x = 15$ or $3x = -15$
$\dfrac{3x}{3} = \dfrac{15}{3}$ or $\dfrac{3x}{3} = \dfrac{-15}{3}$
$x = 5$ or $x = -5$

Solution set: $\{-5, 5\}$

13.
$|-2x| = \dfrac{3}{5}$

Case I   Case II

$-2x = \dfrac{3}{5}$ or $-2x = -\dfrac{3}{5}$

$\left(-\dfrac{1}{2}\right)(-2x) = \left(-\dfrac{1}{2}\right)\dfrac{3}{5}$ or $\left(-\dfrac{1}{2}\right)(-2x) = \left(-\dfrac{1}{2}\right)\left(-\dfrac{3}{5}\right)$

$x = -\dfrac{3}{10}$ or $x = \dfrac{3}{10}$

Solution set: $\left\{-\dfrac{3}{10}, \dfrac{3}{10}\right\}$

15.
$\left|\dfrac{2}{3}x\right| = 6$

Case I   Case II

$\dfrac{2}{3}x = 6$ or $\dfrac{2}{3}x = -6$
$\dfrac{3}{2} \cdot \dfrac{2}{3}x = \dfrac{3}{2} \cdot 6$ or $\dfrac{3}{2} \cdot \dfrac{2}{3}x = \dfrac{3}{2}(-6)$
$x = 9$ or $x = -9$

Solution set: $\{-9, 9\}$

17.
$|6x| = 40$

Case I   Case II

$6x = 40$ or $6x = -40$
$\dfrac{6x}{6} = \dfrac{40}{6}$ or $\dfrac{6x}{6} = \dfrac{-40}{6}$
$x = \dfrac{40}{6}$ or $x = \dfrac{-40}{6}$
$x = \dfrac{20}{3}$ or $x = \dfrac{-20}{3}$

Solution set: $\left\{\dfrac{-20}{3}, \dfrac{20}{3}\right\}$

19.
$|-9x| = 6$

Case I   Case II

$-9x = 6$ or $-9x = -6$
$\dfrac{-9x}{-9} = \dfrac{6}{-9}$ or $\dfrac{-9x}{-9} = \dfrac{-6}{-9}$
$x = \dfrac{-6}{9}$ or $x = \dfrac{6}{9}$
$x = -\dfrac{2}{3}$ or $x = \dfrac{2}{3}$

Solution set: $\left\{-\dfrac{2}{3}, \dfrac{2}{3}\right\}$

21.
$|x| = -1$

No solution since $|x| \geq 0$ and $-1 < 0$.

23.
$\left|-\dfrac{5}{6}x\right| = \dfrac{3}{5}$

Case I   Case II

$-\dfrac{5}{6}x = \dfrac{3}{5}$ or $-\dfrac{5}{6}x = -\dfrac{3}{5}$

$-\dfrac{6}{5}\left(\dfrac{-5}{6}x\right) = \dfrac{-6}{5} \cdot \dfrac{3}{5}$ or $\dfrac{-6}{5}\left(\dfrac{-5}{6}x\right) = \dfrac{-6}{5}\left(-\dfrac{3}{5}\right)$

#23 continued

$$x = -\frac{18}{25} \quad \text{or} \quad x = \frac{18}{25}$$

Solution set: $\left\{-\frac{18}{25}, \frac{18}{25}\right\}$

25.
$|x+3| = 9$

Case I		Case II

$$\begin{array}{lcl} x + 3 = 9 & \text{or} & x + 3 = -9 \\ x + 3 - 3 = 9 - 3 & \text{or} & x + 3 - 3 = -9 - 3 \\ x = 6 & \text{or} & x = -12 \end{array}$$

Solution set: $\{-12, 6\}$

27.
$|x - 2| = 5$

Case I		Case II

$$\begin{array}{lcl} x - 2 = 5 & \text{or} & x - 2 = -5 \\ x - 2 + 2 = 5 + 2 & \text{or} & x - 2 + 2 = -5 + 2 \\ x = 7 & \text{or} & x = -3 \end{array}$$

Solution set: $\{-3, 7\}$

29.
$|8 - x| = 7$

Case I		Case II

$$\begin{array}{lcl} 8 - x = 7 & \text{or} & 8 - x = -7 \\ -8 + 8 - x = 7 - 8 & \text{or} & -8 + 8 - x = -7 - 8 \\ -x = -1 & \text{or} & -x = -15 \\ x = 1 & \text{or} & x = 15 \end{array}$$

Solution set: $\{1, 15\}$

31.
$|4 - x| = \frac{3}{4}$

#31 continued

Case I	Case II

$$\begin{array}{lcl} 4 - x = \frac{3}{4} & \text{or} & 4 - x = -\frac{3}{4} \\ -4 + 4 - x = \frac{3}{4} - 4 & \text{or} & -4 + 4 - x = -\frac{3}{4} - 4 \\ -x = -\frac{13}{4} & \text{or} & -x = -\frac{19}{4} \\ x = \frac{13}{4} & \text{or} & x = \frac{19}{4} \end{array}$$

Solution set: $\left\{\frac{13}{4}, \frac{19}{4}\right\}$

33.
$|9 - x| = -1$

There is no solution since $|9 - x| \geq 0$ and $-1 < 0$.

35.
$|2x + 5| = 9$

Case I		Case II

$$\begin{array}{lcl} 2x + 5 = 9 & \text{or} & 2x + 5 = -9 \\ 2x + 5 - 5 = 9 - 5 & \text{or} & 2x + 5 - 5 = -9 - 5 \\ 2x = 4 & \text{or} & 2x = -14 \\ \frac{2x}{2} = \frac{4}{2} & \text{or} & \frac{2x}{2} = -\frac{14}{2} \\ x = 2 & \text{or} & x = -7 \end{array}$$

Solution set: $\{-7, 2\}$

37.
$|4x + 5| = 8$

Case I		Case II

$$\begin{array}{lcl} 4x + 5 = 8 & \text{or} & 4x + 5 = -8 \\ 4x + 5 - 5 = 8 - 5 & \text{or} & 4x + 5 - 5 = -8 - 5 \\ 4x = 3 & \text{or} & 4x = -13 \\ \frac{4x}{4} = \frac{3}{4} & \text{or} & \frac{4x}{4} = -\frac{13}{4} \\ x = \frac{3}{4} & \text{or} & x = -\frac{13}{4} \end{array}$$

#37 continued

Solution set: $\left\{-\dfrac{13}{4}\,,\,\dfrac{3}{4}\right\}$

39.

$|\,5x - 1\,| = 12$

Case I		Case II
$5x - 1 = 12$	or	$5x - 1 = -12$
$5x - 1 + 1 = 12 + 1$	or	$5x - 1 + 1 = -12 + 1$
$5x = 13$	or	$5x = -11$
$\dfrac{5x}{5} = \dfrac{13}{5}$	or	$\dfrac{5x}{5} = -\dfrac{11}{5}$
$x = \dfrac{13}{5}$	or	$x = -\dfrac{11}{5}$

Solution set: $\left\{-\dfrac{11}{5}\,,\,\dfrac{13}{5}\right\}$

41.

$|\,5 - 2x\,| = \dfrac{1}{6}$

Case I		Case II
$5 - 2x = \dfrac{1}{6}$	or	$5 - 2x = -\dfrac{1}{6}$
$-5 + 5 - 2x = -5 + \dfrac{1}{6}$	or	$-5 + 5 - 2x = -5 - \dfrac{1}{6}$
$-2x = -\dfrac{29}{6}$	or	$-2x = -\dfrac{31}{6}$
$-\dfrac{1}{2}(-2x) = -\dfrac{1}{2}\left(-\dfrac{29}{6}\right)$	or	$-\dfrac{1}{2}(-2x) = -\dfrac{1}{2}\left(-\dfrac{31}{6}\right)$
$x = \dfrac{29}{12}$	or	$x = \dfrac{31}{12}$

Solution set: $\left\{\dfrac{29}{12}\,,\,\dfrac{31}{12}\right\}$

43.

$\left|\dfrac{1}{2}x + 2\right| = \dfrac{1}{3}$

#43 continued

Case I		Case II
$\dfrac{1}{2}x + 2 = \dfrac{1}{3}$	or	$\dfrac{1}{2}x + 2 = -\dfrac{1}{3}$
$\dfrac{1}{2}x + 2 - 2 = \dfrac{1}{3} - 2$	or	$\dfrac{1}{2}x + 2 - 2 = -\dfrac{1}{3} - 2$
$\dfrac{1}{2}x = -\dfrac{5}{3}$	or	$\dfrac{1}{2}x = -\dfrac{7}{3}$
$2\left(\dfrac{1}{2}x\right) = 2\left(-\dfrac{5}{3}\right)$	or	$2\left(\dfrac{1}{2}x\right) = 2\left(-\dfrac{7}{3}\right)$
$x = -\dfrac{10}{3}$	or	$x = -\dfrac{14}{3}$

Solution set: $\left\{-\dfrac{14}{3}\,,\,-\dfrac{10}{3}\right\}$

45.

$|\,2x - 6\,| = 5$

Case I		Case II
$2x - 6 = 5$	or	$2x - 6 = -5$
$2x - 6 + 6 = 5 + 6$	or	$2x - 6 + 6 = -5 + 6$
$2x = 11$	or	$2x = 1$
$\dfrac{2x}{2} = \dfrac{11}{2}$	or	$\dfrac{2x}{2} = \dfrac{1}{2}$
$x = \dfrac{11}{2}$	or	$x = \dfrac{1}{2}$

Solution set: $\left\{\dfrac{1}{2}\,,\,\dfrac{11}{2}\right\}$

47.

$\left|\dfrac{5}{8}x - \dfrac{1}{4}\right| = \dfrac{3}{8}$

Case I		Case II
$\dfrac{5}{8}x - \dfrac{1}{4} = \dfrac{3}{8}$	or	$\dfrac{5}{8}x - \dfrac{1}{4} = -\dfrac{3}{8}$
$\dfrac{5}{8}x - \dfrac{1}{4} + \dfrac{1}{4} = \dfrac{3}{8} + \dfrac{1}{4}$	or	$\dfrac{5}{8}x - \dfrac{1}{4} + \dfrac{1}{4} = -\dfrac{3}{8} + \dfrac{1}{4}$
$\dfrac{5}{8}x = \dfrac{5}{8}$	or	$\dfrac{5}{8}x = -\dfrac{1}{8}$
$\dfrac{8}{5}\left(\dfrac{5}{8}x\right) = \dfrac{8}{5} \cdot \dfrac{5}{8}$	or	$\dfrac{8}{5}\left(\dfrac{5}{8}x\right) = \dfrac{8}{5}\left(-\dfrac{1}{8}\right)$

#47 continued

$$x = 1 \quad \text{or} \quad x = -\frac{1}{5}$$

Solution set: $\left\{-\frac{1}{5}, 1\right\}$

49.

$|5x - 4| = -10$

There is no solution since $|5x - 4| = -10$

51.

$\left|\frac{2}{3}x - \frac{1}{6}\right| = 0$

Case I and Case II are the same.

$$\frac{2}{3}x - \frac{1}{6} = 0$$

$$\frac{2}{3}x - \frac{1}{6} + \frac{1}{6} = \frac{1}{6}$$

$$\frac{2}{3}x = \frac{1}{6}$$

$$\frac{3}{2}\left(\frac{2}{3}x\right) = \frac{3}{2}\left(\frac{1}{6}\right)$$

$$x = \frac{1}{4}$$

Solution set: $\left\{\frac{1}{4}\right\}$

53.

$|18 - x| = 12$

Case I		Case II
$18 - x = 12$	or	$18 - x = -12$
$-18 + 18 - x = 12 - 18$	or	$-18 + 18 - x = -12 - 18$
$-x = -6$	or	$-x = -30$
$\frac{-x}{-1} = \frac{-6}{-1}$	or	$\frac{-x}{-1} = \frac{-30}{-1}$
$x = 6$	or	$x = 30$

Solution set: $\{6, 30\}$

55.

$|20 - 2x| = 9$

Case I		Case II
$20 - 2x = 9$	or	$20 - 2x = -9$
$-20 + 20 - 2x = 9 - 20$	or	$-20 + 20 - 2x = -9 - 20$
$-2x = -11$	or	$-2x = -29$
$\frac{-2x}{-2} = \frac{-11}{-2}$	or	$\frac{-2x}{-2} = \frac{-29}{-2}$
$x = \frac{11}{2}$	or	$x = \frac{29}{2}$

Solution set: $\left\{\frac{11}{2}, \frac{29}{2}\right\}$

57.

$$|x - 8| - 2 = -2$$
$$|x - 8| - 2 + 2 = -2 + 2$$
$$|x - 8| = 0$$
Isolating the absolute value expression

Case I and Case II are the same.

$$x - 8 = 0$$
$$x = 8$$

Solution set: $\{8\}$

59.

$$|x + 3| - 4 = 1$$
$$|x + 3| - 4 + 4 = 1 + 4$$
$$|x + 3| = 5$$
Isolating the absolute value expression

Case I		Case II
$x + 3 = 5$	or	$x + 3 = -5$
$x + 3 - 3 = 5 - 3$	or	$x + 3 - 3 = -5 - 3$
$x = 2$	or	$x = -8$

Solution set: $\{-8, 2\}$

61.

$$|1 - x| + 6 = -3$$
$$|1 - x| + 6 - 6 = -3 - 6$$
$$|1 - x| = -9$$
Isolating the absolute value expression

#61 continued

Solution set: ∅

There is no solution since $|1 - x| \geq 0$.

63.
$$|x + 2| - 7 = -4$$
$$|x + 2| - 7 + 7 = -4 + 7$$
$$|x + 2| = 3$$
Isolating the absolute value expression

Case I		Case II
$x + 2 = 3$	or	$x + 2 = -3$
$x + 2 - 2 = 3 - 2$	or	$x + 2 - 2 = -3 - 2$
$x = 1$	or	$x = -5$

Solution set: {-5, 1}

65.
$$|x + 6| - 10 = -3$$
$$|x + 6| - 10 + 10 = -3 + 10$$
$$|x + 6| = 7$$
Isolating the absolute value expression

Case I		Case II
$x + 6 = 7$	or	$x + 6 = -7$
$x + 6 - 6 = 7 - 6$	or	$x + 6 - 6 = -7 - 6$
$x = 1$	or	$x = -13$

Solution set: {-13, 1}

67.
$$|x - 4| + 11 = 22$$
$$|x - 4| + 11 - 11 = 22 - 11$$
$$|x - 4| = 11$$
Isolating the absolute value expression

Case I		Case II
$x - 4 = 11$	or	$x - 4 = -11$
$x - 4 + 4 = 11 + 4$	or	$x - 4 + 4 = -11 + 4$
$x = 15$	or	$x = -7$

Solution set: {-7, 15}

69.
$$|x + 8| + 2 = 2$$
$$|x + 8| + 2 - 2 = 2 - 2$$
$$|x + 8| = 0$$

Case I and Case II are the same.

$$x + 8 = 0$$
$$x = -8$$

Solution set: {-8}

71.
$$|5 - x| + 3 = 7$$
$$|5 - x| + 3 - 3 = 7 - 3$$
$$|5 - x| = 4$$
Isolating the absolute value expression

Case I		Case II
$5 - x = 4$	or	$5 - x = -4$
$-5 + 5 - x = -5 + 4$	or	$-5 + 5 - x = -5 - 4$
$-x = -1$	or	$-x = -9$
$x = 1$	or	$x = 9$

Solution set: {1, 9}

73.
$$|9 - x| + 3 = 10$$
$$|9 - x| + 3 - 3 = 10 - 3$$
$$|9 - x| = 7$$
Isolating the absolute value expression

Case I		Case II
$9 - x = 7$	or	$9 - x = -7$
$-9 + 9 - x = -9 + 7$	or	$-9 + 9 - x = -9 - 7$
$-x = -2$	or	$-x = -16$
$x = 2$	or	$x = 16$

Solution set: {2, 16}

75.
$$|x - 8| - 2 = 3$$
$$|x - 8| - 2 + 2 = 3 + 2$$
$$|x - 8| = 5$$
Isolating the absolute value expression

#75 continued

Case I   Case II

$$x - 8 = 5 \quad \text{or} \quad x - 8 = -5$$
$$x - 8 + 8 = 5 + 8 \quad \text{or} \quad x - 8 + 8 = -5 + 8$$
$$x = 13 \quad \text{or} \quad x = 3$$

Solution set: {3, 13}

77.
$$|5x - 9| - 2 = -6$$
$$|5x - 9| - 2 + 2 = -6 + 2$$
$$|5x - 9| = -4$$

Solution set: Ø

There is no solution since $|5x - 4| \geq 0$.

79.
$$|3 - 2x| + 4 = 9$$
$$|3 - 2x| + 4 - 4 = 9 - 4$$
$$|3 - 2x| = 5$$
Isolating the absolute value expression

Case I   Case II

$$3 - 2x = 5 \quad \text{or} \quad 3 - 2x = -5$$
$$-3 + 3 - 2x = -3 + 5 \quad \text{or} \quad -3 + 3 - 2x = -3 - 5$$
$$-2x = 2 \quad \text{or} \quad -2x = -8$$
$$\frac{-2x}{-2} = \frac{2}{-2} \quad \text{or} \quad \frac{-2x}{-2} = \frac{-8}{-2}$$
$$x = -1 \quad \text{or} \quad x = 4$$

Solution set: {-1, 4}

81.
$$|6 - 4x| - 3 = 0$$
$$|6 - 4x| = 3$$
Isolating the absolute value expression

Case I   Case II

$$6 - 4x = 3 \quad \text{or} \quad 6 - 4x = -3$$
$$-6 + 6 - 4x = -6 + 3 \quad \text{or} \quad -6 + 6 - 4x = -6 - 3$$
$$-4x = -3 \quad \text{or} \quad -4x = -9$$
$$\frac{-4x}{-4} = \frac{-3}{-4} \quad \text{or} \quad \frac{-4x}{-4} = \frac{-9}{-4}$$
$$x = \frac{3}{4} \quad \text{or} \quad x = \frac{9}{4}$$

#81 continued

Solution set: $\left\{ \frac{3}{4}, \frac{9}{4} \right\}$

83.
$$|2x + 5| + 7 = 13$$
$$|2x + 5| + 7 - 7 = 13 - 7$$
$$|2x + 5| = 6$$
Isolating the absolute value expression

Case I   Case II

$$2x + 5 = 6 \quad \text{or} \quad 2x + 5 = -6$$
$$2x + 5 - 5 = 6 - 5 \quad \text{or} \quad 2x + 5 - 5 = -6 - 5$$
$$2x = 1 \quad \text{or} \quad 2x = -11$$
$$x = \frac{1}{2} \quad \text{or} \quad x = -\frac{11}{2}$$
Divide by 2

Solution set: $\left\{ -\frac{11}{2}, \frac{1}{2} \right\}$

85.
$$|9x - 3| + 8 = 2$$
$$|9x - 3| + 8 - 8 = 2 - 8$$
$$|9x - 3| = -6$$

Solution set: Ø

There is no solution since $|9x - 3| \geq 0$.

87.
$$|5x + 2| + 3 = 11$$
$$|5x + 2| + 3 - 3 = 11 - 3$$
$$|5x + 2| = 8$$

Case I   Case II

$$5x + 2 = 8 \quad \text{or} \quad 5x + 2 = -8$$
$$5x + 2 - 2 = 8 - 2 \quad \text{or} \quad 5x + 2 - 2 = -8 - 2$$
$$5x = 6 \quad \text{or} \quad 5x = -10$$
$$x = \frac{6}{5} \quad \text{or} \quad x = -2$$
Divide by 5

#87 continued

Solution set: $\left\{-2, \dfrac{6}{5}\right\}$

89.

$$\left|\dfrac{1}{3}x+1\right| + 2 = 6$$

$$\left|\dfrac{1}{3}x+1\right| + 2 - 2 = 6 - 2$$

$$\left|\dfrac{1}{3}x+1\right| = 4$$

Isolating the absolute value expression

Case I		Case II
$\dfrac{1}{3}x + 1 = 4$	or	$\dfrac{1}{3}x + 1 = -4$
$\dfrac{1}{3}x + 1 - 1 = 4 - 1$	or	$\dfrac{1}{3}x + 1 - 1 = -4 - 1$
$\dfrac{1}{3}x = 3$	or	$\dfrac{1}{3}x = -5$
$x = 9$	or	$x = -15$

Solution set: $\{-15, 9\}$

91.

$$|6x - 16| - 2 = 38$$
$$|6x - 16| - 2 + 2 = 38 + 2$$
$$|6x - 16| = 40$$

Isolating the absolute value expression

Case I		Case II
$6x - 16 = 40$	or	$6x - 16 = -40$
$6x - 16 + 16 = 40 + 16$	or	$6x - 16 + 16 = -40 + 16$
$6x = 56$	or	$6x = -24$
$\dfrac{6x}{6} = \dfrac{56}{6}$	or	$\dfrac{6x}{6} = \dfrac{-24}{6}$
$x = \dfrac{28}{3}$	or	$x = -4$

Solution set: $\left\{-4, \dfrac{28}{3}\right\}$

93.

$$\left|\dfrac{3}{2}x - \dfrac{1}{4}\right| - \dfrac{1}{8} = \dfrac{3}{8}$$

#93 continued

$$\left|\dfrac{3}{2}x - \dfrac{1}{4}\right| - \dfrac{1}{8} + \dfrac{1}{8} = \dfrac{3}{8} + \dfrac{1}{8}$$

$$\left|\dfrac{3}{2}x - \dfrac{1}{4}\right| = \dfrac{1}{2}$$

Case I		Case II
$\dfrac{3}{2}x - \dfrac{1}{4} = \dfrac{1}{2}$	or	$\dfrac{3}{2}x - \dfrac{1}{4} = -\dfrac{1}{2}$
$\dfrac{3}{2}x - \dfrac{1}{4} + \dfrac{1}{4} = \dfrac{1}{2} + \dfrac{1}{4}$	or	$\dfrac{3}{2}x - \dfrac{1}{4} + \dfrac{1}{4} = -\dfrac{1}{2} + \dfrac{1}{4}$
$\dfrac{3}{2}x = \dfrac{3}{4}$	or	$\dfrac{3}{2}x = -\dfrac{1}{4}$
$\dfrac{2}{3}\left(\dfrac{3}{2}x\right) = \dfrac{2}{3}\left(\dfrac{3}{4}\right)$	or	$\dfrac{2}{3}\left(\dfrac{3}{2}x\right) = \dfrac{2}{3}\left(-\dfrac{1}{4}\right)$
$x = \dfrac{1}{2}$	or	$x = -\dfrac{1}{6}$

Solution set: $\left\{-\dfrac{1}{6}, \dfrac{1}{2}\right\}$

95.

$$\left|\dfrac{3}{2}x + \dfrac{1}{4}\right| - \dfrac{3}{8} = \dfrac{5}{8}$$

$$\left|\dfrac{3}{2}x + \dfrac{1}{4}\right| - \dfrac{3}{8} + \dfrac{3}{8} = \dfrac{5}{8} + \dfrac{3}{8}$$

$$\left|\dfrac{3}{2}x + \dfrac{1}{4}\right| = 1$$

Case I		Case II
$\dfrac{3}{2}x + \dfrac{1}{4} = 1$	or	$\dfrac{3}{2}x + \dfrac{1}{4} = -1$
$\dfrac{3}{2}x + \dfrac{1}{4} - \dfrac{1}{4} = 1 - \dfrac{1}{4}$	or	$\dfrac{3}{2}x + \dfrac{1}{4} - \dfrac{1}{4} = -1 - \dfrac{1}{4}$
$\dfrac{3}{2}x = \dfrac{3}{4}$	or	$\dfrac{3}{2}x = -\dfrac{5}{4}$
$\dfrac{2}{3}\left(\dfrac{3}{2}x\right) = \dfrac{2}{3}\left(\dfrac{3}{4}\right)$	or	$\dfrac{2}{3}\left(\dfrac{3}{2}x\right) = \dfrac{2}{3}\left(-\dfrac{5}{4}\right)$
$x = \dfrac{1}{2}$	or	$x = -\dfrac{5}{6}$

Solution set: $\left\{-\dfrac{5}{6}, \dfrac{1}{2}\right\}$

Exercises 2.6

1.
A = LW, for L

$$\frac{A}{W} = \frac{LW}{W}$$

$$\frac{A}{W} = L \quad \text{or} \quad L = \frac{A}{W}$$

3.
F = ma, for a

$$\frac{F}{m} = \frac{ma}{m}$$

$$\frac{F}{m} = a \quad \text{or} \quad a = \frac{F}{m}$$

5.
C = 2πr, for r

$$\frac{C}{2\pi} = \frac{2\pi r}{2\pi}$$

$$\frac{C}{2\pi} = r \quad \text{or} \quad r = \frac{C}{2\pi}$$

7.
I = PRT, for T

$$\frac{I}{PR} = \frac{PRT}{PR}$$

$$\frac{I}{PR} = T \quad \text{or} \quad T = \frac{I}{PR}$$

9.
V = LWH, for H

$$\frac{V}{LW} = \frac{LWH}{LW}$$

$$\frac{V}{LW} = H \quad \text{or} \quad H = \frac{V}{LW}$$

11.
$A = \frac{1}{2} BH$, for B

$$2A = BH \qquad \text{Multiply by 2}$$

$$\frac{2A}{H} = \frac{BH}{H}$$

$$\frac{2A}{H} = B \quad \text{or} \quad B = \frac{2A}{H}$$

13.
$F = \frac{GmM}{r^2}$ for M

$$r^2 F = GmM \qquad \text{Multiply by } r^2$$

$$\frac{r^2 F}{Gm} = \frac{GmM}{Gm}$$

$$\frac{r^2 F}{Gm} = M \quad \text{or} \quad M = \frac{r^2 F}{Gm}$$

15.
$F = \frac{Mv^2}{r}$, for M

$$rF = Mv^2 \qquad \text{Multiply by } r$$

$$\frac{rF}{v^2} = \frac{Mv^2}{v^2}$$

$$\frac{Fr}{v^2} = M \quad \text{or} \quad M = \frac{Fr}{v^2}$$

17.
PV = nRT, for T

$$\frac{PV}{nR} = \frac{nRT}{nR}$$

$$\frac{PV}{nR} = T \quad \text{or} \quad T = \frac{PV}{nR}$$

19.
y = mx + b, for m

$$y - b = mx \qquad \text{Subtract b}$$

$$\frac{y - b}{x} = \frac{mx}{x}$$

#19 continued

$$\frac{y - b}{x} = m \ \text{ or } \ m = \frac{y - b}{x}$$

21.
$S = 2\pi r^2 + 2\pi rh$, for h

$$S - 2\pi r^2 = 2\pi rh \qquad \text{Subtract } 2\pi r^2$$

$$\frac{S - 2\pi r^2}{2\pi r} = \frac{2\pi rh}{2\pi r}$$

$$\frac{S - 2\pi r^2}{2\pi r} = h \ \text{ or } \ h = \frac{S - 2\pi r^2}{2\pi r}$$

23.
$3x + 2y = 7$, for y

$$2y = -3x + 7 \qquad \text{Subtract } 3x$$

$$\frac{2y}{2} = \frac{-3x + 7}{2}$$

$$y = \frac{7 - 3x}{2}$$

25.
$2x - 7y = 12$, for y

$$-7y = -2x + 12 \qquad \text{Subtract } 2x$$

$$\frac{-7y}{-7} = \frac{-2x + 12}{-7}$$

$$y = \frac{2x - 12}{7}$$

Note: $\dfrac{-1}{-1} \cdot \left(\dfrac{-2x + 12}{-7} \right) = \dfrac{2x - 12}{7}$

27.
$x = \dfrac{y + z}{2}$, for y

$$2x = y + z \qquad \text{Multiply by 2}$$
$$2x - z = y \qquad \text{Subtract } z$$

29.
$a = \dfrac{b - c}{3}$, for b

$$3a = b - c \qquad \text{Multiply by 3}$$
$$3a + c = b \qquad \text{Add } c$$

31.
$t = \dfrac{2s - 5g}{7}$, for g

$$7t = 2s - 5g \qquad \text{Multiply by 7}$$
$$7t - 2s = -5g \qquad \text{Subtract } 2s$$
$$-7t + 2s = 5g \qquad \text{Multiply by -1}$$
$$\frac{2s - 7t}{5} = g \qquad \text{Divide by 5}$$

33.
$S = \pi s(r + R)$, for R

$$S = \pi rs + \pi sR \qquad \text{Distributive property}$$
$$S - \pi rs = \pi sR \qquad \text{Subtract } \pi rs$$
$$\frac{S - \pi rs}{\pi s} = \frac{\pi sR}{\pi s}$$
$$\frac{S - \pi rs}{\pi s} = R$$

35.

$A = \dfrac{1}{2}(b + B)h$, for b

$$2A = (b + B)h \qquad \text{Multiply by 2}$$
$$2A = bh + Bh \qquad \text{Distributive property}$$
$$2A - Bh = bh \qquad \text{Subtract } Bh$$
$$\frac{2A - Bh}{h} = b \qquad \text{Divide by h}$$

37.
$V = r + at$, for a

$$V - r = at \qquad \text{Subtract } r$$
$$\frac{V - r}{t} = a \qquad \text{Divide by t}$$

39.
$V^2 = r^2 + 2as$, for s

$$V^2 - r^2 = 2as \qquad \text{Subtract } r^2$$
$$\frac{V^2 - r^2}{2a} = \frac{2as}{2a}$$
$$\frac{V^2 - r^2}{2a} = s$$

41.
D = RT, for R

$\dfrac{D}{T} = R$

$R = \dfrac{135}{3}$ D = 135, T = 3

R = 45

43.
W = Fs, for s

$\dfrac{W}{F} = s$

$s = \dfrac{20}{15}$ W = 20, F = 15

$s = \dfrac{4}{3}$

45.
$A = \dfrac{1}{2}BH$, for B

2A = BH

$\dfrac{2A}{H} = B$

$B = \dfrac{2(84)}{7}$ A = 84, H = 7

B = 24

47.
V = LWH, for L

$\dfrac{V}{WH} = L$

$L = \dfrac{60}{3 \cdot 4}$ V = 60, W = 3, H = 4

$L = \dfrac{12 \cdot 5}{12}$

L = 5

49.
P = 2L + 2W, for L

P - 2W = 2L

$\dfrac{P - 2W}{2} = L$ Divide by 2

#49 continued

$L = \dfrac{36 - 2(5)}{2}$ P = 36, W = 5

$L = \dfrac{36 - 10}{2}$

L = 13

51.
y = mx + b, for x

y - b = mx

$\dfrac{y-b}{m} = x$

$x = \dfrac{-10 - 2}{3}$ y = -10, b = 2, m = 3

$x = \dfrac{-12}{3}$

x = -4

53.
$F = \dfrac{9}{5}C + 32;$ C = 100

$F = \dfrac{9}{5} \cdot 100 + 32$

F = 180 + 32

F = 212

The boiling point of water is 212° F.

55.
F = Kx, for K

$K = \dfrac{F}{x};$ F = 84, x = 36

$K = \dfrac{84}{36}$

$K = \dfrac{7}{3}$

57.

$b = \dfrac{2A - Bh}{h};$ B = 14, h = 9, A = 99

See exercise 35.

#57 continued

$$b = \frac{2(99) - 14(9)}{9}$$

$$b = \frac{198 - 126}{9}$$

$$b = \frac{72}{9}$$

$$b = 8$$

59.

S = 2WL + 2LH + 2WH, for W

S - 2LH = 2WL + 2WH   Subtract 2LH.

S - 2LH = W(2L + 2H) Distributive property

$$\frac{S - 2LH}{2L + 2H} = \frac{W(2L + 2H)}{2L + 2H}$$

$$\frac{S - 2LH}{2L + 2H} = W; \quad S = 228, L = 6, H = 9$$

$$W = \frac{228 - 2(6 \cdot 9)}{2(6) + 2(9)}$$

$$W = \frac{228 - 108}{12 + 18}$$

$$W = \frac{120}{30} = 4$$

Exercises 2.7

1-19.

The answers do not require worked out solutions.

21.

Let x be the number.
Thus,

$$3x + 1 = 19$$
$$3x + 1 - 1 = 19 - 1$$
$$3x = 18$$
$$\frac{3x}{3} = \frac{18}{3}$$
$$x = 6$$

The number is 6.

23.

Let x be the number.
Thus,

$$8x - 1 = 3$$
$$8x - 1 + 1 = 3 + 1$$
$$8x = 4$$
$$\frac{8x}{8} = \frac{4}{8}$$
$$x = \frac{1}{2}$$

The number is $\frac{1}{2}$.

25.

Let x be the number.
Four minus 3 times a number is 4 - 3x.
Thus,

$$4 - 3x = 10$$
$$-4 + 4 - 3x = -4 + 10$$
$$-3x = 6$$
$$\frac{-3x}{-3} = \frac{6}{-3}$$
$$x = -2$$

The number is -2.

27.

Three consecutive integers are x, x + 1, and x + 2.
Thus,

$$x + (x + 1) + (x + 2) = 39$$
$$3x + 3 = 39$$
$$3x + 3 - 3 = 39 - 3$$
$$3x = 36$$
$$\frac{3x}{3} = \frac{36}{3}$$
$$x = 12$$

The consecutive integers are 12, 13, and 14.

29.

Three consecutive integers are x, x + 1, and x + 2.
Thus,

$$x + (x + 1) + (x + 2) = -63$$
$$3x + 3 = -63$$
$$3x + 3 - 3 = -63 - 3$$

#29 continued

$$3x = -66$$
$$\frac{3x}{3} = \frac{-66}{3}$$
$$x = -22$$

The consective integers are -22, -21, and -20.

31.

Three consecutive odd integers are x, x+2, and x + 4.
Thus,

$$x + (x + 2) + (x + 4) = 93$$
$$3x + 6 = 93$$
$$3x + 6 - 6 = 93 - 6$$
$$3x = 87$$
$$\frac{3x}{3} = \frac{87}{3}$$
$$x = 29$$

The three consecutive odd integers are 29, 31, and 33.

33.

Three consecutive even integers are x, x+2, and x+4.
Thus,

$$x + (x + 2) + (x + 4) = -30$$
$$3x + 6 = -30$$
$$3x + 6 - 6 = -30 - 6$$
$$3x = -36$$
$$\frac{3x}{3} = \frac{-36}{3}$$
$$x = -12$$

The consecutive even integers are -12, -10, and -8.

35.

The length of the second and third pieces are in terms of the first piece.

So,   x = length of first piece
   3x = length of second piece
  x + 4 = length of third piece

The total length is 64 in.

#35 continued

The equation is:

$$x + 3x + (x + 4) = 64$$
$$5x + 4 = 64$$
$$5x + 4 - 4 = 64 - 4$$
$$5x = 60$$
$$x = 12$$
Thus, $$3x = 36$$
$$x + 4 = 16$$

The three pieces have lengths of 12", 36" and 16".

37.

The length of the second piece is in terms of the first piece, and the length of the third piece is in terms of the second piece.
So,   x = length of first piece
    x + 3 = length of second piece
  (x + 3) + 4 = length of third piece

The total length is 37 in.
The equation is:

$$x + (x + 3) + [(x + 3) + 4] = 37$$
$$3x + 10 = 37$$
$$3x + 10 - 10 = 37 + 10$$
$$3x = 27$$
$$\frac{3x}{3} = \frac{27}{3}$$
$$x = 9$$
Thus, $$x + 3 = 12$$
$$(x + 3) + 4 = 16$$

The three pieces have lengths of 9", 12" and 16".

39.

The length of the first and second piece are in terms of the third piece.

So,   x = length of third piece
    2x = length of the first piece
    $\frac{1}{2}x$ = length of the second piece

The total length is 98 cm.

#39. continued.

The equation is:

$$x + 2x + \frac{1}{2}x = 98$$

$$\frac{7}{2}x = 98$$

$$\frac{2}{7}\left(\frac{7}{2x}\right) = \frac{2}{7} \cdot 98$$

$$x = 28$$

Thus, $2x = 56$

$$\frac{1}{2}x = 14$$

The three pieces have lengths of 56 cm, 14 cm, and 28 cm.

41.
The perimeter formula is P = 2W + 2L.
The length is in terms of the width

So, x = measurement of the width
   x + 7 = measurement of the length
       P = 74

Thus,
       74 = 2x + 2 (x + 7)
       74 = 2x + 2x + 14
       74 = 4x + 14
 74 - 14 = 4x + 14 - 14
       60 = 4x
       $\frac{60}{4} = \frac{4x}{4}$
       15 = x
Thus, x = 15
   x + 7 = 22

The dimensions are 15' by 22'.

43.
The perimeter formula is P = 2W + 2L.
The length is in terms of the width.

#43 continued

So, x = measurement of the width
   3x + 4 = measurement of the length
       P = 80

Thus,
       80 = 2x +2(3x + 4)
       80 = 2x + 6x + 8
       80 = 8x + 8
 80 - 8 = 8x + 8 - 8
       72 = 8x
        9 = x
Thus, x = 9
   3x + 4 = 31

The dimensions are 9' by 31'.

45.
The perimeter formula is P = 2W + 2L.
The length is in terms of the width.

So, x = measurement of the width
   2x - 7 = measurement of the length
       P = 64

Thus,
       64 = 2x + 2(2x - 7)
       64 = 2x + 4x - 14
       64 = 6x - 14
 64 + 14 = 6x - 14 + 14
       78 = 6x
       13 = x
Thus, x = 13
   2x - 7 = 19

The dimensions are 13' by 19'.

47.
The number of fives are in terms of tens.

So, x = number of tens
   3x - 1 = number of fives

#47 continued

	Tens	Fives	Total
Number of bills	x	3x - 1	
Value (dollars)	10x	5(3x - 1)	120

The equation is in terms of value.
Thus,

$$10x + 5(3x - 1) = 120$$
$$10x + 15x - 5 = 120$$
$$25x - 5 + 5 = 120 + 5$$
$$25x = 125$$
$$x = 5$$
$$3x - 1 = 14$$

There are 14 fives and 5 tens.

49.
The number of nickels are in terms of quarters.

So,   x = number of quarters
    2x = number of nickels

	Nickels	Quarters	Total
Number of coins	2x	x	
Value (cents)	5(2x)	25(x)	455

The equation is in terms of value.
Thus,

$$5(2x) + 25(x) = 455$$
$$10x + 25x = 455$$
$$35x = 455$$
$$x = 13$$
$$2x = 26$$

There are 26 nickels and 13 quarters.

51.
The number of $1 bills and $5 bills is in terms of $10 bills.

So,   x = number of $10 bills
    x + 4 = number of $1 bills
    2x + 3 = number of $5 bills

	$1	$5	$10	Total
Number of bills	x + 4	2x +3	x	
Value (dollars)	x + 4	5(2x + 3)	10(x)	166

The equation is in terms of value.
Thus,

$$x + 4 + 5(2x + 3) + 10(x) = 166$$
$$x + 4 + 10x + 15 + 10x = 166$$
$$21x + 19 = 166$$
$$21x = 147$$
$$x = 7$$
$$2x + 3 = 17$$
$$x + 4 = 11$$

There are 11 ones, 17 fives, and 7 tens.

53.
The number of dimes is in terms of nickels, and the number of quarters is in terms of dimes.

So,   x = number of nickels
    x - 3 = number of dimes
    (x - 3) + 7 = number of quarters

	5¢	10¢	25¢	Total
Number of coins	x	x - 3	(x - 3) + 7	
Value (cents)	5x	10(x - 3)	25(x + 4)	550

The equation is in terms of value.
Thus,

$$5x + 10(x - 3) + 25(x + 4) = 550$$

#53 continued

$$5x + 10x - 30 + 25x + 100 = 550$$
$$40x + 70 = 550$$
$$40x = 480$$
$$x = 12$$
$$x - 3 = 9$$
$$x + 4 = 16$$

There are 12 nickels, 9 dimes and 16 quarters.

55.
The number of $20 and $10 bills is in terms of $5 bills.

So, x = number of $5 bills
$2x + 1$ = number of $10 bills
$\frac{1}{3}x$ = number of $20 bills

	$5	$10	$20	Total
Number of bills	x	2x + 1	$\frac{1}{3}x$	
Value (dollars)	5x	10(2x + 1)	$20\left(\frac{1}{3}x\right)$	200

The equation is in terms of value.
Thus,

$$5x + 10(2x + 1) + 20\left(\frac{1}{3}x\right) = 200$$
$$5x + 20x + 10 + \frac{20}{3}x = 200$$
$$\frac{95}{3}x + 10 = 200$$
$$\frac{95}{3}x = 190$$
$$x = 6$$
$$2x + 1 = 13$$
$$\frac{1}{3}x = 2$$

There are 6 fives, 13 tens and 2 twenties.

57.
Let x = number of nickels
$36 - x$ = number of quarters

	5¢	25¢	Total
Number of coins	x	36 - x	36
Value (cents)	5x	25(36 - x)	460

The equation is in terms of value.
Thus,
$$5x + 25(36 - x) = 460$$
$$5x + 900 - 25x = 460$$
$$-20x = -440$$
$$x = 22$$
$$36 - x = 14$$

There were 22 nickels and 14 quarters.

59.
Let x = number of $5 bills
$24 - x$ = number of $20 bills
The equation is in terms of value.
Thus,

	$5	$20	Total
Number of bills	x	24 - x	24
Value (dollars)	5x	20(24 - x)	210

$$5x + 20(24 - x) = 210$$
$$5x + 480 - 20x = 210$$
$$-15x = -270$$
$$x = 18$$
$$24 - x = 6$$

There were 18 fives and 6 twenties.

61.
Let x = number of $5 bills
$3x + 1$ = number of $1 bills
$48 - [x + (3x + 1)]$ = number of $10 bills

#61 continued

Note: $48 - [x + (3x + 1)]$
$= 48 - (4x + 1)$
$= 48 - 4x - 1$
$= 47 - 4x$

	$1	$5	$10	Total
Number of bills	3x + 1	x	47 - 4x	48
Value (dollars)	3x + 1	5x	10(47 - 4x)	215

The equation is in terms of value.

$$3x + 1 + 5x + 10(47 - 4x) = 215$$
$$3x + 1 + 5x + 470 - 40x = 215$$
$$-32x = -256$$
$$x = 8$$
$$3x + 1 = 25$$
$$47 - 4x = 15$$

There were 25 ones, 8 fives, and 15 tens.

Exercises 2.8

1.
Let x = number of hours it will take for Debbie and Jim to be 150 miles apart.

	Rate ·	Time =	Distance
Debbie	45	x	45x
Jim	55	x	55x

Since they are traveling in the opposite direction from the same starting point, each one's distance is added for the total miles apart.

$$45x + 55x = 150$$
$$100x = 150$$
$$x = 1.5$$

It took 1.5 hours.

3.
Let x = number of hours for the trains to meet.

	Rate ·	Time =	Distance
Train A	30	x	30x
Train B	35	x	35x

The equation is

$$30x + 35x = 325$$
$$65x = 325$$
$$x = 5$$

They will meet after 5 hours or 5:00 p.m.

5.
Let x = Roscoe's traveling time.
Let $3 - x$ = Junior's traveling time.

	Rate ·	Time =	Distance
Roscoe	35	x	35x
Junior	60	x - 3	60(x - 3)

When Junior caught up with Roscoe, they both had gone the same distance, thus both distances are equal.
So,
$$35x = 60(x - 3)$$
$$35x = 60x - 180$$
$$180 = 25x$$
$$7.2 = x$$
$$x - 3 = 4.2$$

It took Junior 4.2 hours to overtake Roscoe.

7.

Let x   =   Bob's traveling time

x - 4   =   Judy's traveling time

	Rate ·	Time =	Distance
Bob	14	x	14x
Judy	54	x-4	54(x - 4)

When Judy caught up with Bob, they both had gone the same distance, thus both distances are equal.

So,

$$14x = 54(x - 4)$$
$$14x = 54x - 216$$
$$-40x = -216$$
$$x = 5.4$$
$$x-4 = 1.4$$

It took Judy 1.4 hours to catch Bob.

9.

Let x   =   Mike's traveling time

x - 2   =   Sam's traveling time

	Rate ·	Time =	Distance
Mike	42	x	42x
Sam	54	x-2	54(x - 2)

Since they are traveling in the opposite direction from the same starting point, each distance is added for the total miles apart.

$$42x + 54(x - 2) = 276$$
$$42x + 54x - 108 = 276$$
$$96x = 384$$
$$x = 4$$

Mike travels 4 hours. They will be 276 miles apart at 4:00 p.m.

11.

Let x   =   Charles' travel time

x - 1   =   Bernadette's travel time

	Rate ·	Time =	Distance
Charles	32	x	32x
Bernadette	50	x - 1	50(x - 1)

Since they are traveling in opposite directions from the same starting point, each distance is added for the total miles apart.

$$32x + 50(x - 1) = 401$$
$$32x + 50x - 50 = 401$$
$$82x = 451$$
$$x = 5.5$$

Charles travels $5\frac{1}{2}$ hours. They will be 401 miles apart at 5:30 p.m.

13.

Let x = amount invested in the bond

x - 2,000 = amount invested in savings

	P ·	R ·	T =	I
Bond	x	.11	1	.11x
Savings	x - 2,000	.06	1	.06(x - 2,000)

The total interest was $730.

Thus,

Bond Interest	+	Savings Interest	=	Total Interest
.11x	+ .06(x - 2,000)		=	730
11x	+ 6(x - 2,000)		=	73,000
11x	+ 6x - 12,000		=	73,000
	17x		=	85,000
	x		=	5,000
	x - 2,000		=	3,000

#13 continued

The bond was $5,000 and the savings was $3,000.

15.
Let x = amount invested in retirement
4x = amount invested in a bond

	P ·	R .	T =	I
Retirement	x	.05	1	.05x
Bond	4x	.12	1	.12(4x)

The total interest was $1,060.
Thus,

Retirement Interest	+	Bond Interest	=	Total Interest
.05x	+	.12(4x)	=	1,060
5x	+	12(4x)	=	106,000
5x	+	48x	=	106,000
		53x	=	106,000
		x	=	2,000
		4x	=	8,000

The retirement investment was $2,000 and the bond was $8,000.

17.
Let x = amount invested in a CD
x - 1,000 = amount invested in a money market account.

	P ·	R · T		I
CD	x	.07	2	.14x
Money Market	x - 1,000	.04	2	.08(x - 1,000)

Note: x · (.07) · 2 = .14x
(x - 1,000)(.04) · 2 = .08(x - 1,000)

The total interest was $1,240.

#17 continued

Thus,

CD Interest	+	Money Market Interest	=	Total Interest
.14x	+	.08(x - 1,000)	=	1,240
14x	+	8(x - 1,000)	=	124,000
14x	+	8x - 8,000	=	124,000
		22x	=	132,000
		x	=	6,000
		x - 1,000	=	5,000

The CD was $6,000 and the money market was $5,000.

19.
Let x = amount invested in a checking account
7,000 - x = amount invested in a savings account.

	P ·	R · T		I
Checking	x	.05	1	.05x
Savings	7,000 - x	.08	1	.08(7,000- x)

The total interest was $500.

Thus,

Checking Interest	+	Savings Interest	=	Total Interest
.05x	+	.08(7,000 - x)	=	500
5x	+	8(7,000 - x)	=	50,000
5x	+	56,000 - 8x	=	50,000
		-3x	=	-6,000
		x	=	2,000
		7,000 - x	=	5,000

The checking account was $2,000 and the savings was $5,000.

21.

Let x = amount invested in retirement

$12,000 - x = amount invested in Lifeline

	P ·	R · T	=	I
Retire	x	.09	1	.09x
Life	12,000 - x	.05	1	.05(12,000 - x)

The total interest was $960.

Thus,

Retirement Interest	+	Lifeline Interest	=	Total Interest
.09x	+	.05(12,000 - x)	=	960
9x	+	5(12,000 - x)		96,000
9x	+	60,000 - 5x		96,000
4x	+	60,000		96,000
		4x	=	36,000
		x	=	9,000
		12,000 - x	=	3,000

The retirement was $9,000 and the Lifeline was $3,000.

23.

Let x = amount invested in a passbook account

8,000 - x = amount invested in a retirement account

	P ·	R · T	=	I
Passbook	x	.04	2	.08x
Retire	8,000 - x	.09	2	.18(8,000 - x)

Note: x(.04) · 2 = .08x

(8,000 - x)(.09) · 2 = .18(8,000 - x)

The total interest was $840.

#23 continued

Thus,

Passbook Interest	+	Retirement Interest	=	Total Interest
.08x	+	.18(8,000 - x)	=	840
8x	+	18(8,000 - x)	=	84,000
8x	+	144,000 - 18x	=	84,000
		-10x	=	-60,000
		x	=	6,000
		8000 - x	=	2,000

The passbook was $6,000 and the retirement was $2,000.

25.

Let x = amount of Big E

18 - x = amount of Big M

	Amount	Percent	Fluoride
Big E	x	.45	.45x
Big M	18 - x	.36	.36(18 - x)
New Product	18	.42	.42(18)

The equation is:

Big E Pure Fluoride	+	Big M Pure Fluoride	=	N/P Pure Flouride
.45x	+	.36(18 - x)	=	.42(18)
45x	+	36(18 - x)	=	42(18)
45x	+	648 - 36x	=	756
		9x	=	108
		x	=	12
		18 - x	=	6

Use 12oz of Big E and 6oz of Big M.

27.
Let x = amount of "Bald Away"
23 - x = amount of "Old Skin Head"

	Amount	Percent	Alcohol
Bald Away	x	.46	.46x
Old Skin Head	23 - x	.69	.69(23 - x)
New Product	23	.55	.55(23)

The equation is:

"B.A." Pure Alcohol		"O.S.H." Pure Alcohol		N/P Pure Alcohol
.46x	+	.69(23 - x)	=	.55(23)
46x	+	69(23 - x)	=	55(23)
46x	+	1587 - 69x	=	1,265
		-23x	=	-322
		x	=	14
		23 - x	=	9

Use 14oz of "Bald Away" and 9oz of "Old Skin Head".

29.
Let x = amount of "Sweety"
16 - x = amount of "Nectar"

	Amount	Percent	Dextrose
Sweety	x	.32	.32x
Nectar	16 - x	.48	.48(16 - x)
New Product	16	.42	.42(16)

#29 continued

The equation is:

"Sweety" Pure Dextrose		"Nectar" Pure Dextrose		N/P Pure Dextrose
.32x	+ .	.48(16 - x)	=	.42(16)
32x	+	48(16 - x)	=	42(16)
32x	+	768 - 48x	=	672
		-16x	=	-96
		x	=	6
		16 - x	=	10

Use 6oz of "Sweety" and 10oz of "This Might Be Nectar".

31.
Let x = amount of Bobo's
24-x = amount of Red's

	Amount	Cost/Pint	Cost
Bobo's	x	$1.20	$1.20x
Red's	24 - x	$1.80	$1.80(24 - x)
New Product	24	$1.55	$1.55(24)

The equation is:

Bobo Cost		Red Cost		N/P Cost
1.20x	+	1.80(24 - x)	=	1.55(24)
120x	+	180(24 - x)	=	155(24)
120x	+	4320 - 180x	=	3,720
		-60x	=	-600
		x	=	10
		24 - x	=	14

Use 10 pts of Bobo's and 14 pts of Red's.

33.
Let x = amount of Texas peanuts
14 - x = amount of Georgia peanuts

	Amount	Cost/Lb.	Cost
Texas	x	$2.80	$2.80x
Georgia	14 - x	$2.10	$2.10(14 - x)
New Product	14	$2.50	$2.50(14)

The equation is:

Texas Cost	+	Georgia Cost	=	N/P Cost
2.80x	+	2.10(14 - x)	=	2.50(14)
280x	+	210(14 - x	=	250(14)
280x	+	2,940 - 210x	=	3,500
		70x	=	560
		x	=	8
		14 - x	=	6

Use 8 lb. of Texas and 6 lb. of Georgia.

35.
Let x = amount of West Texas
18 - x = amount of Moss County

	Amount	Cost/gal.	Cost
West Texas	x	$0.72	$0.72x
Moss County	18 - x	$0.54	$0.54(18 - x)
New Product	18	$0.67	$0.67(18)

#35 continued

The equation is:

West Texas Cost	+	Moss County Cost	=	N/P Cost
0.72x	+	0.54(18 - x)	=	0.67(18)
72x	+	54(18 - x)	=	67(18)
72x	+	972 - 54x	=	1,206
		18x	=	234
		x	=	13
		18 - x	=	5

Use 13 gal. of West Texas and 5 gal. of Moss County.

Exercises 2.9

The graphs of the solution sets are found in the answer section of the main text book.

1.
$$x + 4 < 6$$
$$x + 4 - 4 < 6 - 4$$
$$x < 2$$

The solution set consists of all numbers x such that x < 2.

3.
$$x - 5 \geq 2$$
$$x - 5 + 5 \geq 2 + 5$$
$$x \geq 7$$

The solution set consists of all numbers x such that x ≥ 7.

5.
$$x + \frac{3}{5} \leq \frac{2}{3}$$
$$x + \frac{3}{5} - \frac{3}{5} \leq \frac{2}{3} - \frac{3}{5}$$
$$x \leq \frac{1}{15}$$

#5 continued

The solution set consists of all numbers x such that $x \leq \dfrac{1}{15}$.

7.
$$3 < x - 7$$
$$3 + 7 < x - 7 + 7$$
$$10 < x \text{ or } x > 10$$

The solution set consists of all numbers x such that x > 10.

9.
$$1 < x + 3 < 5$$
$$1 - 3 < x + 3 - 3 < 5 - 3$$
$$-2 < x < 2$$

The solution set consists of all numbers x such that $-2 < x < 2$.

11.
$$-4 \leq x - 8 \leq 2$$
$$-4 + 8 \leq x - 8 + 8 \leq 2 + 8$$
$$4 \leq x \leq 10$$

The solution set consists of all numbers x such that $4 \leq x \leq 10$.

13.
$$-2 \leq x + \frac{3}{8} \leq -\frac{3}{2}$$
$$-2 - \frac{3}{8} \leq x + \frac{3}{8} - \frac{3}{8} \leq -\frac{3}{2} - \frac{3}{8}$$
$$-\frac{19}{8} \leq x \leq -\frac{15}{8}$$

The solution set consists of all numbers x such that $-\dfrac{19}{8} \leq x \leq -\dfrac{15}{8}$.

15.
$$2x > 6$$
$$\frac{2x}{2} > \frac{6}{2}$$
$$x > 3$$

The solution set consists of all numbers x such that x > 3.

17.
$$-6x \geq 4$$
$$\frac{-6x}{-6} \leq \frac{4}{-6} \quad \text{Reverse inequality}$$
$$x \leq -\frac{2}{3}$$

The solution set consists of all numbers x such that $x \leq -\dfrac{2}{3}$.

19.
$$\frac{2}{3}x > \frac{4}{15}$$
$$\frac{3}{2}\left(\frac{2}{3}x\right) > \frac{3}{2}\left(\frac{4}{15}\right)$$
$$x > \frac{2}{5}$$

The solution set consists of all numbers x such that $x > \dfrac{2}{5}$.

21.
$$3 \leq 5x$$
$$\frac{3}{5} \leq \frac{5x}{5}$$
$$\frac{3}{5} \leq x \text{ or } x \geq \frac{3}{5}$$

The solution set consists of all numbers x such that $x \geq \dfrac{3}{5}$.

The graphs of the solution sets are found in the answer section of the main text book.

The graphs of the solution sets are found in the answer section of the main text book.

23.

$$-4 < 3x < 6$$

$$-\frac{4}{3} < \frac{3x}{3} < \frac{6}{3}$$

$$-\frac{4}{3} < x < 2$$

The solution set consists of all numbers x such that $-\frac{4}{3} < x < 2$.

25.

$$-2 \leq 4x \leq 5$$

$$-\frac{2}{4} \leq \frac{4x}{4} \leq \frac{5}{4}$$

$$-\frac{1}{2} \leq x \leq \frac{5}{4}$$

The solution set consists of all numbers x such that $-\frac{1}{2} \leq x \leq \frac{5}{4}$.

27.

$$-6 < \frac{3}{4}x < -3$$

$$\frac{4}{3}(-6) < \frac{4}{3}\left(\frac{3}{4}x\right) < \frac{4}{3}(-3)$$

$$-8 < x < -4$$

The solution set consists of all numbers x such that $-8 < x < -4$.

29.

$$2x + 5 \geq 11$$

$$2x + 5 - 5 \geq 11 - 5$$

$$2x \geq 6$$

$$\frac{2x}{2} \geq \frac{6}{2}$$

$$x \geq 3$$

The solution set consists of all numbers x such that $x \geq 3$.

31.

$$3x - 7 < 5$$

$$3x - 7 + 7 < 5 + 7$$

$$3x < 12$$

$$\frac{3x}{3} < \frac{12}{3}$$

$$x < 4$$

The solution set consists of all numbers x such that $x < 4$.

33.

$$1 - 6x \leq 3$$

$$-1 + 1 - 6x \leq -1 + 3$$

$$-6x \leq 2$$

$$\frac{-6x}{-6} \geq \frac{2}{-6} \qquad \text{Reverse inequality}$$

$$x \geq -\frac{1}{3}$$

The solution set consists of all numbers x such that $x \geq -\frac{1}{3}$.

35.

$$\frac{4}{5}x + 2 > \frac{2}{5}$$

$$\frac{4}{5}x + 2 - 2 > \frac{2}{5} - 2$$

$$\frac{4}{5}x > -\frac{8}{5}$$

$$\frac{5}{4}\left(\frac{4}{5}x\right) > \frac{5}{4}\left(-\frac{8}{5}\right)$$

$$x > -2$$

The solution set consists of all numbers x such that $x > -2$.

37.

$$10 < 2x + 5$$
$$10 - 5 < 2x + 5 - 5$$
$$5 < 2x$$
$$\frac{5}{2} < \frac{2x}{2}$$
$$\frac{5}{2} < x \text{ or } x > \frac{5}{2}$$

The solution set consists of all numbers x such that $x > \frac{5}{2}$.

39.

$$1 \geq 8x - 3$$
$$1 + 3 \geq 8x - 3 + 3$$
$$4 \geq 8x$$
$$\frac{4}{8} \geq \frac{8x}{8}$$
$$\frac{1}{2} \geq x \text{ or } x \leq \frac{1}{2}$$

The solution set consists of all numbers x such that $x \leq \frac{1}{2}$.

41.

$$6 \geq 1 - 3x$$
$$-1 + 6 \geq -1 + 1 - 3x$$
$$5 \geq -3x$$
$$\frac{5}{-3} \leq \frac{-3x}{-3} \quad \text{Reverse inequality}$$
$$-\frac{5}{3} \leq x \text{ or } x \geq -\frac{5}{3}$$

The solution set consists of all numbers x such that $x \geq -\frac{5}{3}$.

43.

$$1.8x - 3.14 < 4.06$$
$$1.8x - 3.14 + 3.14 < 4.06 + 3.14$$
$$1.8x < 7.20$$
$$\frac{1.8x}{1.8} < \frac{7.2}{1.8}$$
$$x < 4$$

The solution set consists of all numbers x such that $x < 4$.

45.

$$4x - 3 \leq x + 7$$
$$4x - x - 3 \leq x - x + 7$$
$$3x - 3 \leq 7$$
$$3x - 3 + 3 \leq 7 + 3$$
$$3x \leq 10$$
$$\frac{3x}{3} \leq \frac{10}{3}$$
$$x \leq \frac{10}{3}$$

The solution set consists of all numbers x such that $x \leq \frac{10}{3}$.

47.

$$x + 9 > 5x - 7$$
$$x - x + 9 > 5x - x - 7$$
$$9 > 4x - 7$$
$$9 + 7 > 4x - 7 + 7$$
$$16 > 4x$$
$$\frac{16}{4} > \frac{4x}{4}$$
$$4 > x \text{ or } x < 4$$

The solution set consists of all numbers x such that $x < 4$.

| The graphs of the solution sets are found in the answer section of the main text book. | The graphs of the solution sets are found in the answer section of the main text book. |

49.

$$6 - x \geq 7x - 12$$
$$6 - x + x \geq 7x + x - 12$$
$$6 \geq 8x - 12$$
$$6 + 12 \geq 8x - 12 + 12$$
$$18 \geq 8x$$
$$\frac{18}{8} \geq \frac{8x}{8}$$
$$\frac{9}{4} \geq x \text{ or } x \leq \frac{9}{4}$$

The solution set consists of all numbers x such that $x \leq \frac{9}{4}$.

51.

$$4x + 11 + 6x < 6 + 3x - 2$$
$$10x + 11 < 4 + 3x$$
$$10x - 3x + 11 < 4 + 3x - 3x$$
$$7x + 11 < 4$$
$$7x + 11 - 11 < 4 - 11$$
$$7x < -7$$
$$\frac{7x}{7} < \frac{-7}{7}$$
$$x < -1$$

The solution set consists of all numbers x such that $x < -1$.

53.

$$x - 8 + 7x > 5 + 3x - 13$$
$$8x - 8 > 3x - 8$$
$$8x - 8 + 8 > 3x - 8 + 8$$
$$8x > 3x$$
$$8x - 3x > 3x - 3x$$
$$5x > 0$$
$$\frac{5x}{5} > \frac{0}{5}$$
$$x > 0$$

The solution set consists of all numbers x such that $x > 0$.

55.

$$\frac{1}{2}x - 2 + \frac{2}{3}x \geq \frac{5}{4}x - \frac{5}{2}$$
$$\frac{7}{6}x - 2 \geq \frac{5}{4}x - \frac{5}{2}$$
$$\frac{7}{6}x - \frac{7}{6}x - 2 \geq \frac{5}{4}x - \frac{7}{6}x - \frac{5}{2}$$
$$-2 \geq \frac{1}{12}x - \frac{5}{2}$$
$$-2 + \frac{5}{2} \geq \frac{1}{12}x - \frac{5}{2} + \frac{5}{2}$$
$$\frac{1}{2} \geq \frac{1}{12}x$$
$$12\left(\frac{1}{2}\right) \geq 12\left(\frac{1}{12}x\right)$$
$$6 \geq x \text{ or } x \leq 6$$

The solution set consists of all numbers x such that $x \leq 6$.

57.

$$3(4x-1) \leq 7x + 12$$
$$12x - 3 \leq 7x + 12$$
$$12x - 7x - 3 \leq 7x - 7x + 12$$
$$5x - 3 \leq 12$$
$$5x - 3 + 3 \leq 12 + 3$$
$$5x \leq 15$$
$$\frac{5x}{5} \leq \frac{15}{5}$$
$$x \leq 3$$

The solution set consists of all numbers x such that $x \leq 3$.

59.

$$7(x + 3) - 8 > 3x + 11$$
$$7x + 21 - 8 > 3x + 11$$
$$7x + 13 > 3x + 11$$
$$7x - 3x + 13 > 3x - 3x + 11$$
$$4x + 13 > 11$$
$$4x + 13 - 13 > 11 - 13$$
$$4x > -2$$

#59 continued

$$\frac{4x}{4} > \frac{-2}{4}$$

$$x > -\frac{1}{2}$$

The solution set consists of all numbers x such that $x > -\frac{1}{2}$.

61.

$5(4x + 3) - 2x < 6x - 9$

$20x + 15 - 2x < 6x - 9$

$18x + 15 < 6x - 9$

$18x - 6x + 15 < 6x - 6x - 9$

$12x + 15 < -9$

$12x + 15 - 15 < -9 - 15$

$12x < -24$

$$\frac{12x}{12} < \frac{-12}{12}$$

$x < -2$

The solution set consists of all numbers x such that $x < -2$.

63.

$2(1 - 3x) \leq 5(x - 4)$

$2 - 6x \leq 5x - 20$

$2 - 6x + 6x \leq 5x + 6x - 20$

$2 \leq 11x - 20$

$2 + 20 \leq 11x - 20 + 20$

$22 \leq 11x$

$$\frac{22}{11} \leq \frac{11x}{11}$$

$2 \leq x$ or $x \geq 2$

The solution set consists of all numbers x such that $x \geq 2$.

65.

$6 - 3(5x - 4) \geq 9 - 3x$

$6 - 15x + 12 \geq 9 - 3x$

$-15x + 18 \geq 9 - 3x$

#65 continued

$-15x + 15x + 18 \geq 9 - 3x + 15x$

$18 \geq 9 + 12x$

$18 - 9 \geq 9 - 9 + 12x$

$9 \geq 12x$

$$\frac{9}{12} \geq \frac{12x}{12}$$

$$\frac{3}{4} \geq x \text{ or } x \leq \frac{3}{4}$$

The solution set consists of all numbers x such that $x \leq \frac{3}{4}$.

67.

$$\frac{1}{3}(2x - 5) + 2 > \frac{3}{2}(x + 3)$$

$$\frac{2}{3}x - \frac{5}{3} + 2 > \frac{3}{2}x + \frac{9}{2}$$

$$\frac{2}{3}x + \frac{1}{3} > \frac{3}{2}x + \frac{9}{2}$$

$$\frac{2}{3}x - \frac{2}{3}x + \frac{1}{3} > \frac{3}{2}x - \frac{2}{3}x + \frac{9}{2}$$

$$\frac{1}{3} > \frac{5}{6}x + \frac{9}{2}$$

$$\frac{1}{3} - \frac{9}{2} > \frac{5}{6}x + \frac{9}{2} - \frac{9}{2}$$

$$-\frac{25}{6} > \frac{5}{6}x$$

$$\frac{6}{5}\left(\frac{-25}{6}\right) > \frac{6}{5}\left(\frac{5}{6}x\right)$$

$-5 > x$ or $x < -5$

The solution set consists of all numbers x such that $x < -5$.

69.

$$2\left(\frac{3}{4}x + \frac{3}{5}\right) - \frac{1}{2}x < \frac{5}{4}x + 1$$

$$\frac{3}{2}x + \frac{6}{5} - \frac{1}{2}x < \frac{5}{4}x + 1$$

75

The graphs of the solution sets are found in the answer section of the main text book.

#69 continued

$$20\left(\frac{3}{2}x + \frac{6}{5} - \frac{1}{2}x\right) < 20\left(\frac{5}{4}x + 1\right) \quad LCD = 20$$

$$30x + 24 - 10x < 25x + 20$$
$$20x + 24 < 25x + 20$$
$$20x - 20x + 24 < 25x - 20x + 20$$
$$24 < 5x + 20$$
$$24 - 20 < 5x + 20 - 20$$
$$4 < 5x$$
$$\frac{4}{5} < \frac{5x}{5}$$
$$\frac{4}{5} < x \text{ or } x > \frac{4}{5}$$

The solution set consists of all numbers x such that $x > \frac{4}{5}$.

71.
$$-1 < 3x - 7 < 2$$
$$-1 + 7 < 3x - 7 + 7 < 2 + 7$$
$$6 < 3x < 9$$
$$\frac{6}{3} < \frac{3x}{3} < \frac{9}{3}$$
$$2 < x < 3$$

The solution set consists of all numbers x such that $2 < x < 3$.

73.
$$6 \le 5x + 1 \le 13$$
$$6 - 1 \le 5x + 1 - 1 \le 13 - 1$$
$$5 \le 5x \le 12$$
$$\frac{5}{5} \le \frac{5x}{5} \le \frac{12}{5}$$
$$1 \le x \le \frac{12}{5}$$

The solution set consists of all numbers x such that $1 \le x \le \frac{12}{5}$.

The graphs of the solution sets are found in the answer section of the main text book.

75.
$$-7 \le 5 - 2x \le 3$$
$$-7 - 5 \le 5 - 5 - 2x \le 3 - 5$$
$$-12 \le -2x \le -2$$
$$\frac{-12}{-2} \ge \frac{-2x}{-2} \ge \frac{-2}{-2} \quad \text{Reverse inequality}$$
$$6 \ge x \ge 1 \text{ or } 1 \le x \le 6$$

The solution set consists of all numbers x such that $1 \le x \le 6$.

77.
$$4 < 4 - \frac{1}{3}x < 5$$
$$4 - 4 < 4 - 4 - \frac{1}{3}x < 5 - 4$$
$$0 < -\frac{1}{3}x < 1$$
$$-3 \cdot 0 > -3\left(-\frac{1}{3}x\right) > -3(1) \quad \text{Reverse inequality}$$
$$0 > x > -3 \text{ or } -3 < x < 0$$

The solution set consists of all numbers x such that $-3 < x < 0$.

79.
$$-4 < \frac{5}{2}x - \frac{1}{4} < 1$$
$$-4 + \frac{1}{4} < \frac{5}{2}x - \frac{1}{4} + \frac{1}{4} < 1 + \frac{1}{4}$$
$$-\frac{15}{4} < \frac{5}{2}x < \frac{5}{4}$$
$$\frac{2}{5}\left(-\frac{15}{4}\right) < \frac{2}{5}\left(\frac{5}{2}x\right) < \frac{2}{5}\left(\frac{5}{4}\right)$$
$$-\frac{3}{2} < x < \frac{1}{2}$$

The solution set consists of all numbers x such that $-\frac{3}{2} < x < \frac{1}{2}$.

Review Exercises

1.
$3 - 5x$; $x = -4$
$3 - 5(-4) = 3 + 20 = 23$

3.
$(3x+1)(x-9)$; $x = \dfrac{3}{2}$

$$\left(3 \cdot \frac{3}{2} + 1\right)\left(\frac{3}{2} - 9\right) = \left(\frac{9}{2} + 1\right)\left(\frac{3}{2} - 9\right)$$
$$= \frac{11}{2}\left(-\frac{15}{2}\right)$$
$$= -\frac{165}{4}$$

5.
$3(x - 5) - (2x + 4)$; $x = -\dfrac{1}{3}$

$$3\left(-\frac{1}{3} - 5\right) - \left[2\left(-\frac{1}{3}\right) + 4\right]$$
$$= 3\left(-\frac{16}{3}\right) - \left(-\frac{2}{3} + 4\right)$$
$$= 3\left(-\frac{16}{3}\right) - \left(\frac{10}{3}\right)$$
$$= -16 - \frac{10}{3}$$
$$= -\frac{48}{3} - \frac{10}{3}$$
$$= -\frac{58}{3}$$

7.
$$\frac{2x}{x - 4} + \frac{9x}{x + 2}; \quad x = 3$$
$$\frac{2 \cdot 3}{3 - 4} + \frac{9 \cdot 3}{3 + 2}$$
$$= \frac{6}{-1} + \frac{27}{5}$$
$$= -6 + \frac{27}{5}$$
$$= -\frac{30}{5} + \frac{27}{5}$$
$$= -\frac{3}{5}$$

9.
$$\frac{3x + 7y}{\sqrt{8z}}; \quad x = 1, y = -5, z = 2$$
$$\frac{3 \cdot 1 + 7(-5)}{\sqrt{8 \cdot 2}} = \frac{3 - 35}{\sqrt{16}}$$
$$= \frac{-32}{4}$$
$$= -8$$

11.
$5x - 7 + 3x - 4$
$= 5x + 3x - 7 - 4$
$= (5 + 3)x - 11$
$= 8x - 11$

13.
$$\frac{1}{2}x^2 + \frac{2}{3}y^2 - \frac{3}{8}x^2 - \frac{8}{9}y^2$$
$$= \frac{1}{2}x^2 - \frac{3}{8}x^2 + \frac{2}{3}y^2 - \frac{8}{9}y^2$$
$$= \left(\frac{1}{2} - \frac{3}{8}\right)x^2 + \left(\frac{2}{3} - \frac{8}{9}\right)y^2$$
$$= \left(\frac{4}{8} - \frac{3}{8}\right)x^2 + \left(\frac{6}{9} - \frac{8}{9}\right)y^2$$
$$= \frac{1}{8}x^2 - \frac{2}{9}y^2$$

15.
$5(x + 3) + 2(x - 4)$
$= 5x + 15 + 2x - 8$
$= 5x + 2x + 15 - 8$
$= (5+2)x + 7$
$= 7x + 7$

17.
$3(2x + 9y) + 9(3x - 4y)$
$= 6x + 27y + 27x - 36y$
$= 6x + 27x + 27y - 36y$
$= (6 + 27)x + (27 - 36)y$
$= 33x - 9y$

19.
$$\frac{2}{3}(x + 4) - \frac{1}{4}(x + 5)$$
$$= \frac{2}{3}x + \frac{8}{3} - \frac{1}{4}x - \frac{5}{4}$$

#19 continued

$$= \frac{2}{3}x - \frac{1}{4}x + \frac{8}{3} - \frac{5}{4}$$

$$= \left(\frac{2}{3} - \frac{1}{4}\right)x + \frac{8}{3} - \frac{5}{4}$$

$$= \left(\frac{8}{12} - \frac{3}{12}\right)x + \frac{32}{12} - \frac{15}{12}$$

$$= \frac{5}{12}x + \frac{17}{12}$$

21.

$$x + 7 = 2$$
$$x + 7 - 7 = 2 - 7$$
$$x = -5$$

Solution set: {-5}

23.

$$8x + 4 - 7x + 2 = 5 - 17$$
$$x + 6 = -12$$
$$x + 6 - 6 = -12 - 6$$
$$x = -18$$

Solution set: {-18}

25.

$$4x + 9 = 3x + 11$$
$$4x - 3x + 9 = 3x - 3x + 11$$
$$x + 9 = 11$$
$$x + 9 - 9 = 11 - 9$$
$$x = 2$$

Solution set: {2}

27.

$$5x + 2x + 4 = 6x + 4$$
$$7x + 4 = 6x + 4$$
$$7x - 6x + 4 = 6x - 6x + 4$$
$$x + 4 = 4$$
$$x + 4 - 4 = 4 - 4$$
$$x = 0$$

Solution set: {0}

29.

$$3x - \frac{1}{2} - x = x + \frac{3}{4}$$

#29 continued

$$4\left(3x - \frac{1}{2} - x\right) = 4\left(x + \frac{3}{4}\right) \qquad \text{LCD} = 4$$
$$12x - 2 - 4x = 4x + 3$$
$$8x - 2 = 4x + 3$$
$$8x - 4x - 2 = 4x - 4x + 3$$
$$4x - 2 = 3$$
$$4x - 2 + 2 = 3 + 2$$
$$4x = 5$$
$$\frac{4x}{4} = \frac{5}{4}$$
$$x = \frac{5}{4}$$

Solution set: $\left\{\frac{5}{4}\right\}$

31.

$$5x - 3 + 4x + 17 = 2x + 5 + 8x + 1$$
$$9x + 14 = 10x + 6$$
$$9x - 9x + 14 = 10x - 9x + 6$$
$$14 = x + 6$$
$$8 = x$$

Solution set: {8}

33.

$$4(5x - 1) = 21x - 4$$
$$20x - 4 = 21x - 4$$
$$20x - 20x - 4 = 21x - 20x - 4$$
$$-4 = x - 4$$
$$-4 + 4 = x - 4 + 4$$
$$0 = x$$

Solution set: {0}

35.

$$\frac{1}{3}(12x + 9) - 4 = \frac{1}{4}(12x - 8) + 5$$
$$4x + 3 - 4 = 3x - 2 + 5$$
$$4x - 1 = 3x + 3$$
$$4x - 3x - 1 = 3x - 3x + 3$$
$$x - 1 = 3$$
$$x - 1 + 1 = 3 + 1$$
$$x = 4$$

Solution set: {4}

37.

$3(2x + 5) + 2(x - 8) = 7(x - 1)$

$6x + 15 + 2x - 16 = 7x - 7$

$8x - 1 = 7x - 7$

$8x - 7x - 1 = 7x - 7x - 7$

$x - 1 = -7$

$x - 1 + 1 = -7 + 1$

$x = -6$

Solution set: {-6}

39.

$2[4(x + 4) - 2(x + 3)] = 3(x + 5)$

$2(4x + 16 - 2x - 6) = 3x + 15$

$2(2x + 10) = 3x + 15$

$4x + 20 = 3x + 15$

$4x - 3x + 20 = 3x - 3x + 15$

$x + 20 = 15$

$x + 20 - 20 = 15 - 20$

$x = -5$

Solution set: {-5}

41.

$4x = -28$

$\dfrac{4x}{4} = -\dfrac{28}{4}$

$x = -7$

Solution set: {-7}

43.

$\dfrac{2}{3}x = 4$

$\dfrac{3}{2}\left(\dfrac{2}{3}x\right) = \dfrac{3}{2}(4)$

$x = 6$

Solution set: {6}

45.

$\dfrac{5}{4} = \dfrac{1}{10}x$

$10\left(\dfrac{5}{4}\right) = 10\left(\dfrac{1}{10}\right)x$

$\dfrac{25}{2} = x$

#45 continued

Solution set: $\left\{\dfrac{25}{2}\right\}$

47.

$-x = -12$

$-1(-x) = -1(-12)$

$x = 12$

Solution set: {12}

49.

$4x + 1 = 7$

$4x + 1 - 1 = 7 - 1$

$4x = 6$

$\dfrac{4x}{4} = \dfrac{6}{4}$

$x = \dfrac{3}{2}$

Solution set: $\left\{\dfrac{3}{2}\right\}$

51.

$\dfrac{5}{6}x + 2 = 4$

$\dfrac{5}{6}x + 2 - 2 = 4 - 2$

$\dfrac{5}{6}x = 2$

$\dfrac{6}{5}\left(\dfrac{5}{6}x\right) = \dfrac{6}{5} \cdot 2$

$x = \dfrac{12}{5}$

Solution set: $\left\{\dfrac{12}{5}\right\}$

53.

$5x + 4 = 3x$

$5x - 3x + 4 = 3x - 3x$

$2x + 4 = 0$

$2x + 4 - 4 = 0 - 4$

$2x = -4$

#53 continued

$$\frac{2x}{2} = \frac{-4}{2}$$
$$x = -2$$

Solution set: {-2}

55.
$$3x + 14 = 8x - 1$$
$$3x - 3x + 14 = 8x - 3x - 1$$
$$14 = 5x - 1$$
$$14 + 1 = 5x - 1 + 1$$
$$15 = 5x$$
$$\frac{15}{5} = \frac{5x}{5}$$
$$3 = x$$

Solution set: {3}

57.
$$11 + 9x - 3 = 8x - 10 - 2x$$
$$9x + 8 = 6x - 10$$
$$9x - 6x + 8 = 6x - 6x - 10$$
$$3x + 8 = -10$$
$$3x + 8 - 8 = -10 - 8$$
$$3x = -18$$
$$\frac{3x}{3} = \frac{-18}{3}$$
$$x = -6$$

Solution set: {-6}

59.
$$\frac{4}{3}x + \frac{3}{8} - 2x = \frac{1}{9}x + 3$$
$$72\left(\frac{4}{3}x + \frac{3}{8} - 2x\right) = 72\left(\frac{1}{9}x + 3\right) \quad LCD = 72$$
$$96x + 27 - 144x = 8x + 216$$
$$-48x + 27 = 8x + 216$$
$$-48x - 8x + 27 = 8x - 8x + 216$$
$$-56x + 27 = 216$$
$$-56x + 27 - 27 = 216 - 27$$
$$-56x = 189$$
$$\frac{-56x}{-56} = \frac{189}{-56}$$

#59 continued

$$x = -\frac{27}{8} \quad Reduce$$
Solution set: $\left\{-\frac{27}{8}\right\}$

61.
$$3(4x - 1) = 5x + 6$$
$$12x - 3 = 5x + 6$$
$$12x - 5x - 3 = 5x - 5x + 6$$
$$7x - 3 = 6$$
$$7x - 3 + 3 = 6 + 3$$
$$7x = 9$$
$$\frac{7x}{7} = \frac{9}{7}$$
$$x = \frac{9}{7}$$

Solution set: $\left\{\frac{9}{7}\right\}$

63.
$$8 - (3x + 4) = 6x$$
$$8 - 3x - 4 = 6x$$
$$-3x + 4 = 6x$$
$$-3x + 3x + 4 = 6x + 3x$$
$$4 = 9x$$
$$\frac{4}{9} = \frac{9x}{9}$$
$$\frac{4}{9} = x$$

Solution set: $\left\{\frac{4}{9}\right\}$

65.
$$\frac{1}{3} + \frac{3}{4}(5x + 2) = \frac{28}{3}$$
$$12\left[\frac{1}{3} + \frac{3}{4}(5x + 2)\right] = 12 \cdot \frac{28}{3}$$
$$4 + 9(5x + 2) = 112$$
$$4 + 45x + 18 = 112$$
$$45x + 22 = 112$$
$$45x + 22 - 22 = 112 - 22$$
$$45x = 90$$
$$x = 2$$

Solution set: {2}

67.

$$\frac{2}{3}(6x - 4) = \frac{1}{2}(7x - 5)$$

$$6\left[\frac{2}{3}(6x-4)\right] = 6\left[\frac{1}{2}(7x - 5)\right]$$

$$4(6x - 4) = 3(7x - 5)$$
$$24x - 16 = 21x - 15$$
$$24x - 21x - 16 = 21x - 21x - 15$$
$$3x - 16 = -15$$
$$3x - 16 + 16 = -15 + 16$$
$$3x = 1$$
$$x = \frac{1}{3}$$

Solution set: $\left\{\frac{1}{3}\right\}$

69.

$$5(4 - 7x) = 2(3x - 2) - x$$
$$20 - 35x = 6x - 4 - x$$
$$20 - 35x = 5x - 4$$
$$20 - 35x - 5x = 5x - 5x - 4$$
$$20 - 40x = -4$$
$$-20 + 20 - 40x = -4 - 20$$
$$-40x = -24$$
$$\frac{-40x}{-40} = \frac{-24}{-40}$$
$$x = \frac{4}{5}$$

Solution set: $\left\{\frac{4}{5}\right\}$

71.

$$2.6(x + 5) + 4.1(2x - 4) = 3.6x + 0.2$$
$$2.6x + 13 + 8.2x - 16.4 = 3.6x + 0.2$$
$$10.8x - 3.4 = 3.6x + 0.2$$
$$10.8x - 3.6x - 3.4 = 3.6x - 3.6x + 0.2$$
$$7.2x - 3.4 = 0.2$$
$$7.2x - 3.4 + 3.4 = 0.2 + 3.4$$
$$7.2x = 3.6$$
$$x = 0.5$$

Solution set: $\{0.5\}$

73.

$$5(x - 1) - 2(3x + 1) = 4(2x + 1) - 11$$
$$5x - 5 - 6x - 2 = 8x + 4 - 11$$
$$-x - 7 = 8x - 7$$

#73 continued

$$-x + x - 7 = 8x + x - 7$$
$$-7 = 9x - 7$$
$$-7 + 7 = 9x - 7 + 7$$
$$0 = 9x$$
$$0 = x \quad \text{or} \quad x = 0$$

Solution set: $\{0\}$

75.

$$2[5x - (4x - 7)] = 3(5 - x) + 6$$
$$2(5x - 4x + 7) = 15 - 3x + 6$$
$$2(x + 7) = -3x + 21$$
$$2x + 14 = -3x + 21$$
$$2x + 3x + 14 = -3x + 3x + 21$$
$$5x + 14 = 21$$
$$5x + 14 - 14 = 21 - 14$$
$$5x = 7$$
$$x = \frac{7}{5}$$

Solution set: $\left\{\frac{7}{5}\right\}$

77.

$$|x| = 7$$

$$\{-7, 7\}$$

79.

$$\left|\frac{3}{4}x\right| = 9$$

Case I   Case II

$$\frac{3}{4}x = 9 \qquad \text{or} \qquad \frac{3}{4}x = -9$$

$$\frac{4}{3}\left(\frac{3}{4}x\right) = \frac{4}{3}(9) \quad \text{or} \quad \frac{4}{3}\left(\frac{3}{4}x\right) = \frac{4}{3}(-9)$$

$$x = 12 \qquad \text{or} \qquad x = -12$$

Solution set: $\{-12, 12\}$

81.

$$\left|2 - x\right| = \frac{3}{4}$$

#81 continued

Case I   Case II

$2 - x = \dfrac{3}{4}$ or $2 - x = -\dfrac{3}{4}$

$-x = \dfrac{3}{4} - 2$ or $-x = -\dfrac{3}{4} - 2$

$-x = -\dfrac{5}{4}$ or $-x = -\dfrac{11}{4}$

$x = \dfrac{5}{4}$ or $x = \dfrac{11}{4}$

Solution set: $\left\{ \dfrac{5}{4}, \dfrac{11}{4} \right\}$

83.

$|3x + 6| = 5$

Case I   Case II

$3x + 6 = 5$ or $3x + 6 = -5$

$3x + 6 - 6 = 5 - 6$ or $3x + 6 - 6 = -5 - 6$

$3x = -1$ or $3x = -11$

$\dfrac{3x}{3} = \dfrac{-1}{3}$ or $\dfrac{3x}{3} = \dfrac{-11}{3}$

$x = -\dfrac{1}{3}$ or $x = -\dfrac{11}{3}$

Solution set: $\left\{ -\dfrac{11}{3}, -\dfrac{1}{3} \right\}$

85.

$|3x + 4| = -8$

Does not have a solution since
$|3x + 4| > 0$.

87.

$|x - 5| + 2 = 13$

$|x - 5| = 11$ Isolate absolute value

Case I   Case II

$x - 5 = 11$ or $x - 5 = -11$

$x = 11 + 5$ or $x = -11 + 5$

$x = 16$ or $x = -6$

Solution set: $\{-6, 16\}$

89.

$|7 - x| + 3 = 10$

$|7 - x| = 7$ Isolate absolute value

Case I   Case II

$7 - x = 7$ or $7 - x = -7$

$-x = 7 - 7$ or $-x = -7 - 7$

$-x = 0$ or $-x = -14$

$x = 0$ or $x = 14$

Solution set: $\{0, 14\}$

91.

$|4x + 2| - 1 = 4$

$|4x + 2| = 5$ Isolate absolute value

Case I   Case II

$4x + 2 = 5$ or $4x + 2 = -5$

$4x = 5 - 2$ or $4x = -5 - 2$

$4x = 3$ or $4x = -7$

$x = \dfrac{3}{4}$ or $x = -\dfrac{7}{4}$

Solution set: $\left\{ -\dfrac{7}{4}, \dfrac{3}{4} \right\}$

93.

$|2x - 4| - 9 = 0$

$|2x - 4| = 9$ Isolate absolute value

Case I   Case II

$2x - 4 = 9$ or $2x - 4 = -9$

$2x = 9 + 4$ or $2x = -9 + 4$

$2x = 13$ or $2x = -5$

$x = \dfrac{13}{2}$ or $x = -\dfrac{5}{2}$

Solution set: $\left\{ -\dfrac{5}{2}, \dfrac{13}{2} \right\}$

95.

$\left| \dfrac{3}{2}x - \dfrac{1}{8} \right| + \dfrac{1}{4} = \dfrac{3}{4}$

$\left| \dfrac{3}{2}x - \dfrac{1}{8} \right| = \dfrac{1}{2}$

#95 continued

Case I	Case II

$$\frac{3}{2}x - \frac{1}{8} = \frac{1}{2} \quad \text{or} \quad \frac{3}{2}x - \frac{1}{8} = -\frac{1}{2}$$

$$\frac{3}{2}x = \frac{1}{2} + \frac{1}{8} \quad \text{or} \quad \frac{3}{2}x = -\frac{1}{2} + \frac{1}{8}$$

$$\frac{3}{2}x = \frac{5}{8} \quad \text{or} \quad \frac{3}{2}x = -\frac{3}{8}$$

$$\frac{2}{3}\left(\frac{3}{2}x\right) = \frac{2}{3}\left(\frac{5}{8}\right) \quad \text{or} \quad \frac{2}{3}\left(\frac{3}{2}x\right) = \frac{2}{3}\left(-\frac{3}{8}\right)$$

$$x = \frac{5}{12} \quad \text{or} \quad x = -\frac{1}{4}$$

Solution set: $\left\{-\frac{1}{4}, \frac{5}{12}\right\}$

97.

I = PRT, for R

$$\frac{I}{PT} = \frac{PRT}{PT}$$

$$\frac{I}{PT} = R \quad \text{or} \quad R = \frac{I}{PT}$$

99.

$$m = \frac{3n - 4p}{5}, \text{ for } p$$

$$5m = 3n - 4p$$

$$5m - 3n = -4p$$

$$\frac{5m - 3n}{-4} = \frac{-4p}{-4}$$

$$\frac{5m - 3n}{-4} = p$$

$$\frac{3n - 5m}{4} = p$$

Note: $\frac{-1}{-1} \cdot \frac{5m - 3n}{-4} = \frac{-5m + 3n}{4}$

$$= \frac{3n - 5m}{4}$$

101.

A - 2x^2 + 4xy, for y

$$A - 2x^2 = 4xy$$

#101 continued

$$\frac{A - 2x^2}{4x} = \frac{4xy}{4x}$$

$$\frac{A - 2x^2}{4x} = y$$

103.

D = RT, for R

$$\frac{D}{T} = R ; \quad D = 252, \ T = 6$$

$$\frac{252}{6} = R$$

$$42 = R$$

105.

P = 2L + 2W, for L

$$P - 2W = 2L$$

$$\frac{P - 2W}{2} = L; \quad P = 32, \ W = 4$$

$$\frac{32 - 2(4)}{2} = L$$

$$\frac{32 - 8}{2} = L$$

$$\frac{24}{2} = L$$

$$12 = L$$

107.

F = kx, for k

$$\frac{F}{x} = k; \quad F = 120, \ x = 192$$

$$\frac{120}{192} = k$$

$$\frac{5}{8} = k$$

109.

Let x be the number.
Thus,

$$3x + 14 = 8$$

$$3x + 14 - 14 = 8 - 14$$

$$3x = -6$$

83

#109 continued

$$x = -2$$

The number is -2

111.
Three consecutive integers are x, x + 1, and x + 2.
Thus,
$$x + (x + 1) + (x + 2) = 66$$
$$3x + 3 = 66$$
$$3x = 63$$
$$x = 21$$
$$x + 1 = 22$$
$$x + 2 = 23$$

The consecutive integers are 21, 22, and 23.

113.
The length of the second and third pieces are in terms of the first piece.
So,
$$x = \text{length of first piece}$$
$$2x + 1 = \text{length of second piece}$$
$$x - 3 = \text{length of third piece}$$

The total length is 34 in.
The equation is:
$$x + (2x + 1) + (x - 3) = 34$$
$$4x - 2 = 34$$
$$4x = 36$$
$$x = 9$$
$$2x + 1 = 19$$
$$x - 3 = 6$$

The three pieces have lengths 9", 19" and 6".

115.
The perimeter formula is P = 2W + 2L
The length is in terms of the width.
So,
$$x = \text{measurement of the width}$$
$$x + 8 = \text{measurement of the length}$$
$$P = 44m$$

#115 continued

Thus,
$$44 = 2x + 2(x + 8)$$
$$44 = 2x + 2x + 16$$
$$44 = 4x + 16$$
$$28 = 4x$$
$$7 = x$$
$$x = 7$$
$$x + 8 = 15$$

The dimensions are 7m. by 15m.

117.
The number of quarters is in terms of nickels.

So,   x = number of nickels
4x + 5 = number of quarters

	5¢	25¢	Total
Number of coins	x	4x + 5	
Value(cents)	5x	25(4x + 5)	1,175

The equation is in terms of value.
Thus,
$$5x + 25(4x + 5) = 1,175$$
$$5x + 100x + 125 = 1,175$$
$$105x = 1,050$$
$$x = 10$$
$$4x + 5 = 45$$

There are 10 nickels and 45 quarters

119.
Let x = number of dimes
39 - x = number of quarters

#119 continued

	10¢	25¢	Total
Number of coins	x	39 - x	
Value(cents)	10x	25(39 - x)	795

The equation is in terms of value.
Thus,

$$10x + 25(39 - x) = 795$$
$$10x + 975 - 25x = 795$$
$$-15x = -180$$
$$x = 12$$
$$39 - x = 27$$

There are 12 dimes and 27 quarters.

121.
Let x = number of hours it will take for Joey and Amy to be $16\frac{1}{2}$ miles apart.

	Rate ·	Time	= Distance
Joey	12	x	12x
Amy	10	x	10x

Since they are traveling in the opposite direction from the same starting point at the same time, each distance is added for the total miles apart.

$$12x + 10x = 16\frac{1}{2}$$
$$22x = \frac{33}{2}$$
$$\frac{1}{22} \cdot 22x = \frac{1}{22} \cdot \frac{33}{2}$$

#121 continued

$$x = \frac{3}{4}$$

It took $\frac{3}{4}$ hr.

123.
Let x = Susie's traveling time
x - 1 = Earl's traveling time

	Rate ·	Time	= Distance
Susie	42	x	42x
Earl	56	x - 1	56(x - 1)

When Earl caught up with Susie, they both had gone the same distance, thus both distances are equal.
So,

$$42x = 56(x - 1)$$
$$42x = 56x - 56$$
$$42x - 56x = -56$$
$$-14x = -56$$
$$x = 4$$

Susie left at 3:00 p.m. and 4 hours later or at 7:00 p.m,. Earl overtook Susie.

125.
Let x = amount invested in a checking account.
x - 500 = amount invested in savings

	P ·	R ·	T	= I
Checking	x	.05	1	.05x
Savings	x - 500	.06	1	.06(x - 500)

The total interest was $102.

#125 continued

Thus,

Checking Interest	+	Savings Interest	=	Total Interest
.05x	+	.06(x - 500)	=	102
5x	+	6(x - 500)	=	10,200
5x	+	6x - 3,000	=	10,200
		11x	=	13,200
		x	=	1,200
		x - 500	=	700

There was $1,200 in checking and $700 in savings.

127.
Let x = amount invested in the first CD
5,000 - x = amount invested in second CD

	P ·	R ·	T =	I
1st CD	x	.08	1	.08x
2nd CD	5,000 - x	.09	1	.09(5,000 - x)

The total interest was $436.50.
Thus,

1st CD Interest	+	2nd CD Interest	=	Total Interest
.08x	+	.09(5000 - x)	=	436.50
8x	+	9(5,000 - x)	=	43,650
8x	+	45,000 - 9x	=	43,650
		-x	=	-1,350
		x	=	1,350
		5000 - x	=	3,650

The first CD was $1,350 and the second CD was $3,650.

129.
Let x = amount of "Real Stuff"
12 - x = amount of "Sugar Water"

#129 continued

	Amount ·	Percent	= Juice
Real Stuff	x	.18	.18x
Sugar Water	12 - x	.03	.03(12 - x)
NewProduct	12	.10	.10(12)

The equation is

"Real" Pure Fruit Juice	+	"Sugar" Pure Fruit Juice	=	N/P Real Fruit Juice
.18x	+	.03(12 - x)	=	.10(12)
18x	+	3(12 - x)	=	10(12)
18x	+	36 - 3x	=	120
		15x	=	84
		x	=	5.6
		12 - x	=	6.4

Use 5.6 oz of "Real Stuff" and 6.4 oz of "Sugar Water".

131.
Let x = amount of East Connecticut honey
25 - x = amount of Moss County honey

	Amount	Cost/Qty.	Cost
EastConn.	x	$4.00	$4.00x
Moss County	25 - x	$1.50	$1.50(25 - x)
New Product	25	$2.60	$2.60(25)

#131 continued

The equation is

East Conn. Cost	+	Moss County Cost	=	N/P Cost
4.00x	+	1.50(25 - x)	=	2.60(25)
400x	+	150(25 - x)	=	260(25)
400x	+	3750 - 150x	=	6,500
		250x	=	2,750
		x	=	11
		25 - x	=	14

Use 11 quarts of East Connecticut honey and 14 quarts of Moss County honey.

The graphs of the solution sets are found in the answer section of the main textbook.

133.

$$x - 7 \le -4$$
$$x - 7 + 7 \le -4 + 7$$
$$x \le 3$$

The solution set consists of all numbers x such that $x \le 3$.

135.

$$-1 < x + 2 < 3$$
$$-1 - 2 < x + 2 - 2 < 3 - 2$$
$$-3 < x < 1$$

The solution set consists of all numbers x such that $-3 < x < 1$.

137.

$$-5x > 20$$
$$\frac{-5x}{-5} < \frac{20}{-5} \quad \text{Reverse inequality.}$$
$$x < -4$$

The solution set consists of all numbers x such that $x < -4$.

139.

$$-6 \le 2x \le \frac{2}{3}$$

The graphs of the solution sets are found in the answer section of the main textbook.

#139 continued

$$\frac{1}{2}(-6) \le \frac{1}{2}(2x) \le \frac{1}{2}\left(\frac{2}{3}\right)$$
$$-3 \le x \le \frac{1}{3}$$

The solution set consists of all numbers x such that $-3 \le x \le \frac{1}{3}$.

141.

$$5x - 7 > 3$$
$$5x - 7 + 7 > 3 + 7$$
$$5x > 10$$
$$x > 2$$

The solution set consists of all numbers x such that $x > 2$.

143.

$$1 - 9x \ge 4$$
$$-1 + 1 - 9x \ge -1 + 4$$
$$-9x \ge 3$$
$$\frac{-9x}{-9} \le \frac{3}{-9} \quad \text{Reverse inequality.}$$
$$x \le -\frac{1}{3}$$

The solution set consists of all numbers x such that $x \le \frac{1}{3}$.

145.

$$x - 8 < 5x + 2$$
$$x - x - 8 < 5x - x + 2$$
$$-8 < 4x + 2$$
$$-8 - 2 < 4x + 2 - 2$$
$$-10 < 4x$$
$$\frac{-10}{4} < \frac{4x}{4}$$

The graphs of the solution sets are found in the answer section of the main textbook.

145 continued

$$-\frac{5}{2} < x \text{ or } x > -\frac{5}{2}$$

The solution set consists of all numbers x such that $x > -\frac{5}{2}$.

147.
$$2x - 7 - 8x > 3 - 3x - 1$$
$$-6x - 7 > -3x + 2$$
$$-6x + 6x - 7 > -3x + 6x + 2$$
$$-7 > 3x + 2$$
$$-7 - 2 > 3x + 2 - 2$$
$$-9 > 3x$$
$$\frac{-9}{3} > \frac{3x}{3}$$
$$-3 > x \text{ or } x < -3.$$

The solution set consists of all numbers x such that $x < -3$.

149.
$$\frac{2}{3}(5x - 1) \geq (4x - 2)$$
$$3\left[\frac{2}{3}(5x - 1)\right] \geq 3(4x - 2) \quad \text{LCD} = 3$$
$$2(5x - 1) \geq 3(4x - 2)$$
$$10x - 2 \geq 12x - 6$$
$$10x - 10x - 2 \geq 12x - 10x - 6$$
$$-2 \geq 2x - 6$$
$$-2 + 6 \geq 2x - 6 + 6$$
$$4 \geq 2x$$
$$2 \geq x \text{ or } x \leq 2$$

The solution set consists of all numbers x such that $x \leq 2$.

151.
$$-1 < 3x + 8 < 5$$
$$-1 - 8 < 3x + 8 - 8 < 5 - 8$$
$$-9 < 3x < -3$$

#151 continued

$$\frac{-9}{3} < \frac{3x}{3} < \frac{-3}{3}$$
$$-3 < x < -1$$

The solution set consists of all numbers x such that $-3 < x < -1$.

Chapter 2 Test Solutions

1.
$$\frac{4}{5}(x + 2) - (3x - 5); \quad x = -7$$
$$\frac{4}{5}(-7 + 2) - [3(-7) - 5] = \frac{4}{5}(-5) - (-21 - 5)$$
$$= \frac{4}{5}(-5) - (-26)$$
$$= -4 + 26 = 22$$

3.
$$5x^2 - 7x + 3x^2 + 2x = 5x^2 + 3x^2 - 7x + 2x$$
$$= 8x^2 - 5x$$

5.
$$I = pm + V, \text{ for } m$$
$$I - V = pm$$
$$\frac{I - V}{P} = \frac{pm}{p}$$
$$\frac{I - V}{p} = m$$

7.
$$5x - 2 = 11$$
$$5x - 2 + 2 = 11 + 2$$
$$5x = 13$$
$$\frac{5x}{5} = \frac{13}{5}$$
$$x = \frac{13}{5}$$
Solution set: $\left\{\frac{13}{5}\right\}$

9.
$$\frac{4}{5}x + \frac{2}{3} = \frac{5}{6}x + 1$$

88

#9 continued

$$\frac{4}{5}x - \frac{4}{5}x + \frac{2}{3} = \frac{5}{6}x - \frac{4}{5}x + 1$$

$$\frac{2}{3} = \frac{1}{30}x + 1$$

$$\frac{2}{3} - 1 = \frac{1}{30}x$$

$$-\frac{1}{3} = \frac{1}{30}x$$

$$30\left(-\frac{1}{3}\right) = 30\left(\frac{1}{30}\right)x$$

$$-10 = x$$

Solution set: {- 10}

11.

$$|4x - 3| = 5$$

Case I		Case II
$4x - 3 = 5$	or	$4x - 3 = -5$
$4x - 3 + 3 = 5 + 3$	or	$4x - 3 + 3 = -5 + 3$
$4x = 8$	or	$4x = -2$
$x = 2$	or	$x = -\frac{1}{2}$

Solution set: $\left\{-\frac{1}{2}, 2\right\}$

13.

$$2x + 17 \leq 9$$

$$2x + 17 - 17 \leq 9 - 17$$

$$2x \leq -8$$

$$\frac{2x}{2} \leq \frac{-8}{2}$$

$$x \leq -4$$

The solution set consists of all numbers x such that
x ≤ 4.

The graph is found in the answer section of the main textbook.

15.

$$-\frac{5}{2} < 3x - \frac{1}{4} < 2$$

#15 continued

$$-\frac{5}{2} + \frac{1}{4} < 3x - \frac{1}{4} + \frac{1}{4} < 2 + \frac{1}{4}$$

$$-\frac{9}{4} < 3x < \frac{9}{4}$$

$$\frac{1}{3}\left(-\frac{9}{4}\right) < \frac{1}{3} \cdot 3x < \frac{1}{3} \cdot \frac{9}{4}$$

$$-\frac{3}{4} < x < \frac{3}{4}$$

The solution set consists of all numbers x such that

$$-\frac{3}{4} < x < \frac{3}{4}$$

The graph is found in the answer section of the main textbook.

17.

x = number of $5 bills
25 - x = number of $20 bills
5x = value of x number of $5 bills
20(25 - x) = value of (25 - x) number of $20 bills

Thus,

$$5x + 20(25 - x) = 230$$

$$5x + 500 - 20x = 230$$

$$-15x + 500 = 230$$

$$-15x = -270$$

$$x = 18$$

$$(25 - x) = 7$$

Answer: 18 fives and 7 twenties

Test Your Memory

1. - 3.
The answers do not require worked-out solutions.

5.

$$-\frac{5}{4} + \frac{7}{12} \qquad\qquad LCD = 12$$

#5 continued

$$= -\frac{5}{4} \cdot \frac{3}{3} + \frac{7}{12} = -\frac{15}{12} + \frac{7}{12}$$

$$= -\frac{8}{12} \text{ or } -\frac{2}{3}$$

7.

$$5 - |-2 - 7| = 5 - |-9| = 5 - 9 = -4$$

9.

$$2.8 - (-7.5)(0.44) = 2.8 - (-3.3)$$
$$= 2.8 + 3.3 = 6.1$$

11.

$$5(7) + 3(9) = 35 + 27 = 62$$

13.

$$-36 + 4 \cdot 3 = -9 \cdot 3 = -27$$

15.

$$\left(-\frac{3}{10}\right)\left(\frac{35}{6}\right)$$

$$= \left(-\frac{1}{2}\right)\left(\frac{7}{2}\right) \qquad \text{Canceling}$$

$$= -\frac{7}{4}$$

17.

$$\frac{1 - \sqrt{(-1)^2 - 4(-6)}}{2}$$

$$= \frac{1 - \sqrt{1 + 24}}{2} = \frac{1 - \sqrt{25}}{2} = \frac{1 - 5}{2} = -\frac{4}{2} = -2$$

19.

$$(-3)(2)(-5)(-4) = -6(-5)(-4)$$
$$= 30(-4) = -120$$

21.

$$\frac{7}{4} - \left(\frac{5}{6} + \frac{2}{15}\right) = \frac{7}{4} - \left(\frac{25}{30} + \frac{4}{30}\right) \qquad \text{LCD} = 30$$

#21 continued

$$= \frac{7}{4} - \left(\frac{29}{30}\right) = \frac{105}{60} - \left(\frac{58}{60}\right)$$

$$= \frac{47}{60}$$

23.

$$8 - 2[6 + 4(3 - 11)] = 8 - 2[6 + 4(-8)]$$
$$= 8 - 2[6 - 32]$$
$$= 8 - 2[-26]$$
$$= 8 + 52 = 60$$

25.

$$\frac{2}{3}(5x + 2) - (7x - \frac{1}{3}); \quad x = 3$$

$$\frac{2}{3}(5 \cdot 3 + 2) - (7 \cdot 3 - \frac{1}{3}) = \frac{2}{3}(15 + 2) - (21 - \frac{1}{3})$$

$$= \frac{2}{3}(17) - \left(\frac{62}{3}\right)$$

$$= \frac{34}{3} - \frac{62}{3} = -\frac{28}{3}$$

27.

$$8x^2 - 11x + x - 5x^2 \quad = 8x^2 - 5x^2 - 11x + x$$
$$= 3x^2 - 10x$$

29.

$$3x - 4y = 8; \text{ for } y$$
$$-4y = 8 - 3x$$
$$\frac{-4y}{-4} = \frac{8 - 3x}{-4}$$
$$y = \frac{8 - 3x}{-4} \text{ or } \frac{3}{4}x - 2$$

31.

$$3x - 4 = 7$$
$$3x - 4 + 4 = 7 + 4$$
$$3x = 11$$
$$\frac{3x}{3} = \frac{11}{3}$$
$$x = \frac{11}{3}$$

Solution set: $\left\{\frac{11}{3}\right\}$

33.

$$2(4x - 3) = 9x + 7$$
$$8x - 6 = 9x + 7$$
$$8x - 8x - 6 = 9x - 8x + 7$$
$$-6 = x + 7$$
$$-6 - 7 = x + 7 - 7$$
$$-13 = x$$

Solution set: $\{-13\}$

35.

$$7x - 3(2x - 1) = 8$$
$$7x - 6x + 3 = 8$$
$$x + 3 = 8$$
$$x + 3 - 3 = 8 - 3$$
$$x = 5$$

Solution set: $\{5\}$

37.

$$\frac{1}{4}x - \frac{5}{6} = \frac{1}{6}x - \frac{1}{2}$$
$$\frac{1}{4}x - \frac{1}{6}x - \frac{5}{6} = \frac{1}{6}x - \frac{1}{6}x - \frac{1}{2}$$
$$\frac{1}{12}x - \frac{5}{6} = -\frac{1}{2}$$
$$\frac{1}{12}x - \frac{5}{6} + \frac{5}{6} = -\frac{1}{2} + \frac{5}{6}$$
$$\frac{1}{12}x = \frac{1}{3}$$
$$12 \cdot \frac{1}{12}x = 12 \cdot \frac{1}{3}$$
$$x = 4$$

Solution set: $\{4\}$

39.

$$3x + 10 \leq 1$$
$$3x + 10 - 10 \leq 1 - 10$$
$$3x \leq -9$$
$$\frac{3x}{3} \leq \frac{-9}{3}$$
$$x \leq -3$$

The solution set consists of all numbers x such that $x \leq -3$.

The graph is found in the answer section of the main text.

41.

$$\frac{1}{2}x - \frac{1}{3} > 2$$
$$\frac{1}{2}x - \frac{1}{3} + \frac{1}{3} > 2 + \frac{1}{3}$$
$$\frac{1}{2}x > \frac{7}{3}$$
$$2 \cdot \frac{1}{2}x > 2 \cdot \frac{7}{3}$$
$$x > \frac{14}{3}$$

The solution set consists of all numbers x such that $x > \frac{14}{3}$.

The graph is found in the answer section of the main text.

43.

$$-9 < 2x - 1 < 9$$
$$-9 + 1 < 2x - 1 + 1 < 9 + 1$$
$$-8 < 2x < 10$$
$$\frac{-8}{2} < \frac{2x}{2} < \frac{10}{2}$$
$$-4 < x < 5$$

The solution set consists of all numbers x such that $-4 < x < 5$.

The graph is found in the answer section of the main text.

45.

$$P = 4s$$
$$P = 4\left(4\frac{2}{3}\right)$$
$$P = 4\left(\frac{14}{3}\right)$$
$$P = \frac{56}{3} \text{ or } 18\frac{2}{3} \qquad \text{Answer: } 18\frac{2}{3} \text{ ft.}$$

47.

Let x be the measrement of the width.
Then 3x - 1 is the measurement of the length.

#47 continued

$$P = 2W + 2L$$

$$54 = 2x + 2(3x - 1)$$
$$54 = 2x + 6x - 2$$
$$54 = 8x - 2$$
$$54 + 2 = 8x - 2 + 2$$
$$56 = 8x$$
$$\frac{56}{8} = \frac{8x}{8}$$
$$7 = x$$
$$20 = (3x - 1) \qquad \text{Answer: } 7 \text{ X } 20 \text{ ft.}$$

49.

Let x be the number of dimes.
Then 32 - x is the number of quarters.

Let 10x be the value of x number of dimes in cents.
Then 25(32 - x) is the value of (32 - x) number of quarters in cents.

$$\text{Thus, } 10x + 25(32 - x) = 440$$
$$10x + 800 - 25x = 440$$
$$- 15x + 800 = 440$$
$$- 15x + 800 - 800 = 440 - 800$$
$$- 15x = - 360$$
$$\frac{-15x}{-15x} = \frac{-360}{-15}$$
$$x = 24$$
$$32 - x = 8$$

Answer: 24 dimes, 8 quarters

Chapter 2 Study Guide

Self Test Exercises

I. Evaluate the following algebraic expressions for the given values of the variables.

1. $\frac{2}{3}(x - 3) - (2x + 1)$; $x = -6$

2. $\frac{3x - 5y}{\sqrt{4x}}$; $x = 16, y = 2$

II. Simplify the following algebraic expressions by combining like terms.

3. $5x - 7x^2 + 9x + 10x^2$

4. $\frac{3}{5}(2x - 3) - 2\left(x + \frac{6}{7}\right)$

III. Solve the following literal equations for the indicated variable.

5. $A = P + Prt$;   Solve for r.

6. $S = 4\pi r^2$;   Solve for r^2

IV. Solve the following equations.

7. $6x + 2 = -2$

8. $-8x - 14 - 6x = 23x - 88$

9. $\frac{7}{6}y - 8 = \frac{5}{2} + y$

10. $5(x - 4) - 2(x - 2) = 12 + x - 2$

11. $|3x - 5| = 7$

12. $\left|\frac{2}{3}x - 3\right| = 5$

V. Find the solutions of the following inequalities, and graph the solution sets on the number line.

13. $4x - 18 \leq 7$

14. $3 - 2(3x - 4) < -4x + 8$

15. $-5 < 2x + 3 < 5$

16. $-\frac{3}{4} < 5x - \frac{1}{8} < 3$

VI. Use algebraic expressions to find the solutions of the following problems.

17. The perimeter of a rectangle is 28 ft.. The length of the rectangle is 2 less than three times the width. What are the dimensions of the rectangle?

18. Jed has $630 in $10 bills and $20 bills. He has a total of 40 bills. How many $10 bills and how many $20 bills does he have?

19. John leaves home on a bicycle traveling south at a rate of 10 mi/hr. At the same time, Sandy leaves home on a bicycle traveling north at a rate of 6 mi/hr. In how many hours will John and Sandy be $5\frac{1}{3}$ mi. apart?

20. How many gallons of a 15% alcohol solution must be mixed with 8 gallons of a 20% alcohol solution to obtain a 17% alcohol solution?

The worked-out solutions begin on the next page.

Self-Test Solutions

1.

$\frac{2}{3}(x - 3) - (2x + 1)$; $x = -6$

$\frac{2}{3}(-6 - 3) - [2(-6) + 1] = \frac{2}{3}(-9) - (-12 + 1)$

$= \frac{2}{3}(-9) - (-11)$

$= -6 + 11 = 5$

2.

$\frac{3x - 5y}{\sqrt{4x}}$; $x = 16, y = 2$

$\frac{3 \cdot 16 - 5 \cdot 2}{\sqrt{4 \cdot 16}} = \frac{48 - 10}{\sqrt{64}} = \frac{48 - 10}{8} = \frac{38}{8} = \frac{19}{4}$

3.

$5x - 7x^2 + 9x + 10x^2 = 3x^2 + 14x$

4.

$\frac{3}{5}(2x - 3) - 2(x + \frac{6}{7}) = \frac{6}{5}x - \frac{9}{5} - 2x - \frac{12}{7}$

$= \frac{6}{5}x - 2x - \frac{9}{5} - \frac{12}{7}$

$= \frac{6}{5}x - \frac{10}{5}x - \frac{63}{35} - \frac{60}{35}$

$= -\frac{4}{5}x - \frac{123}{35}$

5.

$A = P + Prt$;   Solve for r

$A - P = Prt$ Subtract P

$\frac{A - P}{Pt} = \frac{Prt}{Pt}$ Divide by Pt

$\frac{A - P}{Pt} = r$

6.

$S = 4\pi r^2$;   Solve for r^2

#6 continued

$\frac{S}{4\pi} = \frac{4\pi r^2}{4\pi}$ Divide by 4π

$\frac{S}{4\pi} = r^2$

7.

$6x + 2 = -2$

$6x + 2 - 2 = -2 - 2$

$6x = -4$

$x = -\frac{4}{6}$ or $-\frac{2}{3}$

Solution set: $\left\{ -\frac{2}{3} \right\}$

8.

$-8x - 14 - 6x = 23x - 88$

$-14x - 14 = 23x - 88$

$-14x + 14x - 14 = 23x + 14x - 88$

$-14 = 37x - 88$

$-14 + 88 = 37x - 88 + 88$

$74 = 37x$

$2 = x$

Solution set: { 2 }

9.

$\frac{7}{6}y - 8 = \frac{5}{2} + y$

$\frac{7}{6}y - y - 8 = \frac{5}{2} + y - y$

$\frac{1}{6}y - 8 = \frac{5}{2}$

$\frac{1}{6}y - 8 + 8 = \frac{5}{2} + 8$

$\frac{1}{6}y = \frac{21}{2}$

$6 \cdot \frac{1}{6}y = 6 \cdot \frac{21}{2}$

$y = 63$

Solution set: { 63 }

10.

$5(x - 4) - 2(x - 2) = 10 + x$
$5x - 20 - 2x + 4 = 10 + x$
$3x - 16 = 10 + x$
$3x - x - 16 = 10 + x - x$
$2x - 16 = 10$
$2x - 16 + 16 = 10 + 16$
$2x = 26$
$\frac{2x}{2} = \frac{26}{2}$
$x = 13$

Solution set: $\{ 13 \}$

11.

$|3x - 5| = 7$

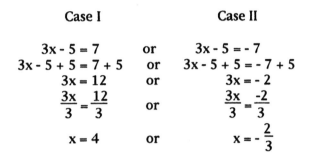

Case I		Case II
$3x - 5 = 7$	or	$3x - 5 = -7$
$3x - 5 + 5 = 7 + 5$	or	$3x - 5 + 5 = -7 + 5$
$3x = 12$	or	$3x = -2$
$\frac{3x}{3} = \frac{12}{3}$	or	$\frac{3x}{3} = \frac{-2}{3}$
$x = 4$	or	$x = -\frac{2}{3}$

Solution set: $\left\{ -\frac{2}{3}, \ 4 \right\}$

12.

$\left| \frac{2}{3}x - 3 \right| = 5$

Case I		Case II
$\frac{2}{3}x - 3 = 5$	or	$\frac{2}{3}x - 3 = -5$
$\frac{2}{3}x - 3 + 3 = 5 + 3$	or	$\frac{2}{3}x - 3 + 3 = -5 + 3$
$\frac{2}{3}x = 8$	or	$\frac{2}{3}x = -2$
$\frac{3}{2} \cdot \frac{2}{3}x = \frac{8}{1} \cdot \frac{3}{2}$	or	$\frac{3}{2} \cdot \frac{2}{3}x = \frac{-2}{1} \cdot \frac{3}{2}$
$x = 12$	or	$x = -3$

Solution set: $\{ 12, -3 \}$

13.

$4x - 18 \leq 7$
$4x - 18 + 18 \leq 7 + 18$
$4x \leq 25$
$x \leq \frac{25}{4}$

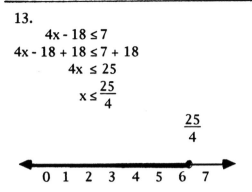

14.

$3 - 2(3x - 4) < -4x + 8$
$3 - 6x + 8 < -4x + 8$
$-6x + 11 < -4x + 8$
$-6x + 4x + 11 < -4x + 4x + 8$
$-2x + 11 < 8$
$-2x + 11 - 11 < 8 - 11$
$-2x < -3$
$x > \frac{3}{2}$

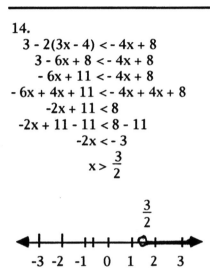

15.

$-5 < 2x + 3 < 5$
$-5 - 3 < 2x + 3 - 3 < 5 - 3$
$-8 < 2x < 2$
$\frac{-8}{2} < \frac{2x}{2} < \frac{2}{2}$
$-4 < x < 1$

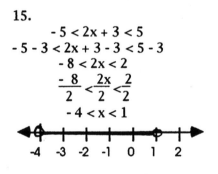

16.

$$-\frac{3}{4} < 5x - \frac{1}{8} < 3$$

$$-\frac{3}{4} + \frac{1}{8} < 5x - \frac{1}{8} + \frac{1}{8} < 3 + \frac{1}{8}$$

$$-\frac{5}{8} < 5x < \frac{25}{8}$$

$$-\frac{1}{8} < x < \frac{5}{8}.$$

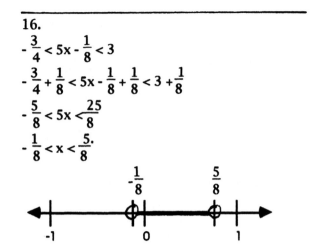

17.
Let x be the measurement of the width.
Then 3x - 2 is the measurement of the length.
Thus;, P = 2W + 2L

$$28 = 2x + 2(3x - 2)$$
$$28 = 2x + 6x - 4$$
$$28 = 8x - 4$$
$$28 + 4 = 8x - 4 + 4$$
$$32 = 8x$$
$$\frac{32}{8} = \frac{8x}{8}$$
$$4 = x$$
$$10 = 3x - 2$$

Answer: 4 by 10 ft.

18.
Let x be the number of $10 bills.
Then 40 - x is the number of $20 bills.
Let 10x be the value of x number of $10 bills.
Then 20(40 - x) is the value of 40 - x number of $20 bills.
Thus,
$$10x + 20(40 - x) = 630$$
$$10x + 800 - 20x = 630$$
$$-10x = -170$$
$$x = 17$$
$$40 - x = 23$$
Answer: 17 $10 bills, 23 $20 bills

19.
Let x = number of hours it will take for John and Sandy to be $5\frac{1}{3}$ miles apart.

	Rate	· Time	= Distance
John	10	x	10x
Sandy	6	x	6x

We add their distances.

$$10x + 6x = 5\frac{1}{3}$$
$$16x = \frac{16}{3}$$
$$x = \frac{1}{3}$$

Answer: $\frac{1}{3}$ of an hr. or 20 min.

20.
Let x be the amount of 15% alcohol solution.

	Amount	Percent of Alcohol	Amount of Alcohol
15% Sol.	x	15% = 0.15	0.15x
20% Sol.	8	20% = 0.2	0.2(8)
Combined	8 + x	17% - 0.17	0.17(8+x)

Thus,
$$0.15x + 0.2(8) = 0.17(8 + x)$$
$$0.15x + 1.6 = 1.36 + 0.17x$$
$$0.24 = 0.02x$$
$$12 = x$$
Answer: 12 gallons

CHAPTER 3   EXPONENTS AND POLYNOMIALS

Solutions to Text Exercises

Exercises 3.1

1.
Base: 8
Exponent: 3

3.
Base: 2
Exponent: 1

5.
Base: x
Exponent: 4

7.
Base: 7
Exponent: 5

9.
Base: (-4)
Exponent: 10

11.
Base: x
Exponent: 8

13. - 25.
Answers do not require worked out solutions.

27.
$-1 \cdot 3 \cdot 3 \cdot 3 \cdot 3 = -81$

29.
$-1 \cdot 5 \cdot 5 \cdot 5 \cdot 5 = -625$

31.
$5^2 + 5^3 = 25 + 125 = 150$

33.
$2^5 - 2^2 = 32 - 4 = 28$

35.
$2 \cdot 2^4 = 2 \cdot 16 = 32$

37.
$3^3 \cdot 3^2 = 27 \cdot 9 = 243$

39.
$(2^4)^2 = (16)^2 = 256$

41.
$\dfrac{5^6}{5^3} = 5^{6-3} = 5^3 = 125$

43.
$\dfrac{7^3}{7} = 7^{3-1} = 7^2 = 49$

45.
$\left(\dfrac{2}{3}\right)^3 = \dfrac{2^3}{3^3} = \dfrac{8}{27}$

47.
$2^3 \cdot 3^2 = 8 \cdot 9 = 72$

49.
$x^8 \cdot x^{13} = x^{8+13} = x^{21}$

51.
$x^{42} \cdot x^{29} = x^{42+29} = x^{71}$

53.
Cannot be simiplified since the bases are not the same, so the answer is $x^5 y^9$.

55.
$x^5 \cdot x^{12} \cdot x^4 = x^{5+12+4} = x^{21}$

57.
$(-7x^3)(2x^6) = (-7 \cdot 2)(x^3 \cdot x^6) = -14x^9$

Product law of exponents

59.
$(x^4)^6 = x^{4 \cdot 6} = x^{24}$

61.
$$\left[(x^3)^6\right]^9 = (x^{3 \cdot 6})^9 = (x^{18})^9$$
$$= x^{18 \cdot 9} = x^{162}$$

Apply the Power to a Power Law twice.

63.

$$\left[\left(x^2\right)^5\right]^7 = \left(x^{10}\right)^7$$

$$= x^{70}$$

Apply the Power to a Power Law twice.

65.

$$\left(x \cdot x^4 \cdot x^7\right)^{10} = \left(x^{1+4+7}\right)^{10}$$

$$= \left(x^{12}\right)^{10}$$

$$= x^{120}$$

Apply the Power to a Power Law.

67.

$$(6xy)^2 = 6^2 x^2 y^2$$

$$= 36 x^2 y^2$$

Power to Power Law

69.

$$\left(pt^7\right)^{12} = p^{12} t^{7 \cdot 12}$$

$$= p^{12} t^{84}$$

Power to Power Law

71.

$$\left(xy^3 z^7\right)^2 = x^2 y^{3 \cdot 2} z^{7 \cdot 2}$$

$$= x^2 y^6 z^{14}$$

Power to Power Law

73.

$$\left(5x^7 y^4\right)^2 = 5^2 x^{7 \cdot 2} y^{4 \cdot 2}$$

$$= 25 x^{14} y^8$$

Power to Power Law

75.

$$\frac{x^{12}}{x^3} = x^{12-3} = x^9$$

77.

$$\frac{x^{31}}{x^8} = x^{31-8} = x^{23}$$

79.

Cannot be simplified since the bases are different, so the answer is $\dfrac{x^8}{y^2}$.

81.

$$\left(\frac{x^9}{x^4}\right)^5 = \left(x^{9-4}\right)^5 = \left(x^5\right)^5 = x^{25}$$

Use the Quotient Law of Exponents first.

83.

$$\frac{\left(x^4\right)^6}{\left(x^3\right)^5} = \frac{x^{24}}{x^{15}} = x^{24-15} = x^9$$

Use Power to Power Law, then Quotient Law of Exponents.

85.

$$\left(\frac{x^9}{y}\right)^5 = \frac{x^{9 \cdot 5}}{y^5} = \frac{x^{45}}{y^5}$$

87.

$$\left(\frac{x^8}{y^2}\right)^4 = \frac{x^{8 \cdot 4}}{y^{2 \cdot 4}} = \frac{x^{32}}{y^8}$$

89.

$$\left(\frac{6x^8}{y}\right)^2 = \frac{6^2 x^{8 \cdot 2}}{y^2} = \frac{36 x^{16}}{y^2}$$

Use Quotient to a Power Law first.

91.

$$\left(3x^7\right)^2\left(2x^6\right)^3 = 3^2x^{14} \cdot 2^3 x^{18}$$

$$= \left(3^2 \cdot 2^3\right)\left(x^{14} \cdot x^{18}\right)$$

$$= 9 \cdot 8 \cdot x^{14+18}$$

$$= 72x^{32}$$

Use Power to a Power Law first, then Product Law of Exponents.

93.

$$\left[\left(x^3 y^5\right)^3\right]^4\left[\left(x^3 y\right)^4\right]^3 = \left(x^9 y^{15}\right)^4\left(x^{12} y^4\right)^3$$

$$= x^{36} y^{60} \cdot x^{36} y^{12}$$

$$= x^{36} \cdot x^{36} \cdot y^{60} \cdot y^{12}$$

$$= x^{72} y^{72}$$

Apply Power to a Power Law twice, then Product Law of Exponents.

95.

$$\frac{-12x^7 y^9}{3x^3 y} = \frac{-12}{3} \cdot \frac{x^7}{x^3} \cdot \frac{y^9}{y}$$

$$= -4x^{7-3} y^{9-1}$$

$$= -4x^4 y^8$$

Use Quotient Law of Exponents.

97.

$$\left(\frac{3x^4}{7y^8}\right)^2 = \frac{3^2 x^8}{7^2 y^{16}} = \frac{9x^8}{49y^{16}}$$

Use Quotient to Power Law.

99.

$$\left(\frac{6x^9 y^{12}}{3x^5 y^4}\right)^2 = \left(2x^4 y^8\right)^2$$

$$= 2^2 x^8 y^{16}$$

$$= 4x^8 y^{16}$$

Use Quotient Law of Exponents, then Power to a Power Law.

101.

$$\left(\frac{16xy^7}{24y^3}\right)^3 = \left(\frac{2xy^4}{3}\right)^3$$

$$= \frac{2^3 x^3 y^{4 \cdot 3}}{3^3}$$

$$= \frac{8x^3 y^{12}}{27} \quad \text{or} \quad \frac{8}{27}x^3 y^{12}$$

Use Quotient Law of Exponents, then Power to a Power Law.

103.

$$\frac{\left(2x^3 y^6\right)^5}{\left(4xy^5\right)^2} = \frac{\left(2x^3 y^6\right)^5}{\left(2^2 xy^5\right)^2}$$

$$= \frac{2^5 x^{15} y^{30}}{2^4 x^2 y^{10}}$$

$$= 2^{5-4} x^{15-2} y^{30-10}$$

$$= 2x^{13} y^{20}$$

Use Power to a Power Law, then the Quotient Law of Exponents.

105.

$$\frac{\left(5xy^3\right)\left(8x^4y^9\right)^2}{10x^3y^7}$$

$$=\frac{5xy^3 \cdot 8^2x^8y^{18}}{10x^3y^7}$$

$$=\frac{5 \cdot 64 \cdot x^9y^{21}}{10x^3y^7}$$

$$=32x^{9-3}y^{21-7}$$

$$=32x^6y^{14}$$

Use Power to Power Law of Exponents, then Quotient Law of Exponents.

107.

$$\frac{\left(9xy^3\right)\left(2x^2y^6\right)^3}{\left(6x^2y\right)^2}$$

$$=\frac{9xy^3 \cdot 2^3x^6y^{18}}{6^2x^4y^2}$$

$$=\frac{9 \cdot 8 \cdot x^7y^{21}}{36x^4y^2}$$

$$=2x^3y^{19}$$

Use Power to Power Law of Exponents, then Quotient Law of Exponents.

109.

$$\frac{\left(2x^2y^3\right)^4\left(3xy^4\right)^2}{\left(4x^3y\right)^2}$$

$$=\frac{2^4x^8y^{12} \cdot 3^2x^2y^8}{4^2x^6y^2}$$

$$=\frac{16 \cdot 9 \cdot x^{10}y^{20}}{16x^6y^2}$$

$$=9x^4y^{18}$$

Use Power to Power Law of Exponents, then Quotient Law of Exponents.

Exercises 3.2

1.
$$3^0 = 1$$

3.
$$-8^0 = -1 \cdot 8^0$$
$$= -1 \cdot 1 = -1$$

5.
$$7^{-1} = \frac{1}{7}$$

7.
$$4^{-2} = \frac{1}{4^2} = \frac{1}{16}$$

9.
$$-3^{-4} = -1 \cdot 3^{-4} = -1 \cdot \frac{1}{3^4}$$
$$= -\frac{1}{81}$$

The exponent applies to the base number 3 only.

11.
$$(-5)^{-3} = \frac{1}{(-5)^3} = -\frac{1}{125}$$

13.
$$-7^{-2} = -1 \cdot 7^{-2} = -1 \cdot \frac{1}{7^2} = -\frac{1}{49}$$

15.

$$\left(\frac{6}{5}\right)^{-2} = \frac{1}{\left(\frac{6}{5}\right)^2} = \frac{1}{\frac{36}{25}}$$

$$= \frac{25}{36}$$

Note: $1 + \frac{36}{25} = 1 \cdot \frac{25}{36} = \frac{25}{36}$

17.

$$-\left(\frac{3}{2}\right)^{-2} = -1 \cdot \left(\frac{3}{2}\right)^{-2} = -1 \cdot \frac{1}{\left(\frac{3}{2}\right)^2} = -1 \cdot \frac{1}{\frac{9}{4}} = -\frac{4}{9}$$

See note on Exercise 15.

19.

$$3^{-1} - 3^{-2} = \frac{1}{3} - \frac{1}{3^2} = \frac{1}{3} - \frac{1}{9} = \frac{2}{9}$$

21.

$$3^0 + 3 + 3^{-1} = 1 + 3 + \frac{1}{3}$$

$$= 4\frac{1}{3} \quad \text{or} \quad \frac{13}{3}$$

23.

$$5^8 \cdot 5^{-6} = 5^{8+(-6)} = 5^2 = 25$$

25.

$$(-4)^{-1} \cdot (-4)^{-2} = \frac{1}{(-4)} \cdot \frac{1}{(-4)^2}$$

$$= \frac{1}{(-4)^3} = -\frac{1}{64}$$

27.

$$\left(\left(\frac{2}{3}\right)^{-1}\right)^{-2} = \left(\frac{2}{3}\right)^{-1(-2)} = \left(\frac{2}{3}\right)^2 = \frac{4}{9}$$

Power to a Power Law.

29.

$$\left(3^{-2}\right)^2 = 3^{-2(2)} = 3^{-4} = \frac{1}{3^4} = \frac{1}{81}$$

First use the Power to a Power Law.

31.

$$\frac{7^{-4}}{7^{-2}} = \frac{1}{7^{-2+4}} = \frac{1}{7^2} = \frac{1}{49}$$

A positive exponent is needed, so the -4 is subtracted.

33.

$$\frac{4}{4^{-2}} = 4^{1-(-2)} = 4^3 = 64$$

A positive exponent is needed, so the -2 is subtracted.

35.

$$\frac{4^2}{3^{-1}} = \frac{4^2}{1} \cdot \frac{1}{3^{-1}} = \frac{4^2}{1} \cdot \frac{3}{1} = 48$$

37.

$$x^7 \cdot x^{-4} = x^{7+(-4)} = x^3$$

Product Law of Exponents

39.

$$x^{-4} \cdot x = x^{-4+1} = x^{-3} = \frac{1}{x^3}$$

41.

$$x^{-2} \cdot x^{-3} \cdot x^4 = x^{-2+(-3)+4} = x^{-1} = \frac{1}{x}$$

43.

$$\left(x^3\right)^{-2} = x^{-6} = \frac{1}{x^6}$$

Power to a Power Law.

45.

$$\left(x^{-3}\right)^{-5} = x^{15}$$

47.

$$\left(\left(x^{-2}\right)^{-3}\right)^{-2} = \left(x^6\right)^{-2} = x^{-12} = \frac{1}{x^{12}}$$

Use Power to Power Law twice.

49.

$$\left(x^{-2}y^{-1}\right)^{-2} = x^{-2(-2)} \cdot y^{-1(-2)} = x^4y^2$$

Power to Power Law.

51.

$$\left(x^{-2}y^{-4}\right)^5 = x^{-2\cdot5} \cdot y^{-4\cdot5} = x^{-10}y^{-20} = \frac{1}{x^{10}y^{20}}$$

Power to Power Law.

53.

$$\frac{x^5}{x^{-2}} = x^{5-(-2)} = x^{5+2} = x^7$$

A positive exponent is needed, so the -2 is subtracted.

55.

$$\frac{x^{-2}}{x^{-2}} = x^{-2-(-2)} = x^0 = 1$$

57.

$$\left(\frac{x^{-3}}{x^{-1}}\right)^4 = \frac{x^{-12}}{x^{-4}} \qquad (1)$$

$$= \frac{1}{x^{-4-(-12)}} \qquad (2)$$

$$= \frac{1}{x^8}$$

(1) Quotient to a Power Law.
(2) A positive exponent is needed, so -12 is subtracted.

59.

$$\left(\frac{x^{-3}}{x4}\right)^{-2} = \frac{x^6}{x^{-8}} \qquad (1)$$

$$= x^{6-(-8)} \qquad (2)$$

$$= x^{14}$$

(1) Quotient to a Power Law.
(2) A positive exponent is needed, so -8 is subtracted.

61.

$$\frac{x^{-8}}{y^{-3}} = \frac{x^{-8}}{1} \cdot \frac{1}{y^{-3}} = \frac{1}{x^8} \cdot \frac{y^3}{1} = \frac{y^3}{x^8}$$

63.

$$\frac{6x^{-2}}{y^3} = \frac{6}{y^3} \cdot \frac{x^{-2}}{1} = \frac{6}{y^3} \cdot \frac{1}{x^2} = \frac{6}{x^2y^3}$$

65.

$$\frac{12x}{16x^9} = \frac{3}{4x^9 \cdot x^{-1}} = \frac{3}{4x^8}$$

67.

$$\frac{20x^4}{5x^2} = 4x^{4-2} = 4x^2$$

Use Quotient Law of Exponents.

69.

$$\frac{24x^{-3}y^{-5}}{20x^{-1}y^{-11}} = \frac{24}{20} \cdot \frac{x^{-3}}{x^{-1}} \cdot \frac{y^{-5}}{y^{-11}}$$

$$= \frac{6}{5} \cdot \frac{1}{x^{-1-(-3)}} \cdot \frac{y^{-5-(-11)}}{1} = \frac{6y^6}{5x^2}$$

71.

$$\left(-5xy^{-3}\right)^{-2}\left(3x^{-4}y^{-1}\right)^2$$

$$=(-5)^{-2}x^{-2}y^6 \cdot 3^2x^{-8}y^{-2}$$

$$=(-5)^{-2} \cdot 3^2 \cdot x^{-2} \cdot x^{-8} \cdot y^6 \cdot y^{-2}$$

$$=(-5)^{-2} \cdot 3^2x^{-10}y^4$$

$$=\frac{3^2y^4}{(-5)^2x^{10}}$$

$$=\frac{9y^4}{25x^{10}}$$

Use Power to a Power Law, then Product Law of Exponents.

73.

$$\left(6^2 x^{-3} y^8\right)^0 \left(5x^0 y^4\right)^3$$

$$= 1 \cdot 5^3 x^0 y^{12}$$

$$= 1 \cdot 5^3 \cdot 1 \cdot y^{12}$$

$$= 125 y^{12}$$

All numbers or expressions, except 0, to the power of zero are equal to 1.

75.

$$\left(\frac{2x^{-2} y^4}{3x^5 y^{-5}}\right)^3 = \frac{2^3 x^{-6} y^{12}}{3^3 x^{15} y^{-15}}$$

$$= \frac{8}{27} \cdot \frac{x^{-6}}{x^{15}} \cdot \frac{y^{12}}{y^{-15}}$$

$$= \frac{8}{27} \cdot \frac{1}{x^{15-(-6)}} \cdot \frac{y^{12-(-15)}}{1}$$

$$= \frac{8 y^{27}}{27 x^{21}}$$

Use Quotient to a Power Law, then Quotient Law of Exponents.

77.

$$\frac{\left(4x^{-2} y^{-3}\right)^2}{\left(3x^4 y^{-1}\right)^3} = \frac{4^2 x^{-4} y^{-6}}{3^3 x^{12} y^{-3}} = \frac{16}{27} \cdot \frac{x^{-4}}{x^{12}} \cdot \frac{y^{-6}}{y^{-3}}$$

$$= \frac{16}{27} \cdot \frac{1}{x^{12-(-4)}} \cdot \frac{1}{y^{-3-(-6)}} = \frac{16}{27 x^{16} y^3}$$

Use Quotient to a Power Law, then Quotient Law of Exponents.

79.

$$\frac{\left(2x^3 y^{-4}\right)^3}{\left(4x^{-2} y^{-5}\right)^{-2}} = \frac{2^3 x^9 y^{-12}}{4^{-2} x^4 y^{10}} = \frac{2^3}{4^{-2}} \cdot \frac{x^9}{x^4} \cdot \frac{y^{-12}}{y^{10}}$$

$$= \frac{2^3 \cdot 4^2}{1} \cdot \frac{x^{9-4}}{1} \cdot \frac{1}{y^{10-(-12)}} = \frac{128 x^5}{y^{22}}$$

Use Quotient to a Power Law, then Quotient Law of Exponents.

81.

$$\frac{\left(2x^{-2} y^{-3}\right)^2 \left(3x^4 y^{-2}\right)^{-2}}{\left(5x^4 y^{-2}\right)^{-1}} = \frac{2^2 x^{-4} y^{-6} \cdot 3^{-2} x^{-8} y^4}{5^{-1} x^{-4} y^2}$$

$$= \frac{2^2 \cdot 3^{-2} x^{-12} y^{-2}}{5^{-1} x^{-4} y^2} \qquad \text{Simplify the numerator}$$

$$= \frac{2^2 \cdot 3^{-2}}{5^{-1}} \cdot \frac{x^{-12}}{x^{-4}} \cdot \frac{y^{-2}}{y^2} = \frac{4 \cdot 5}{3^2} \cdot \frac{1}{x^{-4-(-12)}} \cdot \frac{1}{y^{2-(-2)}}$$

$$= \frac{20}{9 x^8 y^4}$$

83.

$$5,376,000 = 5.\overset{\frown}{376000} \times 10^6$$

Six places to the left
$$= 5.376 \times 10^6$$

85.

$$75,100,000 = 7.\overset{\frown}{5100000} \times 10^7$$

Seven places to the left.

$$= 7.51 \times 10^7$$

87.

$$1023 = 1.\overset{\frown}{023} \times 10^3$$

Three places to the left.

$$= 1.023 \times 10^3$$

89.

$101,000,000 = 1.\underset{\smile}{01000000} \times 10^8$

Eight places to the left.

$= 1.01 \times 10^8$

91.

$0.00079 = \underset{\smile}{0000}7.9 \times 10^{-4}$

Four places to the right.

$= 7.9 \times 10^{-4}$

93.

$0.000000605 = \underset{\smile}{0000000}6.05 \times 10^{-7}$

Seven places to the right

$= 6.05 \times 10^{-7}$

95.

$0.024 = \underset{\smile}{00}2.4 \times 10^{-2}$

Two places to the right.

$= 2.4 \times 10^{-2}$

97.

$0.004009 = \underset{\smile}{000}4.009 \times 10^{-3}$

Three places to the right

$= 4.009 \times 10^{-3}$

99.

$$\frac{(42,000,000)(1,500,000)}{45,000,000,000}$$

$$= \frac{\left(4.2 \times 10^7\right)\left(1.5 \times 10^6\right)}{4.5 \times 10^{10}}$$

#99 continued

$$= \frac{4.2 \times 1.5 \times 10^{13}}{4.5 \times 10^{10}}$$

$$= \frac{6.3 \times 10^{13-10}}{4.5}$$

$$= \frac{6.3}{4.5} \times 10^3$$

$$= 1.4 \times 10^3 = 1400$$

101.

$$\frac{(0.00000054)(0.00041)}{0.000000018}$$

$$= \frac{(5.4 \times 10^{-7})(4.1 \times 10^{-4})}{1.8 \times 10^{-8}}$$

$$= \frac{(5.4 \times 4.1)(10^{-7} \times 10^{-4})}{1.8 \times 10^{-8}}$$

$$= \frac{5.4 \times 4.1 \times 10^{-11}}{1.8 \times 10^{-8}}$$

$$= \frac{5.4 \times 4.1 \times 10^{-11-(-8)}}{\underset{1}{1.8}}$$

$$= 12.3 \times 10^{-3} = 1.23 \times 10^{-2}$$

103.

$$\frac{(32,000)(5,100)}{(160,000)(1,700,000)}$$

$$= \frac{(3.2 \times 10^4)(5.1 \times 10^3)}{(1.6 \times 10^5)(1.7 \times 10^6)}$$

$$= \frac{(3.2 \times 5.1)(10^4 \times 10^3)}{(1.6 \times 1.7)(10^5 \times 10^6)}$$

$$= \frac{3.2 \times 5.1 \times 10^7}{1.6 \times 1.7 \times 10^{11}}$$

$$= \frac{\overset{2}{\cancel{3.2}} \times \overset{3}{\cancel{5.1}} \times 10^{7-11}}{\underset{1}{\cancel{1.6}} \times \underset{1}{\cancel{1.7}}} = 6 \times 10^{-4} = 0.0006$$

$$= 6 \times 10^{-4} = 0.0006$$

105.

$$\frac{(0.000000072)(0.000084)}{(0.0000000096)(0.0000042)}$$

$$= \frac{\left(7.2 \times 10^{-8}\right)\left(8.4 \times 10^{-5}\right)}{\left(9.6 \times 10^{-9}\right)\left(4.2 \times 10^{-6}\right)}$$

$$= \frac{(7.2 \times 8.4)\left(10^{-8} \times 10^{-5}\right)}{(9.6 \times 4.2)\left(10^{-9} \times 10^{-6}\right)}$$

$$= \frac{7.2 \times 8.4 \times 10^{-13}}{9.6 \times 4.2 \times 10^{-15}}$$

$$= \frac{\overset{1}{\cancel{7.2}} \times \overset{\overset{1}{1.5}\ \ \cancel{2}}{\cancel{8.4}} \times 10^{-13-(-15)}}{\underset{1}{\underset{\cancel{4.8}}{\cancel{9.6} \times \cancel{4.2}}}}$$

$$= 1.5 \times 10^{2}$$

107.

$$\frac{(124,000,000)(0.0000054)}{(0.0000000009)(3,100,000)}$$

$$= \frac{\left(1.24 \times 10^{8}\right)\left(5.4 \times 10^{-6}\right)}{\left(9 \times 10^{-10}\right)\left(3.1 \times 10^{6}\right)}$$

$$= \frac{(1.24 \times 5.4)\left(10^{8} \times 10^{-6}\right)}{(9 \times 3.1)\left(10^{-10} \times 10^{6}\right)}$$

$$= \frac{\overset{.4}{\cancel{1.24}} \times \overset{.6}{\cancel{5.4}} \times 10^{2-(-4)}}{\underset{1}{\cancel{9}} \times \underset{1}{\cancel{3.1}}}$$

$$= .24 \times 10^{6} = 2.4 \times 10^{5}$$

109.

$$\frac{(0.00000532)(605,000,000)}{(9,680,000)(0.000133)}$$

$$= \frac{\left(5.32 \times 10^{-6}\right)\left(6.05 \times 10^{8}\right)}{\left(9.68 \times 10^{6}\right)\left(1.33 \times 10^{-4}\right)}$$

$$= \frac{(5.32 \times 6.05)\left(10^{-6} \times 10^{8}\right)}{(9.68 \times 1.33)\left(10^{6} \times 10^{-4}\right)}$$

$$= \frac{5.32 \times 6.05 \times 10^{2}}{9.68 \times 1.33 \times 10^{2}}$$

$$= \frac{\overset{4}{\cancel{5.32}} \times \overset{.0625}{\cancel{6.05}}}{\underset{1}{\cancel{9.68}} \times \underset{1}{\cancel{1.33}}}$$

$$= 2.5$$

111.
We translate to scientific notation, then multiply.

$$602,300,000,000,000,000,000,000$$
$$= 6.02 \times 10^{23}$$

$\left(\begin{array}{c}\text{Number of}\\ \text{molecules}\end{array}\right)$	$\left(\begin{array}{c}\text{Number of}\\ \text{moles}\end{array}\right)$	$\left(\begin{array}{c}\text{Total number}\\ \text{molecules}\end{array}\right)$
$\left(6.02 \times 10^{23}\right) \cdot$	850	$= (6.02 \cdot 850) \times 10^{23}$
		$= 5117 \times 10^{23}$
		$= 5.117 \times 10^{26}$
		$= 5.12 \times 10^{26}$
		molecules

113.
Speed of light: 186,300 mi/sec.
There are 31,536,000 seconds in one year.
Light year distance of Proxima Centauri: 4.3

 # 113 Continued

$$\begin{aligned}
\text{Miles Dist.} &= (186,300)(31,536,000)(4.3) \\
&= (1.863 \times 10^5)(3.1536 \times 10^7(4.3) \\
&= (1.863 \cdot 3.1536 \cdot 4.3) \times 10^{12} \\
&= 25.263 \times 10^{12} \\
&= 2.5263 \times 10^{13}
\end{aligned}$$

115.

$$\begin{pmatrix} \text{Family} \\ \text{Members} \end{pmatrix} \begin{pmatrix} \text{Hare} \\ \text{Families} \end{pmatrix} \begin{pmatrix} \text{Number} \\ \text{of Hares} \end{pmatrix}$$

$$\begin{aligned}
12 \quad \times \quad 289,000 \ &= 12(2.89 \times 10^5) \\
&= 34.68 \times 10^5 \\
&= 3.468 \times 10^6 \text{ hares}
\end{aligned}$$

Exercises 3.3

1.

$5x - 7x^2 + 11$

Standard form: $-7x^2 + 5x + 11$

Degree: 2nd   Type: Trinomial

3.

$6x - 8 + 7x$

Standard form: $13x - 8$

Degree: 1st   Type: Binomial

5.

$$\begin{aligned}
7xy &+ 4x^2 - 3xy + y^3 - 7 \\
&= 4x^2 + 7xy - 3xy + y^3 - 7 \quad \text{Commutative law} \\
&= 4x^2 + (7 - 3)xy + y^3 - 7 \quad \text{Distributive law} \\
&= 4x^2 + 4xy + y^3 - 7 \quad \text{Simplified}
\end{aligned}$$

Degree: 3rd   Type: Neither

This is a four term polynomial.

7.

9

Degree: 0   Type: Monomial

9.

$$\begin{aligned}
5xy &+ x^4 - 5xy - 4y^2 \\
&= x^4 - 4y^2 + 5xy - 5xy \quad &\text{Commutative law} \\
&= x^4 - 4y^2 + (5 - 5)xy \quad &\text{Distributive law} \\
&= x^4 - 4y^2 + 0xy \\
&= x^4 - 4y^2
\end{aligned}$$

Degree: 4th   Type: Binomial

---

11. $\left(\dfrac{2}{3}\right) x^6 y^2 z$

Degree: 9th   Type: Monomial

13.

$$\begin{aligned}
5x^2 &+ 7x - 4 - 9x^2 - 2x - 8 \\
&= 5x^2 - 9x^2 + 7x - 2x - 4 - 8 \quad &\text{Commutative law} \\
&= (5 - 9)x^2 + (7 - 2)x - 4 - 8 \quad &\text{Associative law} \\
&= -4x^2 + 5x - 12 \quad &\text{Simplified}
\end{aligned}$$

Degree: 2nd   Type: Trinomial

15.

$$\frac{2}{3}x^2 - \frac{1}{9}x + 2 + \frac{1}{2}x^2 + \frac{1}{3}x - \frac{4}{5}$$

$$= \frac{2}{3}x^2 + \frac{1}{2}x^2 - \frac{1}{9}x + \frac{1}{3}x + 2 - \frac{4}{5}$$

$$= \left(\frac{2}{3} + \frac{1}{2}\right)x^2 + \left(-\frac{1}{9} + \frac{1}{3}\right)x + 2 - \frac{4}{5}$$

$$= \frac{7}{6}x^2 + \frac{2}{9}x + \frac{6}{5}$$

Degree: 2nd   Type: Trinomial

17.

$-1.21x^3 + 3.24z^2 - 6.48x^3 - 0.002z^2$

$= -1.21x^3 - 6.48x^3 + 3.24z^2 - 0.002z^2$

$= (-1.21 - 6.48)x^3 + (3.24 - 0.002)z^2$

$= -7.69x^3 + 3.238z^2$

Degree: 3rd   Type: Binomial

19.

$2x - 3y + 4z - 8 + 7x - 8y + 11$

$= 2x + 7x - 3y - 8y + 4z - 8 + 11$

$= (2 + 7)x + (-3 - 8)y + 4z - 8 + 11$

$= 9x - 11y + 4z + 3$

Degree: 1st   Type: Neither

This is a   4 term polynomial.

21.

$-3x^3y - 16x^4y^2 - 18x^3y + 12x^4y^2$

$= -3x^3y - 18x^3y - 16x^4y^2 + 12x^4y^2$

$= (-3 - 18)x^3y + (-16 + 12)x^4y^2$

$= -21x^3y - 4x^4y^2$

Degree: 6th   Type: Binomial

23.

$\dfrac{5}{8}xz^2 - \dfrac{2}{3}x^2y^2 + \dfrac{1}{4}xz^2 + \dfrac{4}{6}x^2y^2$

$= \dfrac{5}{8}xz^2 + \dfrac{1}{4}xz^2 - \dfrac{2}{3}x^2y^2 + \dfrac{4}{6}x^2y^2$

$= \left(\dfrac{5}{8} + \dfrac{1}{4}\right)xz^2 + \left(-\dfrac{2}{3} + \dfrac{4}{6}\right)x^2y^2$

$= \dfrac{7}{8}xz^2 + 0x^2y^2 = \dfrac{7}{8}xz^2$

Degree: 3rd   Type: Monomial

25.

$-7 - 2x + 5x^2$

Standard form: $5x^2 - 2x - 7$

Leading Coefficient: 5

27.

$3x^2 - 4x^4 - x^9 - 2$

Standard form: $-x^9 - 4x^4 + 3x^2 - 2$

Leading coefficient: -1

29.

$112x^5 + 19x^{47}$

Standard form: $19x^{47} + 112x^5$

Leading coefficient: 19

31.

$\dfrac{x^7}{5} + \dfrac{3x^9}{4} + \dfrac{3x^3}{7} + 1$

$= \dfrac{1}{5}x^7 + \dfrac{3}{4}x^9 + \dfrac{3}{7}x^3 + 1$

Standard form: $\dfrac{3}{4}x^9 + \dfrac{1}{5}x^7 + \dfrac{3}{7}x^3 + 1$

Leading coefficient: $\dfrac{3}{4}$

33.

$4x - 9 - 3 - 6x$

$= 4x - 6x - 9 - 3$

$= -2x - 12$ Standard form

Leading coefficent: -2

35.

$3x^2 - 2x - 9 + 5x^2 + 2x - 14$

$= 3x^2 + 5x^2 - 2x + 2x - 9 - 14$

$= 8x^2 + 0x - 23$

$= 8x^2 - 23$ Standard form

Leading coefficient: 8

37.

$2x - 5 + 3x^2 + x^3 + 4x - 9$

$= x^3 + 3x^2 + 2x + 4x - 5 - 9$

$= x^3 + 3x^2 + 6x - 14$ Standard form

Leading coefficient: 1

39.

$$\frac{1}{5}x + \frac{1}{9} - \frac{2}{3}x + \frac{1}{6}$$

$$= \frac{1}{5}x - \frac{2}{3}x + \frac{1}{9} + \frac{1}{6}$$

$$= -\frac{7}{15}x + \frac{5}{18} \qquad \text{Standard form}$$

Leading coefficient: $\qquad -\dfrac{7}{15}$

41.

$$\frac{1}{2}x^2 - \frac{1}{3}x - \frac{3}{4} + \frac{3}{10}x^2 - \frac{5}{6}x - \frac{1}{2}$$

$$= \frac{1}{2}x^2 + \frac{3}{10}x^2 - \frac{1}{3}x - \frac{5}{6}x - \frac{3}{4} - \frac{1}{2}$$

$$= \frac{8}{10}x^2 - \frac{7}{6}x - \frac{5}{4}$$

$$= \frac{4}{5}x^2 - \frac{7}{6}x - \frac{5}{4} \qquad \text{Standard form}$$

Leading coefficient: $\qquad \dfrac{4}{5}$

43.

$$8 + 48x^5 - 18x - 27x + 60$$
$$= 48x^5 - 18x - 27x + 8 + 60$$
$$= 48x^5 - 45x + 68 \qquad \text{Standard form}$$
Leading coefficient: $\quad 48$

45.

$$1.24 - 1.21x^2 - 2.36x^2 - 3.06x^2 + 9.56x^4$$
$$= 1.24 - 6.63x^2 + 9.56x^4$$
$$= 9.56x^4 - 6.63x^2 + 1.24 \qquad \text{Standard form}$$

Leading coefficient: $\qquad 9.56$

47.

$$\left(x^2 + 7x - 4\right) + \left(5x^2 - 11x - 12\right)$$
$$= x^2 + 7x - 4 + 5x^2 - 11x - 12$$
$$= x^2 + 5x^2 + 7x - 11x - 4 - 12$$
$$= 6x^2 - 4x - 16$$

49.

$$\left(x^4 + 5x^2 - 3x + 4\right) + \left(9x^3 - 6x^2 - 19\right)$$
$$= x^4 + 5x^2 - 3x + 4 + 9x^3 - 6x^2 - 19$$
$$= x^4 + 9x^3 + 5x^2 - 6x^2 - 3x + 4 - 19$$
$$= x^4 + 9x^3 - x^2 - 3x - 15$$

51.

$$\left(6x^2 - 7xy - y^2\right) + \left(9x^2 + 7xy + y^2\right)$$
$$= 6x^2 - 7xy - y^2 + 9x^2 + 7xy + y^2$$
$$= 6x^2 + 9x^2 - 7xy + 7xy - y^2 + y^2$$
$$= 15x^2 + 0 + 0$$
$$= 15x^2$$

53.

$$\left(2x^3 + 11x^2 + x - 4\right) - \left(5x^3 - 4x - 12\right)$$
$$= 2x^3 + 11x^2 + x - 4 - 5x^3 + 4x + 12 \qquad \text{Adding Opposites}$$
$$= 2x^3 - 5x^3 + 11x^2 + x + 4x - 4 + 12$$
$$= -3x^3 + 11x^2 + 5x + 8$$

55.

$$\left(8x^2 + 2xy - 4y^2\right) - \left(5x^2 + 2xy + 9y^2\right)$$
$$= 8x^2 + 2xy - 4y^2 - 5x^2 - 2xy - 9y^2$$
$$\qquad \qquad \text{Adding Opposites}$$
$$= 8x^2 - 5x^2 + 2xy - 2xy - 4y^2 - 9y^2$$
$$= 3x^2 + 0 - 13y^2$$
$$= 3x^2 - 13y^2$$

57.

$$\left(\frac{2}{3}x^2 - \frac{1}{2}x + \frac{3}{5}\right) + \left(\frac{3}{4}x^2 - \frac{1}{8}x + \frac{7}{10}\right)$$

$$= \frac{2}{3}x^2 - \frac{1}{2}x + \frac{3}{5} + \frac{3}{4}x^2 - \frac{1}{8}x + \frac{7}{10}$$

$$= \frac{2}{3}x^2 + \frac{3}{4}x^2 - \frac{1}{2}x - \frac{1}{8}x + \frac{3}{5} + \frac{7}{10}$$

$$= \frac{17}{12}x^2 - \frac{5}{8}x + \frac{13}{10}$$

59.

$$\frac{3}{4}x^2 + \frac{7}{8}x + \frac{1}{3}$$
$$-\frac{1}{3}x^2 - \frac{1}{2}x + \frac{2}{5}$$

$$\left(\frac{1}{2}x^2 - \frac{1}{3}x - \frac{1}{4}\right) - \left(\frac{2}{3}x^2 - \frac{3}{4}x - \frac{5}{6}\right)$$

$$= \frac{1}{2}x^2 - \frac{1}{3}x - \frac{1}{4} - \frac{2}{3}x^2 + \frac{3}{4}x + \frac{5}{6} \quad \text{Adding opposites}$$

$$= \frac{1}{2}x^2 - \frac{2}{3}x^2 - \frac{1}{3}x + \frac{3}{4}x - \frac{1}{4} + \frac{5}{6}$$

$$= -\frac{1}{6}x^2 + \frac{5}{12}x + \frac{7}{12}$$

61.

$$\left(1.12x^2 - 3.04x - 4.79\right) + \left(6.19x^2 + 1.42x - 6.34\right)$$

$$= 1.12x^2 - 3.04x - 4.79 + 6.19x^2 + 1.42x - 6.34$$

$$= 1.12x^2 + 6.19x^2 - 3.04x + 1.42x - 4.79 - 6.34$$

$$= 7.31x^2 - 1.62x - 11.13$$

63.

$$\begin{array}{ll} 5x^2 + 3x - 4 & \text{Note: } (5-2) = 3 \\ + \underline{-2x^2 - 7x + 12} & (3-7) = -4 \\ 3x^2 - 4x + 8 & (-4+12) = 8 \end{array}$$

65.

$$\begin{array}{ll} -8x^2 + 7xy - 4y^2 & \text{Note: } (-8-2) = -10 \\ + \underline{-2x^2 - 9xy + 16y^2} & (7-9) = -2 \\ -10x^2 - 2xy + 12y^2 & (-4+16) = 12 \end{array}$$

67.

$$\begin{array}{l} 4x^3 - 5x^2 - 2x - 3 \\ - \underline{\quad -3x^2 - 4x + 1} \end{array}$$

Change the signs and add.

$$\begin{array}{l} 4x^3 - 5x^2 - 2x - 3 \\ + \underline{\quad 3x^2 + 4x - 1} \\ 4x^3 - 2x^2 + 2x - 4 \end{array}$$

69.

$$\begin{array}{l} \frac{4}{3}x^2 + \frac{1}{2}x - \frac{1}{4} \\ + \underline{\frac{1}{6}x^2 - \frac{3}{4}x - \frac{5}{8}} \\ \frac{3}{2}x^2 - \frac{1}{4}x - \frac{7}{8} \end{array}$$

Note: $\dfrac{4}{3} + \dfrac{1}{6} = \dfrac{8}{6} + \dfrac{1}{6} = \dfrac{9}{6} = \dfrac{3}{2}$

71.

$$\begin{array}{l} \frac{3}{4}x^2 + \frac{7}{8}x + \frac{1}{3} \\ -\frac{1}{3}x^2 - \frac{1}{2}x + \frac{2}{5} \end{array}$$

Change the signs and add.

$$\begin{array}{l} \frac{3}{4}x^2 + \frac{7}{8}x + \frac{1}{3} \\ + \underline{\frac{1}{3}x^2 + \frac{1}{2}x - \frac{2}{5}} \\ \frac{5}{12}x^2 + \frac{11}{8}x - \frac{1}{15} \end{array}$$

73.

$$\begin{array}{l} 1.24x^2 - 3.96x + 4.09 \\ + \underline{5.61x^2 + 1.45x - 6.34} \\ 6.85x^2 - 2.51x - 2.25 \end{array}$$

75.

$5(3x + 4) + 6(2x - 7)$

$= 15x + 20 + 12x - 42$
$= 15x + 12x + 20 - 42$
$= 27x - 22$

77.

$-7(3x + 4) - 2(x - 3)$

$= -21x - 28 - 2x + 6$
$= -21x - 2x - 28 + 6$
$= -23x - 22$

79.

$2(-4x^2 - 3x + 7) - 5(2x^2 + x - 2)$

$= -8x^2 - 6x + 14 - 10x^2 - 5x + 10$
$= -8x^2 - 10x^2 - 6x - 5x + 14 + 10$
$= -18x^2 - 11x + 24$

81.

$4(x^3 - 3x^2 + 7) - 5(-2x^2 + 3x - 8)$

$= 4x^3 - 4 \cdot 3x^2 + 4 \cdot 7 - 5(-2)x^2 - 5 \cdot 3x - 5(-8)$
$= 4x^3 - 12x^2 + 28 + 10x^2 - 15x + 40$
$= 4x^3 - 12x^2 + 10x^2 - 15x + 28 + 40$
$= 4x^3 - 2x^2 - 15x + 68$

83.

$-2(4x^2 + 3xy - 9y^2) - (2x^2 + 7xy + y^2)$

$= -2(4x^2 + 3xy - 9y^2) - 1 \cdot (2x^2 + 7xy + y^2)$

$= -2 \cdot 4x^2 - 2 \cdot 3xy - 2(-9y)^2 - 1 \cdot 2x^2 -$
$\qquad\qquad\qquad\qquad 1 \cdot 7xy - 1 \cdot y^2$
$= -8x^2 - 6xy + 18y^2 - 2x^2 - 7xy - y^2$
$= -8x^2 - 2x^2 - 6xy - 7xy + 18y^2 - y^2$
$= -10x^2 - 13xy + 17y^2$

85.

$\frac{1}{2}(7x - 3) - \frac{1}{3}(4x - 2)$

$= \frac{7}{2}x - \frac{3}{2} - \frac{4}{3}x - \frac{2}{3} = \frac{7}{2}x - \frac{4}{3}x - \frac{3}{2} - \frac{2}{3}$

$\qquad\qquad = \frac{13}{6}x - \frac{13}{6}$

87.

$-\frac{1}{2}\left(4x^2 - 7x + 2\right) - \frac{2}{3}\left(3x^2 + 5x - 6\right)$

$= -\frac{4}{2}x^2 + \frac{7}{2}x - \frac{2}{2} - \frac{6}{3}x^2 - \frac{10}{3}x + \frac{12}{3}$

$= -2x^2 + \frac{7}{2}x - 1 - 2x^2 - \frac{10}{3}x + 4$ Simplified

$= -2x^2 - 2x^2 + \frac{7}{2}x - \frac{10}{3}x - 1 + 4$

$= -4x^2 + \frac{1}{6}x + 3$

89.

$\qquad 3.6(1.9x - 4.3) + 2.7(5.5x - 9.3)$
$= \ 6.84x - 15.48 + 14.85x - 25.11$
$= \ 6.84x + 14.85x - 15.48 - 25.11$
$= \ 21.69x - 40.59$

$\boxed{\text{Exercises } 3.4}$

1.
$5x^2 \cdot 8x^6 = 5 \cdot 8 \cdot x^2 \cdot x^6$
$\qquad\qquad\qquad$ Commutative property
$\qquad = (5 \cdot 8)(x^2 \cdot x^6)$
$\qquad\qquad\qquad$ Associative property
$\qquad = 40x^8$ $\quad$ Product law of exponents

3.
$\left(-9x^3\right)\left(7x^2\right) = (-9)7 \cdot x^3 \cdot x^2$
$\qquad\qquad\qquad$ Commutative property
$\qquad = (-9 \cdot 7) \cdot \left(x^3 \cdot x^2\right)$
$\qquad\qquad\qquad$ Associative property
$\qquad = -63x^5$ $\quad$ Product law of exponents

5.
$$(-8xy^3)(6x^3y^6) = (-8 \cdot 6)(x \cdot x^3)(y^3 \cdot y^6)$$
$$= -48x^4y^9$$
Product law of exponents

7.
$$(5x^2y^4)(2x^3y^6)(4x^3y^5)$$
$$= (5 \cdot 2 \cdot 4)(x^2 \cdot x^3 \cdot x^3)(y^4y^6y^5)$$
Commutative and associative properties
$$= 40x^8y^{15}$$ Product law of exponents

9.
$$\left(-5xy^3z\right)\left(3x^4y^2z^2\right)\left(3x^2y^3z^6\right)$$
$$= (-5 \cdot 3 \cdot 3)\left(x \cdot x^4 \cdot x^2\right)\left(y^3 \cdot y^2 \cdot y^3\right)\left(z \cdot z^2 \cdot z^6\right)$$
Commutative and associative properties
$$= -45x^7y^8z^9$$ Product law of exponents

11.
$$\left(6xy^2\right)^2\left(3xy^4\right)$$

$$= \left(6^2x^2y^4\right)\left(3xy^4\right)$$ Power to a

power law

$$= \left(6^2 \cdot 3\right)\left(x^2 \cdot x\right)\left(y^4 \cdot y^4\right)$$ Commutative

and associative
properties

$$= 108x^3y^8$$ Product law
of exponents

13.

$$5x(2x + 4) = 5x \cdot 2x + 5x \cdot 4$$
Distributive property
$$= 5 \cdot 2 \cdot x \cdot x + 5 \cdot 4 \cdot x$$
Commutative property
$$= 10x^2 + 20x$$
Product law of exponents

15.

$$-6x^3(4x - 8)$$
$$= (-6x^3)4x - (-6x^3) \cdot 8$$ Distributive property
$$= -6 \cdot 4 \cdot x^3 \cdot x - (-6) \cdot 8 \cdot x^3$$
Commutative property

#15 continued

$$= -24x^4 + 48x^3$$ Product law of exponents

17.

$$8x^2y(2x + 3y) = 8x^2y \cdot 2x + 8x^2y \cdot 3y$$
$$= 8 \cdot 2 \cdot x^2 \cdot x \cdot y + 8 \cdot 3 \cdot x^2 \cdot y \cdot y$$
$$= 16x^3y + 24x^2y^2$$

19.

$$-2x^2(x^2 + 4x - 4)$$
$$= (-2x^2)x^2 + (-2x^2)4x - (-2x^2) \cdot 4$$
$$= -2x^2 \cdot x^2 + (-2)4 \cdot x^2 \cdot x - (-2) \cdot 4 \cdot x^2$$
$$= -2x^4 - 8x^3 + 8x^2$$

21.

$$-4xy^2(1 + 2x - 5xy)$$
$$= \left(-4xy^2\right) \cdot 1 + \left(-4xy^2\right)2x - \left(-4xy^2\right) \cdot 5xy$$
$$= -4xy^2 + (-4)2 \cdot x \cdot x \cdot y^2 - (-4) \cdot 5x \cdot x \cdot y^2 \cdot y$$
$$= -4xy^2 - 8x^2y^2 + 20x^2y^3$$

23.
Distribute the left factor, (3x+4), once to the 2x and once to the 6 by multiplication.

$$(3x + 4)(2x + 6)$$
$$= (3x + 4)2x + (3x + 4)6$$
$$= 3x \cdot 2x + 4 \cdot 2x + 3x \cdot 6 + 4 \cdot 6$$
Distributive property
$$= 6x^2 + 8x + + 18x + 24$$
$$= 6x^2 + 26x + 24$$ Combine like terms

25.

$$(x + 5)(x - 3)$$

$$= (x + 5) \cdot x + (x + 5)(-3)$$
$$= x^2 + 5x + x(-3) + 5(-3)$$
$$= x^2 + 5x - 3x - 15$$
$$= x^2 + 2x - 15$$ Combine like terms

111

27.

$\overline{(x-8)(x-1)}$

$= (x - 8) \cdot x - (x - 8) \cdot 1$

$= x(x - 8) - 1 \cdot (x - 8)$ Use commutative property to avoid sign error

$= x \cdot x - x \cdot 8 - x - (-1)8$

$= x^2 - 8x - x + 8 = x^2 - 9x + 8$

29.

$\overline{(2x+3)(x-5)}$

$= (2x + 3) \cdot x - (2x + 3) \cdot 5$

$= x(2x +3) - 5(2x + 3)$ Use commutative property to avoid sign error

$= x (2x) + x \cdot 3 - 5(2x) + (-5) \cdot 3$

$= 2x^2 + 3x - 10x - 15 = 2x^2 - 7x - 15$

31.

$\overline{(2x-3)(2x-3)}$

$= (2x - 3) \cdot 2x - (2x - 3) \cdot 3$

$= 2x(2x - 3) - 3(2x - 3)$ Commutative property

$= 2x \cdot 2x - 2x \cdot 3 - 3 \cdot 2x - (-3) \cdot 3$

$= 4x^2 - 6x - 6x + 9 = 4x^2 - 12x + 9$

33.

$\overline{(4x+7)(4x+7)}$

$= (4x + 7) \cdot 4x + (4x + 7) \cdot 7$

$= 4x \cdot 4x + 7 \cdot 4x + 4x \cdot 7 + 7 \cdot 7$

$= 16x^2 + 28x + 28x + 49 = 16x^2 + 56x + 49$

35.

$\overline{(2x-3)(2x-4)}$

$= (2x - 3) \cdot 2x - (2x - 3) \cdot 4$

$= 2x(2x - 3) - 4(2x - 3)$ Commutative property

$= 2x \cdot 2x - 2x \cdot 3 - 4 \cdot 2x - (-4) \cdot 3$

$= 4x^2 - 6x - 8x + 12 = 4x^2 - 14x + 12$

37.

$\overline{(6x+7)(6x-7)}$

$= (6x + 7) \cdot 6x - (6x + 7) \cdot 7$

$= 6x (6x + 7) - 7(6x + 7)$

Commutative property

#37 continued

$= 6x \cdot 6x + 6x \cdot 7 - 7 \cdot 6x - 7 \cdot 7$

$= 36x^2 + 42x - 42x - 49 = 36x^2 - 49$

39.

$\overline{(x+y)(x+y)}$

$= (x + y) \cdot x + (x + y) \cdot y$

$= x \cdot x + y \cdot x + x \cdot y + y \cdot y$

$= x^2 + xy + xy + y^2 = x^2 + 2xy + y^2$

41.

$\overline{(2x-y)(x-3y)}$

$= (2x - y) \cdot x^- (2x - y) \cdot 3y$

$= x(2x - y) - 3y(2x - y)$ Commutative property

$= x \cdot 2x - x \cdot y - 3y \cdot 2x - (-3y) \cdot y$

$= 2x^2 - xy - 6xy + 3y^2 = 2x^2 - 7xy + 3y^2$

43.

$\overline{(3x+2y)(5x+4y)}$

$= (3x + 2y) \cdot 5x + (3x + 2y) \cdot 4y$

$= 3x \cdot 5x + 2y \cdot 5x + 3x \cdot 4y + 2y \cdot 4y$

$= 15x^2 + 10xy + 12xy + 8y^2$

$= 15x^2 + 22xy + 8y^2$

45.

$\overline{(5x-2y)(2x+4y)}$

$= (5x - 2y) \cdot 2x + (5x - 2y) \cdot 4y$

$= 5x \cdot 2x - 2y \cdot 2x + 5x \cdot 4y - 2y \cdot 4y$

$= 10x^2 - 4xy + 20xy - 8y^2$

$= 10x^2 + 16xy - 8y^2$

47.

$\overline{(x-2y)(3x-4y)}$

$= (x - 2y) \cdot 3x - (x - 2y) \cdot 4y$

$= 3x(x - 2y) - 4y(x - 2y)$ Commutative property

$= 3x \cdot x - 3x \cdot 2y - 4y \cdot x - (-4y)(2y)$

$= 3x^2 - 6xy - 4xy + 8y^2 = 3x^2 - 10xy + 8y^2$

49.

$$(-x + 4y)(3x - 2y)$$

$= (-x + 4y) \cdot 3x - (-x + 4y) \cdot 2y$

$= 3x(-x + 4y) - 2y(-x + 4y)$

$= 3x(-x) + 3x \cdot 4y - 2y(-x) + (-2y) \cdot 4y$

$= -3x^2 + 12xy + 2xy - 8y^2$

$= -3x^2 + 14xy - 8y^2$

51.

$$\begin{array}{r} 6x - 2y \\ \underline{x - 4y} \\ -24xy + 8y^2 \end{array}$$ Multiply by $-4y$

$\underline{6x^2 - 2xy}$ Multiply by x.

$6x^2 - 26xy + 8y^2$ Add.

53.

$$\begin{array}{r} 4x + 5y \\ \underline{4x + 5y} \\ 20xy + 25y^2 \end{array}$$ Multiply by $5y$

$\underline{16x^2 + 20xy}$ Multiply by $4x$

$16x^2 + 40xy + 25y^2$ Add.

55.

$(6x^2 - 7)(3x^2 - 1)$

$= (6x^2 - 7) \cdot 3x^2 - (6x^2 - 7) \cdot 1$

$= 3x^2(6x^2 - 7) - 1 \cdot (6x^2 - 7)$

$= 3x^2 \cdot 6x^2 - 3x^2 \cdot 7 - 6x^2 - (-1) \cdot 7$

$= 18x^4 - 21x^2 - 6x^2 + 7 = 18x^4 - 27x^2 + 7$

57.

$(5x^2 + 3)(2x - 4)$

$= (5x^2 + 3) \cdot 2x - (5x^2 + 3) \cdot 4$

$= 2x(5x^2 + 3) - 4(5x^2 + 3)$

$= 2x \cdot 5x^2 + 2x \cdot 3 - 4 \cdot 5x^2 - 4 \cdot 3$

$= 10x^3 + 6x - 20x^2 - 12$

$= 10x^3 - 20x^2 + 6x - 12$

59.

$(8x^2 + 4x)(2x + 9)$

$= (8x^2 + 4x) \cdot 2x + (8x^2 + 4x) \cdot 9$

$= 8x^2 \cdot 2x + 4x \cdot 2x + 8x^2 \cdot 9 + 4x \cdot 9$

$= 16x^3 + 8x^2 + 72x^2 + 36x$

$= 16x^3 + 80x^2 + 36x$

61.

$$\left(2x^2 + 5\right)\left(2x^2 - 5\right)$$

$= \left(2x^2 + 5\right) \cdot 2x^2 - \left(2x^2 + 5\right) \cdot 5$

$= 2x^2\left(2x^2 + 5\right) - 5\left(2x^2 + 5\right)$

$= 2x^2 \cdot 2x^2 + 2x^2 \cdot 5 - 5 \cdot 2x^2 - 5 \cdot 5$

$= 4x^4 + 10x^2 - 10x^2 - 25$

$= 4x^4 - 25$

63.

$$\left(6x^2 + 1\right)\left(6x^2 + 1\right)$$

$= \left(6x^2 + 1\right) \cdot 6x^2 + \left(6x^2 + 1\right) \cdot 1$

$= 6x^2 \cdot 6x^2 + 1 \cdot 6x^2 + 6x^2 \cdot 1 + 1 \cdot 1$

$= 36x^4 + 6x^2 + 6x^2 + 1$

$= 36x^4 + 12x^2 + 1$

65. Horizontal Method

$(3x + 2)(5x^2 - 7x + 4)$

$= (3x + 2) \cdot 5x^2 - (3x + 2)7x + (3x + 2) \cdot 4$

$= 5x^2(3x + 2) - 7x(3x + 2) + 4(3x + 2)$

$= 15x^3 + 10x^2 - 21x^2 - 14x + 12x + 8$

$= 15x^3 - 11x^2 - 2x + 8$

65. Vertical Method

$$\begin{array}{r} 5x^2 - 7x + 4 \\ \underline{3x + 2} \\ 10x^2 - 14x + 8 \end{array}$$ Multiply by 2.

$\underline{15x^3 - 21x^2 + 12x}$ Multiply by $3x$.

$15x^3 - 11x^2 - 2x + 8$ Add.

67.

$$\begin{array}{r} x^2 - 2x - 7 \\ \underline{2x - 1} \\ -x^2 + 2x + 7 \end{array}$$ Multiply by -1

$\underline{2x^3 - 4x^2 - 14x}$ Multiply by $2x$

$2x^3 - 5x^2 - 12x + 7$ Add.

69.

$$x^2 + 2x + 4$$
$$\underline{x - 2}$$
$-2x^2 - 4x - 8 \qquad$ Multiply by -2
$\underline{x^3 + 2x^2 + 4x } \qquad$ Multiply by x
$x^3 - 8 \qquad$ Add.

The answer is $x^3 - 8$.

71.

$$-2x^2 - 6x + 7$$
$$\underline{3x + 4}$$
$-8x^2 - 24x + 28 \qquad$ Multiply by 4
$\underline{-6x^3 - 18x^2 + 21x } \qquad$ Multiply by $3x$
$-6x^3 - 26x^2 - 3x + 28 \qquad$ Add.

73.

$$2x^2 + 4x - 9$$
$$\underline{x^2 + 1}$$
$2x^4 + 4x^3 - 9x^2 \qquad$ Multiply by x^2
$\underline{2x^2 + 4x - 9} \qquad$ Multiply by 1
$2x^4 + 4x^3 - 7x^2 + 4x - 9 \qquad$ Add.

75.

$$x^2 - 2x + 3$$
$$\underline{4x^2 - 3x}$$
$4x^4 - 8x^3 + 12x^2 \qquad$ Multiply by $4x^2$
$\underline{- 3x^3 + 6x^2 - 9x} \qquad$ Multiply by $-3x$
$4x^4 - 11x^3 + 18x^2 - 9x \qquad$ Add

77.

$$2x^3 - 7x + 3$$
$$\underline{5x + 2}$$
$10x^4 - 35x^2 + 15x \qquad$ Multiply by $5x$
$\underline{+ 4x^3 - 14x + 6} \qquad$ Multiply by 2
$10x^4 + 4x^3 - 35x^2 + x + 6 \qquad$ Add.

79.

$$4x^3 - 3x^2 + 2$$
$$\underline{4x - 1}$$
$16x^4 - 12x^3 + 8x$
$\underline{- 4x^3 + 3x^2 - 2}$
$16x^4 - 16x^3 + 3x^2 + 8x - 2$

81.

$$2x^2 - 8xy + y^2$$
$$\underline{6x - 5y}$$
$12x^3 - 48x^2y + 6xy^2 \qquad$ Multiply by $6x$
$\underline{- 10x^2y + 40xy^2 - 5y^3} \qquad$ Multiply by $-5y$
$12x^3 - 58x^2y + 46xy^2 - 5y^3$

83.

$$x^2 - 4xy + 16y^2$$
$$\underline{x + 4y}$$
$x^3 - 4x^2y + 16xy^2 \qquad$ Multiply by x
$\underline{4x^2y - 16xy^2 + 64y^3} \qquad$ Multiply by $4y$
$x^3 + 64y^3$

The answer is $x^3 + 64y^3$.

85.

$$x^2 + 6xy + 3y^2$$
$$\underline{4x + 7y}$$
$7x^2y + 42xy^2 + 21y^3 \qquad$ Multiply by $7y$
$\underline{4x^3 + 24x^2y + 12xy^2 } \qquad$ Multiply by $4x$
$4x^3 + 31x^2y + 54xy^2 + 21y^3 \qquad$ Add.

87.

$$3x^2 - 4xy - 4y^2$$
$$\underline{x - 2y}$$
$-6x^2y + 8xy^2 + 8y^3 \qquad$ Multiply by $-2y$
$\underline{3x^3 - 4x^2y - 4xy^2 } \qquad$ Multiply by x
$3x^3 - 10x^2y + 4xy^2 + 8y^3$

89.

$$5x^3 - 7x^2 + 3x - 2$$
$$2x + 4$$
$$10x^4 - 14x^3 + 6x^2 - 4x$$
$$20x^3 - 28x^2 + 12x - 8$$
$$10x^4 + 6x^3 - 22x^2 + 8x - 8$$

91.

$$2x^3 + 6x^2 - 7x - 1$$
$$x - 3$$
$$2x^4 + 6x^3 - 7x^2 - x$$
$$- 6x^3 - 18x^2 + 21x + 3$$
$$2x^4 \qquad - 25x^2 + 20x + 3$$

93.

$$x^3 - 6x^2 - 6x + 5$$
$$3x - 2$$
$$3x^4 - 18x^3 - 18x^2 + 15x$$
$$- 2x^3 + 12x^2 + 12x - 10$$
$$3x^4 - 20x^3 - 6x^2 + 27x - 10$$

95.

$$2x^4 + 3x^2 - 2x - 5$$
$$4x + 2$$
$$8x^5 \qquad + 12x^3 - 8x^2 - 20x$$
$$4x^4 \qquad + 6x^2 - 4x - 10$$
$$8x^5 + 4x^4 + 12x^3 - 2x^2 - 24x - 10$$

97.

$$x^2 - 7x + 4$$
$$x^2 + 3x + 5$$
$$x^4 - 7x^3 + 4x^2$$ Multiply by x^2
$$+ 3x^3 - 21x^2 + 12x$$ Multiply by $3x$
$$+ 5x^2 - 35x + 20$$ Multiply by 5
$$x^4 - 4x^3 - 12x^2 - 23x + 20$$

99.

$$2x^2 + 4x - 5$$
$$3x^2 + 5x - 1$$
$$6x^4 + 12x^3 - 15x^2$$
$$10x^3 + 20x^2 - 25x$$
$$- 2x^2 - 4x + 5$$
$$6x^4 + 22x^3 + 3x^2 - 29x + 5$$

101.

$$x^2 + x + 3$$
$$-3x^2 - 6x + 4$$
$$-3x^4 - 3x^3 - 9x^2$$
$$- 6x^3 - 6x^2 - 18x$$
$$4x^2 + 4x + 12$$
$$-3x^4 - 9x^3 - 11x^2 - 14x + 12$$

Exercises 3.5

1.

$$F \quad O \quad I \quad L$$
$$= x \cdot x + x \cdot 2 + 5 \cdot x + 5 \cdot 2$$
$$= x^2 + 2x + 5x + 10$$
$$= x^2 + 7x + 10$$

3.

$$(x - 9)(x + 1)$$

$$F \quad O \quad I \quad L$$
$$= x \cdot x + x \cdot 1 - 9 \cdot x - 9 \cdot 1$$
$$= x^2 + x - 9x - 9$$
$$= x^2 - 8x - 9$$

5.

$$\begin{array}{cccc} F & O & I & L \end{array}$$
$$= 5x \cdot x + 5x(-2) + 6 \cdot x + 6(-2)$$
$$= 5x^2 - 10x + 6x - 12$$
$$= 5x^2 - 4x - 12$$

7.

$$\begin{array}{cccc} F & O & I & L \end{array}$$
$$= 8x \cdot 3x + 8x \cdot 1 + 4 \cdot 3x + 4 \cdot 1$$
$$= 24x^2 + 8x + 12x + 4$$
$$= 24x^2 + 20x + 4$$

9.

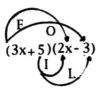

$$\begin{array}{cccc} F & O & I & L \end{array}$$
$$= 3x \cdot 2x + 3x(-3) + 5 \cdot 2x + 5(-3)$$
$$= 6x^2 - 9x + 10x - 15$$
$$= 6x^2 + x - 15$$

11.

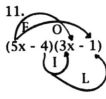

$$\begin{array}{cccc} F & O & I & L \end{array}$$
$$= 5x \cdot 3x + 5x(-1) - 4 \cdot 3x - 4(-1)$$
$$= 15x^2 - 5x - 12x + 4 \quad = 15x^2 - 17x + 4$$

13.

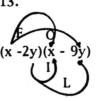

$$\begin{array}{cccc} F & O & I & L \end{array}$$
$$= x \cdot x + x(-9y) - 2y \cdot x - 2y(-9y)$$
$$= x^2 - 9xy - 2xy + 18y^2 = x^2 - 11xy + 18y^2$$

15.

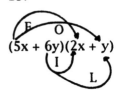

$$\begin{array}{cccc} F & O & I & L \end{array}$$
$$= 5x \cdot 2x + 5x \cdot y + 6y \cdot 2x + 6y \cdot y$$
$$= 10x^2 + 5xy + 12xy + 6y^2$$
$$= 10x^2 + 17xy + 6y^2$$

17.

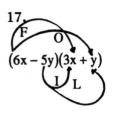

$$\begin{array}{cccc} F & O & I & L \end{array}$$
$$= 6x \cdot 3x + 6x \cdot y - 5y \cdot 3x - 5y \cdot y$$
$$= 18x^2 + 6xy - 15xy - 5y^2$$
$$= 18x^2 - 9xy - 5y^2$$

19.

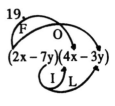

$$\begin{array}{cccc} F & O & I & L \end{array}$$
$$= 2x \cdot 4x + 2x(-3y) - 7y \cdot 4x - 7y(-3y)$$
$$= 8x^2 - 6xy - 28xy + 21y^2$$
$$= 8x^2 - 34xy + 21y^2$$

21.

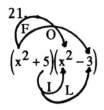

$$(x^2+5)(x^2-3)$$

F   O   I   L

$$= x^2 \cdot x^2 + x^2(-3) + 5 \cdot x^2 + 5(-3)$$
$$= x^4 - 3x^2 + 5x^2 - 15 = x^4 + 2x^2 - 15$$

23.

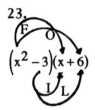

$$(x^2-3)(x+6)$$

F   O   I   L

$$= x^2 \cdot x + x^2 \cdot 6 - 3 \cdot x - 3 \cdot 6$$
$$= x^3 + 6x^2 - 3x - 18$$

25.

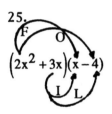

$$(2x^2+3x)(x-4)$$

F   O   I   L

$$= 2x^2 \cdot x + 2x^2(-4) + 3x \cdot x + 3x(-4)$$
$$= 2x^3 - 8x^2 + 3x^2 - 12x = 2x^3 - 5x^2 - 12x$$

27.

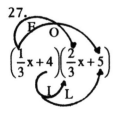

$$\left(\tfrac{1}{3}x+4\right)\left(\tfrac{2}{3}x+5\right)$$

F   O   I   L

$$= \tfrac{1}{3}x \cdot \tfrac{2}{3}x + \tfrac{1}{3}x \cdot 5 + 4 \cdot \tfrac{2}{3}x + 4 \cdot 5$$

$$= \tfrac{2}{9}x^2 + \tfrac{5}{3}x + \tfrac{8}{3}x + 20 = \tfrac{2}{9}x^2 + \tfrac{13}{3}x + 20$$

29.

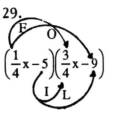

$$\left(\tfrac{1}{4}x-5\right)\left(\tfrac{3}{4}x-9\right)$$

F   O   I   L

$$= \tfrac{1}{4}x \cdot \tfrac{3}{4}x + \tfrac{1}{4}x(-9) - 5 \cdot \tfrac{3}{4}x - 5(-9)$$

$$= \tfrac{3}{16}x^2 - \tfrac{24}{4}x + 45 = \tfrac{3}{16}x^2 - 6x + 45$$

31.

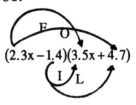

$$(2.3x-1.4)(3.5x+4.7)$$

F   O   I   L

$$= 2.3x(3.5x) + 2.3x(4.7) - 1.4(3.5x) - 1.4(4.7)$$
$$= 8.05x^2 + 10.81x - 4.9x - 6.58$$
$$= 8.05x^2 + 5.91x - 6.58$$

33.

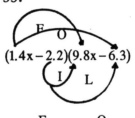

$$(1.4x-2.2)(9.8x-6.3)$$

F   O   I   L

$$= 1.4x(9.8x) + 1.4x(-6.3) - 2.2(9.8x) - 2.2(-6.3)$$
$$= 13.72x^2 - 8.82x - 21.56x + 13.86$$
$$= 13.72x^2 - 30.38x + 13.86$$

35.
$$(x+5)(x-5)$$

This is the product of two conjugates. The product is the difference of the squares of the first and last terms. Thus,

$$(x+5)(x-5) = x^2 - 5^2$$
$$= x^2 - 25$$

37.
$(8 - x)(8 + x)$

Refer to the comment on exercise 35.
$$(8 - x)(8 + x) = 8^2 - x^2$$
$$= 64 - x^2$$

39.
$(2x - 6)(2x + 6)$

Refer to the comment on exercise 35.
$$(2x - 6)(2x + 6) = (2x)^2 - 6^2$$
$$= 4x^2 - 36$$

41.
$(3x + 5)(3x - 5)$

Refer to the comment on exercise 35.
$$(3x + 5)(3x - 5) = (3x)^2 - 5^2$$
$$= 9x^2 - 25$$

43.
$(9x - 1)(9x + 1)$

Refer to the comment on exercise 35.
$$(9x + 1)(9x - 1) = (9x)^2 - 1^2$$
$$= 81x^2 - 1$$

45.
$(x - 3y)(x + 3y)$

Refer to the comment on exercise 35.
$$(x - 3y)(x + 3y) = (x)^2 - (3y)^2$$
$$= x^2 - 9y^2$$

47.
$(x + y)(x - y)$

Refer to the comment on exercise 35.
$$(x + y)(x - y) = x^2 - y^2$$

49.
$(2x + y)(2x - y)$

Refer to the comment on exercise 35.
$$(2x + y)(2x - y) = (2x)^2 - (y)^2$$
$$= 4x^2 - y^2$$

51.
$(4x - 3y)(4x + 3y)$

Refer to the comment on exercise 35.
$$(4x - 3y)(4x + 3y) = (4x)^2 - (3y)^2$$
$$= 16x^2 - 9y^2$$

53.
$(9x - 2y)(9x + 2y)$

Refer to the comment on exercise 35.
$$(9x - 2y)(9x + 2y) = (9x)^2 - (2y)^2$$
$$= 81x^2 - 4y^2$$

55.
$$\left(\frac{1}{2}x + 5\right)\left(\frac{1}{2}x - 5\right)$$

Refer to the comment on exercise 35.
$$\left(\frac{1}{2}x + 5\right)\left(\frac{1}{2}x - 5\right) = \left(\frac{1}{2}x\right)^2 - 5^2$$

$$= \frac{1}{4}x^2 - 25$$

57.
$(1 - 5x)(1 + 5x)$

Refer to the comment on exercise 35.
$$(1 - 5x)(1 + 5x) = 1^2 - (5x)^2$$
$$= 1 - 25x^2$$

59.
$(3x^2 - 1)(3x^2 + 1)$

Refer to the comment on exercise 35.
$$(3x^2 - 1)(3x^2 + 1) = (3x^2)^2 - 1^2 = 9x^4 - 1$$

61.
$[(x + 2) + 5][(x + 2) - 5]$

Refer to the comment on exercise 35.
$$[(x + 2)][(x + 2) - 5] = (x + 2)^2 - 5^2$$
$$= (x^2 + 4x + 4) - 25$$
$$= x^2 + 4x - 21$$

63.

$[(x - 4) - 8][(x - 4) + 8]$

Refer to the comment on exercise 35.

$[(x - 4) - 8][(x - 4) + 8] = (x - 4)^2 - 8^2$
$= (x^2 - 8x + 16) - 64$
$= x^2 - 8x - 48$

65.

$[(2y + 1) + x][(2y + 1) - x]$

Refer to the comment on exercise 35.

$[(2y + 1) + x][(2y + 1) - x] = (2y + 1)^2 - x^2$
$= 4y^2 + 4y + 1 - x^2$

67.

$\left(\dfrac{2}{3}x - \dfrac{3}{4}y\right)\left(\dfrac{2}{3}x + \dfrac{3}{4}y\right)$

Refer to the comment on exercise 35.

$\left(\dfrac{2}{3}x - \dfrac{3}{4}y\right)\left(\dfrac{2}{3}x + \dfrac{3}{4}y\right) = \left(\dfrac{2}{3}x\right)^2 - \left(\dfrac{3}{4}y\right)^2$

$= \dfrac{4}{9}x^2 - \dfrac{9}{16}y^2$

69.

$(1.2x + 7)(1.2x - 7)$

Refer to the comment on exercise 35.

$(1.2x + 7)(1.2x - 7) = (1.2x)^2 - 7^2$

$= 1.44x^2 - 49$

71.

$(6.5x - 2.1y)(6.5x + 2.1y)$

Refer to the comment on exercise 35.

$(6.5x - 2.1y)(6.5x + 2.1y) = (6.5x)^{2\cdot} - (2.1y)^2$

$= 42.25x^{2\cdot} - 4.41y^2$

73.

$(x + 8)^2$

To square the binomial:
(1) Square the first term.
(2) Add twice the product of the two terms.

#73 continued

3) Square the last term.

$$(x + 8)^2 \overset{(1)}{=} x^2 + \overset{(2)}{2(8x)} + \overset{(3)}{8^2}$$

$= x^2 + 16x + 64$

75.

$(x - 1)^2$

Refer to the comment on exercise 73.

$$(x - 1)^2 \overset{(1)}{=} x^2 + \overset{(2)}{2(-1 \cdot x)} + \overset{(3)}{(-1)^2}$$

$= x^2 - 2x + 1$

77.

$(2x - 5)^2$

Refer to the comment on exercise 73.

$$(2x - 5)^2 \overset{(1)}{=} (2x)^2 + \overset{(2)}{2(-5 \cdot 2x)} + \overset{(3)}{(-5)^2}$$

$= 4x^2 - 20x + 25$

79.

$(3x + 7)^2$

Refer to the comment on exercise 73.

$$(3x + 7)^2 \overset{(1)}{=} (3x)^2 + \overset{(2)}{2(3x \cdot 7)} + \overset{(3)}{7^2}$$

$= 9x^2 + 42x + 49$

81.

$(2x + 10)^2$

Refer to the comment on exercise 73.

$$(2x + 10)^2 \overset{(1)}{=} (2x)^2 + \overset{(2)}{2(2x \cdot 10)} + \overset{(3)}{10^2}$$

$= 4x^2 + 40x + 100$

83.

$(x - 2y)^2$

Refer to the comment on exercise 73.

#83 continued

$$(x-2y)^2 \overset{(1)}{=} x^2 \overset{(2)}{+} 2(-2y \cdot x) \overset{(3)}{+} (-2y)^2$$

$$= x^2 - 4xy + 4y^2$$

85.
$(x + 7y)^2$

Refer to the comment on exercise 73.

$$(x+7y)^2 \overset{(1)}{=} x^2 \overset{(2)}{+} 2(x \cdot 7y) \overset{(3)}{+} (7y)^2$$

$$= x^2 + 14xy + 49y^2$$

87.
$(4x - y)^2$

Refer to the comment on exercise 73

$$(4x-y)^2 \overset{(1)}{=} (4x)^2 \overset{(2)}{+} 2(-y \cdot 4x) \overset{(3)}{+} (-y)^2$$

$$= 16x^2 - 8xy + y^2$$

89.
$(3x + 2y)^2$

Refer to the comment on exercise 73.

$$(3x+2y)^2 \overset{(1)}{=} (3x)^2 \overset{(2)}{+} 2(3x \cdot 2y) \overset{(3)}{+} (2y)^2$$

$$= 9x^2 + 12xy + 4y^2$$

91.
$$\left(\frac{1}{5}x - \frac{3}{4}y\right)^2$$

Refer to the comment on exercise 73.

$$\left(\frac{1}{5}x - \frac{3}{4}y\right)^2 \overset{(1)}{=} \left(\frac{1}{5}x\right)^2 \overset{(2)}{+} 2\left(-\frac{3}{4}y \cdot \frac{1}{5}x\right) \overset{(3)}{+} \left(-\frac{3}{4}y\right)^2$$

$$= \frac{1}{25}x^2 - \frac{3}{10}xy + \frac{9}{16}y^2$$

93.
$$\left(\frac{1}{2}x + \frac{2}{3}y\right)^2$$

Refer to the comment on exercise 73.

$$\left(\frac{1}{2}x + \frac{2}{3}y\right)^2 \overset{(1)}{=} \left(\frac{1}{2}x\right)^2 \overset{(2)}{+} 2\left(\frac{1}{2}x \cdot \frac{2}{3}y\right) \overset{(3)}{+} \left(\frac{2}{3}y\right)^2$$

$$= \frac{1}{4}x^2 + \frac{2}{3}xy + \frac{4}{9}y^2$$

95.
$(1.2x - 3)^2$

Refer to the comment on exercise 73.

$$(1.2x-3)^2 \overset{(1)}{=} (1.2x)^2 \overset{(2)}{+} 2(-3 \cdot 1.2x) \overset{(3)}{+} (-3)^2$$

$$= 1.44x^2 - 7.2x + 9$$

97.
$(7.6x + 3.4y)^2$

Refer to the comment on exercise 73.

$$(7.6x+3.4y)^2 \overset{(1)}{=} (7.6x)^2 \overset{(2)}{+} 2(7.6x \cdot 3.4y) \overset{(3)}{+} (3.4y)^2$$

$$= 57.76x^2 + 51.68xy + 11.56y^2$$

Exercises 3.6

1.
$$\frac{7x+35}{7} = \frac{7x}{7} + \frac{35}{7}$$

$$= x + 5 \qquad \text{Reduce}$$

3.
$$\frac{5a^2 + 20ab - 40}{5} = \frac{5a^2}{5} + \frac{20ab}{5} - \frac{40}{5}$$

$$= a^2 + 4ab - 8$$

5.

$$\frac{9p^3 - 18p - 27}{-9} = \frac{9p^3}{-9} + \frac{-18p}{-9} + \frac{-27}{-9}$$

$$= -p^3 + 2p + 3$$

7.

$$\frac{4m^4 - 11m^2 + m}{m} = \frac{4m^4}{m} - \frac{11m^2}{m} + \frac{m}{m}$$

$$= 4m^3 - 11m + 1$$

Simplify by subtracting exponents.

9.

$$\frac{x^5 + 9x^4 - 2x^3}{x^2} = \frac{x^5}{x^2} + \frac{9x^4}{x^2} - \frac{2x^3}{x^2}$$

$$= x^3 + 9x^2 - 2x$$

Simplify by subtracting exponents.

11.

$$\frac{16n^3 - 4n^3p + n^2p^2}{n^2} = \frac{16n^3}{n^2} - \frac{4n^3p}{n^2} + \frac{n^2p^2}{n^2}$$

$$= 16n - 4np + p^2$$

Simplify by subtracting exponents.

13.

$$\frac{6r^4 - 8r^3 + 2r^2}{2r^2} = \frac{6r^4}{2r^2} - \frac{8r^3}{2r^2} + \frac{2r^2}{2r^2}$$

$$= 3r^2 - 4r + 1$$

Simplify by dividing the coefficient and subtracting the exponents.

15.

$$\frac{-20y^5 + 28y^3 - 4y^2}{4y} = \frac{-20y^5}{4y} + \frac{28y^3}{4y} - \frac{4y^2}{4y}$$

$$= -5y^4 + 7y^2 - y$$

#15 continued

Simplify by dividing the coefficients and subtracting the exponents.

17.

$$\frac{15k^8 + 3k^7 - 30k^5 + 36k^4}{3k^3}$$

$$= \frac{15k^8}{3k^3} + \frac{3k^7}{3k^3} - \frac{30k^5}{3k^3} + \frac{36k^4}{3k^3}$$

$$= 5k^5 + k^4 - 10k^2 + 12k$$

19.

$$\frac{8q^5t^2 - 32q^4t^3 - 40q^3t^4}{8qt^2}$$

$$= \frac{8q^5t^2}{8qt^2} - \frac{32q^4t^3}{8qt^2} - \frac{40q^3t^4}{8qt^2}$$

$$= q^4 - 4q^3t - 5q^2t^2$$

Simplify by dividing the coefficient and subtracting the exponents.

21.

$$\frac{22x^3 - 55y^2}{11x^3y^2}$$

$$= \frac{22x^3}{11x^3y^2} - \frac{55y^2}{11x^3y^2}$$

$$= \frac{2}{y^2} - \frac{5}{x^3}$$

Simplify by dividing the coefficient and subtracting the exponents.

23.

$$\frac{6p^3 - 15p^2q + 8q^4}{18p^2q}$$

$$= \frac{6p^3}{18p^2q} - \frac{15p^2q}{18p^2q} + \frac{8q^4}{18p^2q}$$

$$= \frac{p}{3q} - \frac{5}{6} + \frac{4q^3}{9p^2}$$

Simplify by dividing the coefficients and subtracting the exponents.

25.

$$\frac{5b^7 - 45b^6 - 30b^5 - 12b^4}{15b^4}$$

$$= \frac{5b^7}{15b^4} - \frac{45b^6}{15b^4} - \frac{30b^5}{15b^4} - \frac{12b^4}{15b^4}$$

$$= \frac{b^3}{3} - 3b^2 - 2b - \frac{4}{5}$$

Simplify by dividing the coefficients and subtracting the exponents.

27.

$$\frac{42r^5 + 36r^4t^2 + 8r^3t^3}{6r^2}$$

$$= \frac{42r^5}{6r^2} + \frac{36r^4t^2}{6r^2} + \frac{8r^3t^3}{6r^2}$$

$$= 7r^3 + 6r^2t^2 + \frac{4rt^3}{3}$$

Simplify by dividing the coefficients and subtracting the exponents.

29.

$$\frac{24x^3m + 27m^4}{12xm} = \frac{24x^3m}{12xm} + \frac{27m^4}{12xm}$$

$$= 2x^2 + \frac{9m^3}{4x}$$

Simplify by dividing the coefficients and subtracting the exponents.

31.

$$\frac{26y^8p^3 - 39y^6p^5 - 52y^4p^7}{-13y^2p^3}$$

$$= \frac{26y^8p^3}{-13y^2p^3} + \frac{-39y^6p^5}{-13y^2p^3} + \frac{-52y^4p^7}{-13y^2p^3}$$

$$= -2y^6 + 3y^4p^2 + 4y^2p^4$$

Simplify by dividing the coefficients and subtracting the exponents.

33.

$$\frac{-45x^2 + 30xy - y^4}{-5xy^3}$$

$$= \frac{-45x^2}{-5xy^3} + \frac{30xy}{-5xy^3} + \frac{-y^4}{-5xy^3}$$

$$= \frac{9x}{y^3} - \frac{6}{y^2} + \frac{y}{5x}$$

Simplify by dividing the coefficients and subtracting the exponents.

35.

$$\frac{9a^2b^5c^4 + 54a^3b^4c^2 - 6a^4b^3c}{18ab^3c^2}$$

$$= \frac{9a^2b^5c^4}{18ab^3c^2} + \frac{54a^3b^4c^2}{18ab^3c^2} - \frac{6a^4b^3c}{18ab^3c^2}$$

$$= \frac{ab^2c^2}{2} + 3a^2b - \frac{a^3}{3c}$$

#35 continued

Simplify by dividing the coefficients and subtracting the exponents.

37.

$$\frac{4mn^8p^4+10m^2n^6p^3+m^3n^4p^2}{2m^2n^4p^3}$$

$$=\frac{4mn^8p^4}{2m^2n^4p^3}+\frac{10m^2n^6p^3}{2m^2n^4p^3}+\frac{m^3n^4p^2}{2m^2n^4p^3}$$

$$=\frac{2n^4p}{m}+5n^2+\frac{m}{2p}$$

Simplify by dividing the coefficients and subtracting the exponents.

39.

$$\frac{12k^5r^2-2r^4q^3+64k^2q^8}{-8k^3rq^6}$$

$$=\frac{12k^5r^2}{-8k^3rq^6}-\frac{2r^4q^3}{-8k^3rq^6}+\frac{64k^2q^8}{-8k^3rq^6}$$

$$=-\frac{3k^2r}{2q^6}+\frac{r^3}{4k^3q^3}-\frac{8q^2}{kr}$$

Simplify by dividing the coefficients and subtracting the exponents.

41.

$$\frac{3x^2+13x-6}{x+5}$$

$$x+5\overline{)3x^2+13x-6} \quad\quad 3x-2$$

$$\phantom{x+5\overline{)}}{}^-\underline{3x^2\mp15x}\quad\text{Change signs and add.}$$

$$-2x-6$$

$${}^\pm\underline{2x\pm10}\quad\text{Change signs and add.}$$

$$4$$

Answer: $3x-2+\dfrac{4}{x+5}$

43.

$$\frac{5x^2-18x-9}{x-4}$$

$$=x-4\overline{)5x^2-18x-9}\quad\quad 5x+2$$

$$\phantom{=x-4\overline{)}}{}^{--}\underline{5x^2\pm20x}\quad\text{Change signs and add.}$$

$$2x-9$$

$${}^-\underline{2x\pm8}\quad\text{Change signs and add.}$$

$$-1$$

Answer: $5x+2-\dfrac{1}{x-4}$

45.

$$\frac{5x^2+9x-18}{x+3}$$

$$x+3\overline{)5x^2+9x-18}\quad\quad 5x-6$$

$$\phantom{x+3\overline{)}}{}^-\underline{5x^2\mp15x}\quad\text{Change signs and add.}$$

$$-6x-18$$

$${}^\pm\underline{6x\pm18}$$

Answer: $5x-6$

47.

$$\frac{2x^2-117}{x-8}$$

$$x-8\overline{)2x^2+0x-117}\quad\quad 2x+16$$

$$\phantom{x-8\overline{)}}{}^-\underline{2x^2\pm16x}\quad\text{Change signs and add.}$$

$$16x-117$$

$${}^-\underline{16x\pm128}\quad\text{Change signs and add.}$$

$$11$$

Answer: $2x+16+\dfrac{11}{x-8}$

49.

$$\frac{10x^2 + 13x - 1}{2x + 3}$$

$$
\begin{array}{r}
5x - 1 \\
2x+3\overline{\smash{\big)}10x^2 + 13x - 1}
\end{array}
$$

$\underline{^-10x^2 \mp 15x}$ Change signs and add.

$-2x - 1$

$\underline{\pm 2x \mp 3}$ Change signs and add.

2

Answer: $5x - 1 + \dfrac{2}{2x + 3}$

51.

$$\frac{20x^2 - 3x - 9}{5x - 2}$$

$$
\begin{array}{r}
4x + 1 \\
5x-2\overline{\smash{\big)}20x^2 - 3x - 9}
\end{array}
$$

$\underline{^-20x^2 \pm 8x}$ Change signs and add.

$5x - 9$

$\underline{^-5x \pm 2}$ Change signs and add.

-7

Answer: $4x + 1 - \dfrac{7}{5x - 2}$

53.

$$\frac{8x^2 + 22x + 5}{4x + 1}$$

$$
\begin{array}{r}
2x + 5 \\
4x+1\overline{\smash{\big)}8x^2 + 22x + 5}
\end{array}
$$

$\underline{^-8x^2 \mp 2x}$ Change signs and add.

$20x + 5$

$\underline{^-20x \mp 5}$ Change signs and add.

0

Answer: $2x + 5$

55.

$$\frac{18x^2 + 6}{3x - 1}$$

#55 continued

$$
\begin{array}{r}
6x + 2 \\
3x-1\overline{\smash{\big)}18x^2 + 0x + 6}
\end{array}
$$

$\underline{^-18x^2 \pm 6x}$ Change signs and add.

$6x + 6$

$\underline{^-6x \pm 2}$ Change signs and add.

8

Answer: $6x + 2 + \dfrac{8}{3x - 1}$

57.

$$\frac{x^3 + 2x^2 - 13x + 8}{x - 2}$$

$$
\begin{array}{r}
x^2 + 4x - 5 \\
x-2\overline{\smash{\big)}x^3 + 2x^2 - 13x + 8}
\end{array}
$$

$\underline{^-x^3 \pm 2x^2}$ Change signs and add.

$4x^2 - 13x$

$\underline{^-4x^2 \pm 8x}$ Change signs and add.

$-5x + 8$

$\underline{\pm 5x \mp 10}$ Change signs and add.

-2

Answer: $x^2 + 4x - 5 - \dfrac{2}{x - 2}$

59.

$$\frac{2x^3 + 13x^2 + 21x + 7}{x + 4}$$

$$
\begin{array}{r}
2x^2 + 5x + 1 \\
x+4\overline{\smash{\big)}2x^3 + 13x^2 + 21x + 7}
\end{array}
$$

$\underline{^-2x^3 \mp 8x^2}$ Change signs.

$5x^2 + 21x$

$\underline{^-5x^2 \mp 20x}$ Change signs.

$x + 7$

$\underline{^-x \mp 4}$ Change signs.

3

Answer: $2x^2 + 5x + 1 + \dfrac{3}{x + 4}$

61.

$$\frac{4x^3 - 33x + 4}{x + 3}$$

$$\begin{array}{r} 4x^2 - 12x + 3 \\ x+3\overline{\smash{\big)}\,4x^3 + 0x^2 - 33x + 4} \end{array}$$

$\underline{\pm 4x^3 \mp 12x^2}$ Change signs.

$-12x^2 - 33x$

$\underline{\pm 12x^2 \pm 36x}$ Change signs.

$3x + 4$

$\overline{}\underline{3x \mp 9}$ Change signs.

-5

Answer: $4x^2 - 12x + 3 - \dfrac{5}{x + 3}$

63.

$$\frac{x^3 - 8x^2 + 8x - 7}{x - 7}$$

$$\begin{array}{r} x^2 - x + 1 \\ x-7\overline{\smash{\big)}\,x^3 - 8x^2 + 8x - 7} \end{array}$$

$\overline{}\underline{x^3 \pm 7x^2}$ Change signs and add.

$-x^2 + 8x$

$\underline{\pm\ x^2 \mp 7x}$ Change signs and add.

$x - 7$

$\overline{}\underline{x \pm 7}$ Change signs and add.

0

Answer: $x^2 - x + 1$

65.

$$\frac{12x^3 + 5x^2 - 34x + 10}{4x - 1}$$

#65 continued.

$$\begin{array}{r} 3x^2 + 2x - 8 \\ 4x-1\overline{\smash{\big)}\,12x^3 + 5x^2 - 34x + 10} \end{array}$$

$\overline{}\underline{12x^3 \pm 3x^2}$ Change signs.

$8x^2 - 34x$

$\overline{}\underline{8x^2 \pm 2x}$ Change signs.

$-32x + 10$

$\underline{\pm 32x \mp 8}$ Change signs.

2

Answer: $3x^2 + 2x - 8 + \dfrac{2}{4x - 1}$

67.

$$\frac{14x^3 + 37x^2 + 9x + 6}{2x + 5}$$

$$\begin{array}{r} 7x^2 + x + 2 \\ 2x+5\overline{\smash{\big)}\,14x^3 + 37x^2 + 9x + 6} \end{array}$$

$\overline{}\underline{14x^3 \mp 35x^2}$ Change signs.

$2x^2 + 9x$

$\overline{}\underline{2x^2 \mp 5x}$ Change signs.

$4x + 6$

$\overline{}\underline{4x \mp 10}$ Change signs.

-4

Answer: $7x^2 + x + 2 - \dfrac{4}{2x + 5}$

69.

$$\frac{5x^3 + 21x^2 - 16}{5x - 4}$$

$$\begin{array}{r} x^2 + 5x + 4 \\ 5x-4\overline{\smash{\big)}\,5x^3 + 21x^2 + 0x - 16} \end{array}$$

$\overline{}\underline{5x^3 \pm 4x^2}$ Change signs.

$25x^2 + 0x$

$\underline{-25x^2 \pm 20x}$ Change signs.

$20x - 16$

$\underline{-20x \pm 16}$ Change signs.

0

#69 continued

Answer: $x^2 + 5x + 4$

71.

$$\frac{27x^3 + 2}{3x - 2}$$

$$3x-2\overline{)27x^3 + 0x^2 + 0x + 2}$$
$$\quad 9x^2 + 6x + 4$$

$$\underline{^-27x^3 \pm 18x^2}\qquad \text{Change signs.}$$
$$18x^2 + 0x$$
$$\underline{^-18x^2 \mp 12x}\qquad \text{Change signs}$$
$$12x + 2$$
$$\underline{^-12x \pm 8}\quad \text{Change signs}$$
$$10$$

Answer: $9x^2 + 6x + 4 + \dfrac{10}{3x - 2}$

73.

$$\frac{2x^3 + 7x^2 - 11x + 10}{x^2 + 5x + 2}$$

$$x^2+5x+2\overline{)2x^3 + 7x^2 - 11x + 10}$$
$$\quad 2x - 3$$

$$\underline{-2x^3 \mp 10x^2 \mp 4x}\qquad \text{Change signs.}$$
$$-3x^2 - 15x + 10$$
$$\underline{\pm 3x^2 \pm 15x \pm 6}\quad \text{Change signs.}$$
$$16$$

Answer: $2x - 3 + \dfrac{16}{x^2 + 5x + 2}$

75.

$$\frac{4x^3 - 15x^2 - 27x - 3}{x^2 - 4x - 6}$$

$$x^2-4x-6\overline{)4x^3 - 15x^2 - 27x - 3}$$
$$\quad 4x + 1$$

$$\underline{^-4x^3 \pm 16x^2 \pm 24x}\quad \text{Change signs.}$$
$$x^2 - 3x - 3$$
$$\underline{^-x^2 \pm 4x \pm 6}\quad \text{Change signs.}$$
$$x + 3$$

Answer: $4x + 1 + \dfrac{x + 3}{x^2 - 4x - 6}$

77.

$$\frac{6x^3 + 7x^2 + 10x + 25}{2x^2 - x + 5}$$

$$2x^2-x+5\overline{)6x^3 + 7x^2 + 10x + 25}$$
$$\qquad 3x + 5$$

$$\underline{^-6x^3 \pm 3x^2 \mp 15x}\qquad \text{Change signs.}$$
$$10x^2 - 5x + 25$$
$$\underline{^-10x^2 \mp 5x \mp 25}\quad \text{Change signs.}$$
$$0$$

Answer: $3x + 5$

79.

$$\frac{8x^3 - 16x - 8}{4x^2 + 6x + 1}$$

$$4x^2+6x+1\overline{)8x^3 + 0x^2 - 16x - 8}$$
$$\qquad 2x - 3$$

$$\underline{^-8x^3 \mp 12x^2 \mp 2x}\qquad \text{Change signs.}$$
$$-12x^2 - 18x - 8$$
$$\underline{\pm 12x^2 \mp 18x \pm 3}\quad \text{Change signs.}$$
$$-5$$

Answer: $2x - 3 - \dfrac{5}{4x^2 + 6x + 1}$

81.

$$\frac{3x^4 + 19x^3 - 4x^2 - 24x + 13}{x^2 + 6x - 2}$$

$$x^2+6x-2\overline{)3x^4 + 19x^3 - 4x^2 - 24x + 13}$$
$$\qquad 3x^2 + x - 4$$

$$\underline{^-3x^4 \mp 18x^3 \pm 6x^2}\qquad \text{Change signs.}$$
$$x^3 + 2x^2 - 24x$$
$$\underline{^-x^3 \mp 6x^2 \pm 2x}\qquad \text{Change signs.}$$
$$-4x^2 - 22x + 13$$
$$\underline{^-4x^2 \pm 24x \mp 8}\quad \text{Change signs.}$$
$$2x + 5$$

Answer: $3x^2 + x - 4 + \dfrac{2x + 5}{x^2 + 6x - 2}$

83.

$$\frac{2x^4 - 19x^3 + 17x^2 + 4x - 10}{2x^2 - 3x + 1}$$

$$\begin{array}{r} x^2 - 8x - 4 \\ 2x^2 - 3x + 1 \overline{\smash{\big)}\ 2x^4 - 19x^3 + 17x^2 + 4x - 10} \end{array}$$

$$\underline{-2x^4 \pm 3x^3 \mp x^2} \qquad \text{Change.}$$
$$-16x^3 + 16x^2 + 4x$$
$$\underline{\pm 16x^3 \mp 24x^2 \pm 8x} \qquad \text{Change.}$$
$$- 8x^2 + 12x - 10$$
$$\underline{\pm 8x^2 \mp 12x \pm 4} \qquad \text{Change.}$$
$$- 6$$

Answer: $x^2 - 8x - 4 - \dfrac{6}{2x^2 - 3x + 1}$

85.

$$\frac{6x^4 + 2x^3 - 29x^2 - 2}{3x^2 - 5x - 3}$$

$$\begin{array}{r} 2x^2 + 4x - 1 \\ 3x^2 - 5x - 3 \overline{\smash{\big)}\ 6x^4 + 2x^3 - 29x^2 + 0x - 2} \end{array}$$

$$\underline{{}^-6x^4 \pm 10x^3 \pm 6x^2}$$
$$12x^3 - 23x^2 + 0x$$
$$\underline{{}^-12x^3 \pm 20x^2 \pm 12x}$$
$$- 3x^2 + 12x - 2$$
$$\underline{\pm 3x^2 \mp 5x \mp 3}$$
$$7x - 5$$

Answer: $2x^2 + 4x - 1 + \dfrac{7x - 5}{3x^2 - 5x - 3}$

87.

$$\frac{5x^4 + 19x^3 + 18x + 25}{x^2 + 5x + 4}$$

#87 continued

$$\begin{array}{r} 5x^2 - 6x + 10 \\ x^2 + 5x + 4 \overline{\smash{\big)}\ 5x^4 + 19x^3 + 0x^2 + 18x + 25} \end{array}$$

$$\underline{{}^-5x^4 \mp 25x^3 \mp 20x^2} \qquad \text{Change.}$$
$$- 6x^3 - 20x^2 + 18x$$
$$\underline{\pm 6x^3 \pm 30x^2 \pm 24x} \qquad \text{Change.}$$
$$10x^2 + 42x + 25$$
$$\underline{{}^-10x^2 \mp 50x \mp 40} \qquad \text{Change.}$$
$$- 8x - 15$$

Answer: $5x^2 - 6x + 10 + \dfrac{-8x - 15}{x^2 + 5x + 4}$

89.

$$\frac{5x^3 - 17x^2 + 26x - 3}{-5x + 2}$$

$$\begin{array}{r} -x^2 + 3x - 4 \\ -5x + 2 \overline{\smash{\big)}\ 5x^3 - 17x^2 + 26x - 3} \end{array}$$

$$\underline{{}^-5x^3 \pm 2x^2} \qquad \text{Change signs.}$$
$$-15x^2 + 26x$$
$$\underline{\pm 15x^2 \mp 6x} \qquad \text{Change signs.}$$
$$20x - 3$$
$$\underline{{}^-20x \pm 8} \quad \text{Change signs.}$$
$$5$$

Answer: $-x^2 + 3x - 4 + \dfrac{5}{2 - 5x}$

91.

$$\frac{8x^3 - 6x^2y - 31xy^2 + 9y^3}{4x - 9y}$$

#91 continued

$$\begin{array}{r} 2x^2 + 3xy - y^2 \\ 4x - 9y \overline{\smash{\big)}\ 8x^3 - 6x^2y - 31xy^2 + 9y^3} \end{array}$$

$-\underline{8x^3 \overset{+}{-} 18x^2y}$ Change.

$12x^2y - 31xy^2$

$-\underline{12x^2y \overset{+}{-} 27xy^2}$ Change.

$- 4xy^2 + 9y^3$

$\underline{\overset{+}{-} 4xy^2 \overset{-}{+} 9y^3}$ Change.

0

Answer: $2x^2 + 3xy - y^2$

#95 continued

$$\frac{-3x^4 - 8x^3 + 3x^2 + 10x - 4}{-3x^2 - 2x + 4}$$

$$\begin{array}{r} x^2 + 2x - 1 \\ -3x^2 - 2x + 4 \overline{\smash{\big)}\ -3x^4 - 8x^3 + 3x^2 + 10x - 4} \end{array}$$

$\underline{\overset{+}{-}3x^4 \overset{+}{-} 2x^3 \overset{-}{+} 4x^2}$ Change.

$-6x^3 - x^2 + 10x$

$\underline{\overset{+}{-}6x^3 \overset{+}{-} 4x^2 \overset{-}{+} 8x}$ Change.

$3x^2 + 2x - 4$

$\underline{-3x^2 \overset{-}{+} 2x \overset{+}{-} 4}$ Change.

0

Answer: $x^2 + 2x - 1$

93.

Change the numerator and denominator to polynomials in descending order and divide.

$$\frac{3x^4 - 14x^3 - 2x^2 + 19x}{x^2 - 5x + 2}$$

$$\begin{array}{r} 3x^2 + x - 3 \\ x^2 - 5x + 2 \overline{\smash{\big)}\ 3x^4 - 14x^3 - 2x^2 + 19x} \end{array}$$

$-\underline{3x^4 \overset{+}{-} 15x^3 \overset{-}{+} 6x^2}$

$x^3 - 8x^2 + 19x$

$-\underline{x^3 \overset{+}{-} 5x^2 \overset{-}{+} 2x}$

$-3x^2 - 17x$

$\underline{\overset{+}{-}3x^2 \overset{-}{+} 15x \overset{+}{-} 6}$

$2x + 6$

Answer: $3x^2 + x - 3 + \dfrac{2x + 6}{x^2 - 5x + 2}$

95.

Change the denominator to a polynomial in descending order and divide.

97.

$$\frac{2x^4 - 8x^3 + 13x^2 - 36x + 3}{x^2 + 5}$$

$$\begin{array}{r} 2x^2 - 8x + 3 \\ x^2 + 5 \overline{\smash{\big)}\ 2x^4 - 8x^3 + 13x^2 - 36x + 3} \end{array}$$

$-\underline{2x^4 \overset{-}{+} 10x^2}$ Change.

$-8x^3 + 3x^2 - 36x$

$\underline{\overset{+}{-}8x^3 \overset{+}{-} 40x}$ Change.

$3x^2 + 4x + 3$

$-\underline{3x^2 \overset{-}{+} 15}$ Change.

$4x - 12$

Answer: $2x^2 - 8x + 3 + \dfrac{4x - 12}{x^2 + 5}$

99.

$$\frac{12x^4 - 21x^3 - 26x^2 + 35x + 10}{3x^2 - 5}$$

128

#99 continued

$$4x^2 - 7x - 2$$
$$3x^2 - 5 \overline{)12x^4 - 21x^3 - 26x^2 + 35x + 10}$$

$$\underline{-12x^4 \qquad\quad \pm 20x^2} \qquad \text{Change.}$$
$$-21x^3 - 6x^2 + 35x$$

$$\underline{\pm 21x^3 \qquad\quad \mp 35x} \qquad \text{Change.}$$
$$-6x^2 \qquad + 10$$

$$\underline{\pm 6x^2 \qquad\quad \mp 10} \quad \text{Change.}$$
$$0$$

Answer: $4x^2 - 7x - 2$

Review Exercises
1.
$$-7^2 = -\left(7^2\right) = -49$$

3.
$$4^3 \cdot 4^2 = 64 \cdot 16 = 1024$$

5.
$$\frac{3^6}{3^2} = 3^{6-2} = 3^4 = 81$$

7.
$$k^5 \cdot k^8 = k^{5+8} = k^{13}$$

9.
$$\left(y^4\right)^6 = y^{4 \cdot 6} = y^{24}$$

11.
$$\left(2x^3y^5\right)^4 = 2^4 x^{12} y^{20}$$
$$= 16x^{12} y^{20}$$

13.
$$\frac{\left(x^2\right)^6}{\left(x^3\right)^3} = \frac{x^{12}}{x^9} = x^{12-9} = x^3$$

15.
$$\left(6xy^4\right)^2 \left[\left(x^5 y^2\right)^3\right]^2$$
$$= \left(6xy^4\right)^2 \left(x^5 y^2\right)^6$$
$$= 6^2 x^2 y^8 \cdot x^{30} y^{12}$$
$$= 36x^{2+30} y^{8+12}$$
$$= 36x^{32} y^{20}$$

17.
$$\left(\frac{8x^5 y^6}{2x^2 y^2}\right)^3 = \left(4x^{5-2} y^{6-2}\right)^3$$
$$= \left(4x^3 y^4\right)^3$$
$$= 4^3 x^9 y^{12}$$
$$= 64x^9 y^{12}$$

19.
$$\left(\frac{15x^6 y^9}{3xy^3}\right)^3 = \left(5x^{6-1} y^{9-3}\right)^3$$
$$= \left(5x^5 y^6\right)^3$$
$$= 5^3 x^{15} y^{18}$$
$$= 125x^{15} y^{18}$$

21.
$$-5^0 = -\left(5^0\right) = -(1) = -1$$

23.
$$\left(\frac{5}{2}\right)^{-3} = \frac{1}{\left(\frac{5}{2}\right)^3} = \frac{1}{\frac{125}{8}} = \frac{8}{125}$$

25.
$$2^{-4} \cdot 2^{-1} = 2^{-4+(-1)} = 2^{-5} = \frac{1}{2^5} = \frac{1}{32}$$

27.
$$\frac{3^{-1}}{3^{-5}} = 3^{-1-(-5)} = 3^4 = 81$$

29.

$$\left(x^{-9}\right)^2 = x^{-18} = \frac{1}{x^{18}}$$

31.

$$\frac{x^{-5}}{x^2} = x^{-5-(2)} = x^{-7} = \frac{1}{x^7}$$

33.

$$\frac{16x^3}{12x^5} = \frac{4 \cdot 4}{3 \cdot 4 \cdot x^{5-3}}$$

$$= \frac{4}{3x^2}$$

35.

$$\left(4x^{-2}y^{-7}\right)^{-3}\left(6x^3y^{-2}\right)^{-2}$$

$$= 4^{-3}x^6y^{21} \cdot 6^{-2}x^{-6}y^4$$

$$= 4^{-3} \cdot 6^{-2} \cdot x^6 \cdot x^{-6} \cdot y^{21} \cdot y^4$$

$$= \frac{x^{6-6} \cdot y^{21+4}}{4^3 \cdot 6^2}$$

$$= \frac{y^{25}}{2304}$$

37.

$$\frac{\left(3x^{-8}y^{-2}\right)^{-3}\left(2x^{-2}y^{-1}\right)}{\left(8x^{-2}y\right)^{-2}}$$

$$= \frac{3^{-3}x^{24}y^6 \cdot 2x^{-2}y^{-1}}{8^{-2}x^4y^{-2}}$$

$$= \frac{2 \cdot 8^2 x^{24-2}y^{6-1}}{3^3 x^4 y^{-2}}$$

$$= \frac{128x^{22}y^5}{27x^4y^{-2}}$$

$$= \frac{128x^{22-4}y^{5-(-2)}}{27}$$

$$= \frac{128x^{18}y^7}{27}$$

39.

$$0.0000206 = 2.06 \times 10^{-5}$$

41.

$$\frac{(0.00042)(11200)}{(0.00000035)(160000)}$$

$$= \frac{\left(4.2 \times 10^{-4}\right)\left(1.12 \times 10^4\right)}{\left(3.5 \times 10^{-7}\right)\left(1.6 \times 10^5\right)}$$

$$= \frac{4.2 \times 1.12 \times 10^{-4} \times 10^4}{3.5 \times 1.6 \times 10^{-7} \times 10^5}$$

$$= \frac{4.704 \times 10^0}{5.6 \times 10^{-2}}$$

$$= .84 \times 10^2 = 84$$

43.

$$14x - 6y + 3x + 6y$$

$$= 14x + 3x - 6y + 6y$$

$$= 17x$$

Degree: First
Type: Monomial

45.

$$5.04xz^3 - 3.6x^2y^4 - 2.8xz^3$$

$$= 5.04xz^3 - 2.8xz^3 - 3.6x^2y^4$$

$$= 2.24xz^3 - 3.6x^2y^4$$

$$= -3.6x^2y^4 + 2.24xz^3$$

Degree: Sixth
Type: Binomial

47.

$$2.6x - x + 3.05x^2 + 5 - 5x^2$$

$$= 1.6x - 1.95x^2 + 5$$

$$= -1.95x^2 + 1.6x + 5 \qquad \text{Standard form}$$

Leading coefficient: -1.95

49.

$$\frac{7}{8}x - \frac{11}{12} - \frac{3}{2}x - \frac{1}{4}$$

$$= \frac{7}{8}x - \frac{3}{2}x - \frac{11}{12} - \frac{1}{4}$$

$$= -\frac{5}{8}x - \frac{7}{6} \qquad \text{Standard form}$$

Leading coefficient: $-\dfrac{5}{8}$

51.

$$\left(5x^2 - x + 6\right) - \left(x^2 - 8x - 2\right)$$

$= 5x^2 - x + 6 - x^2 + 8x + 2$ Add opposites

$= 5x^2 - x^2 - x + 8x + 6 + 2$

$= 4x^2 + 7x + 8$

53.

$$\left(x^3 - 15x^2 + 3x - 7\right) + \left(4x^3 - 8x^2 + 5x + 1\right)$$

$= x^3 + 4x^3 - 15x^2 - 8x^2 + 3x + 5x - 7 + 1$

$= 5x^3 - 23x^2 + 8x - 6$

55.

$$\left(\frac{4}{3}x^2 + xy - \frac{7}{15}y^2\right) - \left(\frac{2}{15}x^2 - \frac{4}{5}xy + \frac{1}{3}y^2\right)$$

$= \frac{4}{3}x^2 + xy - \frac{7}{15}y^2 - \frac{2}{15}x^2 + \frac{4}{5}xy - \frac{1}{3}y^2$

$= \frac{4}{3}x^2 - \frac{2}{15}x^2 + xy + \frac{4}{5}xy - \frac{7}{15}y^2 - \frac{1}{3}y^2$

$= \frac{18}{15}x^2 + \frac{9}{5}xy - \frac{12}{15}y^2$

$= \frac{6}{5}x^2 + \frac{9}{5}xy - \frac{4}{5}y^2$

57.

$$5x^2 + 17x - 8$$
$$\underline{- \quad x^3 - \quad x^2 + \quad 2x - 9}$$

Change signs and add.

$$5x^2 + 17x - 8$$
$$\underline{-x^3 + \quad x^2 - \quad 2x + 9}$$
$$-x^3 + 6x^2 + 15x + 1$$

59.

$$\frac{2}{3}x^2 + \frac{1}{5}x - \frac{5}{4}$$
$$+ \quad \frac{1}{2}x^2 - \frac{7}{20}x - \frac{9}{10}$$
$$\underline{ \frac{7}{6}x^2 - \frac{3}{20}x - \frac{43}{20}}$$

Note: $\frac{2}{3} + \frac{1}{2} = \frac{4}{6} + \frac{3}{6} = \frac{7}{6}$

$\frac{1}{5} - \frac{7}{20} = \frac{4}{20} - \frac{7}{20} = -\frac{3}{20}$

$-\frac{5}{4} - \frac{9}{10} = -\frac{25}{20} - \frac{18}{20} = \frac{43}{20}$

61.

$$5\left(2x^3 - 6x + 3\right) - 4\left(4x^2 + 3x - 7\right)$$

$= 10x^3 - 30x + 15 - 16x^2 - 12x + 28$

$= 10x^3 - 16x^2 - 42x + 43$

63.

$2.4(4x + 7) + 3.5(1.6x - 5.2)$

$= 9.6x + 16.8 + 5.6x - 18.2$

$= 15.2x - 1.4$

65.

$$\left(2x^4\right)\left(7x^2\right)\left(-3x^8\right)$$

$= 2 \cdot 7(-3) \cdot x^4 x^2 x^8$

$= -42 \cdot x^{4+2+8}$

$= -42x^{14}$

67.

$$-8x^2\left(3x^2 - 5\right)$$

$= -8x^2\left(3x^2\right) - 8x^2(-5)$ Distributive property

$= -24x^4 + 40x^2$

69.

$$4x\left(-2x^2 + 8x - 1\right)$$

$$= 4x\left(-2x^2\right) + 4x(8x) + 4x(-1) \qquad \text{Distributive}$$
$$\qquad\qquad\qquad\qquad\qquad\qquad\qquad\qquad \text{property}$$

$$= \text{-}8x^3 + 32x^2 - 4x$$

71.

$$(3x+8)(6x-5)$$

$$= (3x+8)\cdot 6x - (3x+8)\cdot 5$$
$$= 6x(3x+8) - 5(3x+8)$$
$$= 6x\cdot 3x + 6x\cdot 8 - 5\cdot 3x - 5\cdot 8$$
$$= 18x^2 + 48x - 15x - 40$$
$$= 18x^2 + 33x - 40$$

73.

$$(5x-2y))\,(8x+7y)$$

$$= (5x-2y)\cdot 8x + (5x-2y)\cdot 7y$$
$$= 5x\cdot 8x - 2y\cdot 8x + 5x\cdot 7y - 2y\cdot 7y$$
$$= 40x^2 - 16xy + 35xy - 14y^2$$
$$= 40x^2 + 19xy - 14y^2$$

75.

$$\begin{aligned} & x + 6y \\ & \underline{3x + 10y} \\ & 3x^2 + 18xy \qquad\qquad \text{Multiply by 3x.} \\ & \underline{\quad\; + 10xy + 60y^2} \quad \text{Multiply by 10y.} \\ & 3x^2 + 28xy + 60y^2 \quad \text{Add.} \end{aligned}$$

77.

$$\begin{aligned} & 3x^2 - x + 4 \\ & \underline{\quad 2x + 5} \\ & 6x^3 - 2x^2 + 8x \qquad\quad \text{Multiply by 2x.} \\ & \underline{\quad 15x^2 - 5x + 20} \quad \text{Multiply by 5.} \\ & 6x^3 + 13x^2 + 3x + 20 \end{aligned}$$

79.

$$\begin{aligned} & x^3 + 2x^2 - 2 \\ & \underline{\quad 5x + 6} \\ & 5x^4 + 10x^3 \qquad\; - 10x \qquad \text{Multiply by 5x.} \\ & \underline{\quad\; 6x^3 + 12x^2 \qquad - 12} \quad \text{Multiply by 6.} \\ & 5x^4 + 16x^3 + 12x^2 - 10x - 12 \end{aligned}$$

81.

$$\begin{aligned} & 4x^2 - 5xy - y^2 \\ & \underline{\quad 6x - 7y} \\ & 24x^3 - 30x^2y - 6xy^2 \\ & \underline{\quad\; -28x^2y + 35xy^2 + 7y^3} \\ & 24x^3 - 58x^2y + 29xy^2 + 7y^3 \end{aligned}$$

83.

$$\begin{aligned} & 3x^2 + 6x - 5 \\ & \underline{\quad 2x^2 - x + 5} \\ & 6x^4 + 12x^3 - 10x^2 \\ & \quad\;\; -3x^3 - 6x^2 + 5x \\ & \underline{\quad\qquad\;\; +15x^2 + 30x - 25} \\ & 6x^4 + 9x^3 - x^2 + 35x - 25 \end{aligned}$$

85.

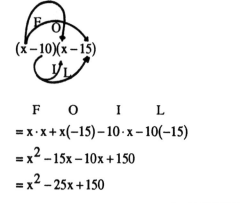

$$(x-10)(x-15)$$

$$\begin{array}{cccc} \text{F} & \text{O} & \text{I} & \text{L} \end{array}$$
$$= x\cdot x + x(-15) - 10\cdot x - 10(-15)$$
$$= x^2 - 15x - 10x + 150$$
$$= x^2 - 25x + 150$$

87.

$(5x+1)(4x+9)$

$$\quad\mathrm{F}\qquad\mathrm{O}\qquad\mathrm{I}\qquad\mathrm{L}$$
$$= 5x\cdot 4x + 5x\cdot 9 + 1\cdot 4x + 1\cdot 9$$
$$= 20x^2 + 45x + 4x + 9$$
$$= 20x^2 + 49x + 9$$

89.

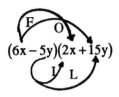

$(6x-5y)(2x+15y)$

$$\quad\mathrm{F}\qquad\mathrm{O}\qquad\mathrm{I}\qquad\mathrm{L}$$
$$= 6x\cdot 2x + 6x\cdot 15y - 5y\cdot 2x - 5y\cdot 15y$$
$$= 12x^2 + 90xy - 10xy - 75y^2$$
$$= 12x^2 + 80xy - 75y^2$$

91.

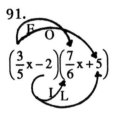

$\left(\dfrac{3}{5}x-2\right)\left(\dfrac{7}{6}x+5\right)$

$$\quad\mathrm{F}\qquad\mathrm{O}\qquad\mathrm{I}\qquad\mathrm{L}$$
$$= \frac{3}{5}x\cdot\frac{7}{6}x + \frac{3}{5}x\cdot 5 - 2\cdot\frac{7}{6}x - 2\cdot 5$$
$$= \frac{7}{10}x^2 + 3x - \frac{7}{3}x - 10$$
$$= \frac{7}{10}x^2 + \frac{2}{3}x - 10$$

93.

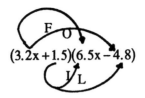

$(3.2x+1.5)(6.5x-4.8)$

$$\quad\mathrm{F}\qquad\mathrm{O}\qquad\mathrm{I}\qquad\mathrm{L}$$
$$= 3.2x\cdot 6.5x + 3.2x(-4.8) + 1.5\cdot 6.5x + 1.5(-4.8)$$
$$= 20.8x^2 - 15.36x + 9.75x - 7.2$$
$$= 20.8x^2 - 5.61x - 7.2$$

95.

$(x-11y)(x+11y)$

This is a product of two conjugates. The product is the difference of the squares of the first and last terms.
Thus,

$$(x-11y)(x+11y) = x^2 - (11y)^2$$
$$= x^2 - 121y^2$$

97.

$\left(2+\dfrac{3}{8}x\right)\left(2-\dfrac{3}{8}x\right)$

Refer to the comment on exercise 95 above.

$$\left(2+\frac{3}{8}x\right)\left(2-\frac{3}{8}x\right) = 2^2 - \left(\frac{3}{8}x\right)^2$$
$$= 4 - \frac{9}{64}x^2$$

99.

$(1.3x-5)(1.3x+5)$

Refer to the comment on exercise 95 above.

$$(1.3x-5)(1.3x+5) = (1.3x)^2 - 5^2$$
$$= 1.69x^2 - 25$$

101.

$(4x-7)^2$

To square a binomial:
(1) Square the first term.
(2) Add twice the product of the two terms.
(3) Square the last term.

#101 continued

$$(4x-7)^2 \overset{(1)\quad\ \ (2)\quad\ \ (3)}{=16x^2 + 2(-7\cdot4x)+(-7)^2}$$
$$= 16x^2 - 56x + 49$$

103.

$$\left(x+\frac{1}{3}\right)^2$$

Refer to the comment on exercise 101 above.

$$\left(x+\frac{1}{3}\right)^2 \overset{(1)\quad (2)\quad\ (3)}{= x^2 + 2\left(\frac{1}{3}x\right)+\left(\frac{1}{3}\right)^2}$$
$$= x^2 + \frac{2}{3}x + \frac{1}{9}$$

105.

$$(3.5x-4)^2$$

Refer to the comment on exercise 101 above.

$$(3.5x-4)^2 \overset{(1)\qquad\quad (2)\qquad\quad (3)}{= (3.5x)^2 + 2(-4\cdot3.5x)+(-4)^2}$$
$$= 12.25x^2 - 28x + 16$$

107.

$$\frac{5p^6 + 12p^4 - 3p^2}{p^2} = \frac{5p^6}{p^2}+\frac{12p^4}{p^2}-\frac{3p^2}{p^2}$$
$$= 5p^4 + 12p^2 - 3$$

109.

$$\frac{7x^5 + 35x^4 - 14x^3 + 42x^2}{7x}$$
$$= \frac{7x^5}{7x}+\frac{35x^4}{7x}-\frac{14x^3}{7x}+\frac{42x^2}{7x}$$
$$= x^4 + 5x^3 - 2x^2 + 6x$$

111.

$$\frac{18m^2y^2 - 6m^4y}{9my^2} = \frac{18m^2y^2}{9my^2}-\frac{6m^4y}{9my^2}$$
$$= 2m - \frac{2m^3}{3y}$$

113.

$$\frac{24x^5yz + 6x^4y^2z^3 - 4x^3y^3z^5}{3x^4yz^2}$$
$$= \frac{24x^5yz}{3x^4yz^2}+\frac{6x^4y^2z^3}{3x^4yz^2}-\frac{4x^3y^3z^5}{3x^4yz^2}$$
$$= \frac{8x}{z}+2yz-\frac{4y^2z^3}{3x}$$

115.

$$\frac{4x^2 + 35x + 24}{x+8}$$

$$\begin{array}{r} 4x+3 \\ x+8\overline{)4x^2 + 35x + 24} \end{array}$$

$$\underline{-4x^2 \mp 32x} \qquad \text{Change signs and add.}$$
$$3x + 24$$
$$\underline{-3x \mp 24} \qquad \text{Change signs and add.}$$
$$0$$

Answer: $4x+3$

117.

$$\frac{3x^3 - 7x^2 - 25x + 11}{x-4}$$

$$\begin{array}{r} 3x^2 + 5x - 5 \\ x-4\overline{)3x^3 - 7x^2 - 25x + 11} \end{array}$$

$$\underline{^{-}3x^3 \pm 12x^2} \qquad\qquad \text{Change signs.}$$
$$5x^2 - 25x$$
$$\underline{^{-}5x^2 \pm 20x} \qquad\qquad \text{Change signs.}$$
$$-5x + 11$$
$$\underline{\pm 5x \mp 20} \qquad\qquad \text{Change signs.}$$
$$-9$$

Answer: $3x^2 + 5x - 5 - \dfrac{9}{x-4}$

119.

$$\frac{6x^3 + x^2 - 20x - 8}{3x+2}$$

#119 continued

$$3x+2 \overline{)\begin{array}{r} 2x^2 - x - 6 \\ 6x^3 + x^2 - 20x - 8 \end{array}}$$

$$\underline{^{-}6x^3 \mp 4x^2} \qquad \text{Change signs.}$$
$$-3x^2 - 20x$$
$$\underline{\pm 3x^2 \pm 2x} \qquad \text{Change signs.}$$
$$-18x - 8$$
$$\underline{\pm 18x \pm 12} \quad \text{Change signs.}$$
$$4$$

Answer: $2x^2 - x - 6 + \dfrac{4}{3x+2}$

121.

$$\frac{6x^3 - 23x^2 - 9x + 56}{3x^2 - x - 8}$$

$$3x^2 - x - 8 \overline{)\begin{array}{r} 2x - 7 \\ 6x^3 - 23x^2 - 9x + 56 \end{array}}$$

$$\underline{^{-}6x^3 \pm 2x^2 \pm 16x} \qquad \text{Change signs.}$$
$$-21x^2 + 7x + 56$$
$$\underline{\pm 21x^2 \mp 7x \mp 56} \quad \text{Change signs.}$$
$$0$$

Answer: $2x - 7$

123.

$$\frac{2x^4 + 9x^3y - 3x^2y^2 + 21xy^3 - 4y^4}{x^2 + 5xy - y^2}$$

$$x^2 + 5xy - y^2 \overline{)\begin{array}{r} 2x^2 - xy + 4y^2 \\ 2x^4 + 9x^3y - 3x^2y^2 + 21xy^3 - 4y^4 \end{array}}$$

$$\underline{^{-}2x^4 \mp 10x^3y \pm 2x^2y^2}$$
$$- x^3y - x^2y^2 + 21xy^3$$
$$\underline{\pm x^3y \pm 5x^2y^2 \mp xy^3}$$
$$4x^2y^2 + 20xy^3 - 4y^4$$
$$\underline{^{-}4x^2y^2 \mp 20xy^3 \pm 4y^4}$$
$$0$$

Answer : $2x^2 - xy + 4y^2$

125.

$$\frac{10x^4 + 4x^3 + 11x^2 + 6x}{2x^2 + 3}$$

$$2x^2 + 3 \overline{)\begin{array}{r} 5x^2 + 2x - 2 \\ 10x^4 + 4x^3 + 11x^2 + 6x + 0 \end{array}}$$

$$\underline{-10x^4 \qquad \mp 15x^2} \qquad \text{Change signs.}$$
$$4x^3 - 4x^2 + 6x$$
$$\underline{-4x^3 \qquad \mp 6x} \qquad \text{Change signs.}$$
$$- 4x^2 \qquad + 0$$
$$\underline{\pm 4x^2 \qquad \pm 6} \quad \text{Change signs.}$$
$$6$$

Answer: $5x^2 + 2x - 2 + \dfrac{6}{2x^2 + 3}$

Chapter 3 Test Solutions

1.
$$-10^2 = -1 \cdot 10 \cdot 10 = -1 \cdot 100 = -100$$

3.
$$\left(\frac{4}{5}\right)^{-2} = \frac{1}{\left(\frac{4}{5}\right)^2} = \frac{1}{\frac{16}{25}} = \frac{25}{16}$$

5.
$$\frac{4^{-3}}{4} = 4^{-3-1} = 4^{-4} = \frac{1}{4^4} = \frac{1}{256}$$

7. $\dfrac{4x^{-3}}{x^5} = 4 \cdot x^{-3-5} = 4x^{-8} = \dfrac{4}{x^8}$

9. $\dfrac{6x^{-2}y^4}{4x^6y^{-2}} = \dfrac{6}{4} \cdot \dfrac{x^{-2}}{x^6} \cdot \dfrac{y^4}{y^{-2}}$
$$= \frac{3}{2} \cdot \frac{1}{x^{6-(-2)}} \cdot \frac{y^{4-(-2)}}{1}$$
$$= \frac{3}{2} \cdot \frac{1}{x^8} \cdot \frac{y^6}{1} = \frac{3y^6}{2x^8}$$

11.
$$(4x^3y^{-2})^{-2} (x^5y^3) = 4^{-2}x^{-6}y^4 \cdot x^5y^3$$
$$= 4^{-2}x^{-6}x^5y^4y^3$$

#11 continued

$$= 4^{-2}x^{-1}y^7$$

$$= \frac{1}{4^2} \cdot \frac{1}{x} \cdot \frac{y^7}{1} = \frac{y^7}{4^2x} = \frac{y^7}{16x}$$

13.

$$\frac{(8000)(450000)}{24000} = \frac{8 \times 10^3 \times 4.5 \times 10^5}{2.4 \times 10^4}$$

$$= \frac{8 \times 4.5 \times 10^3 \times 10^5}{2.4 \times 10^4}$$

$$= \frac{36 \times 10^8}{2.4 \times 10^4}$$

$$= 15 \times 10^{8-4}$$

$$= 15 \times 10^4 \text{ or } 150{,}000$$

15.

$$\frac{3}{2}x - \frac{4}{5}x^3 + \frac{5}{6}x - x^2 + 2x^3$$

$$= 2x^3 - \frac{4}{5}x^3 - x^2 + \frac{5}{6}x + \frac{3}{2}x$$

$$= \frac{6}{5}x^3 - x^2 + \frac{7}{3}x$$

Third degree, trinomial

17.

$$\frac{6x^3 + 31x^2 + 23x - 23}{2x + 7}$$

$$\begin{array}{r} 3x^2 + 5x - 6 \\ 2x+7\overline{)6x^3 + 31x^2 + 23x - 23} \\ \underline{{}^-6x^3 \mp 21x^2} \\ 10x^2 + 23x \\ \underline{{}^-10x^2 \mp 35x} \\ -12x - 23 \\ \underline{\pm 12x \pm 42} \\ 19 \end{array}$$

Answer: $3x^2 + 5x - 6 + \dfrac{19}{2x + 7}$

19.

$$(3x + 5)(x - 4)$$

$$\begin{array}{cccc} \text{F} & \text{O} & \text{I} & \text{L} \end{array}$$

$$= 3x^2 - 12x + 5x - 20$$

$$= 3x^2 - 7x - 20$$

21.

$$\left(\frac{5}{8}x^2 - \frac{1}{2}xy - \frac{4}{9}y^2\right) + \left(\frac{3}{2}x^2 + xy - \frac{1}{6}y^2\right)$$

$$= \frac{5}{8}x^2 + \frac{3}{2}x^2 - \frac{1}{2}xy + xy - \frac{4}{9}y^2 - \frac{1}{6}y^2$$

$$= \frac{5}{8}x^2 + \frac{12}{8}x^2 - \frac{1}{2}xy + xy - \frac{8}{18}y^2 - \frac{3}{18}y^2$$

$$= \frac{17}{8}x^2 + \frac{1}{2}xy - \frac{11}{18}y^2$$

23.

$$\begin{array}{r} 5x^3 - 7x^2 + 3x - 3 \\ - \quad \underline{2x^2 - 6x + 9} \end{array}$$

Change the signs and add.

$$\begin{array}{r} 5x^3 - 7x^2 + 3x - 3 \\ + \quad \underline{-2x^2 + 6x - 9} \\ 5x^3 - 9x^2 + 9x - 12 \end{array}$$

25.

$$\frac{18x^4 + 27x^2y - 15y^2}{3x^2y} = \frac{18x^4}{3x^2y} + \frac{27x^2y}{3x^2y} - \frac{15y^2}{3x^2y}$$

$$= \frac{6x^2}{y} + 9 - \frac{5y}{x^2}$$

27.

$$\left(\frac{3}{4}x + \frac{6}{5}\right)^2 = \left(\frac{3}{4}x\right)^2 + 2\left(\frac{3}{4}x\right)\left(\frac{6}{5}\right) + \left(\frac{6}{5}\right)^2$$

$$= \frac{9}{16}x^2 + \frac{9}{5}x + \frac{36}{25}$$

Test Your Memory

1.

$$\frac{11}{8}\left(\frac{3}{7} + \frac{5}{21}\right) = \frac{11}{8}\left(\frac{9}{21} + \frac{5}{21}\right) \qquad \text{LCD} = 21$$

$$= \frac{11}{8}\left(\frac{14}{21}\right) = \frac{11}{4} \cdot \frac{1}{3} \qquad \text{Canceling}$$

$$= \frac{11}{12}$$

3.

$$\sqrt{(-1 - 5)^2 + (10 - 2)^2} = \sqrt{(-6)^2 + (8)^2}$$

$$= \sqrt{36 + 64}$$

$$= \sqrt{100} = 10$$

5.

$60 + 10 + (-2) = 6 + (-2) = -3$

7.

$-4.08 + 3.9 - 0.625 = -0.18 - 0.625 = -0.805$

9.

$$\frac{7}{6} - 3 - \frac{4}{15} = \frac{35}{30} - \frac{90}{30} - \frac{8}{30} \qquad \text{LCD} = 30$$

$$= \frac{35 - 90 - 8}{30}$$

$$= \frac{-63}{30} = -\frac{21}{10} \qquad \text{Reduce}$$

11.

$$5 + 3(x - 2) = 7 - 2(2x - 4)$$
$$5 + 3x - 6 = 7 - 4x + 8$$
$$3x - 1 = -4x + 15$$
$$3x + 4x - 1 = -4x + 4x + 15$$
$$7x - 1 = 15$$
$$7x - 1 + 1 = 15 + 1$$
$$7x = 16$$
$$x = \frac{16}{7}$$

Solution set: $\left\{\frac{16}{7}\right\}$

13.

$$3(x - 4) = 2(x - 7) + 2$$
$$3x - 12 = 2x - 14 + 2$$
$$3x - 12 = 2x - 12$$
$$3x - 2x - 12 = 2x - 2x - 12$$
$$x - 12 = -12$$
$$x - 12 + 12 = -12 + 12$$
$$x = 0$$

Solution set: $\{0\}$

15.

$$\frac{1}{3}x + \frac{3}{2} = \frac{2}{3}x - \frac{1}{4}$$

$$\frac{1}{3}x - \frac{1}{3}x + \frac{3}{2} = \frac{2}{3}x - \frac{1}{3}x - \frac{1}{4}$$

$$\frac{3}{2} = \frac{1}{3}x - \frac{1}{4}$$

$$\frac{3}{2} + \frac{1}{4} = \frac{1}{3}x - \frac{1}{4} + \frac{1}{4}$$

$$\frac{7}{4} = \frac{1}{3}x$$

$$3 \cdot \frac{7}{4} = 3 \cdot \frac{1}{3}x$$

#15 continued

$$\frac{21}{4} = x$$

Solution set: $\left\{\frac{21}{4}\right\}$

17.

$$-4x + 2 \leq 26$$
$$-4x + 2 - 2 \leq 26 - 2$$
$$-4x \leq 24$$
$$x \geq -6 \qquad \text{Divide by } -4$$

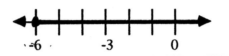

19.

$$2 + 2(x + 3) < 5x + 4$$
$$2 + 2x + 6 < 5x + 4$$
$$2x + 8 < 5x + 4$$
$$2x - 2x + 8 < 5x - 2x + 4$$
$$8 < 3x + 4$$
$$8 - 4 < 3x + 4 - 4$$
$$4 < 3x$$
$$\frac{4}{3} < \frac{3x}{3}$$
$$\frac{4}{3} < x \text{ or } x > \frac{4}{3}$$

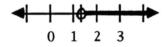

21.

$$-7 \leq 2x + 3 \leq 13$$
$$-7 - 3 \leq 2x + 3 - 3 \leq 13 - 3$$
$$-10 \leq 2x \leq 10$$
$$-5 \leq x \leq 5 \qquad \text{Divide by } 2$$

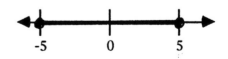

23.

$$6^{-2} = \frac{1}{6^2} = \frac{1}{36}$$

25.

$$\left(\frac{3}{4}\right)^{-3} = \frac{1}{\left(\frac{3}{4}\right)^3} = \frac{1}{\frac{27}{64}} = \frac{64}{27}$$

27.

$$\left(2^3\right)^{-2} = (8)^{-2} = \frac{1}{8^2} = \frac{1}{64}$$

29.

$$\frac{x^4}{8x^{-2}} = \frac{x^{4-(-2)}}{8} = \frac{x^6}{8}$$

31.

$$\frac{10x^3 y^{-5}}{18x^{-4} y^{10}} = \frac{10}{18} \cdot \frac{x^3}{x^{-4}} \cdot \frac{y^{-5}}{y^{10}}$$

$$= \frac{5}{9} \cdot \frac{x^{3-(-4)}}{1} \cdot \frac{1}{y^{10-(-5)}} = \frac{5x^7}{9y^{15}}$$

33.

$$(3x^4 y^{-2})^{-2}(x^3 y^{-3}) = (3^{-2} x^{-8} y^4)(x^3 y^{-3})$$
$$= 3^{-2} \cdot x^{-8} \cdot x^3 \cdot y^4 \cdot y^{-3}$$
$$= 3^{-2} \cdot x^{-5} \cdot y$$
$$= \frac{y}{3^2 x^5} = \frac{y}{9x^5}$$

35.

$$3x^2 y(5x^2 - 6xy - y^2)$$
$$= 15x^4 y - 18x^3 y^2 - 3x^2 y^3$$

37.

$$(4x^2 + 5x - 1)(3x - 2)$$

$$
\begin{array}{r}
4x^2 + 5x - 1 \\
3x - 2 \\
\hline
12x^3 + 15x^2 - 3x \\
- 8x^2 - 10x + 2 \\
\hline
12x^3 + 7x^2 - 13x + 2
\end{array}
$$

39.

$$\frac{6x^3 + 7x^2 - 17x + 20}{3x + 5}$$

$$
\begin{array}{r}
2x^2 - x - 4 \\
3x + 5 \overline{\smash{\big)}\ 6x^3 + 7x^2 - 17x + 20} \\
\underline{-6x^3 \mp 10x^2} \\
-3x^2 - 17x \\
\underline{\pm 3x^2 \pm 5x} \\
-12x + 20 \\
\underline{\pm 12x \pm 20} \\
40
\end{array}
$$

Answer: $2x^2 - x - 4 + \dfrac{40}{3x + 5}$

41.

$$(4x - 3y)(5x + 2y)$$

$$
\begin{array}{cccc}
F & O & I & L
\end{array}
$$
$$= 4x \cdot 5x + 4x \cdot 2y - 3y \cdot 5x - 3y \cdot 2y$$
$$= 20x^2 + 8xy - 15xy - 6y^2$$
$$= 20x^2 - 7xy - 6y^2$$

43.

$$(-5x^3)(4x^2)(6x^4) = -5 \cdot 4 \cdot 6 \cdot x^3 \cdot x^2 \cdot x^4$$
$$= -120 x^9$$

45.

$$P = 3\left(5\frac{3}{4}\right)$$

$$P = 3\left(\frac{23}{4}\right)$$

$$P = \frac{69}{4} \text{ or } 17\frac{1}{4} \qquad \text{Answer: } 17\frac{1}{4} \text{ feet}$$

47.

Let x = measurement of width.
Then $4x + 3$ = measurement of the length.

$$
\begin{aligned}
P &= 2w + 2\ell \\
96 &= 2x + 2(4x + 3) \\
96 &= 2x + 8x + 6 \\
96 &= 10x + 6 \\
90 &= 10x \qquad \text{Subtract 6} \\
9 &= x \qquad \text{Divide by 10} \\
39 &= 4x + 3
\end{aligned}
$$

47 Continued

Answer: 9 X 39 km

49.
Let x = number of 22¢ stamps.
Then 14 - x = number of 39 ¢ stamps.
Let 22x = value of x number of 22¢ stamps in cents.
Then 39(14 - x) = value of (14 - x) number of 39¢ stamps in cents.
Thus,
$$22x + 39(14 - x) = 410$$
$$22x + 546 - 39x = 410$$
$$-17x = -136$$
$$x = 8$$
$$14 - x = 6$$
Answer: 8, 22¢ stamps; 6, 39¢ stamps

CHAPTER 3 STUDY GUIDE

Self -Test Exercises

I. Evaluate each of the following.

1. -13^2

2. $3^{-3} - 3^2$

3. $\left(\dfrac{2}{3}\right)^{-3}$

4. $7^7 \cdot 7^{-5}$

5. $\dfrac{3}{3^{-4}}$

6. -25^0

II. Simplify each of the following. (Remember to write your answers with only positive exponents.)

7. $\dfrac{5y^{-4}}{y^7}$

8. $(3x^{-3})^2$

9. $\dfrac{15x^{-3}y^5}{9x^5y^{-1}}$

10. $\left(\dfrac{x^{-3}}{x^7}\right)^3$

11. $(3x^4y^{-3})^{-2}(x^6y^2)$

12. $\left(\dfrac{3x^3y^5}{2x^{-2}y^0}\right)^{-2}$

III.   Use scientific notation to evaluate the following expression.

13. $\dfrac{(540000)(33000)}{9000}$

IV.   Simplify each of the following polynomials by combining like terms, if possible. After simplifying, state the degree and specify which are monomials, binomials, and trinomials.

14. $3x^2 + 2xy - y^2 - 9x^2 + 4xy$

15. $\dfrac{5}{3}x - x^3 - \dfrac{7}{18}x + x^2 + 3x^3 - \dfrac{2}{5}x^2$

V.   Perform the indicate operations.

16. $5x^3(7x^2 + x - 9)$

17. $\dfrac{6y^3 - 5y^2 - 7y + 6}{2y - 1}$

18. $(2x - 7)(x + 1)$

19. $(3x - 5)^2$

20. $\begin{array}{r} 2x^2 - 6x - 5 \\ \underline{x \qquad\ 3x - 2} \end{array}$

21. $(3x + 12y)(3x - 12y)$

22. $(5x^3)(-3x^2)(4x^5)$

The worked-out solutions begin on the next page.

Self-Test Solutions

1.
$-13^2 = -1 \cdot 13 \cdot 13 = -1 \cdot 169 = -169$

2.
$3^{-3} - 3^2 = \dfrac{1}{3^3} - 3^2$

$= \dfrac{1}{27} - 9 = \dfrac{1}{27} - \dfrac{243}{27} = -\dfrac{242}{27}$

3.
$\left(\dfrac{2}{3}\right)^{-3} = \dfrac{1}{\left(\dfrac{2}{3}\right)^3} = \dfrac{1}{\dfrac{8}{27}} = \dfrac{27}{8}$

4.
$7^7 \cdot 7^{-5} = 7^2 = 49$

5.
$\dfrac{3}{3^{-4}} = 3^{1-(-4)} = 3^5 = 243$

6.
$-25^0 = -1 \cdot 1 = -1$

7.
$\dfrac{5y^{-4}}{y^7} = 5 \cdot y^{-4-7} = 5y^{-11} = \dfrac{5}{y^{11}}$

8.
$(3x^{-3})^2 = 3^2 x^{-6} = \dfrac{9}{x^6}$

9.
$\dfrac{15x^{-3}y^5}{9x^5y^{-1}} = \dfrac{15}{9} \cdot \dfrac{x^{-3}}{x^5} \cdot \dfrac{y^5}{y^{-1}} = \dfrac{5}{3} \cdot \dfrac{1}{x^{5-(-3)}} \cdot \dfrac{y^{5-(-1)}}{1}$

$= \dfrac{5y^6}{3x^8}$

10.
$\left(\dfrac{x^{-3}}{x^7}\right)^{-3} = \dfrac{x^9}{x^{-21}} = x^{9-(-21)} x^{30}$

11.
$(3x^4y^{-3})^{-2}(x^6y^2) = 3^{-2}x^{-8}y^6 \cdot x^6y^2$

$= 3^{-2}x^{-8}x^6y^6y^2$

$= 3^{-2}x^{-2}y^8 = \dfrac{y^8}{3^2x^2} = \dfrac{y^8}{9x^2}$

12.
$\left(\dfrac{3x^3y^5}{2x^{-2}y^0}\right)^{-2} = \dfrac{3^{-2}x^{-6}y^{-10}}{2^{-2}x^4y^0}$

$= \dfrac{3^{-2}}{2^{-2}} \cdot \dfrac{x^{-6}}{x^4} \cdot \dfrac{y^{-10}}{y^0}$

$= \dfrac{2^2}{3^2} \cdot \dfrac{1}{x^{4-(-6)}} \cdot \dfrac{1}{y^{0-(-10)}} = \dfrac{4}{9x^{10}y^{10}}$

13.
$\dfrac{(540000)(33000)}{9000} = \dfrac{5.4 \times 10^5 \times 3.3 \times 10^4}{9 \times 10^3}$

$= \dfrac{5.4 \times 3.3 \times 10^5 \times 10^4}{9 \times 10^3}$

$= \dfrac{17.82 \times 10^9}{9 \times 10^3}$

$= 1.98 \times 10^{9-3} = 1.98 \times 10^6$

$= 1{,}980{,}000$

14.
$3x^2 + 2xy - y^2 - 9x^2 + 4xy$

$= 3x^2 - 9x^2 - y^2 + 2xy + 4xy$

$= -6x^2 - y^2 + 6xy \qquad$ 2nd degree, trinomial

15.
$\dfrac{5}{3}x - x^3 - \dfrac{7}{18}x + x^2 + 3x^3 - \dfrac{2}{5}x^2$

$= -x^3 + 3x^3 + x^2 - \dfrac{2}{5}x^2 + \dfrac{5}{3}x - \dfrac{7}{18}x$

$= 2x^3 + \dfrac{3}{5}x^2 + \dfrac{23}{18}x \qquad$ 3rd degree, trinomial

16.

$5x^3(7x^2 + x - 9) = 35x^5 + 5x^4 - 45x^3$

22.

$(5x^3)(-3x^2)(4x^5)$
$= 5(-3)(4) \cdot x^3 \cdot x^2 \cdot x^5 = -60x^{10}$

17.

$$\dfrac{6y^3 - 5y^2 - 7y + 6}{2y - 1}$$

$$
\begin{array}{r}
3y^2 - y - 4 \\
2y - 1 \overline{\smash{\big)}\, 6y^3 - 5y^2 - 7y + 6} \\
\underline{-6y^3 \pm 3y^2} \\
-2y^2 - 7y \\
\underline{\pm 2y^2 \mp y} \\
-8y + 6 \\
\underline{\pm 8y \mp 4} \\
2
\end{array}
$$

Answer: $3y^2 - y - 4 + \dfrac{2}{2y - 1}$

18.

$$
\begin{array}{cccc}
 & F & O \quad I & L \\
(2x - 7)(x + 1) = & 2x \cdot x + 2x \cdot 1 - 7 \cdot x - 7 \cdot 1
\end{array}
$$
$$= 2x^2 + 2x - 7x - 7$$
$$= 2x^2 - 5x - 7$$

19.

$(3x - 5)^2 = (3x)^2 - 2(3x)(5) + 5^2$
$\qquad\quad = 9x^2 - 30x + 25$

20.

$$
\begin{array}{r}
2x^2 - 6x - 5 \\
\underline{x \qquad 3x - 2} \\
6x^3 - 18x^2 - 15x \\
\underline{-4x^2 + 12x + 10} \\
6x^3 - 22x^2 - 3x + 10
\end{array}
$$

21.

$(3x + 12y)(3x - 12y) = (3x)^2 - (12y)^2$
$\qquad\qquad\qquad\quad = 9x^2 - 144y^2$

CHAPTER 4 FACTORING POLYNOMIALS

Solutions to Text Exercises

1.

$18 = 2 \cdot 9$
$\quad = 2 \cdot 3 \cdot 3$
$\quad = 2 \cdot 3^2$

3.

$175 = 5 \cdot 35$
$\quad\quad = 5 \cdot 5 \cdot 7$
$\quad\quad = 5^2 \cdot 7$

5.

$100 = 2 \cdot 50$
$\quad\quad = 2 \cdot 2 \cdot 25$
$\quad\quad = 2 \cdot 2 \cdot 5 \cdot 5$
$\quad\quad = 2^2 \cdot 5^2$

7.

$441 = 3 \cdot 147$
$\quad\quad = 3 \cdot 3 \cdot 49$
$\quad\quad = 3 \cdot 3 \cdot 7 \cdot 7$
$\quad\quad = 3^2 \cdot 7^2$

9.

$105 = 3 \cdot 35$
$\quad\quad = 3 \cdot 5 \cdot 7$

11.

$90 = 2 \cdot 45$
$\quad = 2 \cdot 3 \cdot 15$
$\quad = 2 \cdot 3 \cdot 3 \cdot 5$
$\quad = 2 \cdot 3^2 \cdot 5$

13.

$525 = 3 \cdot 175$
$\quad\quad = 3 \cdot 5 \cdot 35$
$\quad\quad = 3 \cdot 5 \cdot 5 \cdot 7$
$\quad\quad = 3 \cdot 5^2 \cdot 7$

15.

$252 = 2 \cdot 126$
$\quad\quad = 2 \cdot 2 \cdot 63$
$\quad\quad = 2 \cdot 2 \cdot 3 \cdot 21$
$\quad\quad = 2 \cdot 2 \cdot 3 \cdot 3 \cdot 7$
$\quad\quad = 2^2 \cdot 3^2 \cdot 7$

17.

$18 = 2 \cdot 3^2$
$90 = 2 \cdot 3^2 \cdot 5$
$252 = 2^2 \cdot 3^2 \cdot 7$

The common factors are 2 and 3. Use their smallest power in any of the factored numbers.
$$\text{gcf} = 2^1 \cdot 3^2 = 18$$

19.

$98 = 2 \cdot 7^2$
$441 = 3^2 \cdot 7^2$
$70 = 2 \cdot 5 \cdot 7$

The common factor is 7. Use the smallest power in any of the factored numbers.
$$\text{gcf} = 7^1 = 7$$

21.

$1225 = 5^2 \cdot 7^2$
$441 = 3^2 \cdot 7^2$
$525 = 3 \cdot 5^2 \cdot 7$

#21 continued

The common factor is 7. Use the smallest power in any of the factored numbers.
$$\text{gcf} = 7^1 = 7$$

23.

$90 = 2 \cdot 3^2 \cdot 5$
$60 = 2^2 \cdot 3 \cdot 5$
$315 = 3^2 \cdot 5 \cdot 7$

The common factors are 3 and 5. Use their smallest power in any of the factored numbers.
$$\text{gcf} = 3^1 \cdot 5^1 = 15$$

25.

$8x^4 = 2^3 \, x^4$
$12x^3 = 2^2 \cdot 3 \cdot x^3$
Common factors: 2 and x.
$$\text{gcf} = 2^2 \cdot x^3 = 4x^3$$

27.

$18x^2y^2 = 2 \cdot 3^2 \cdot x^2 \cdot y^2$
$27xy^3 = 3^3 \cdot x \cdot y^3$
Common factors: 3, x and y.
$$\text{gcf} = 3^2 \cdot x \cdot y^2 = 9xy^2$$

29.

$12xy = 2^2 \cdot 3 \cdot x \cdot y$
$18yz = 2 \cdot 3^2 \cdot y \cdot z$
$24xz = 2^3 \cdot 3 \cdot x \cdot z$
Common factors: 2 and 3. $\quad$ gcf: $2^1 \cdot 3^1 = 6$

31.

$x^4 = x^4$
$3x^3 = 3 \cdot x^3$
$15x^2 = 3 \cdot 5 \cdot x^2$
Common factor: x. $\quad\quad$ gcf $= x^2$

33.

$15x^3y^2 = 3 \cdot 5 \cdot x^3 \cdot y^2$
$6x^3 \cdot y^3 = 2 \cdot 3 \cdot x^3 y^3$
$3x^2y = 3 \cdot x^2 \cdot y$
Common factors: 3, x and y.
$$\text{gcf} = 3^1 \cdot x^2 \cdot y = 3x^2y$$

35.

$$21x^3 = 3 \cdot 7 \cdot x^3$$
$$28x^3y = 2^2 \cdot 7 \cdot x^3 \cdot y$$
$$35x^2z = 5 \cdot 7 \cdot x^2 \cdot z$$
$$21x^2yz^2 = 3 \cdot 7 \cdot x^2 \cdot y \cdot z^2$$

Common factors: 7 and x. $gcf = 7 \cdot x^2 = 7x^2$

37.

$$24x^3y^4 = 2^3 \cdot 3 \cdot x^3 \cdot y^4$$
$$16x^4y^4 = 2^4 \cdot x^4 \cdot y^4$$
$$40x^3y^3 = 2^3 \cdot 5 \cdot x^3 \cdot y^3$$
$$8x^2y^4 = 2^3 \cdot x^2 \cdot y^4$$

Common factors: 2, x and y.
$$gcf = 2^3 \cdot x^2 \cdot y^3 = 8x^2y^3$$

39.

$12x - 28y$ $gcf = 4$
$= 4 \cdot 3x - 4 \cdot 7y$
$= 4(3x - 7y)$

41.

$15x^2 + 20x$ $gcf = 5x$
$= 5x \cdot 3x + 5x \cdot 4$
$= 5x(3x + 4)$

43.

$14x^3 - 7x^2$ $gcf = 7x^2$
$= 7x^2 \cdot 2x - 7x^2 \cdot 1$
$= 7x^2(2x - 1)$

45.

$16x^4y + 24x^2y^3$ $gcf = 8x^2y$
$= 8x^2y \cdot 2x^2 + 8x^2y \cdot 3y^2$
$= 8x^2y(2x^2 + 3y^2)$

47.

$10xy^3 - 11ab^2$ $gcf = 1$ prime

49.

$36abcd + 8abc$ $gcf = 4abc$
$= 4abc \cdot 9d + 4abc \cdot 2$
$= 4abc(9d + 2)$

51.

$12x^2 + 42x + 12$ $gcf = 6$
$= 6 \cdot 2x^2 + 6 \cdot 7x + 6 \cdot 2$
$= 6(2x^2 + 7x + 2)$

53.

$4x^2 - 24x - 4$ $gcf = 4$
$= 4 \cdot x^2 - 4 \cdot 6x - 4 \cdot 1$
$= 4(x^2 - 6x - 1)$

55.

$5x^3 + 2x^2y - 3x^2z$ $gcf = x^2$
$= x^2 \cdot 5x + x^2 \cdot 2y - x^2 \cdot 3z$
$= x^2(5x + 2y - 3z)$

57.

$10x^3y - 6x^2y^2 - 8xy$ $gcf = 2xy$
$= 2xy \cdot 5x^2 - 2xy \cdot 3xy - 2xy \cdot 4$
$= 2xy(5x^2 - 3xy - 4)$

59.

$8x^2y^2 + 9yz^2 - 4xz$ $gcf = 1$ prime

61.

$-12x^2 + 18y^2 - 24z^2$ $gcf = 6$
$= 6(-2x^2) + 6 \cdot 3y^2 - 6 \cdot 4z^2$
$= 6(-2x^2 + 3y^2 - 4z^2)$

63.

$15x^3y - 9x^2y^2 + 21x^2y$ $gcf = 3x^2y$
$= 3x^2y \cdot 5x - 3x^2y \cdot 3y + 3x^2y \cdot 7$
$= 3x^2y(5x - 3y + 7)$

65.

$12x^2y^3 - 4x^3y^3 - 8x^3y^2$ $gcf = 4x^2y^2$
$= 4x^2y^2 \cdot 3y - 4x^2y^2 \cdot xy - 4x^2y^2 \cdot 2x$
$= 4x^2y^2(3y - xy - 2x)$

67.

$10x^3y^2 + 15x^3y^3 + 5x^2y$ $gcf = 5x^2y$
$= 5x^2y \cdot 2xy + 5x^2y \cdot 3xy^2 + 5x^2y \cdot 1$
$= 5x^2y(2xy + 3xy^2 + 1)$

69.

$6x^3y^2 - 8y^3 + 15x^2$ $gcf = 1$ prime

71.

$8x^4y^2 + 8x^3y^3 - 40x^2y^4$ $gcf = 8x^2y^2$
$= 8x^2y^2 \cdot x^2 + 8x^2y^2 \cdot xy - 8x^2y^2 \cdot 5y^2$
$= 8x^2y^2(x^2 + xy - 5y^2)$

73.

$-6xy^3z - 30x^2y^2z + 18xyz^3$ $gcf = 6xyz$
$= 6xyz(-y^2) - 6xyz \cdot 5xy + 6xyz \cdot 3z^2$
$= 6xyz(-y^2 - 5xy + 3z^2)$

75.

$18x - 81y + 27z + 63$ $gcf = 9$
$= 9 \cdot 2x - 9 \cdot 9y + 9 \cdot 3z + 9 \cdot 7$
$= 9(2x - 9y + 3z + 7)$

77.

$6x^5 - 8x^4 + 14x^3 - 2x^2$ gcf $= 2x^2$

$= 2x^2 \cdot 3x^3 - 2x^2 \cdot 4x^2 + 2x^2 \cdot 7x - 2x^2 \cdot 1$

$= 2x^2(3x^3 - 4x^2 + 7x - 1)$

79.

$8x^3y - 10x^2y^2 - 12xy^3 - 8x^3y^3$ gcf $= 2xy$

$= 2xy \cdot 4x^2 - 2xy \cdot 5xy - 2xy \cdot 6y^2 - 2xy \cdot 4x^2y^2$

$= 2xy(4x^2 - 5xy - 6y^2 - 4x^2y^2)$

81.

$15x^3y^2 + 6x^3y - 12x^2y^2 + 3x^2y^3$ gcf $= 3x^2y$

$= 3x^2y \cdot 5xy + 3x^2y \cdot 2x - 3x^2y \cdot 4y + 3x^2y \cdot y^2$

$= 3x^2y(5xy + 2x - 4y + y^2)$

83.

$\underbrace{5a(x+2)}+\underbrace{3b(x+2)}$

　　Two terms　　　　　　gcf $= (x + 2)$

$5a(x + 2) + 3b(x + 2)$　　Factor

$= (5a + 3b)(x + 2)$ or $(x + 2)(5a + 3b)$

85.

$\underbrace{8x(x-3)}+\underbrace{5(x-3)}$

　　Two terms　　　　　　gcf $= (x - 3)$

$= 8x(x - 3) + 5(x - 3)$　　Factor

$= (8x + 5)(x - 3)$

87.

$\underbrace{2a(x-1)}-\underbrace{7b(x-1)}$

　　Two terms　　　　　　gcf $= (x - 1)$

$= 2a(x - 1) - 7b(x - 1)$　　Factor

$= (2a - 7b)(x - 1)$

89.

$\underbrace{3x(x+2)}-\underbrace{7(x+2)}$

　　Two terms　　　　　　gcf $= (x + 2)$

$3x(x + 2) - 7(x + 2)$　　Factor

$= (3x - 7)(x + 2)$

91.

$\underbrace{5y(3y+4)}+\underbrace{(3y+4)}$

　　Two terms　　　　　　gcf $= (3y + 4)$

$5y(3y + 4) + 1(3y + 4)$

$= (5y + 1)(3y + 4)$

93.

$\underbrace{4x^2(x-6)}+\underbrace{8x(x-6)}$

　　Two terms　　　　　　gcf $= 4x(x - 6)$

$4x^2(x - 6) + 8x(x - 6)$

$= 4x(x - 6) \cdot x + 4x(x - 6) \cdot 2$

$= 4x(x - 6)(x + 2)$

95.

$\underbrace{9x^2(2x-1)}-\underbrace{3x(2x-1)}$

　　Two terms　　　　　　gcf $= 3x(2x - 1)$

$9x^2(2x - 1) - 3x(2x - 1)$

$= 3x(2x - 1) \cdot 3x - 3x(2x - 1)$

$= 3x(2x - 1)(3x - 1)$

Exercises 4.2

1.

$x^2 - 100 = x^2 - 10^2 = (x + 10)(x - 10)$

3.

$y^2 - 36 = y^2 - 6^2 = (x + 6)(x - 6)$

5.

$t^2 - 1 = (t + 1)(t - 1)$

7.

$m^2 - 6$　　　　　　　　　　　　Prime

9.

$x^4 - 81 = (x^2)^2 - 9^2$　$= (x^2 + 9)(x^2 - 9)$

　　　　　　　　　　$= (x^2 + 9)(x^2 - 3^2)$

　　　　　　　　　　$= (x^2 + 9)(x + 3)(x - 3)$

11.

$36 - x^2 = 6^2 - x^2 = (6 + x)(6 - x)$

13.

$121 - y^2 = 11^2 - y^2 = (11 + y)(11 - y)$

15.

$4 - t^2 = 2^2 - t^2 = (2 + t)(2 - t)$

17.

$x^2 + 49$

Prime. Sum of squares does not factor (over reals).

19.

$x^2 - y^2 = (x + y)(x - y)$

21.

$9x^2 - y^2 = (3x)^2 - y^2 = (3x + y)(3x - y)$

23.
$x^2 - 49y^2 = x^2 - (7y)^2 = (x + 7y)(x - 7y)$

25.
$100x^2 - 9y^2 = (10x)^2 - (3y)^2$
$= (10x + 3y)(10x - 3y)$

27.
$4x^2 + 81y^2$
Prime. Sum of squares does not factor (over reals).

29.
$49m^2 - 121n^2 = (7m)^2 (11n)^2$
$= (7m + 11n)(7m - 11n)$

31.
$x^4 - 81y^4 = (x^2)^2 - (9y^2)^2$
$= (x^2 + 9y^2)(x^2 - 9y^2)$
$= (x^2 + 9y^2)(x + 3y)(x - 3y)$

33.
$16x^4 - 625y^4 = (4x^2)^2 - (25y^2)^2$
$= (4x^2 + 25y^2)(4x^2 - 25y^2)$
$= (4x^2 + 25y^2)(2x + 5y)(2x - 5y)$

35.
$5x^2 - 45 = 5(x^2 - 9)$ Factor out 5,
$= 5(x + 3)(x - 3)$ the gcf.

37.
$4x^2 + 100 = 4(x^2 + 25)$ Factor out 4, the gcf.

39.
$6x^2 - 600y^2 = 6(x^2 - 100y^2)$ Factor out
$= 6(x + 10y)(x - 10y)$ 6, the gcf.

41.
$32x^2 - 50y^2$
$= 2(16x^2 - 25y^2)$ Factor out 2,
$= 2(4x + 5y)(4x - 5y)$ the gcf.

43.
$147x^2 - 12y^2$
$= 3(49x^2 - 4y^2)$ Factor out 3,
$= 3(7x + 2y)(7x - 2y)$ the gcf.

45.
$2x^3 - 50x = 2x(x^2 - 25)$ Factor out 2x,
$= 2x(x + 5)(x - 5)$ the gcf.

47.
$9x^3 + 81x = 9x(x^2 + 9)$

49.
$4x^3 y - 9xy^3$

#49 continued
$= xy(4x^2 - 9y^2)$ Factor out xy, the gcf.
$= xy(2x + 3y)(2x - 3y)$

51.
$75x^3 y^3 - 3xy^5$
$= 3xy^3(25x^2 - y^2)$ Factor out $3xy^3$, the gcf.
$= 3xy^3(5x + y)(5x - y)$

53.
$18x^4 y^3 - 50x^2 y^5$
$= 2x^2 y^3(9x^2 - 25y^2)$ Factor out $2x^2 y^3$
$= 2x^2 y^3(3x + 5y)(3x - 5y)$ the gcf.

55.
$32x^4 y - 162y^5$
$= 2y(16x^4 - 81y^3)$ Factor out 2y, the gcf.
$= 2y (4x^2 + 9y^2)(4x^2 - 9y^2)$
$= 2y(4x^2 + 9y^2)(2x + 3y)(2x - 3y)$

57.
$3x^6 y - 48x^2 y^5$
$= 3x^2 y(x^4 - 16y^4)$ Factor out $3x^2 y$, the gcf.
$= 3x^2 y(x^2 + 4y^2)(x^2 - 4y^2)$
$= 3x^2 y(x^2 + 4y^2)(x + 2y)(x - 2y)$

59.
$x^2 + 6x + 9 = (x)^2 + 2(x \cdot 3) + 3^2 = (x + 3)^2$

61.
$x^2 + 20x + 25$ Prime

63.
$y^2 - 14y + 49 = (y)^2 - 2(y \cdot 7) + 7^2 = (y - 7)^2$

65.
$y^2 - 2y + 1 = (y)^2 - 2(y \cdot 1) + (1)^2 = (y - 1)^2$

67.
$4x^2 + 20x + 25 = (2x)^2 + 2(2x \cdot 5) + (5)^2$
$= (2x + 5)^2$

69.
$9x^2 + 6x + 1 = (3x)^2 + 2(3x \cdot 1) + (1)^2$
$= (3x + 1)^2$

71.
$16x^2 - 24x + 9$
$= (4x)^2 - 2(4x \cdot 3) + (3)^2 = (4x - 3)^2$

73.
$4x^2 + 10x + 25 = (2x)^2 + 2(2x \cdot 5) + 5^2$

146

#73 continued

The middle term is not twice the product of the quantities being squared. The trinomial is prime.

75.
$25x^2 - 20x + 4 = (5x)^2 - 2(5x \cdot 2) + (2)^2$
$= (5x - 2)^2$

77.
$x^2 + 4xy + 4y^2 = (x)^2 + 2(x \cdot 2y) + (2y)^2$
$= (x + 2y)^2$

79.
$x^2 + 14xy + 49y^2 = (x)^2 + 2(x \cdot 7y) + (7y)^2$
$= (x + 7y)^2$

81.
$x^2 - 12xy + 36y^2 = (x)^2 - 2(x \cdot 6y) + (6y)^2$
$= (x - 6y)^2$

83.
$x^2 - 20xy + 100y^2 = (x)^2 - 2(x \cdot 10y) + (10y)^2$
$= (x - 10y)^2$

85.
$9x^2 + 6xy + y^2 = (3x)^2 + 2(3x \cdot y) + (y)^2$
$= (3x + y)^2$

87.
$16x^2 + 40xy + 25y^2$
$= (4x)^2 + 2(4x \cdot 5y) + (5y)^2$
$= (4x + 5y)^2$

89.
$25x^2 - 15xy + 9y^2 = (5x)^2 - (5x \cdot 3y) + (3y)^2$

The middle term is not twice the product of the quantities being squared. The trinomial is prime.

91.
$25x^2 - 20xy + 4y^2 = (5x)^2 - 2(5x \cdot 2y) + (2y)^2$
$= (5x - 2y)^2$

93.
$16x^2 - 40xy + 25y^2$
$= (4x)^2 - 2(4x \cdot 5y) + (5y)^2$
$= (4x - 5y)^2$

95.
$3x^2 + 12x + 12$
$= 3(x^2 + 4x + 4)$ Factor out 3, the gcf.
$= 3(x^2 + 2(2x) + 2^2)$
$= 3(x + 2)^2$

97.
$4x^2 - 48x + 144$
$= 4(x^2 - 12x + 36)$ Factor out 4, the gcf.
$= 4(x^2 - 2(x \cdot 6) + 6^2) = 4(x - 6)^2$

99.
$75x^2 + 30xy + 3y^2$
$= 3(25x^2 + 10xy + y^2)$ Factor out 3, the gcf.
$= 3\left[(5x)^2 + 2(5x \cdot y) + y^2\right]$
$= 3(5x + y)^2$

101.
$16x^2 - 48xy + 36y^2$
$= 4(4x^2 - 12xy + 9y^2)$ Factor out 4, the gcf.
$= 4\left[(2x)^2 - 2(2x \cdot 3y) + (3y)^2\right]$
$= 4(2x - 3y)^2$

103.
$2x^3 - 16x^2 + 32x$
$= 2x(x^2 - 8x + 16)$ Factor out 2x, the gcf.
$= 2x(x^2 - 2(x \cdot 4) + 4^2)$
$= 2x(x - 4)^2$

105.
$6x^4y + 24x^3y^2 + 24x^2y^3$
$= 6x^2y(x^2 + 4xy + 4y^2)$ Factor out $6x^2y$,
$= 6x^2y\left[x^2 + 2(x \cdot 2y) + (2y)^2\right]$ the gcf.
$= 6x^2y(x + 2y)^2$

107.
$16x^2y + 24xy^2 + 36y^3$
$= 4y(4x^2 + 6xy + 9y^2)$ Factor out 4y, the gcf.
$= 4y\left[(2x)^2 + 2(2x \cdot 3y) + (3y)^2\right]$

The middle term is not twice the product of the quantities being squared, so the answer is $4y(4x^2 + 6xy + 9y^2)$.

109.
$12x^3y - 36x^2y^2 + 27xy^3$
$= 3xy(4x^2 - 12xy + 9y^2)$
$= 3xy\left[(2x)^2 - 2(2x \cdot 3y) + (3y)^2\right]$
$= 3xy(2x - 3y)^2$

111.
$64x^4y^2 + 32x^3y^3 + 4x^2y^4$
$= 4x^2y^2(16x^2 + 8xy + y^2)$
$= 4x^2y^2\left[(4x)^2 + 2(4x \cdot y) + y^2\right]$
$= 4x^2y^2(4x + y)^2$

Exercises 4.3

1.

$x^2 + 9x + 18 = (x + m)(x + n)$

We look for two factors of 18 that have a sum of 9.

Factors of 18	Sum of the Factors
$m \cdot n = 18$	$m + n = 9$
$1 \cdot 18 = 18$	$1 + 18 = 19$
$2 \cdot 9 = 18$	$2 + 9 = 11$
$3 \cdot 6 = 18$	$3 + 6 = 9$

Thus, $m = 3$ and $n = 6$.

So, $x^2 + 9x + 18 = (x + 3)(x + 6)$.

3.

$x^2 - 11x + 18 = (x + m)(x + n)$

We look for two factors of 18 that have a sum of -11.

Factors of 18	Sum of the Factors
$m \cdot n = 18$	$m + n = -11$
$1 \cdot 18 = 18$	$1 + 18 = 19$
$-2(-9) = 18$	$-2 + (-9) = -11$

Thus, $m = -2$ and $n = -9$.

So, $x^2 - 11x + 18 = (x - 2)(x - 9)$

Note: All factors need not be listed. Stop when the correct combination is reached.

5.

$x^2 - 2x + 1 = (x + m)(x + n)$

We look for two factors of 1 that have a sum of -2.

Factors of 1	Sum of the Factors
$m \cdot n = 1$	$m + n = -2$
$1 \cdot 1 = 1$	$1 + 1 = 2$
$-1(-1) = 1$	$-1 + (-1) = -2$

Thus, $m = -1$ and $n = -1$.

So, $x^2 - 2x + 1 = (x - 1)(x - 1)$

7.

$x^2 - 5x - 6 = (x + m)(x + n)$

We look for two factors of -6 that have a sum of -5.

Factors of -6	Sum of the Factors
$m \cdot n = -6$	$m + n = -5$
$2(-3) = -6$	$2 + (-3) = -1$
$-2(3) = -6$	$-2 + 3 = 1$
$6(-1) = -6$	$6 + (-1) = 5$
$-6 \cdot 1 = -6$	$-6 + 1 = -5$

Thus, $m = -6$ and $n = 1$.

#7 continued

So, $(x^2 - 5x - 6) = (x - 6)(x + 1)$

9.

$x^2 + 8x + 6$ Prime.

11.

$x^2 - 5x - 36 = (x + m)(x + n)$

We look for two factors of -36 that have a sum of -5.

Factors of -36	Sum of the Factors
$m \cdot n = -36$	$m + n = -5$
$-6 \cdot 6 = -36$	$-6 + 6 = 0$
$-4 \cdot 9 = -36$	$-4 + 9 = 5$
$4(-9) = -36$	$4 + (-9) = -5$

Thus, $m = 4$ and $n = -9$.

So, $x^2 - 5x - 36 = (x + 4)(x - 9)$

Note: All factors need not be listed. Stop when the correct combination is reached.

13.

$x^2 + 11x - 12 = (x + m)(x + n)$

We look for two factors of -12 that have a sum of 11.

Factors of -12	Sum of the Factors
$m \cdot n = -12$	$m + 4 = 11$
$-2 \cdot 6 = -12$	$-2 + 6 = 4$
$-3 \cdot 4 = -12$	$-3 + 4 = 1$
$12(-1) = -12$	$12 + (-1) = 11$

Thus, $m = 12$ and $n = -1$.

So, $x^2 + 11x - 12 = (x + 12)(x - 1)$

15.

$x^2 + x - 72 = (x + m)(x + n)$

We look for two factors of -72 that have a sum of 1.

Factors of -72	Sum of the Factors
$m \cdot n = -72$	$m + n = 1$
$-9 \cdot 8 = -72$	$-9 + 8 = -1$
$9(-8) = -72$	$9 + (-8) = 1$

Thus, $m = 9$ and $n = -8$.

So, $x^2 + x - 72 = (x + 9)(x - 8)$

17.

$x^2 - 3x - 54 = (x + m)(x + n)$

We look for two factors of -54 that have a sum of -3.

148

#17 continued

Factors of -54	Sum of the Factors
$m \cdot n = -54$	$m + n = -3$
$-18 \cdot 3 = -54$	$-18 + 3 = -15$
$-6 \cdot 9 = -54$	$-6 + 9 = 3$
$6(-9) = -54$	$6 + (-9) = -3$

Thus, $m = 6$ and $n = -9$.

So, $x^2 - 3x - 54 = (x + 6)(x - 9)$

19.

$x^2 + 12x + 11 = (x + m)(x + n)$

Two integer factors of 11 whose sum is 12 are 11 and 1
Thus, $m = 11$ and $n = 1$.

So, $x^2 + 12x + 11 = (x + 11)(x + 1)$

21.

$x^2 - 6x + 9 = (x + m)(x + n)$

Two integer factors of 9 whose sum is -6 are -3 and -3. Thus, $m = -3$ and $n = -3$.

So, $x^2 - 6x + 9 = (x - 3)(x - 3)$ or $(x - 3)^2$

23.

$x^2 + x - 2 = (x + m)(x + n)$

Two integer factors of -2 whose sum is 1 are 2 and -1. Thus, $m = 2$ and $n = -1$.

So, $x^2 + x - 2 = (x + 2)(x - 1)$

25.

$x^2 - x - 12 = (x + m)(x + n)$

Two integer factors of -12 whose sum is -1 are -4 and 3. Thus, $m = -4$ and $n = 3$.

So, $x^2 - x - 12 = (x - 4)(x + 3)$

27.

$3x^2 + 9x - 30 = 3(x^2 + 3x - 10)$
$= 3(x + m)(x + n)$

Two integer factors of -10 whose sum is 3 are 5 and -2. Thus, $m = 5$ and $n = -2$.

So, $3x^2 + 9x - 30 = 3(x^2 + 3x - 10)$
$\qquad\qquad\qquad\quad = 3(x + 5)(x - 2)$

29.

$2x^2 - 14x + 24 = 2(x^2 - 7x + 12)$
$\qquad\qquad\qquad = 2(x + m)(x + n)$

Two integer factors of 12 whose sum is -7 are -3 and -4. Thus, $m = -3$ and $n = -4$.

So, $2x^2 - 14x + 24 = 2(x^2 - 7x + 12)$
$\qquad\qquad\qquad\qquad = 2(x - 3)(x - 4)$

31.

$4x^2 - 36x + 32 = 4(x^2 - 9x + 8)$
$\qquad\qquad\qquad = 4(x + m)(x + n)$

Two integer factors of 8 whose sum is -9 are -1 and -8. Thus, $m = -1$ and $n = -8$.

So, $4x^2 - 36x + 32 = 4(x^2 - 9x + 8)$
$\qquad\qquad\qquad\qquad = 4(x - 1)(x - 8)$

33.

$3x^2 - 3x - 90 = 3(x^2 - x - 30)$
$\qquad\qquad\qquad = 3(x + m)(x + n)$

Two integer factors of -30 whose sum is -1 are -6 and 5. Thus, $m = -6$ and $n = 5$.

So, $3x^2 - 3x - 90 = 3(x^2 - x - 30)$
$\qquad\qquad\qquad\quad = 3(x - 6)(x + 5)$

35.

$x^4 - 3x^3 - 18x^2 = x^2(x^2 - 3x - 18)$
$\qquad\qquad\qquad\quad = x^2(x + m)(x + n)$

Two integer factors of -18 whose sum is -3 are -6 and 3. Thus, $m = -6$ and $n = 3$.

So, $x^4 - 3x^3 - 18x^2 = x^2(x^2 - 3x - 18)$
$\qquad\qquad\qquad\qquad\quad = x^2(x - 6)(x + 3)$

37.

$4x^3 + 36x^2 + 32x = 4x(x^2 + 9x + 8)$
$\qquad\qquad\qquad\quad = 4x(x + m)(x + n)$

Two integer factors of 8 whose sum is 9 are 1 and 8. Thus, $m = 1$ and $n = 8$.

So, $4x^3 + 36x^2 + 32x = 4x(x^2 + 9x + 8)$
$\qquad\qquad\qquad\qquad\quad = 4x(x + 1)(x + 8)$

39.

$2x^3y + 8x^2y - 64xy = 2xy(x^2 + 4x - 32)$
$\qquad\qquad\qquad\qquad = 2xy(x + m)(x + n)$

Two integer factors of -32 whose sum is 4 are -4 and 8. Thus, $m = -4$ and $n = 8$.

So, $2x^3y + 8x^2y - 64xy = 2xy(x^2 + 4x - 32)$
$\qquad\qquad\qquad\qquad\qquad = 2xy(x - 4)(x + 8)$

41.

$3x^3y^2 - 12x^2y^2 + 12xy^2$
$= 3xy^2(x^2 - 4x + 4)$
$= 3xy^2(x + m)(x + n)$

Two integer factors of 4 whose sum is -4 are -2 and -2. Thus, $m = -2$ and $n = -2$.

So,
$3x^3y^2 - 12x^2y^2 + 12xy^2 = 3xy^2(x^2 - 4x + 4)$
$\qquad\qquad\qquad\qquad\qquad\quad = 3xy^2(x - 2)(x - 2)$

43.

$-x^2 + 4x + 21 = -(x^2 - 4x - 21)$
$= -(x + m)(x + n)$

Two integer factors of -21 whose sum is -4 are -7 and 3. Thus, m = -7 and n = 3.

So, $-x^2 + 4x + 21 = -(x^2 - 4x - 21)$
$= -(x - 7)(x + 3)$

45.

$-x^2 + 8x - 12 = -(x^2 - 8x + 12)$
$= - (x + m)(x + n)$

Two integer factors of 12 whose sum is -8 are -2 and -6. Thus, m = -2 and n = -6.

So, $-x^2 + 8x - 12 = -(x^2 - 8x + 12)$
$= -(x - 2)(x - 6)$

47.

$-x^2 - 8x - 15 = -(x^2 + 8x + 15)$
$= -(x + m)(x + n)$

Two integer factors of 15 whose sum is 8 are 3 and 5. Thus, m = 3 and n = 5.

So, $-x^2 - 8x - 15 = -(x^2 + 8x + 15)$
$= -(x + 3)(x + 5)$

49.

$-x^2 - 11x + 12 = -(x^2 + 11x - 12)$
$= -(x + m)(x + n)$

Two integer factors of -12 whose sum is 11 are 12 and -1. Thus, m = 12 and n = -1.

So, $-x^2 - 11x + 12 = -(x^2 + 11x - 12)$
$= -(x + 12)(x - 1)$

51.

$-2x^2 - 16x - 30 = -2(x^2 + 8x + 15)$
$= -2(x + m)(x + n)$

Two integers whose product is 15 and sum is 8 are 3 and 5. Thus, m = 3 and n = 5.

So, $-2x^2 - 16x - 30 = -2(x^2 + 8x + 15)$
$= -2(x + 3)(x + 5)$

53.

$-3x^2 + 27x - 60 = -3(x^2 - 9x + 20)$
$= -3(x + m)(x + n)$

Two integers whose product is 20 and sum is -9 are -4 and -5. Thus, m = -4 and n = -5.

So, $-3x^2 + 27x - 60 = -3(x^2 - 9x + 20)$
$= -3(x - 4)(x - 5)$

55.

$x^2 - 5xy - 24y^2 = (x + my)(x + ny)$

#55 continued

Two integers whose product is -24 and sum is -5 are 3 and -8. Thus, m = 3 and n = -8.

So, $x^2 - 5xy - 24y^2 = (x + 3y)(x - 8y)$

57.

$x^2 + 9xy + 8y^2 = (x + my)(x + ny)$

Two integers whose product is 8 and sum is 9 are 8 and 1. Thus, m = 8 and n = 1.

So, $x^2 + 9xy + 8y^2 = (x + 8y)(x + y)$

59.

$x^2 + 14xy + 45y^2 = (x + my)(x + ny)$

Two integers whose product is 45 and sum is 14 are 5 and 9. Thus, m = 5 and n = 9.

So, $x^2 + 14xy + 45y^2 = (x + 5y)(x + 9y)$

61.

$x^2 + 3xy - 40y^2 = (x + my)(x + ny)$

Two integers whose product is -40 and sum is 3 are 8 and -5. Thus, m = 8 and n = -5.

So, $x^2 + 3xy - 40y^2 = (x + 8y)(x - 5y)$

63.

$x^2 - 2xy + y^2 = (x + my)(x + ny)$

Two integers whose product is 1 and sum is -2 are -1 and -1. Thus, m = -1 and n = -1.

So, $x^2 - 2xy + y^2 = (x - y)(x - y)$ or $(x - y)^2$

65.

$x^2 + 3xy + 2y^2 = (x + my)(x + ny)$

Two integers whose product is 2 and sum is 3 are 2 and 1. Thus, m = 2 and n = 1.

So, $x^2 + 3xy + 2y^2 = (x + 2y)(x + y)$

67.

$x^2 + 7xy - 8y^2 = (x + my)(x + ny)$

Two integers whose product is -8 and sum is 7 are 8 and -1. Thus, m = 8 and n = -1.

So, $x^2 + 7xy - 8y^2 = (x + 8y)(x - y)$

69.

$x^2 + 3xy - 18y^2 = (x + my)(x + ny)$

Two integers whose product is -18 and sum is 3 are 6 and -3. Thus, m = 6 and n = -3.

So, $x^2 + 3xy - 18y^2 = (x + 6y)(x - 3y)$

71.

$x^2 - 7xy + 10y^2 = (x + my)(x + ny)$

#71 continued

Two integers whose product is 10 and sum is -7 are -2 and -5. Thus, m = -2 and n = -5.

So, $x^2 - 7xy + 10y^2 = (x - 2y)(x - 5y)$

73.

$x^2 - 6xy - 3y^2 = (x + my)(x + ny)$

There are no two integers whose product is -3 and sum is -6. The trinomial is prime.

75.

$x^2 - 7xy - 18y^2 = (x + my)(x + ny)$

Two integers whose product is -18 and sum is -7 are -9 and 2. Thus, m = -9 and n = 2.

So, $x^2 - 7xy - 18y^2 = (x - 9y)(x + 2y)$

77.

$x^2 + 3xy - 4y^2 = (x + my)(x + ny)$

Two integers whose product is -4 and sum is 3 are 4 and -1. Thus, m = 4 and n = -1.

$x^2 + 3xy - 4y^2 = (x + 4y)(x - y)$

79.

$2x^2 + 6xy - 36y^2 = 2(x^2 + 3xy - 18y^2)$
$= 2(x + my)(x + ny)$

Two integers whose product is -18 and sum is 3 are 6 and -3. Thus, m = 6 and n = -3.

So, $2x^2 + 6xy - 36y^2 = 2(x^2 + 3xy - 18y^2)$
$= 2(x + 6y)(x - 3y)$

81.

$3x^2 - 27xy + 60y^2 = 3(x^2 - 9xy + 20y^2)$
$= 3(x + my)(x + ny)$

Two integers whose product is 20 and sum is -9 are -4 and -5. Thus m = -4 and n = -5.

So, $3x^2 - 27xy + 60y^2 = 3(x^2 - 9xy + 20y^2)$
$= 3(x - 4y)(x - 5y)$

83.

$4x^3 + 24x^2y + 32xy^2 = 4x(x^2 + 6xy + 8y^2)$
$= 4x(x + my)(x + ny)$

Two integers whose product is 8 and sum is 6 are 4 and 2. Thus, m = 4 and n = 2.

So, $4x^3 + 24x^2y + 32xy^2 = 4x(x^2 + 6xy + 8y^2)$
$= 4x(x + 4y)(x + 2y)$

85.

$2x^3y + 16x^2y^2 - 18xy^3$

#85 continued

$= 2xy(x^2 + 8xy - 9y^2)$
$= 2xy(x + my)(x + ny)$

Two integers whose product is -9 and sum is 8 are 9 and -1. Thus, m = 9 and n = -1.

So,
$2x^3y + 16x^2y^2 - 18xy^3 = 2xy(x^2 + 8xy - 9y^2)$
$= 2xy(x + 9y)(x - y)$

87.

$-3x^2 + 12xy + 36y^2 = -3(x^2 - 4xy - 12y^2)$
$= -3(x + my)(x + ny)$

Two integers whose product is -12 and sum is -4 are 2 and -6. Thus, m = 2 and n = -6.

So, $-3x^2 + 12xy + 36y^2 = -3(x^2 - 4xy - 12y^2)$
$= -3(x + 2y)(x - 6y)$

89.

$-2x^3 + 8x^2y - 8xy^2 = -2x(x^2 - 4xy + 4y^2)$
$= -2(x + my)(x + ny)$

Two integers whose product is 4 and sum is -4 are -2 and -2. Thus, m = -2 and n = -2.

So, $-2x^3 + 8x^2y - 8xy^2 = -2x(x^2 - 4xy + 4y^2)$
$= -2x(x - 2y)(x - 2y)$
$= -2x(x - 2y)^2$

Exercises 4.4

1.

$2x^2 + 15x + 7 = (2x + h)(x + k)$

The possible factors of 7 are 7 and 1.

Thus, $(2x + 7)(x + 1) = 2x^2 + 9x + 7$
$(2x + 1)(x + 7) = 2x^2 + 15x + 7$

So, $2x^2 + 15x + 7 = (2x+1)(x + 7)$

3.

$5x^2 + 16x + 11 = (5x + h)(x + k)$

The possible factors of 11 are 11 and 1.

Thus, $(5x + 1)(x + 11) = 5x^2 + 56x + 11$
$(5x + 11)(x + 1) = 5x^2 + 16x + 11$

So, $5x^2 + 16x + 11 = (5x + 11)(x + 1)$

5.

$7x^2 + 20x - 3 = (7x + h)(x + k)$

Possible Factor Combinations

	h · k = -3		
-1 · 3	-3 · 1	Signs are
1(-3)	3(-1)	opposite

#5 continued

The factor combination that gives the correct product is -1 and 3.

Thus, $(7x - 1)(x + 3) = 7x^2 + 20x - 3$

So, $7x^2 + 20x - 3 = (7x - 1)(x + 3)$

7.

$3x^2 + 4x - 7 = (3x + h)(x + k)$

Possible Factor Combinations
$h \cdot k = -7$

1(-7)	-7 · 1	Signs are
7(-1)	-1 · 7	opposite

The factor combination that gives the correct product is 7 and -1.

Thus, $(3x + 7)(x - 1) = 3x^2 + 4x - 7$

So, $3x^2 + 4x - 7 = (3x + 7)(x - 1)$

9.

$7x^2 + 3x - 5 = (7x + h)(x + k)$

Possible Factor Combinations
$h \cdot k = -5$

1(-5)	-5 · 1	Signs are
5(-1)	-1 · 5	opposite

There are no combinations of factors that will give the correct product.
The trinomial is prime.

11.

$5x^2 - 26x + 5 = (5x + h)(x + k)$

Possible Factor Combinations
$h \cdot k = 5$

-5(-1)	-1(-5)	Both signs
		are negative

The factor combination that gives the correct product is -1 and -5.

Thus, $(5x - 1)(x - 5) = 5x^2 - 26x + 5$

So, $5x^2 - 26x + 5 = (5x - 1)(x - 5)$

13.

$3x^2 - 8x + 5 = (3x + h)(x + k)$

Possible Factor Combinations
$h \cdot k = 5$

-1(-5)	-5(-1)	Both signs
		are negative

The factors combination that gives the correct product is -5 and -1.

Thus, $(3x - 5)(x - 1) = 3x^2 - 8x + 5$

So, $3x^2 - 8x + 5 = (3x - 5)(x - 1)$

15.

$7x^2 - 2x - 5 = (7x + h)(x + k)$

Possible Factor Combinations
$h \cdot k = -5$

-1 · 5	-5 · 1	Signs are
5(-1)	1(-5)	opposite

The factor combination that gives the correct product is 5 and -1.

Thus, $(7x + 5)(x - 1) = 7x^2 - 2x - 5$

So, $7x^2 - 2x - 5 = (7x + 5)(x - 1)$

17.

$5x^2 - 8x - 13 = (5x + h)(x + k)$

Possible Factor Combinations
$h \cdot k = -13$

1(-13)	-1 · 13	Signs are
-13 · 1	13(-1)	opposite

The factor combination that gives the correct product is -13 and 1.

Thus, $(5x - 13)(x + 1) = 5x^2 - 8x - 13$

So, $5x^2 - 8x - 13 = (5x - 13)(x + 1)$

19.

$5x^2 - 22x + 8 = (5x + h)(x + k)$

Possible Factor Combinations
$h \cdot k = 8$

-1(-8)	-2(-4)	Both signs
-8(-1)	-4(-2)	are negative

The factor combination that gives the correct product is -2 and -4.

Thus, $(5x - 2)(x - 4) = 5x^2 - 22x - 8$

So, $5x^2 - 22x + 8 = (5x - 2)(x - 4)$

21.

$7x^2 - 71x + 10 = (7x + h)(x + k)$

Possible Factor Combinations
$h \cdot k = 10$

-1(-10)	-2(-5)	Both signs
-10(-1)	-5(-2)	are negative

The factor combination that gives the correct product is -1 and -10.

Thus, $(7x - 1)(x - 10) = 7x^2 - 71x + 10$

So, $7x^2 - 71x + 10 = (7x - 1)(x - 10)$

23.

$6x^2 + x - 12 \quad = (6x + h)(x + k)$
$$\text{or}$$
$$= (2x + h)(3x + k)$$

Possible Factor Combinations
$h \cdot k = -12$

1(-12)	12(-1)	6(-2)	2(-6)
-12 · 1	-1 · 12	-2 · 6	-6(2)
3(-4)	-4 · 3	-3 · 4	4(-3)

The signs are opposites.

After a little or much trial and error, the factor combination that gives the correct product is 3 and -4 with $(2x + h)(3x + k)$

Thus, $(2x + 3)(3x - 4) = 6x^2 + x - 12$

So, $6x^2 + x - 12 = (2x + 3)(3x - 4)$

25.

$4x^2 - 7x + 6 \quad = (4x + h)(x + k)$
$$\text{or}$$
$$= (2x + h)(2x + k)$$

Possible Factor Combinations
$h \cdot k = 6$

-1(-6)	-2(-3)	Both signs
-6(-1)	-3(-2)	are negative

After trying all possible factor combinations, there are none that will give the correct product. The trinomial is prime.

27.

$8x^2 - 26x + 15 \quad = (8x + h)(x + k)$
$$\text{or}$$
$$= (2x + h)(4x + k)$$

Possible Factor Combinations
$h \cdot = 15$

-3(-5)	-1(-15)	Both signs
-5(-3)	-15(-1)	are negative

After little or much trial and error, the factor combination that gives the correct product is -5 and -3 with $(2x + h)(4x + k)$.

Thus, $(2x - 5)(4x - 3) = 8x^2 - 26x + 15$
So, $8x^2 - 26x + 15 = (2x - 5)(4x - 3)$

29.

$25x^2 - 20x + 4 \quad = (25x + h)(x + k)$
$$\text{or}$$
$$= (5x + h)(5x + k)$$

#29 continued

Possible Factor Combinations
$h \cdot k = 4$

-4(-1)	-2(-2)	Both signs
-1(-4)		are negative

After little or much trial and error, the factor combination that gives the correct product is -2 and -2 with $(5x + h)(5x + k)$

Thus, $(5x - 2)(5x - 2) = 25x^2 - 20x + 4$
So, $25x^2 - 20x + 4 = (5x - 2)(5x - 2)$
Note: $(5x)^2 - 2(5x)(2) + 2^2 = (5x - 2)^2$

31.

$14x^2 - 19x + 6 \quad = (14x + h)(x + k)$
$$\text{or}$$
$$= (2x + h)(7k + k)$$

Possible Factor Combinations
$h \cdot k = 6$

-6(-1)	-2(-3)	Both signs
-1(-6)	-3(-2)	are negative

After little or much trial and error, the factor combination is -1 and -6 with $(2x + h)(7x + k)$

Thus, $(2x - 1)(7x - 6) = 14x^2 - 19x + 6$.
So, $14x^2 - 19x + 6 = (2x - 1)(7x - 6)$

33.

$9x^2 + 9x - 10 \quad = (9x + h)(x + k)$
$$\text{or}$$
$$= (3x + h)(3x - k)$$

Possible Factor Combinations
$h \cdot k = -10$

1 · (-10)	10(-1)	-2 · 5	-5 · 2	Signs are
-10 · 1	-1 · 10	5(-2)	2(-5)	opposite

After little or much trial and error, the factor combination that gives the correct product is 5 and -2 with $(3x + h)(3x + h)$.

Thus, $(3x + 5)(3x - 2) = 9x^2 + 9x - 10$
So, $9x^2 + 9x - 10 = (3x + 5)(3x - 2)$

35.

$6x^2 + 17x + 10 \quad = (6x + h)(x + k)$
$$\text{or}$$
$$= (2x + h)(3x + k)$$

Possible Factor Combinations
$h \cdot k = 10$

1 · 10	2 · 5	Signs are
10 · 1	5 · 2	positive

#35 continued

After little or much trial and error, the factor combination that gives the correct product is 5 and 2 with $(6x + h)(x + k)$.

Thus, $(6x + 5)(x + 2) = 6x^2 + 17x + 10$

So, $6x^2 + 17x + 10 = (6x + 5)(x + 2)$

37.

$$8x^2 + 9x - 14 = (8x + h)(x + k)$$
$$\text{or}$$
$$= (2x + h)(4x + k)$$

Possible Factor Combinations

$$h \cdot k = -14$$

$-1 \cdot 14$	$1 \cdot (-14)$	$2(-7)$	Signs are
$14(-1)$	$-14 \cdot 1$	$-7 \cdot 2$	opposite
$-2 \cdot 7$	$7(-2)$		

After little or much trial and error, the factor combination that gives the correct product is -7 and 2 with $(8x + h)(x + k)$.

Thus, $(8x - 7)(x + 2) = 8x^2 + 9x - 14$

So, $8x^2 + 9x - 14 = (8x - 7)(x + 2)$

39.

$$12x^2 + 34x + 10 = 2(6x^2 + 17x + 5)$$
$$= 2(6x + h)(x + k)$$
$$= 2(3x + h)(2x + k)$$

Possible Factor Combinations

$$h \cdot k = 5$$

$1 \cdot 5$	$5 \cdot 1$	Both signs are positive

After little or much trial and error, the factor combination that gives the correct product is 1 and 5 with $2(3x + h)(2x + k)$.

Thus, $2(3x + 1)(2x + 5) = 2(6x^2 + 17x + 5)$
$$= 12x^2 + 34x + 10$$

So, $12x^2 + 34x + 10 + 2(3x + 1)(2x + 5)$

41.

$$24x^2 - 44x - 40 = 4(6x^2 - 11x - 10)$$
$$= 4(6x + h)(x + k)$$
$$\text{or}$$
$$= 4(2x + h)(3x + k)$$

Possible Factor Combinations

$$h \cdot k = -10$$

$1(-10)$	$-1 \cdot 10$	$2(-5)$	$-2 \cdot 5$	Signs are
$-10 \cdot 1$	$10(-1)$	$-5 \cdot 2$	$5(-2)$	opposite

#41 continued

After little or much trial and error, the factor combination that gives the correct product is -5 and 2 with $4(2x + h)(3x + k)$.

Thus, $4(2x - 5)(3x + 2) = 4(6x^2 - 11x - 10)$
$$= 24x^2 - 44x - 40$$

So, $24x^2 - 44x - 40 = 4(2x - 5)(3x + 2)$

43.

$$12x^2 - 48x + 45 = 3(4x^2 - 16x + 15)$$
$$= 3(4x + h)(x + k)$$
$$\text{or}$$
$$= 3(2x + h)(2x + k)$$

Possible Factor Combinations

$$h \cdot k = 15$$

$-1(-15)$	$-5(-3)$	Both signs
$-15(-1)$	$-3(-5)$	are negative

After little or much trial and error, the factor combination that gives the correct product is -3 and -5 with $3(2x + h)(2x + h)$.

Thus, $3(2x - 3)(2x - 5) = 3(4x^2 - 16x + 15)$
$$= 12x^2 - 48x + 45$$

So, $12x^2 - 48x + 45 = 3(2x - 3)(2x - 5)$

45.

$$10x^2 + 25x - 35 = 5(2x^2 + 5x - 7)$$
$$= 5(2x + h)(x + k)$$

Possible Factor Combinations

$$h \cdot k = -7$$

$-7 \cdot 1$	$-1 \cdot 7$	Signs are
$1(-7)$	$7(-1)$	opposite

The factor combination that gives the correct product is 7 and -1.

Thus, $5(2x + 7)(x - 1) = 5(2x^2 + 5x - 7)$
$$= 10x^2 + 25x - 35$$

So, $10x^2 + 25x - 35 = 5(2x + 7)(x - 1)$

47.

$$12x^2 - 32x + 4 = 4(3x^2 - 8x + 1)$$
$$= 4(3x + h)(x + k)$$

Possible Factor Combinations

$$h \cdot k = 1$$
$$-1 \cdot (-1) = 1$$

There is no combination of factors that will give the correct product.

Thus, $12x^2 - 32x + 4 = 4(3x^2 - 8x + 1)$

49.

$-3x^2 + 8x - 5 = -(3x^2 - 8x + 5)$
$\qquad\qquad\qquad = -(3x + h)(x + k)$

Possible Factor Combinations
$$h \cdot k = 5$$

-5(-1)	Both signs
-1(-5)	are negative

The factor combination that gives the correct product is -5 and -1.

Thus, $-(3x - 5)(x - 1) = -(3x^2 - 8x + 5)$
$\qquad\qquad\qquad\qquad = -3x^2 + 8x - 5$

So, $\qquad -3x^2 + 8x - 5 = -(3x - 5)(x - 1)$

51.

$-6x^2 - 11x + 10 = -(6x^2 + 11x - 10)$
$\qquad\qquad\qquad\quad = -(6x + h)(x + k)$
$\qquad\qquad\qquad\quad$ or
$\qquad\qquad\qquad\quad = -(2x + h)(3x + k)$

Possible Factor Combinations
$$h \cdot k = -10$$

1(-10)	-1 · 10	2(-5)	Signs are
-10 · 1	10(-1)	-5 · 2	opposite
	-2 · 5	5(-2)	

After little or much trial and error, the factor combination that gives the correct product is 5 and -2 with $-(2x + h)(3x + k)$.

Thus, $-(2x + 5)(3x - 2) = -(6x^2 + 11x - 10)$
$\qquad\qquad\qquad\qquad = -6x^2 - 11x + 10$

So, $-6x^2 - 11x + 10 = -(2x + 5)(3x - 2)$

53.

$-5x^2 - 14x - 8 = -(5x^2 + 14x + 8)$
$\qquad\qquad\qquad = -(5x + h)(x + k)$

Possible Factor Combinations
$$h \cdot k = 8$$

8 · 1	2 · 4	Both signs
1 · 8	4 · 2	are positive

The factor combination that gives the correct product is 4 and 2.

Thus, $-(5x + 4)(x + 2) = -(5x^2 + 14x + 8)$
$\qquad\qquad\qquad\qquad = -5x^2 - 14x - 8$

So, $\qquad -5x^2 - 14x - 8 = -(5x + 4)(x + 2)$

55.

$-9x^2 + 3x + 20 = -(9x^2 - 3x - 20)$
$\qquad\qquad\qquad\quad = -(9x + h)(x + k)$
$\qquad\qquad\qquad\quad$ or
$\qquad\qquad\qquad\quad = -(3x + h)(3x + k)$

Possible Factor Combinations
$$h \cdot k = -20$$

1 · (-20)	-1 · 20	4(-5)	
-20 · 1	20(-1)	-5 · 4	Signs are
2(-10)	-2 · 10	-4 · 5	opposite
-10 · 2	10(-2)	5(-4)	

After little or much trial and error, the factor combination that gives the correct product is 4 and -5 with $-(3x + h)(3x + k)$.

Thus, $-(3x + 4)(3x - 5) = -(9x^2 - 3x - 20)$
$\qquad\qquad\qquad\qquad = -9x^2 + 3x + 20$

So, $\qquad -9x^2 + 3x + 20 = -(3x + 4)(3x - 5)$

57.

$-24x^2 - 34x - 12 = -2(12x^2 + 17x + 6)$
$\qquad\qquad\qquad\qquad = -2(12x + h)(x + k)$
$\qquad\qquad\qquad\qquad$ or
$\qquad\qquad\qquad\qquad = -2(6x + h)((2x + k)$
$\qquad\qquad\qquad\qquad$ or
$\qquad\qquad\qquad\qquad = -2(3x + h)(4x + k)$

Possible Factor Combinations
$$h \cdot k = 6$$

2 · 3	6 · 1	Both signs
3 · 2	1 · 6	are positive

After little or much trial and error, the factor combination that gives the correct product is 2 and 3 with $-2(3x + h)(4x + k)$.

Thus, $-2(3x + 2)(4x + 3) = -2(12x^2 + 17x + 6)$
$\qquad\qquad\qquad\qquad = -24x^2 - 34x - 12$

So, $\qquad -24x^2 - 34x - 12 = -2(3x + 2)(4x + 3)$

59.

$-70x^2 - 95x + 15 = -5(14x^2 + 19x - 3)$
$\qquad\qquad\qquad\qquad = -5(14x + h)(x + k)$
$\qquad\qquad\qquad\qquad$ or
$\qquad\qquad\qquad\qquad = -5(2x + h)(7x + k)$

Possible Factor Combinations
$$h \cdot k = -3$$

-1 · 3	Signs are
3(-1)	opposite

#59 continued

After little or much trial and error, the factor combination that gives the correct product is 3 and -1 with $-5(2x + h)(7x + k)$.

Thus, $-5(2x + 3)(7x - 1) = -5(14x^2 + 19x - 3)$
$= -70x^2 - 95x + 15$

So, $-70x^2 - 95x + 15 = -5(2x + 3)(7x - 1)$

61.
$8x^4 - 14x^3 + 5x^2 = x^2(8x^2 - 14x + 5)$
$= x^2(8x + h)(x + k)$
or
$= x^2(2x + h)(4x + k)$

Possible Factor Combinations
$h \cdot k = 5$

$-1 \cdot (-5)$	Both signs
$-5 \cdot (-1)$	are negative

After little or much trial and error, the factor combination that gives the correct product is -1 and -5 with $x^2(2x + h)(4x + k)$.

Thus,

$x^2(2x - 1)(4x - 5) = x^2(8x^2 - 14x + 5)$
$= 8x^4 - 14x^3 + 5x^2$

So, $8x^4 - 14x^3 + 5x^2 = x^2(2x - 1)(4x - 5)$

63.
$10x^3y - 7x^2y - 12xy = xy(10x^2 - 7x - 12)$
$= xy(10x + h)(x + k)$
or
$= xy(2x + h)(5x + k)$

Possible Factor Combinations
$h \cdot k = -12$

$1 \cdot (-12)$	$-1 \cdot 12$	$-6 \cdot 2$	
$-12 \cdot 1$	$12(-1)$	$2(-6)$	Signs are
$-3 \cdot 4$	$3(-4)$	$-2 \cdot 6$	opposite
$4(-3)$	$-4 \cdot 3$	$6(-2)$	

After little or much trial and error, the factor combination that gives the correct product is -3 and 4 with $xy(2x + h)(5x + k)$.

Thus,

$xy(2x - 3)(5x + 4) = xy(10x^2 - 7x - 12)$
$= 10x^3y - 7x^2y - 12xy$

So, $10x^3y - 7x^2y - 12xy = xy(2x - 3)(5x + 4)$

65.
$5x^2 + 8xy + 3y^2 = (5x + hy)(x + ky)$

#65 continued

Possible Factor Combinations
$h \cdot k = 3$

$1 \cdot 3$	$3 \cdot 1$	Both signs
		are positive

The factor combination that gives the correct product is 3 and 1.

Thus, $(5x + 3y)(x + y) = 5x^2 + 8xy + 3y^2$
So, $5x^2 + 8xy + 3y^2 = (5x + 3y)(x + y)$

67.
$9x^2 + 62xy - 7y^2 = (9x + hy)(x + ky)$
or
$= (3x + hy)(3x + ky)$

Possible Factor Combinations
$h \cdot k = -7$

$-1 \cdot 7$	$-7 \cdot 1$	Signs are
$7(-1)$	$1 \cdot (-7)$	opposite

After little or much trial and error, the factor combination that gives the correct product is -1 and 7 with $(9x + hy)(x + ky)$.

Thus, $(9x - y)(x + 7y) = 9x^2 + 62xy - 7y^2$
So, $9x^2 + 62xy - 7y^2 = (9x - y)(x + 7y)$

69.
$7x^2 - 17xy + 6y^2 = (7x + hy)(x + ky)$

Possible Factor Combinations
$h \cdot k = 6$

$-1(-6)$	$-2(-3)$	Both signs
$-6(-1)$	$-3(-2)$	are negative

The factor combination that gives the correct product is -3 and -2.

Thus, $(7x - 3y)(x - 2y) = 7x^2 - 17xy + 6y^2$

So, $7x^2 - 17xy + 6y^2 = (7x - 3y)(x - 2y)$

71.
$6x^2 + 7xy - 5y^2 = (6x + hy)(x + ky)$
or
$= (3x + hy)(2x + ky)$

Possible Factor Combinations
$h \cdot k = -5$

$-1 \cdot 5$	$-5 \cdot 1$	Signs are
$5(-1)$	$1 \cdot (-5)$	opposite

After little or much trial and error, the factor combination that gives the correct product is 5 and -1 with $(3x + hy)(2x + ky)$.

Thus, $(3x + 5y)(2x - y) = 6x^2 + 7xy - 5y^2$

#71 continued

So, $6x^2 + 7xy - 5y^2 = (3x + 5y)(2x - y)$

73.

$9x^2 - 12xy + 4y^2 \quad = (9x + hy)(x + ky)$
or
$\qquad\qquad = (3x + hy)(3x + ky)$

Possible Factor Combinations
$$h \cdot k = -4$$

-1(4)	-2(2)	Both signs
	4(-1)	are negative

After little or much trial and error, the factor combination that gives the correct product is -2 and -2 with $(3x + hy)(3x + ky)$.

Thus, $(3x - 2y)(3x - 2y) = 9x^2 - 12xy + 4y^2$

So, $9x^2 - 12xy + 4y^2 = (3x - 2y)(3x - 2y)$

75.

$8x^2 + 26xy + 15y^2 \quad = (8x + hy)(x + ky)$
or
$\qquad\qquad = (4x + hy)(2x + ky)$

Possible Factor Combinations
$$h \cdot k = 15$$

1 · 15	3 · 5	Both signs
15 · 1	5 · 3	are positive

After little or much trial and error, the factor combination that gives the correct product is 3 and 5 with $(4x + hy)(2x + ky)$.

Thus,
$(4x + 3y)(2x + 5y) = 8x^2 + 26xy + 15y^2$
So, $8x^2 + 26xy + 15y^2 = (4x + 3y)(2x + 5y)$

77.

$9x^2 - 35xy - 4y^2 \quad = (9x + hy)(x + ky)$
or
$\qquad\qquad = (3x + hy)(3x + ky)$

Possible Factor Combinations
$$h \cdot k = -4$$

4(-1)	-4 · 1	2(-2)	Signs are
-1 · 4	1 · (-4)	-2 · 2	opposite

After little or much trial and error, the factor combination that gives the correct product is 1 and -4 with $(9x + hy)(x + ky)$.

Thus,
$(9x + y)(x - 4y) = 9x^2 - 35xy - 4y^2$
So, $9x^2 - 35xy - 4y^2 = (9x + y)(x - 4y)$

79.

$10x^2 - 31xy - 14y^2 \quad = (10x + hy)(x + ky)$
or
$\qquad\qquad = (2x + hy)(5x + ky)$

Possible Factor Combinations
$$h \cdot k = -14$$

14(-1)	1 · (-14)	-2 · 7	-7 · 2	Signs are
-1 · 14	-14 · 1	7 (-2)	2(-7)	opposite

After little or much trial and error, the factor combination that gives the correct product is -7 and 2 with $(2x + hy)(5x + ky)$.

Thus,
$(2x - 7y)(5x + 2y) = 10x^2 - 31xy - 14y^2$
So, $10x^2 - 31xy - 14y^2 = (2x - 7y)(5x + 2y)$

81.

$-6x^2 + 11xy - 4y^2 \quad = -(6x^2 - 11xy + 4y^2)$
$\qquad\qquad = -(6x + hy)(x + ky)$
or
$\qquad\qquad = -(2x + hy)(3x + ky)$

Possible Factor Combinations
$$h \cdot k = 4$$

-1(-4)	-4(-1)	-2(-2)	Both signs
-4(-1)	-1(-4)		are negative

After little or much trial and error, the factor combination that gives the correct product is -1 and -4 with $-(2x + hy)(3x + ky)$.

Thus,
$-(2x - y)(3x - 4y) = -(6x^2 - 11xy + 4y^2)$
$\qquad\qquad = -6x^2 + 11xy - 4y^2$
So, $-6x^2 + 11xy - 4y^2 = -(2x - y)(3x - 4y)$

83.

$-12x^2 + 17xy + 7y^2 \quad = - (12x^2 - 17xy - 7y^2)$
$\qquad\qquad = -(12x + hy)(x + ky)$
or
$\qquad\qquad = -(6x + hy)(2x + ky)$
or
$\qquad\qquad = -(3x + hy)(4x + ky)$

Possible Factor Combinations
$$h \cdot k = -7$$

-7 · 1	7(-1)	Signs are
1(-7)	-1 · 7	opposite

After little or much trial and error, the factor combination that gives the correct product is 1 and -7 with $-(3x + hy)(4x + ky)$.

#83 continued

Thus,

$$-(3x + y)(4x - 7y) = -(12x^2 - 17xy - 7y^2)$$
$$= -12x^2 + 17xy + 7y^2$$

So,

$$-12x^2 + 17xy + 7y^2 = -(3x + y)(4x - 7y)$$

85.

$$16x^2 + 12xy - 10y^2 = 2(8x^2 + 6xy - 5y^2)$$
$$= 2(8x + hy)(x + ky)$$
$$\text{or}$$
$$= 2(4x + hy)(2x + ky)$$

Possible Factor Combinations
$$h \cdot k = -5$$

5(-1)	1(-5)	Signs are
-1 · 5	-5 · 1	opposite

After little or much trial and error, the factor combination that gives the correct product is 5 and -1 with $2(4x + hy)(2x + ky)$.

Thus,

$$2(4x + 5y)(2x - y) = 2(8x^2 + 6xy - 5y^2)$$
$$= 16x^2 + 12xy - 10y^2$$

So,

$$16x^2 + 12xy - 10y^2 = 2(4x + 5y)(2x - y)$$

87.

$$21x^2 + 36xy + 15y^2 = 3(7x^2 + 12xy + 5y^2)$$
$$= 3(7x + hy)(x + ky)$$

Possible Factor Combinations
$$h \cdot k = 5$$

1 · 5	5 · 1	Both signs are
		positive

The factor combination that gives the correct product is 5 and 1.

Thus, $3(7x + 5y)(x + y) = 3(7x^2 + 12xy + 5y^2)$
So, $21x^2 + 36xy + 15y^2 = 3(7x + 5y)(x + y)$

89.

$$6x^2 + 21xy - 18y^2 = 3(2x^2 + 7xy - 6y^2)$$
$$= 3(2x + hy)(x + ky)$$

There are no combinations of factors that will give the correct product.

So, $6x^2 + 21xy - 18y^2 = 3(2x^2 + 7xy - 6y^2)$

91.

$$8x^4 - 26x^3y + 15x^2y^2 = x^2(8x^2 - 26xy + 15y^2)$$
$$= x^2(8x + hy)(x + ky)$$
$$= x^2(4x + hy)(2x + ky)$$

#91 continued

Possible Factor Combinations
$$h \cdot k = 15$$

-1(-15)	-3(-5)	Both signs
-15(-1)	-5(-3)	are negative

After little or much trial and error, the factor combination that gives the correct product is -3 and -5 with x^2 $(4x + hy)(2x + ky)$.

Thus,

$$x^2(4x - 3y)(2x - 5y) = x^2(8x^2 - 26xy + 15y^2)$$
$$= 8x^4 - 26x^3y + 15x^2y^2$$

So,

$$8x^4 - 26x^3y + 15x^2y^2 = x^2(4x - 3y)(2x - 5y)$$

93.

$$4x^3y - 2x^2y^2 - 42xy^3 = 2xy(2x^2 - xy - 21y^2)$$
$$= 2xy(2x + hy)(x + ky)$$

Possible Factor Combinations
$$h \cdot k = -21$$

1(-21)	21(-1)	
-21 · 1	-1 · 21	Signs are
-3 · 7	7(-3)	opposite
3(-7)	-7 · 3	

The factor combination that gives the correct product is -7 and 3.

Thus,

$$2xy(2x - 7y)(x + 3y) = 2xy(2x^2 - xy - 21y^2)$$
$$= 4x^3y - 2x^2y^2 - 42xy^3$$

So,

$$4x^3y - 2x^2y^2 - 42xy^3 = 2xy(2x - 7y)(x + 3y)$$

95.

$$15x^4y + 24x^3y^2 + 9x^2y^3$$
$$= 3x^2y(5x^2 + 8xy + 3y^2)$$
$$= 3x^2y(5x + hy)(x + ky)$$

Possible Factor Combinations
$$h \cdot k = 3$$

3 · 1	1 · 3	Both signs are
		positive

The factor combination that gives the correct product is 3 and 1.

Thus,

$$3x^2y(5x + 3y)(x + y) = 3x^2y(5x^2 + 8xy + 3y^2)$$
$$= 15x^4y + 24x^3y^2 + 9x^2y^3$$

#95 continued

So,

$15x^4y + 24x^3y^2 + 9x^2y^3$
$$= 3x^2y(5x + 3y)(x + y)$$

97.

$-60x^3 + 76x^2y - 24xy^2$
$$= -4x(15x^2 - 19xy + 6y^2)$$
$$= -4x(15x + hy)(x + ky)$$
$$= -4x(3x + hy)(5x + ky)$$

Possible Factor Combinations
$$h \cdot k = 6$$

-6(-1)	-3(-2)	Both signs
-1(-6)	-2(-3)	are negative

After little or much trial and error, the factor combination that gives the correct product is -2 and -3 with $-4x(3x + hy)(5x + ky)$.

Thus, $-4x(3x - 2y)(5x - 3y)$
$$= -4x(15x^2 - 19xy + 6y^2)$$
$$= -60x^3 + 76x^2y - 24xy^2$$

So, $-60x^3 + 76x^2y - 24xy^2$
$$= -4x(3x - 2y)(5x - 3y)$$

99.

$-16x^3y^2 + 20x^2y^3 + 14xy^4$
$$= -2xy^2(8x^2 - 10xy - 7y^2)$$
$$= -2xy^2(8x + hy)(x + ky)$$
$$= -2xy^2(2x + hy)(4x + ky)$$

Possible Factor Combinations
$$h \cdot k = -7$$

1(-7)	7(-1)	Signs are
-7 · 1	-1 · 7	opposite

After little or much trial and error, the factor combination that gives the correct product is 1 and -7 with $-2xy^2(2x + hy)(4x + ky)$.

Thus,

$-2xy^2(2x + y)(4x - 7y)$
$$= -2xy^2(8x^2 - 10xy - 7y^2)$$
$$= -16x^3y^2 + 20x^2y^3 + 14xy^4$$

So, $-16x^3y^2 + 20x^2y^3 + 14xy^4$
$$= -2xy^2(2x + y)(4x - 7y)$$

101.

$3x^4 - 10x^2 - 8 = (3x^2 + h)(x^2 + k)$

#101 continued

Possible Factor Combinations
$$h \cdot k = -8$$

1(-8)	2(-4)	Signs are
-1(8)	-2(4)	opposite

The factor combination that gives the correct product is 2 and -4.

Thus,

$(3x^2 + 2)(x^2 - 4) = 3x^2 - 10x^2 - 8$

So, $3x - 10x^2 - 8 = (3x^2 + 2)(x^2 - 4)$
$$= (3x^2 + 2)(x + 2)(x - 2)$$

103.

$4x^4 - 13x^2 + 9 = (4x^2 + h)(x^2 + k)$
or
$$= (2x^2 + h)(2x + k)$$

Possible Factor Combinations
$$h \cdot k = 9$$

-1(-9)	-3(-3)	Both signs are
		negative

The factor combination that gives the correct product is -1 and -9.

Thus, $(4x^2 - 9)(x^2 - 1) = 4x^4 - 13x^2 + 9$

So,

$4x^4 - 13x^2 + 9 = (4x^2 - 9)(x^2 - 1)$
$$= (2x + 3)(2x - 3)(x + 1)(x - 1)$$

Exercises 4.5

1.

$xy + 2x + 5y + 10$

$= (xy + 2x) + (5y + 10)$	Step 1
$= x(y + 2) + 5(y + 2)$	Step 1
$= (x + 5)(y + 2)$	

3.

$2xy + 8x + 7y + 28$

$(2xy + 8x) + (7y + 28)$	Step 1
$= 2x(y + 4) + 7(y + 4)$	Step 1
$= (2x + 7)(y + 4)$	

5.

$6xy + 4x + 9y + 6$

$= (6xy + 4x) + (9y + 6)$	Step 1
$= 2x(3y + 2) + 3(3y + 2)$	Step 1
$= (2x + 3)(3y + 2)$	

7.

$12xy + 42x + 10y + 35$

#7 continued

$(12xy + 42x) + (10y + 35)$ Step 1
$= 6x(2y + 7) +5(2y + 7)$ Step 1
$= (6x + 5)(2y + 7)$

9.

$xy - 6x + 3y - 18$
$= (xy - 6x) + (3y -18)$ Step 1
$= x(y - 6) + 3(y - 6)$ Step 1
$= (x + 3)(y - 6)$

11.

$3xy - 9x + 4y - 12$
$= (3xy - 9x) + (4y - 12)$ Step 1
$= 3x(y - 3) + 4(y - 3)$ Step 1
$= (3x + 4)(y - 3)$

13.

$xy - 6y + 2x -12$
$(xy - 6y)+(2x - 12)$ Step 1
$= y(x - 6) +2(x - 6)$ Step 1
$= (y + 2)(x - 6)$

15.

$4xy - 14y + 6x - 21$
$= (4xy - 14y) + (6x - 21)$ Step 1
$=2y(2x - 7) +3(2x - 7)$ Step 1
$= (2y + 3)(2x - 7)$

17.

$2xy + 3y + 2x + 3$
$= (2xy + 3y) + (2x + 3)$ Step 1
$= y(2x + 3) +1(2x + 3)$ Step 1
$= (y + 1)(2x + 3)$

19.

$10xy - 35x + 2y -7$
$= (10xy - 35x) + (2y - 7)$ Step 1
$= 5x(2y - 7) +1 (2y - 7)$ Step 1
$= (5x + 1)(2y -7)$

21.

$xy + 5y - x -5$
$= (xy + 5y) + (-x -5)$
$= y(x + 5) -1 (x + 5)$ Factor out -1
$= (y - 1)(x + 5)$

23.

$6xy - 8y - 3x + 4$
$= (6xy - 8y) + (-3x + 4)$ Step 1
$= 2y(3x - 4) -1(3x - 4)$ Factor out -1
$= (2y - 1)(3x - 4)$

25.

$2xy + 3y - 10x - 15$

#25 continued

$= (2xy + 3y) + (-10x - 15)$ Step 1
$= y(2x + 3) -5(2x + 3)$ Factor out -5
$= (y - 5)(2x + 3)$

27.

$10xy - 15x - 4y + 6$
$= (10xy - 15x) + (-4y + 6)$ Step 1
$= 5x(2y - 3) -2(2y - 3)$ Factor out -2
$= (5x - 2)(2y -3)$

29.

$3ax + 4ay - 3bx - 4by$
$= (3ax + 4ay) + (-3bx - 4by)$ Step 1
$= a(3x + 4y) -b(3x + 4y)$ Factor out -b
$= (a - b)(3x + 4y)$

31.

$6ax - 4bx - 21ay + 14by$
$=(6ax - 4bx) + (-21ay + 14by)$
$= 2x(3a - 2b) -7y(3a - 2b)$ Factor out -7
$= (2x - 7y)(3a - 2b)$

33.

$xy + 32 + 8x + 4y$
$= xy + 4y + 8x + 32$ Step 3
$= (xy + 4y) + (8x + 32)$
$= y(x + 4) +8(x + 4)$
$= (y + 8)(x + 4)$

35.

$15xy + 8 + 20x + 6y$
$= 15xy + 6y + 20x + 8$ Step 1
$= (15xy + 6y) + (20x + 8)$
$= 3y(5x + 2) +4(5x + 2)$
$= (3y + 4)(5x + 2)$

37.

$2x^2 -12y - 8xy + 3x$
$= 2x^2 - 8xy + 3x - 12y$ Step 3
$= (2x^2 - 8xy) + (3x - 12y)$
$= 2x(x - 4y)+3(x - 4y)$
$= (2x + 3)(x - 4y)$

39.

$6x^2 + 14y - 4xy -21x$
$= 6x^2 - 4xy -21x + 14y$ Step 3
$= (6x^2 - 4xy) + (-21x + 14y)$
$= 2x(3x - 2y) -7(3x - 2y)$ Factor out -7
$= (2x - 7)(3x - 2y)$

41.

$12x^2 +2y - 8xy - 3x$
$= 12x^2 -3x- 8xy + 2y$ Step 3
$= (12x^2 - 3x) + (-8xy + 2y)$

#41 continued

$= 3x(4x - 1) -2y(4x - 1)$
$= (3x - 2y)(4x - 1)$

43.
$x^3 + 7x^2 - 9x - 63$
$= x^3 -9x + 7x^2 - 63$ Step 3
$= (x^3 - 9x) + (7x^2 - 63)$
$= x(x^2 - 9) + 7(x^2 - 9)$
$= (x + 7)(x^2 - 9)$ Factor difference
$= (x + 7)(x + 3)(x -3)$ of two squares

45.
$3x^3 + 7x^2 - 3x - 7$
$= (3x^3 + 7x^2) + (-3x - 7)$
$= x^2(3x + 7) - 1(3x + 7)$ Factor out -1
$= (x^2 -1)(3x + 7)$ Factor difference
$= (x + 1)(x - 1)(3x + 7)$ of two squares

47.
$20x^3 + 32x^2 - 45x - 72$
$= (20x^3 + 32x^2) + (-45x - 72)$
$= 4x^2(5x + 8) -9(5x + 8)$ Factor out -9
$= (4x^2 - 9)(5x + 8)$ Factor diference
$= (2x - 3)(2x + 3)(5x + 8)$ of two squares

49.
$75x^3 - 50x^2 -3x + 2$
$= (75x^3 - 50x^2) + (-3x + 2)$
$= 25x^2(3x - 2) - 1(3x - 2)$ Factor out -1
$= (25x^2 - 1)(3x - 2)$ Factor difference
$= (5x + 1)(5x - 1)(3x - 2)$ of two squares

51.
$10x^2 + 19x + 6$

<u>Product</u>
$10 \cdot 6 = 60$

<u>Factors</u>	<u>Sum</u>
$4 \cdot 15 = 60$	$4 + 15 = 19$

Replace 19x with 4x + 15x

So, $10x^2 + 19x + 6$
$= 10x^2 + 4x + 15x +6$
$= (10x^2 + 4x) + (15x + 6)$ Factor by
$= 2x(5x + 2) + 3(5x + 2)$ grouping
$= (2x + 3)(5x + 2)$

53.
$5x^2 + 6x - 8$

<u>Product</u>
$5(-8) = -40$

<u>Factors</u>	<u>Sum</u>
$10(-4) = -40$	$10 - 4 = 6$

Replace 6x with 10x - 4x

#53 continued

So, $5x^2 + 6x - 8$
$= 5x^2 + 10x - 4x - 8$
$= (5x^2 + 10x) + (-4x - 8)$ Factor by grouping
$= 5x(x + 2) -4(x + 2)$
$= (5x - 4)(x + 2)$

55.
$9x^2 - 29x + 6$

<u>Product</u>
$9 \cdot 6 = 54$

<u>Factors</u>	<u>Sum</u>
$-2(-27) = 54$	$-2 -27 = -29$

Replace -29x with -2x -27x

So, $9x^2 - 29x + 6$
$= 9x^2 - 2x - 27x + 6$
$= (9x^2 - 2x) + (-27x + 6)$ Factor by grouping
$= x(9x - 2) -3(9x - 2)$
$= (x - 3)(9x - 2)$

57.
$4x^2 + 4x - 35$

<u>Product</u>
$4(-35) = -140$

<u>Factors</u>	<u>Sum</u>
$14(-10) = 140$	$14 - 10 = 4$

Replace 4x with 14x - 10x

So, $4x^2 + 4x - 35$
$= 4x^2 + 14x -10x -35$
$= (4x^2 + 14x) + (-10x -35)$ Factor by
$= 2x(2x + 7) -5(2x + 7)$ grouping
$= (2x - 5)(2x + 7)$

59.
$10x^2 - 27x + 5$

<u>Product</u>
$10 \cdot 5 = 50$

<u>Factors</u>	<u>Sum</u>
$-2(-25) = 50$	$-2 - 25 = -27$

Replace -27x with -2x - 25x

So, $10x^2 - 27x + 5$
$= 10x^2 - 2x - 25x + 5$
$= (10x^2 - 2x) + (-25x + 5)$ Factor by
$= 2x(5x - 1) -5(5x - 1)$ grouping
$= (2x - 5)(5x - 1)$

61.
$6x^2 + 23x + 10$

#61 continued

Product
$6 \cdot 10 = 60$

Factors	Sum
$3 \cdot 20 = 60$	$3 + 20 = 23$

Replace 23x with 3x + 20x

So, $6x^2 + 23x + 10$
$= 6x^2 + 3x + 20x + 10$
$= (6x^2 + 3x) + (20x + 10)$ Factor by
$= 3x(2x + 1) + 10(2x + 1)$ grouping
$= (3x + 10)(2x + 1)$

63.
$12x^2 - 17x + 6$

Product
$12 \cdot 6 = 72$

Factors	Sum
$-9(-8) = 72$	$-9 - 8 = -17$

Replace -17x with -9x -8x

So, $12x^2 - 17x + 6$
$= 12x^2 - 9x - 8x + 6$
$= (12x^2 - 9x) + (-8x + 6)$
$= 3x(4x - 3) - 2(4x - 3)$
$= (3x - 2)(4x - 3)$

65.
$12x^2 + 20x + 7$

Product
$12 \cdot 7 = 84$

Factors	Sum
$6 \cdot 14 = 84$	$6 + 14 = 20$

Replace 20x with 6x + 14x

So, $12x^2 + 20x + 7$
$= 12x^2 + 6x + 14x + 7$
$= 6x(2x + 1) + 7(2x + 1)$
$= (6x + 7)(2x + 1)$

67.
$4x^2 + 20x + 25$

Product
$4 \cdot 25 = 100$

Factors	Sum
$10 \cdot 10 = 100$	$10 + 10 = 20$

Replace 20x with 10x + 10x

So, $4x^2 + 20x + 25$
$= 4x^2 + 10x + 10x + 25$
$= (4x^2 + 10x) + (10x + 25)$
$= 2x(2x + 5) + 5(2x + 5)$
$= (2x + 5)(2x + 5)$

69.
$6x^2 + x - 7$

Product
$6(-7) = -42$

Factors	Sum
$7(-6) = -42$	$7 - 6 = 1$

Replace x with 7x - 6x

So, $6x^2 + x - 7$
$= 6x^2 + 7x - 6x - 7$
$= (6x^2 + 7x) + (-6x - 7)$
$= x(6x + 7) - 1(6x + 7)$
$= (x - 1)(6x + 7)$

71.
$8x^2 - 10x - 3$

Product
$8(-3) = -24$

Factors	Sum
$-12 \cdot 2 = -24$	$-12 + 2 = -10$

Replace -10x with -12x + 2x

So, $8x^2 - 10x - 3$
$= 8x^2 - 12x + 2x - 3$
$= (8x^2 - 12x) + (2x - 3)$
$= 4x(2x - 3) + 1(2x - 3)$
$= (4x + 1)(2x - 3)$

73.
$4x^2 + 5x - 9$

Product
$4(-9) = -36$

Factors	Sum
$9(-4) = -36$	$9 - 4 = 5$

Replace 5x with 9x - 4x

So, $4x^2 + 5x - 9$
$= 4x^2 + 9x - 4x - 9$
$= (4x^2 + 9x) + (-4x - 9)$
$= x(4x + 9) - 1(4x + 9)$
$= (x - 1)(4x + 9)$

75.
$5x^2 + 16x + 12$

Product
$5 \cdot 12 = 60$

Factors	Sum
$6 \cdot 10 = 60$	$6 + 10 = 16$

Replace 16x with 6x + 10x

So, $5x^2 + 16x + 12$
$= 5x^2 + 6x + 10x + 12$
$= (5x^2 + 6x) + (10x + 12)$
$= x(5x + 6) + 2(5x + 6)$

#75 continued

$= (x + 2)(5x + 6)$

77.
$9x^2 - 18x + 8$

$$\underline{\text{Product}}$$
$$9 \cdot 8 = 72$$

Factors	Sum
$-6(-12) = 72$	$-6 -12 = -18$

Replace $-18x$ with $-6x - 12x$

So, $9x^2 - 18x + 8$
$= 9x^2 - 6x - 12x + 8$
$= (9x^2 - 6x) + (-12x + 8)$
$= 3x(3x - 2) - 4(3x - 2)$
$= (3x - 4)(3x - 2)$

79.
$9x^2 - 6x - 8$

$$\underline{\text{Product}}$$
$$9(-8) = -72$$

Factors	Sum
$-12 \cdot 6 = -72$	$-12 + 6 = -6$

Replace $-6x$ with $-12x + 6x$

So, $9x^2 - 6x - 8$
$= 9x^2 - 12x + 6x - 8$
$= (9x^2 - 12x) + (6x - 8)$
$= 3x(3x - 4) + 2(3x - 4)$
$= (3x + 2)(3x - 4)$

81.
$8x^2 + 35x + 12$

$$\underline{\text{Product}}$$
$$8 \cdot 12 = 96$$

Factors	Sum
$3 \cdot 32 = 96$	$3 + 32 = 35$

Replace $35x$ with $3x + 32x$

So, $8x^2 + 35x + 12$
$= 8x^2 + 3x + 32x + 12$
$= (8x^2 + 3x) + (32x + 12)$
$= x(8x + 3) + 4(8x + 3)$
$= (x + 4)(8x + 3)$

83.
$4x^2 - 12x + 9$

$$\underline{\text{Product}}$$
$$4 \cdot 9 = 36$$

Factors	Sum
$-6(-6) = 36$	$-6 -6 = -12$

Replace $12x$ with $-6x - 6x$

#83 continued

So, $4x^2 - 12x + 9$
$= 4x^2 - 6x - 6x + 9$
$= (4x^2 - 6x) + (-6x + 9)$
$= 2x(2x - 3) - 3(2x - 3)$
$= (2x - 3)(2x - 3)$

85.
$5x^2 - 13x + 6$

$$\underline{\text{Product}}$$
$$5 \cdot 6 = 30$$

Factors	Sum
$-10(-3) = 30$	$-10 -3 = -13$

Replace $13x$ with $-10x - 3x$

So, $5x^2 - 13x + 6$
$= 5x^2 - 10x - 3x + 6$
$= (5x^2 - 10x) + (-3x + 6)$
$= 5x(x - 2) - 3(x - 2)$
$= (5x - 3)(x - 2)$

Exercises 4.6

1.
$x^3 + 27 = (x)^3 + (3)^3$
$= (x + 3)(x^2 - 3 \cdot x + 3^2)$
$= (x + 3)(x^2 - 3x + 9)$

3.
$125 - t^3 = (5)^3 - (t)^3$
$= (5-t)(5^2 + 5 \cdot t + t^2)$
$= (5 - t)(25 + 5t + t^2)$

5.
$r^3 - 216 = (r)^3 - (6)^3$
$= (r - 6)(r^2 + 6 \cdot r + 6^2)$
$= (r - 6)(r^2 + 6r + 36)$

7.
$x^3 - 8 = (x)^3 - 2^3$
$= (x - 2)(x^2 + 2 \cdot x + 2^2)$
$= (x - 2)(x^2 + 2x + 4)$

9.
$x^3 + 1 = (x)^3 + (1)^3$
$= (x + 1)(x^2 - 1 \cdot x + 1^2)$
$= (x + 1)(x^2 - x + 1)$

11.
$8x^3 + 125 = (2x)^3 + (5)^3$
$= (2x + 5)[(2x)^2 - 2x \cdot 5 + 5^2]$
$= (2x + 5)(4x^2 - 10x + 25)$

13.

$64y^3 - 1 = (4y)^3 - (1)^3$
$= (4y - 1)[(4y)^2 + 4y \cdot 1 + 1^2]$
$= (4y - 1)(16y^2 + 4y + 1)$

15.

$27 - 64m^3 = (3)^3 - (4m)^3$
$= (3 - 4m)[3^2 + 3 \cdot 4m + (4m)^2]$
$= (3 - 4m)(9 + 12m + 16m^2)$

17.

$x^3 + 8y^3 = (x)^3 + (2y)^3$
$= (x + 2y)[x^2 - x \cdot 2y + (2y)^2]$
$= (x + 2y)(x^2 - 2xy + 4y^2)$

19.

$125x^3 - 8y^3$
$= (5x)^3 - (2y)^3$
$= (5x - 2y)[(5x)^2 + 5x \cdot 2y + (2y)^2]$
$= (5x - 2y)(25x^2 + 10xy + 4y^2)$

21.

$125p^3 + q^3 = (5p)^3 + (q)^3$
$= (5p + q)[(5p)^2 - 5p \cdot q + q^2]$
$= (5p + q)(25p^2 - 5pq + q^2)$

23.

$x^3 - 64y^3 = (x)^3 - 4(y)^3$
$= (x - 4y)[x^2 + x \cdot 4y + (4y)^2]$
$= (x - 4y)(x^2 + 4xy + 16y^2)$

25.

$64m^3 + 125n^3$
$= (4m)^3 + (5n)^3$
$= (4m + 5n)[(4m)^2 - 4m \cdot 5n + (5n)^2]$
$= (4m + 5n)(16m^2 - 20mn + 25n^2)$

27.

$2x^3 + 250 = 2(x^3 + 125)$
$= 2(x + 5)(x^2 - 5x + 5^2)$
$= 2(x + 5)(x^2 - 5x + 25)$

29.

$3r^3 - 81 = 3(r^3 - 27)$
$= 3(r - 3)(r^2 + 3 \cdot r + 3^2)$
$= 3(r - 3)(r^2 + 3r + 9)$

31.

$32m^3 - 108n^3$
$= 4(8m^3 - 27n^3)$
$= 4(2m - 3n)[(2m)^2 + 2m \cdot 3n + (3n)^2]$

#31 continued

$= 4(2m - 3n)(4m^2 + 6mn + 9n^2)$

33.

$432p^3 + 2q^3 = 2(216p^3 + q^3)$
$= 2(6p + q)[(6p)^2 - 6p \cdot q + q^2]$
$= 2(6p + q)(36p^2 - 6pq + q^2)$

35.

$128x^4y - 2xy^4$
$= 2xy(64x^3 - y^3)$
$= 2xy(4x - y)[(4x)^2 + 4x \cdot y + y^2]$
$= 2xy(4x - y)(16x^2 + 4xy + y^2)$

37.

$81x^4y^2 - 375xy^5$
$= 3xy^2(27x^3 - 125y^3)$
$= 3xy^2(3x - 5y)[(3x)^2 + 3x \cdot 5y + (5y)^2]$
$= 3xy^2(3x - 5y)(9x^2 + 15xy + 25y^2)$

39.

$8x^6 - y^3 = (2x^2)^3 - (y)^3$
$= (2x^2 - y)[(2x^2)^2 + 2x^2 \cdot y + y^2]$
$= (2x^2 - y)(4x^4 + 2x^2y + y^2)$

41.

$27x^3 + 8y^6$
$= (3x)^3 + (2y^2)^3$
$= (3x + 2y^2)[(3x)^2 - 3x \cdot 2y^2 + (2y^2)^2]$
$= (3x + 2y^2)(9x^2 - 6xy^2 + 4y^4)$

43.

$8x^9 + y^3 = (2x^3)^3 + (y)^3$
$= (2x^3 + y)[(2x^3)^2 - 2x^3 \cdot y + y^2]$
$= (2x^3 + y)(4x^6 - 2x^3y + y^2)$

45.

$x^9 - y^6 = (x^3)^3 - (y^2)^3$
$= (x^3 - y^2)[(x^3)^2 + x^3 \cdot y^2 + (y^2)^2]$
$= (x^3 - y^2)(x^6 + x^3y^2 + y^4)$

47.

$81x^6 + 192y^3$
$= 3(27x^6 + 64y^3)$
$= (3x^2 + 4y)[(3x^2)^2 - 3x^2 \cdot 4y + (4y)^2]$
$= 3(3x^2 + 4y)(9x^4 - 12x^2y + 16y^2)$

49.

$128x^7y + 250xy^7$
$= 2xy(64x^6 - 125y^6)$
$= 2xy(4x^2 - 5y^2)[(4x^2)^2 + 4x^2 \cdot 5y^2 + (5y^2)^2]$
$= 2xy(4x^2 - 5y^2)(16x^4 + 20x^2y^2 + 25y^4)$

Exercises 4.7

1.
$x^2 - 3x + 2 = (x + m)(x + n)$

-2 and -1 are two integers whose product is 2 and sum is -3. Thus, m = -2 and n = -1

So, $x^2 - 3x + 2 = (x-2)(x - 1)$

3.
$x^2 - 2x - 3 = (x + m)(x + n)$

-3 and 1 are two integers whose product is -3 and sum is -2. Thus, m = -3 and n = 1.

So, $x^2 - 2x - 3 = (x - 3)(x + 1)$

5.
$2x^2 + 7x + 3 = (2x + h)(x + k)$

<u>Possible Factor Combinations</u>

h · k = 3	
1 · 3	Both factors
3 · 1	are positive

The factor combination that gives the correct product is 1 and 3
Thus, $(2x + 1)(x + 3) = 2x^2 + 7x + 3$

So, $2x^2 + 7x + 3 = (2x + 1)(x + 3)$

7.
$9x^2 - 1 = (3x)^2 - (1)^2 = (3x + 1)(3x - 1)$

This is the difference of two squares that results in the product of two conjugates.

9.
$x^2 + 11x + 24 = (x + m)(x + n)$

3 and 8 are two integers whose product is 24 and sum is 11. Thus, m = 3 and n = 8.

So, $x^2 + 11x + 24 = (x + 3)(x + 8)$

11.
$x^2 + 10x + 25 = (x)^2 + 2(x \cdot 5) + 5^2 = (x + 5)^2$

The trinomial is a perfect square.

13.
$y^2 - y - 12 = (y + m)(y + n)$

-4 and 3 are two integers whose product is -12 and sum is -1. Thus, m = -4 and n = 3

So, $y^2 - y - 12 = (y - 4)(y + 3)$

15.
$x^2 - 36 = (x)^2 - (6)^2 = (x - 6)(x + 6)$

17.
$x^2 - 7x + 8$

There are no two integers whose product is 8 and sum is -7. The trinomial is prime.

19.
$2x^2 - 4x = 2x(x - 2)$ 	GCF = 2x

21.
$9x^2 - 42x + 49$
$= (3x)^2 - 2(3x \cdot 7) + (7)^2 = (3x - 7)^2$

The trinomial is a perfect square.

23.
$x^2 + 9x - 22 = (x + m)(x + n)$

11 and -2 are two integers whose product is -22 and sum is 9. Thus, m = 11 and n = -2

So, $x^2 + 9x - 22 = (x + 11)(x - 2)$

25.
$y^2 - 11y + 28 = (y + m)(y + n)$

-4 and -7 are two integers whose product is 28 and sum is -11. Thus, m = -4 and n = -7.

So, $y^2 - 11y + 28 = (y - 4)(y - 7)$

27.
$2x^3 + 5x^2 + 6x + 15$
$= (2x^3 + 5x^2) + (6x + 15)$
$= x^2(2x + 5) + 3(2x + 5)$
$= (x^2 + 3)(2x + 5)$ 	Factor

29.
$21 + 4x - x^2 \quad = -x^2 + 4x + 21$
$= - (x^2 - 4x - 21)$
$= -(x - 7)(x + 3)$
$= (-x + 7)(x + 3)$
$= (7 - x)(3 + x)$

31.
$64x^2 - 81 = (8x)^2 - (9)^2 = (8x + 9)(8x - 9)$

33.
$4x^2 + 4x - 15 \quad = (4x + h)(x + k)$
or
$= (2x + h)(2x + k)$

<u>Possible Factor Combinations</u>

	h · k = -15	
1 · (-15)	15(-1)	
-15 · 1	-1 · 15	Signs are
3(-5)	5(-3)	opposite
-5 · 3	-3 · 5	

#33 continued

The factor combination that gives the correct product is -3 and 5 with $(2x + h)(2x + k)$.

Thus, $(2x - 3)(2x + 5) = 4x^2 + 4x - 15$

So, $4x^2 + 4x - 15 = (2x - 3)(2x + 5)$

35.

$$24x^2 - 37x - 5 = (24x + h)(x + k)$$
$$\text{or}$$
$$= (6x + h)(4x + k)$$
$$\text{or}$$
$$= (12x + h)(2x + k)$$
$$\text{or}$$
$$= (8x + h)(3x + k)$$

Possible Factor Combinations

$$h \cdot k = -5$$

| $1 \cdot (-5)$ | $5(-1)$ | Signs are |
| $-5 \cdot 1$ | $-1 \cdot 5$ | opposite |

The factor combination that gives the correct product is 1 and -5 with $(8x + h)(3x + k)$.

Thus, $(8x + 1)(3x - 5) = 24x^2 - 37x - 5$

So, $24x^2 - 37x - 5 = (8x + 1)(3x - 5)$

37.

$$5x^5 - 80xy^4 = 5x(x^4 - 16y^4)$$
$$= 5x(x^2 + 4y^2)(x^2 - 4y^2)$$
$$= 5x(x^2 + 4y^2)(x - 2y)(x + 2y)$$

39.

$2x^2 + 4x + 1$ is prime since no two integers have a product 2 and sum 4.

41.

$$3x^2(2x - 7) + 6x(2x - 7)$$
$$= (3x^2 + 6x)(2x - 7) \qquad \text{Factor}$$
$$= 3x(x + 2)(2x - 7) \qquad \text{Factor}$$

43.

$$8y^2 - 6y - 9 = (8y + h)(y + k)$$
$$\text{or}$$
$$= (4y + h)(2y + k)$$

Possible Factor Combinations

$$h \cdot k = -9$$

| $-9 \cdot 1$ | $9(-1)$ | Signs are |
| $3(-3)$ | $-1 \cdot 9$ | opposite |

The factor combination that gives the correct product is 3 and -3 with $(4y + h)(2y + k)$.

Thus, $(4y + 3)(2y - 3) = 8y^2 - 6y - 9$

#43 continued

So, $8y^2 - 6y - 9 = (4y + 3)(2y - 3)$

45.

$$18a^3b^3 - 12ab^5 - 6ab^3 = 6ab^3(3a^2 - 2b^2 - 1)$$
$$\text{Factor out GCF}$$

47.

$$6x^2 + x - 12 = (6x + h)(x + k)$$
$$\text{or}$$
$$= (2x + h)(3x + k)$$

Possible Factor Combinations

$$h \cdot k = -12$$

$12 \cdot (-1)$	$-12 \cdot 1$	
$-1 \cdot 12$	$1(-12)$	Signs are
$3(-4)$	$-3 \cdot 4$	opposite
$-4 \cdot 3$	$4(-3)$	

The factor combination that gives the correct product is 3 and -4 with $(2x + h)(3x + k)$.

Thus, $(2x + 3)(3x - 4) = 6x^2 + x - 12$

So, $6x^2 + x - 12 = (2x + 3)(3x - 4)$

49.

$$x^2 + 20xy + 75y^2 = (x + my)(x + ny)$$

5 and 15 are two integers whose product is 75 and sum is 20. Thus, m = 5 and n = 15

So, $x^2 + 20xy + 75y^2 = (x + 5y)(x + 15y)$

51.

$$3x^3 + x^2y - 3x - y$$
$$= (3x^3 + x^2y) + (-3x - y) \qquad \text{Group}$$
$$= x^2(3x + y) - 1 \cdot (3x + y) \qquad \text{Factor}$$
$$= (x^2 - 1)(3x + y) \qquad \text{Factor}$$
$$= (x + 1)(x - 1)(3x + y) \qquad \text{Factor}$$

53.

$$6x^3 + 27x^2 + 12x = 3x(2x^2 + 9x + 4)$$
$$= 3x(2x + h)(x + k)$$

Possible Factor Combinations

$$h \cdot k = 4$$

| $4 \cdot 1$ | $1 \cdot 4$ | $2 \cdot 2$ | Signs are |
| | | | positive |

The factor combination that gives the correct product is 1 and 4.

Thus, $3x(2x + 1)(x + 4) = 3x(2x^2 + 9x + 4)$
$$= 6x^3 + 27x^2 + 12x$$

So, $6x^3 + 27x^2 + 12x = 3x(2x + 1)(x + 4)$

55.

$$12x^2 + 11x - 15 \quad = (12x + h)(x + k)$$
$$\text{or}$$
$$= (6x + h)(2x + k)$$
$$\text{or}$$
$$= (4x + h)(3x + k)$$

Possible Factor Combinations
$$h \cdot k = -15$$

-15 · 1	15(-1)	3(-5)	Signs are
1 · (-15)	-1 · 15	-5 · 3	opposite
-3 · 5	5(-3)		

The factor combination that gives the correct product is -3 and 5 with $(4x + h)(3x + k)$

Thus, $(4x - 3)(3x + 5) = 12x^2 + 11x - 15)$

So, $12x^2 + 11x - 15 = (4x - 3)(3x + 5)$

57.

$$5x^3 - 125x = 5x(x^2 - 25) = 5x(x + 5)(x - 5)$$

59.

$$2x^3 + 5x^2 - 2x - 5$$
$$= (2x^3 + 5x^2) - 1 \cdot (2x + 5) \qquad \text{Group}$$
$$= x^2(2x + 5) - 1 \cdot (2x + 5) \qquad \text{Factor}$$
$$= (x^2 - 1)(2x + 5) \qquad \text{Factor}$$
$$= (x + 1)(x - 1)(2x + 5) \qquad \text{Factor}$$

61.

$$x^4 + 19x^2 - 20 = (x^2 + h)(x^2 + k)$$

Possible Factor Combinations
$$h \cdot k = -20$$

1 · (-20)	20(-1)	5(-4)	
-20 · 1	-1 · 20	-4 · 5	Signs are
2(-10)	-2 · 10	-5 · 4	opposite
-10 · 2	10(-20)	4(-5)	

The factor combination that gives the correct product is 20 and -1.

Thus, $(x^2 + 20)(x^2 - 1) = x^4 + 19x^2 - 20$
Notice that $(x^2 - 1)$ can be factored.

So, $x^4 + 19x^2 - 20 = (x^2 + 20)(x + 1)(x - 1)$

63.

$$x^4 + 3x^3y - 40x^2y^2 = x^2(x^2 + 3xy - 40y^2)$$
$$= x^2(x + my)(x + ny)$$

8 and -5 are two integers whose product is -40 and sum is 3. Thus, m = 8 and n = -5.

So, $x^4 + 3x^3y - 40x^2y^2 = x^2(x + 8y)(x - 5y)$

65.

$$8x^3 + 27y^3$$
$$= (2x)^3 + (3y)^3$$
$$= (2x + 3y)((2x)^2 - 2x \cdot 3y + (3y)^2)$$
$$= (2x + 3y)(4x^2 - 6xy + 9y^2)$$

67.

$$2x^5 - 16x^2 \quad = 2x^2(x^3 - 8)$$
$$= 2x^2((x)^3 - (2)^3)$$
$$= 2x^2(x - 2)(x^2 + 2x + 4)$$

69.

$$x^6 + 7x^3 - 8 = (x^3 + m)(x^3 + n)$$

8 and -1 are two integers whose product is -8 and sum is 7.

Thus, $(x^3 + 8)(x^3 - 1) = x^6 + 7x^3 - 8$

Notice that the two factors are the sum and difference of perfect cubes.

So, $x^6 + 7x^3 - 8 = (x^3 + 8)(x^3 - 1)$
$= (x + 2)(x^2 - 2x + 4)(x - 1)(x^2 + x + 1)$

Exercises 4.8

1.

$$(x + 3)(x + 7) = 0$$
$$x + 3 = 0 \quad \text{or} \quad x + 7 = 0 \qquad \text{Zero factor law}$$

Thus, x = -3 or x = -7; and the solution set is {-3, -7}.

3.

$$(2x - 5)(x + 12) = 0$$

$2x - 5 = 0$	or	$x + 12 = 0$	Zero
$2x = 5$		$x = -12$	factor law
$x = \dfrac{5}{2}$			

Thus, $x = \dfrac{5}{2}$ or $x = -12$ and the solution set is $\left\{ \dfrac{5}{2}, -12 \right\}$.

5.

$$(4x - 1)(4x - 1) = 0$$

$4x - 1 = 0$	or	$4x - 1 = 0$	Zero
$4x = 1$		$4x = 1$	factor law
$x = \dfrac{1}{4}$		$x = \dfrac{1}{4}$	

Thus, $x = \dfrac{1}{4}$ and the solution set is $\left\{ \dfrac{1}{4} \right\}$.

7.

$(3x - 4)(2x + 7) = 0$

$3x - 4 = 0$ or $2x + 7 = 0$ Zero factor

   $3x = 4$ $2x = -7$ law

     $x = \dfrac{4}{3}$ $x = -\dfrac{7}{2}$

Thus, $x = \dfrac{4}{3}$ or $x = -\dfrac{7}{2}$

Solution set: $\left\{\dfrac{4}{3}, -\dfrac{7}{2}\right\}$.

9.

$5x(x - 2)(x + 3) = 0$

$5x = 0$ or $x - 2 = 0$ or $x + 3 = 0$

 $x = 0$ $x = 2$ $x = -3$

Thus, $x = 0$, or $x = 2$, or $x = -3$

Solution set: $\{0, 2, -3\}$.

11.

$(2x + 7)(x - 3)(3x + 1) = 0$

$2x + 7 = 0$ or $x - 3 = 0$ or $3x + 1 = 0$

  $2x = -7$ $x = 3$ $3x = -1$

   $x = -\dfrac{7}{2}$ $x = -\dfrac{1}{3}$

Thus, $x = -\dfrac{7}{2}$, or $x = 3$ or $x = -\dfrac{1}{3}$

Solution set: $\left\{-\dfrac{7}{2}, 3, -\dfrac{1}{3}\right\}$.

13.

$(x + 1)(x - 7)(x + 8)(x - 1) = 0$

$x + 1 = 0$ or $x - 7 = 0$ or $x + 8 = 0$ or $x - 1 = 0$

  $x = -1$ $x = 7$ $x = 8$ $x = 1$

A check verifies that the solution set is $\{-1, 7, 8, 1\}$.

15.

$x^2 - 9x + 18 = 0$

$(x - 3)(x - 6) = 0$ Step 2

$x - 3 = 0$ or $x - 6 = 0$ Step 3

   $x = 3$ $x = 6$ Step 4

A check verifies that the solution set is $\{3, 6\}$.

17.

$2x^2 + x - 15 = 0$

$(2x - 5)(x + 3) = 0$ Step 2

#17 continued

$2x - 5 = 0$ or $x + 3 = 0$ Step 3

   $2x = 5$ $x = -3$ Step 4

    $x = \dfrac{5}{2}$

A check verifies that the solution set is $\left\{\dfrac{5}{2}, -3\right\}$.

19.

$5x^2 + 15x = 0$

$5x(x + 3) = 0$ Step 2

$5x = 0$ or $x + 3 = 0$ Step 3

  $x = 0$ $x = -3$ Step 4

A check verifies that the solution set is $\{0, -3\}$.

21.

$x^2 - 4x - 12 = 0$

$(x - 6)(x + 2) = 0$ Step 2

$x - 6 = 0$ or $x + 2 = 0$ Step 3

   $x = 6$ $x = -2$ Step 4

A check verifies that the solution set is $\{6, -2\}$.

23.

$6x^2 - 3x = 0$

$3x(2x - 1) = 0$ Step 2

$3x = 0$ or $2x - 1 = 0$ Step 3

  $x = 0$ $2x = 1$ Step 4

           $x = \dfrac{1}{2}$

A check verifies that the solution set is $\left\{0, \dfrac{1}{2}\right\}$.

25.

$10x^2 - 19x + 7 = 0$

$(2x - 1)(5x - 7) = 0$ Step 2

$2x - 1 = 0$ or $5x - 7 = 0$ Step 3

  $2x = 1$ $5x = 7$ Step 4

   $x = \dfrac{1}{2}$ $x = \dfrac{7}{5}$

A check verifies that the solution set is $\left\{\dfrac{1}{2}, \dfrac{7}{5}\right\}$.

27.

$9x^2 + 12x + 4 = 0$

$(3x + 2)(3x + 2) = 0$ Step 2

#27 continued

$$3x + 2 = 0 \quad \text{or} \quad 3x + 2 = 0 \qquad \text{Step 3}$$
$$3x = -2 \qquad\qquad\qquad\qquad \text{Step 4}$$
$$x = -\frac{2}{3}$$

A check verifies that the solution set is
$\left\{-\dfrac{2}{3}\right\}.$

29.

$$8x^2 - 22x - 21 = 0$$
$$(4x + 3)(2x - 7) = 0 \qquad\qquad \text{Step 2}$$
$$4x + 3 = 0 \quad \text{or} \quad 2x - 7 = 0 \qquad \text{Step 3}$$
$$4x = -3 \qquad\qquad 2x = 7 \qquad \text{Step 4}$$
$$x = -\frac{3}{4} \qquad\qquad x = \frac{7}{2}$$

A check verifies that the solution set
is $\left\{-\dfrac{3}{4}, \dfrac{7}{2}\right\}.$

31.

$$6x^2 + 13x + 5 = 0$$
$$(2x + 1)(3x + 5) = 0 \qquad\qquad \text{Step 2}$$
$$2x + 1 = 0 \quad \text{or} \quad 3x + 5 = 0 \qquad \text{Step 3}$$
$$2x = -1 \qquad\qquad 3x = -5 \qquad \text{Step 4}$$
$$x = -\frac{1}{2} \qquad\qquad x = -\frac{5}{3}$$

A check verifies that the solution set
is $\left\{-\dfrac{1}{2}, -\dfrac{5}{3}\right\}.$

33.

$$16x^2 - 1 = 0$$
$$(4x + 1)(4x - 1) = 0 \qquad\qquad \text{Step 2}$$
$$4x + 1 = 0 \quad \text{or} \quad 4x - 1 = 0 \qquad \text{Step 3}$$
$$4x = -1 \qquad\qquad 4x = 1 \qquad \text{Step 4}$$
$$x = -\frac{1}{4} \qquad\qquad x = \frac{1}{4}$$

A check verifies that the solution set
is $\left\{-\dfrac{1}{4}, \dfrac{1}{4}\right\}.$

35.

$$x^2 + 11x + 38 = 8$$
$$x^2 + 11x + 30 = 0 \qquad\qquad \text{Step 1}$$
$$(x + 5)(x + 6) = 0 \qquad\qquad \text{Step 2}$$
$$x + 5 = 0 \quad \text{or} \quad x + 6 = 0 \qquad \text{Step 3}$$
$$x = -5 \quad \text{or} \quad x = -6 \qquad \text{Step 4}$$

A check verifies that the solution set
is {-5, -6}.

37.

$$8x^2 + x - 10 = 2x - 3$$
$$8x^2 - x - 7 = 0 \qquad\qquad \text{Step 1}$$
$$(8x + 7)(x - 1) = 0 \qquad\qquad \text{Step 2}$$
$$8x + 7 = 0 \quad \text{or} \quad x - 1 = 0 \qquad \text{Step 3}$$
$$8x = -7 \qquad\qquad x = 1 \qquad \text{Step 4}$$
$$x = -\frac{7}{8}$$

A check verifies that the solution set is
$\left\{-\dfrac{7}{8}, 1\right\}.$

39.

$$13x^2 - 13x + 9 = x^2 + 6$$
$$12x^2 - 13x + 3 = 0 \qquad\qquad \text{Step 1}$$
$$(3x - 1)(4x - 3) = 0 \qquad\qquad \text{Step 2}$$
$$3x - 1 = 0 \quad \text{or} \quad 4x - 3 = 0 \qquad \text{Step 3}$$
$$3x = 1 \qquad\qquad 4x = 3 \qquad \text{Step 4}$$
$$x = \frac{1}{3} \qquad\qquad x = \frac{3}{4}$$

A check verifies that the solution set
is $\left\{\dfrac{1}{3}, \dfrac{3}{4}\right\}.$

41

$$7x^2 + 4x + 1 = 2x^2 - 4x + 5$$
$$5x^2 + 8x - 4 = 0 \qquad\qquad \text{Step 1}$$
$$(5x - 2)(x + 2) \qquad\qquad \text{Step 2}$$
$$5x - 2 = 0 \quad \text{or} \quad x + 2 = 0 \qquad \text{Step 3}$$
$$5x = 2 \qquad\qquad x = -2 \qquad \text{Step 4}$$
$$x = \frac{2}{5}$$

A check verifies that the solution set
is $\left\{\dfrac{2}{5}, -2\right\}.$

43.

$$x^2 + 12x + 2 = 3(x - 4)$$
$$x^2 + 12x + 2 = 3x - 12$$
$$x^2 + 9x + 14 = 0 \qquad\qquad \text{Step 1}$$
$$(x + 7)(x + 2) = 0 \qquad\qquad \text{Step 2}$$
$$x + 7 = 0 \quad \text{or} \quad x + 2 = 0 \qquad \text{Step 3}$$
$$x = -7 \qquad\qquad x = -2 \qquad \text{Step 4}$$

A check verifies that the solution set is
{-7, -2}.

45.

$$5x^2 - x - 50 = 2(2x^2 + 3)$$
$$5x^2 - x - 50 = 4x^2 + 6$$
$$x^2 - x - 56 = 0 \qquad\qquad \text{Step 1}$$

#45 continued

$(x - 8)(x + 7) = 0$	Step 2
$x - 8 = 0$ or $x + 7 = 0$	Step 3
$x = 8$ $x = -7$	Step 4

A check verifies that the solution set is $\{8, -7\}$.

47.

$3x^2 - 18x + 21 = x(x-5)$

$3x^2 - 18x + 21 = x^2 - 5x$

$2x^2 - 13x + 21 = 0$	Step 1
$(2x - 7)(x - 3) = 0$	Step 2
$2x - 7 = 0$ or $x - 3 = 0$	Step 3
$2x = 7$ $x = 3$	Step 4

$x = \dfrac{7}{2}$

A check verifies that the solution set is $\left\{\dfrac{7}{2}, 3\right\}$.

49.

$13x^2 + 17x + 2 = 4x(x + 2)$

$13x^2 + 17x + 2 = 4x^2 + 8x$

$9x^2 + 9x + 2 = 0$	Step 1
$(3x + 1)(3x + 2) = 0$	Step 2
$3x + 1 = 0$ or $3x + 2 = 0$	Step 3
$3x = -1$ $3x = -2$	Step 4

$x = -\dfrac{1}{3}$ $x = -\dfrac{2}{3}$

A check verifies that the solution set is $\left\{-\dfrac{1}{3}, -\dfrac{2}{3}\right\}$.

51.

$10x^2 - 2x - 64 = (x + 3)(x - 5)$

$10x^2 - 2x - 64 = x^2 - 2x - 15$

$9x^2 - 49 = 0$

$(3x - 7)(3x + 7) = 0$

$3x - 7 = 0$ or $3x + 7 = 0$

$3x = 7$ $3x = -7$

$x = \dfrac{7}{3}$ $x = -\dfrac{7}{3}$

A check verifies that the solution set is $\left\{\dfrac{7}{3}, -\dfrac{7}{3}\right\}$.

53.

$9x^2 - 71x + 5 = (2x + 1)(x - 4)$

$9x^2 - 71x + 5 = 2x^2 - 7x - 4$

$7x^2 - 64x + 9 = 0$	Step 1

#53 continued

$(7x - 1)(x - 9) = 0$	Step 2
$7x - 1 = 0$ or $x - 9 = 0$	Step 3
$7x = 1$ $x = 9$	Step 4

$x = \dfrac{1}{7}$

A check verifies that the solution set is $\left\{\dfrac{1}{7}, 9\right\}$.

55.

$(2x + 3)(x + 1) = 1$

$2x^2 + 5x + 3 = 1$

$2x^2 + 5x + 2 = 0$	Step 1
$(x + 2)(2x + 1) = 0$	Step 2
$x + 2 = 0$ or $2x + 1 = 0$	Step 3
$x = -2$ $x = -\dfrac{1}{2}$	

A check verifies that the solution set is $\left\{-2, -\dfrac{1}{2}\right\}$.

57.

$(2x - 7)(x - 5) = 2$

$2x^2 - 17x + 35 = 2$

$2x^2 - 17x + 33 = 0$	Step 1
$(x - 3)(2x - 11) = 0$	Step 2
$x - 3 = 0$ or $2x - 11 = 0$	Step 3
$x = 3$ $2x = 11$	Step 4

$x = \dfrac{11}{2}$

A check verifies that the solution set is $\left\{3, \dfrac{11}{2}\right\}$.

59.

$(3x + 1)(2x - 5) = 39$

$6x^2 - 13x - 5 = 39$

$6x^2 - 13x - 44 = 0$	Step 1
$(x - 4)(6x + 11) = 0$	Step 2
$x - 4 = 0$ or $6x + 11 = 0$	Step 3
$x = 4$ $6x = -11$	Step 4

$x = -\dfrac{11}{6}$

A check verifies that the solution set is $\left\{4, -\dfrac{11}{6}\right\}$.

61.

$(2x + 5)(4x - 5) = -13$

$8x^2 + 10x - 25 = -13$

#61 continued

$$8x^2 + 10x - 12 = 0$$
$$4x^2 + 5x - 6 = 0 \qquad \text{Divide by 2}$$
$$(x + 2)(4x - 3) = 0 \qquad \text{Step 2}$$
$$x + 2 = 0 \quad \text{or} \quad 4x - 3 = 0 \qquad \text{Step 3}$$
$$x = -2 \qquad\qquad 4x = 3 \qquad \text{Step 4}$$
$$x = \frac{3}{4}$$

A check verifies that the solution set

is $\left\{-2, \dfrac{3}{4}\right\}$.

63.

$$(x - 9)(2x + 3) = (x - 9)(x + 4)$$
$$2x^2 - 15x - 27 = x^2 - 5x - 36$$
$$x^2 - 10x + 9 = 0 \qquad \text{Step 1}$$
$$(x - 1)(x - 9) = 0 \qquad \text{Step 2}$$
$$x - 1 = 0 \quad \text{or} \quad x - 9 = 0 \qquad \text{Step 3}$$
$$x = 1 \qquad\qquad x = 9 \qquad \text{Step 4}$$

A check verifies that the solution set is {1, 9}.

65.

$$(2x - 3)(3x + 4) = (3x + 4)(x - 5)$$
$$6x^2 - x - 12 = 3x^2 - 11x - 20$$
$$3x^2 + 10x + 8 = 0 \qquad \text{Step 1}$$
$$(3x + 4)(x + 2) = 0 \qquad \text{Step 2}$$
$$3x + 4 = 0 \quad \text{or} \quad x + 2 = 0 \qquad \text{Step 3}$$
$$3x = -4 \qquad\qquad x = -2 \qquad \text{Step 4}$$
$$x = -\frac{4}{3}$$

A check verifies that the solution set

is $\left\{-\dfrac{4}{3}, -2\right\}$.

67.

$$(2x - 1)(x + 3) = (x + 2)(x - 5)$$
$$2x^2 + 5x - 3 = x^2 - 3x - 10$$
$$x^2 + 8x + 7 = 0 \qquad \text{Step 1}$$
$$(x + 7)(x + 1) = 0 \qquad \text{Step 2}$$
$$x + 7 = 0 \quad \text{or} \quad x + 1 = 0 \qquad \text{Step 3}$$
$$x = -7 \qquad\qquad x = -1 \qquad \text{Step 4}$$

A check verifies that the solution set
is {-7, -1}.

69

$$(3x + 4)(x - 2) = (x - 1)(x + 8)$$
$$3x^2 - 2x - 8 = x^2 + 7x - 8$$
$$2x^2 - 9x = 0 \qquad \text{Step 1}$$
$$x(2x - 9) = 0 \qquad \text{Step 2}$$

#69 continued

$$x = 0 \quad \text{or} \quad 2x - 9 = 0 \qquad \text{Step 3}$$
$$2x = 9 \qquad \text{Step 4}$$
$$x = \frac{9}{2}$$

A check verifies that the solution set

is $\left\{0, \dfrac{9}{2}\right\}$.

71.

$$2x^3 + 4x^2 - 6x = 0$$
$$2x(x^2 + 2x - 3) = 0$$
$$2x(x + 3)(x - 1) = 0 \qquad \text{Step 2}$$
$$2x = 0 \quad \text{or} \quad x + 3 = 0 \quad \text{or} \quad x - 1 = 0 \qquad \text{Step 3}$$
$$x = 0 \qquad\qquad x = -3 \qquad\qquad x = 1 \qquad \text{Step 4}$$

A check verifies that the solution set
is {0, -3, 1}.

73.

$$27x^3 + 18x^2 + 3x = 0$$
$$3x(9x^2 + 6x + 1) = 0$$
$$3x(3x + 1)(3x + 1) = 0 \qquad \text{Step 2}$$
$$3x = 0 \quad \text{or} \quad 3x + 1 = 0 \quad \text{or} \quad 3x + 1 = 0$$
$$x = 0 \qquad\qquad 3x = -1$$
$$x = -\frac{1}{3}$$

A check verifies that the solution set

is $\left\{0, -\dfrac{1}{3}\right\}$.

75.

$$30x^3 - 65x^2 + 30x = 0$$
$$5x(6x^2 - 13x + 6) = 0$$
$$5x(2x - 3)(3x - 2) = 0 \qquad \text{Step 2}$$
$$5x = 0 \quad \text{or} \quad 2x - 3 = 0 \quad \text{or} \quad 3x - 2 = 0$$
$$x = 0 \qquad\qquad 2x = 3 \qquad\qquad 3x = 2$$
$$x = \frac{3}{2} \qquad\qquad x = \frac{2}{3}$$

A check verifies that the solution set

is $\left\{0, \dfrac{3}{2}, \dfrac{2}{3}\right\}$.

77.

$$(x + 2)(12x^2 - 32x + 5) = 0$$
$$(x + 2)(2x - 5)(6x - 1) = 0$$

#77 continued

$x + 2 = 0$ or $2x - 5 = 0$ or $6x - 1 = 0$
$\qquad x = -2 \qquad\qquad 2x = 5 \qquad\qquad 6x = 1$
$\qquad\qquad\qquad\qquad x = \dfrac{5}{2} \qquad\qquad x = \dfrac{1}{6}$

A check verifies that the solution set
is $\left\{-2, \dfrac{5}{2}, \dfrac{1}{6}\right\}$.

79.
$(2x - 5)(3x^2 - 29x + 18) = 0$
$(2x - 5)(3x - 2)(x - 9) = 0$

$2x - 5 = 0$ or $3x - 2 = 0$ or $x - 9 = 0$
$\quad 2x = 5 \qquad\qquad 3x = 2 \qquad\qquad x = 9$
$\quad x = \dfrac{5}{2} \qquad\qquad x = \dfrac{2}{3}$

A check verifies that the solution set
is $\left\{\dfrac{5}{2}, \dfrac{2}{3}, 9\right\}$.

81.
$(x^2 - 4)(4x^2 - 4x - 15) = 0$
$(x + 2)(x - 2)(2x - 5)(2x + 3) = 0$

$x + 2 = 0$ or $x - 2 = 0$ or $2x - 5 = 0$ or $2x + 3 = 0$
$\quad x = -2 \qquad x = 2 \qquad 2x = 5 \qquad 2x = -3$
$\qquad\qquad\qquad\qquad\qquad x = \dfrac{5}{2} \qquad x = -\dfrac{3}{2}$

A check verifies that the solution set
is $\left\{-2, 2, \dfrac{5}{2}, -\dfrac{3}{2}\right\}$.

Exercises 4.9

1.

Let x = 1st integer.
The two consecutive integers are x and $x + 1$.

The equation is
$x(x + 1) = 72$
$x^2 + x = 72$
$x^2 + x - 72 = 0$
$(x + 9)(x - 8) = 0$

$(x + 9) = 0$ or $x - 8 = 0$
$\quad x = -9 \qquad\qquad x = 8$

There are two possible sets of consecutive
integers: -9 and -8 or 8 and 9.
Note: If $\quad x = 9$ If $\quad x = 8$
$\qquad\qquad x + 1 = 8 \qquad x + 1 = 9$

3.

Let x = 1st even integer.

The two consecutive even integers are x
and $x + 2$.

The even integer product is $x(x + 2)$.

Eight times the larger integer plus 24
is $8(x + 2) + 24$.

The equation is
$x(x + 2) = 8(x + 2) + 24$
$x^2 + 2x = 8x + 16 + 24$
$x^2 + 2x = 8x + 40$
$x^2 - 6x - 40 = 0$
$(x - 10)(x + 4) = 0$

$x - 10 = 0$ or $x + 4 = 0$
$\quad x = 10 \qquad\qquad x = -4$

There are two possible sets of consecutive even
integers: 10 and 12 or -4 and -2.
Note: If $\quad x = 10$ If $\quad x = -4$
$\qquad\qquad x + 2 = 12 \qquad x + 2 = -2$

5.

Let x = 1st odd integer.

The two consecutive odd integers
are x and $x + 2$.

The odd integer product is $x(x + 2)$.

Twice their sum plus 95 is $2[x + (x + 2)] + 95$.

The equation is
$x(x + 2) = 2(x + (x + 2)) + 95$
$x^2 + 2x = 2(2x + 2) + 95$
$x^2 + 2x = 4x + 4 + 95$
$x^2 + 2x = 4x + 99$
$x^2 - 2x - 99 = 0$
$(x - 11)(x + 9) = 0$
$x - 11 = 0$ or $x + 9 = 0$
$\quad x = 11 \qquad\qquad x = -9$

There are two possible sets of consecutive odd
integers: 11 and 13 or -9 and -7. See note on
Exercise 3.

7.

Let x = 1st integer.

The two consecutive integers are x and $x + 1$.

The sum of the squares of each integer
is $x^2 + (x + 1)^2$.

Twelve times the larger plus 1
is $12(x + 1) + 1$.

#7 continued

The equation is
$$x^2 + (x+1)^2 = 12(x+1) + 1$$
$$x^2 + x^2 + 2x + 1 = 12x + 12 + 1$$
$$2x^2 + 2x + 1 = 12x + 13$$
$$2x^2 - 10x - 12 = 0$$
$$x^2 - 5x - 6 = 0 \qquad \text{Divide by 2}$$
$$(x-6)(x+1) = 0$$

$x - 6 = 0$ or $x + 1 = 0$
$\quad x = 6 \qquad\qquad x = -1$

There are two possible sets of consecutive integers: 6 and 7 or -1 and 0.
See note on Exercise 3.

9.

Let x = 1st number.

Let $6 - x$ = 2nd number.

The sum of the squares of each number is $x^2 + (6 - x)^2$.

The equation is
$$x^2 + (6 - x)^2 = 180$$
$$x^2 + 36 - 12x + x^2 = 180$$
$$2x^2 - 12x - 144 = 0$$
$$x^2 - 6x - 72 = 0 \qquad \text{Divide by 2}$$
$$(x+6)(x-12) = 0$$

$x + 6 = 0$ or $x - 12 = 0$
$\quad x = -6 \qquad\qquad x = 12$

The two integers are -6 and 12.

11.

Let x = 1st number.

Let $10 - x$ = 2nd number.

The difference of the squares of each number is $x^2 - (10 - x)^2$.

Their product less 4 is $x(10 - x) - 4$.

The equation is
$$x^2 - (10 - x)^2 = x(10 - x) - 4$$
$$x^2 - (100 - 20x + x^2) = 10x - x^2 - 4$$
$$x^2 - 100 + 20x - x^2 = 10x - x^2 - 4$$
$$x^2 + 10x - 96 = 0$$
$$(x-6)(x+16) = 0$$

$x - 6 = 0$ or $x + 16 = 0$
$\quad x = 6 \qquad\qquad x = -16$

There are two possible sets of numbers:
$x = 6$ and $10 - x = 4$
$x = -16$ and $10 - x = 26$

The numbers are 6 and 4 or -16 and 26.

13.

Let x = width measurement.

Let $x + 5$ = length measurement.

The area formula for a rectangle is $A = L \cdot W$.

The equation is
$$x(x + 5) = 36$$
$$x^2 + 5x = 36$$
$$x^2 + 5x - 36 = 0$$
$$(x-4)(x+9) = 0$$
$x - 4 = 0$ or $x + 9 = 0$
$\quad x = 4 \qquad\qquad x = -9$

There are two solutions, 4 and -9. Since measurement is not negative, -9 is rejected. The dimensions of the rectangle are 4 ft. by 9 ft. Note: $x + 5 = 4 + 5 = 9$.

15.

Let x = width measurement.

Let $2x + 3$ = length measurement.

The area formula for a rectangle is $A = L \cdot W$.

The equation is
$$x(2x + 3) = 77$$
$$2x^2 + 3x = 77$$
$$2x^2 + 3x - 77 = 0$$
$$(2x - 11)(x + 7) = 0$$
$2x - 11 = 0$ or $x + 7 = 0$
$\quad 2x = 11 \qquad\qquad x = -7$
$\quad x = \dfrac{11}{2}$

There are two solutions, $\dfrac{11}{2}$ and -7. Since measurement is not negative, -7 is rejected. The dimensions are
$5\dfrac{1}{2}$ yd. by 14 yd.

Note: $2x + 3 = 2 \cdot \dfrac{11}{2} + 3 = 11 + 3 = 14$.

17.

Let x = length measurement.

Let $\dfrac{1}{2}x - 4$ = width measurement.

173

#17 continued

$$\tfrac{1}{2}x - 4$$

The area formula for a rectangle is $A = L \cdot W$.

The equation is

$$x\left(\tfrac{1}{2}x - 4\right) = 90$$
$$\tfrac{1}{2}x^2 - 4x - 90 = 0$$
$$x^2 - 8x - 180 = 0 \qquad \text{Multiply by 2}$$
$$(x - 18)(x + 10) = 0$$
$$x - 18 = 0 \quad \text{or} \quad x + 10 = 0$$
$$x = 18 \qquad\qquad x = -10$$

There are two solutions, 18 and -10.
Since measurement is not negative, -10 is rejected.
The dimensions are 5m by 18m.

Note: $\tfrac{1}{2}x - 4 = \tfrac{1}{2} \cdot 18 - 4 = 9 - 4 = 5$

19.

Let x = height measurement.

Let $x + 5$ = base measurement.

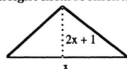

$$x + 5$$

The area formula for a triangle is $A = \tfrac{1}{2}bh$

The equation is

$$\tfrac{1}{2}(x + 5) \cdot x = 42$$
$$\tfrac{1}{2}x^2 + \tfrac{5}{2}x = 42$$
$$x^2 + 5x = 84 \qquad \text{Multiply by 2}$$
$$x^2 + 5x - 84 = 0$$
$$(x - 7)(x + 12) = 0$$
$$x - 7 = 0 \quad \text{or} \quad x + 12 = 0$$
$$x = 7 \qquad\qquad x = -12$$

There are two solutions, 7 and -12. Since measurement is not negative, -12 is rejected.
The height is 7 ft. and base is 12 ft.
Note: $x + 5 = 7 + 5 = 12$

21.

Let x = height measurement.

Let $2x - 3$ = base measurement.

#21 continued

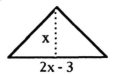

$$2x - 3$$

The area formula for a triangle is $A = \tfrac{1}{2}bh$.

The equation is

$$\tfrac{1}{2}(2x - 3) \cdot x = 22$$
$$x^2 - \tfrac{3}{2}x - 22 = 0$$
$$2x^2 - 3x - 44 = 0 \qquad \text{Multiply by 2}$$
$$(2x - 11)(x + 4) = 0$$
$$2x - 11 = 0 \quad \text{or} \quad x + 4 = 0$$
$$2x = 11 \qquad\qquad x = -4$$
$$x = \tfrac{11}{2}$$

Measurement is positive, so -4 is rejected. The
base is 8 m. and height is $5\tfrac{1}{2}$ m.

Note: $2x - 3 = 2 \cdot \tfrac{11}{2} - 3 = 11 - 3 = 8$

23.

Let x = base measurement.

Let $2x + 1$ = height measurement.

$$2x + 1$$

$$x$$

The area formula for a triangle is $A = \tfrac{1}{2}bh$.

The equation is

$$\tfrac{1}{2} \cdot x \cdot (2x + 1) = 10\tfrac{5}{2}$$
$$x^2 + \tfrac{1}{2}x - \tfrac{105}{2} = 0$$
$$2x^2 + x - 105 = 0 \qquad \text{Multiply by 2}$$
$$(x - 7)(2x + 15) = 0$$
$$x - 7 = 0 \quad \text{or} \quad 2x + 15 = 0$$
$$x = 7 \qquad\qquad 2x = -15$$
$$x = -\tfrac{15}{2}$$

Measurement is positive, so $-\tfrac{15}{2}$ is rejected.

The base is 7 in. and height is 15 in.

#23 continued

Note: $2x + 1 = 2 \cdot 7 + 1 = 15$.

25.

Let x = shorter leg measurement.

Let $2x - 3$ = hypotenuse measurement.

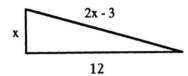

The Pythagorean formula for a triangle is $a^2 + b^2 = c^2$.

The equation is
$$x^2 + 12^2 = (2x - 3)^2$$
$$x^2 + 144 = 4x^2 - 12x + 9$$
$$0 = 3x^2 - 12x - 135$$
$$0 = x^2 - 4x - 45 \qquad \text{Divide by 3}$$
$$0 = (x - 9)(x + 5)$$
$$x - 9 = 0 \quad \text{or} \quad x + 5 = 0$$
$$x = 9 \qquad\qquad x = -5$$

Measurement is positive, so -5 is rejected. The dimension is 9 in.

27.

The measurements are in terms of the shorter leg.

Let x = shorter leg measurement.

Let $3x + 3$ = longer leg measurement.

Let $4x - 3$ = hypotenuse measurement.

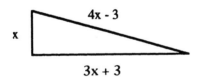

The Pythagorean formula is $a^2 + b^2 = c^2$.

The equation is
$$x^2 + (3x + 3)^2 = (4x - 3)^2$$
$$x^2 + 9x^2 + 18x + 9 = 16x^2 - 24x + 9$$
$$-6x^2 + 42x = 0$$
$$x^2 - 7x = 0 \qquad \text{Divide by -6}$$
$$x(x - 7) = 0$$
$$x = 0 \quad \text{or} \quad x - 7 = 0$$

The dimensions are 7 ft., 24 ft., and 25 ft. Note: 0 is rejected since measurement is positive.
$$3x + 3 = 3 \cdot 7 + 3 = 21 + 3 = 24$$
$$4x - 3 = 4 \cdot 7 - 3 = 28 - 3 = 25$$

29.

Let x = shorter leg measurement.

Let $x + 1$ = longer leg measurement.

Let $x + 2$ = hypotenuse measurement.

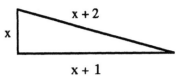

The Pythagorean formula is $a^2 + b^2 = c^2$.

The equation is
$$x^2 + (x + 1)^2 = (x + 2)^2$$
$$x^2 + x^2 + 2x + 1 = x^2 + 4x + 4$$
$$x^2 - 2x - 3 = 0$$
$$(x - 3)(x + 1) = 0$$
$$x - 3 = 0 \quad \text{or} \quad x + 1 = 0$$
$$x = 3 \qquad\qquad x = -1$$

The dimensions are 3 in., 4 in., and 5 in. Note: -1 is rejected since measurement is positive.

31.

$h = -16t^2 + 128t$
Solve for t when h = 240.

$$240 = -16t^2 + 128t$$
$$0 = -16t^2 + 128t - 240$$
$$0 = t^2 - 8t + 15 \qquad \text{Divide by -16}$$
$$0 = (t - 3)(t - 5)$$
$$t - 3 = 0 \quad \text{or} \quad t - 5 = 0$$
$$t = 3 \qquad\qquad t = 5$$

The object will be 240 feet above the ground at 3 seconds on its way up, and at 5 seconds on its way down.

33.

$h = -16t^2 + 64t + 960$
Solve for t when h = 0.

$$0 = -16t^2 + 64t + 960$$
$$0 = t^2 - 4t - 60$$
$$0 = (t - 10)(t + 6)$$
$$t - 10 = 0 \quad \text{or} \quad t + 60 = 0$$
$$t = 10 \qquad\qquad t = -6$$

The object will hit the ground after 10 seconds. Note: Time is not negative in this case, therefore -6 is rejected.

35.

$h = -16t^2 + 100$

#35 continued

Solve for t when h = 0.

$$0 = -16t^2 + 100$$
$$0 = 16t^2 - 100$$
$$0 = (4t - 10)(4t + 10)$$

$4t - 10 = 0$ or $4t + 10 = 0$
$$4t = 10 \qquad\qquad 4t = -10$$
$$t = \frac{10}{4} \qquad\qquad t = -\frac{10}{4}$$
$$t = \frac{5}{2} \qquad\qquad t = -\frac{5}{2}$$

The object will hit the ground after $2\frac{1}{2}$

seconds. Note that $t \geq 0$.

37.

$$E = \frac{1}{2}mv^2$$

Find v when m = 5, E = 10.

$$10 = \frac{1}{2} \cdot 5 \cdot v^2$$
$$10 = \frac{5}{2} \cdot v^2$$
$$4 = v^2$$
$$2 = v$$

The velocity is 2 m/sec.

39.

Let x = width of the border.

Let 130 + 2x = length of pool and border.

Let 60 + 2x = width of pool and border.

Let 7800 = area of pool.

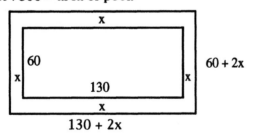

The equation is

Area Of Outer Rectangle		Area Of Inner Rectangle		Area Of Border
(130 + 2x)(60 + 2x)	−	7800	=	2000
7800 + 380x + 4x²	−	7800	=	2000
4x² + 380x	−	2000	=	0
x² + 95x	−	500	=	0
(x - 5)(x + 100)			=	0

#39 continued

$x - 5 = 0$ or $x + 100 = 0$
$$x = 5 \qquad\qquad x = -100$$

The border is 5m. Note: -100 is rejected since measurement is positive.

Review Exercises

1.
$$231 = 3 \cdot 77$$
$$= 3 \cdot 7 \cdot 11$$

3.
$$56 = 2^3 \cdot 7$$
$$42 = 2 \cdot 3 \cdot 7$$
$$70 = 2 \cdot 5 \cdot 7$$

The common factors are 2 and 7. Use their smallest power in any of the factored numbers.
$$\text{gcf} = 2^1 \cdot 7^1 = 14$$

5.
$$20x^3y^2 = 2^2 \cdot 5x^3y^2$$
$$56x^4y^5 = 2^3 \cdot 7x^4y^5$$

Common factors: 2, x and y
$$\text{gcf} = 2^2 \cdot x^3 \cdot y^2 = 4x^3y^2$$

7.
$$15xy^3 = 3 \cdot 5xy^3$$
$$27x^2y^2 = 3^3x^2y^2$$
$$33x^3y^2 = 3 \cdot 11x^3y^2$$

Common factors: 3, x and y
$$\text{gcf} = 3^1x^1y^2 = 3xy^2$$

9.
$$6x^3 + 12xy \qquad\qquad \text{gcf} = 6x$$
$$= 6x(x^2) + 6x(2y)$$
$$= 6x(x^2 + 2y)$$

11.
$$3x^3y - 15xy^2 + 6y \qquad\qquad \text{gcf} = 3y$$
$$= 3y(x^3) - 3y(5xy) + 3y(2)$$
$$= 3y(x^3 - 5xy + 2)$$

13.
$$-20x^3 + 25xy - 15x^2 \qquad\qquad \text{gcf} = -5x$$
$$= -5x(4x^2) - 5x(-5y) - 5x(3x)$$
$$= -5x(4x^2 - 5y + 3x)$$

15.
$$8x^4y + 12x^4y^2 + 4x^3y \qquad\qquad \text{gcf} = 4x^3y$$
$$= 4x^3y(2x) + 4x^3y(3xy) + 4x^3y(1)$$
$$= 4x^3y(2x + 3xy + 1)$$

17.

$3x^3y^2 + 9x^2y^3 - 3xy^3 - 12xy^4$ gcf = $3xy^2$

$= 3xy^2(x^2) + 3xy^2(3xy) + 3xy^2(-y) + 3xy^2(-4y^2)$

$= 3xy^2(x^2 + 3xy - y - 4y^2)$

19.

$3x(2x + 5) - y(2x + 5)$ gcf = $(2x + 5)$

$= (3x - y)(2x + 5)$

21.

$x^2 - 81 = (x)^2 - (9)^2 = (x + 9)(x - 9)$

23.

$16x^2 - y^2 = (4x)^2 - (y)^2 = (4x + y)(4x - y)$

25.

$36x^2 - 121y^2 = (6x)^2 - (11y)^2$

$= (6x + 11y)(6x - 11y)$

27.

$54x^2 - 24y^2 = 6(9x^2 - 4y^2)$ Factor out

$= 6(3x + 2y)(3x - 2y)$ the gcf

29.

$y^4 - 256 = (y^2 + 16)(y^2 - 16)$

$= (y^2 + 16)(y + 4)(y - 4)$

31.

$16x^4y - 4x^2y = 4x^2y(4x^2 - 1)$

$= 4x^2y(2x + 1)(2x - 1)$

33.

$y^2 + 12y + 36 = (y)^2 + 2(y \cdot 6) + (6)^2$

$= (y + 6)^2$

35.

$25x^2 + 10x + 1 = (5x)^2 + 2(5x \cdot 1) + (1)^2$

$= (5x + 1)^2$

37.

$x^2 - 26xy + 169y^2$

$= (x)^2 - 2(x \cdot 13y) + (13y)^2$

$= (x - 13y)^2$

39.

$4x^2 + 10xy + 25y^2$

$(2x)^2 + 2(10xy) + (5y)^2$

The middle term is not twice the product of the quantities being squared. The trinomial is prime.

41.

$81x^2 + 18xy + y^2 = (9x)^2 + 2(9x \cdot y) + (y)^2$

$= (9x + y)^2$

43.

$18x^3y - 24x^2y^2 + 8xy^3$

$= 2xy(9x^2 - 12xy + 4y^2)$

$= 2xy((3x)^2 - 2(3x \cdot 2y) + (2y)^2)$

$= 2xy(3x - 2y)^2$

45.

$x^2 + 6x + 8 = (x + m)(x + n)$

Two integer factors of 8 whose sum is 6 are 2 and 4.

Thus, m = 2 and n = 4.

So, $x^2 + 6x + 8 = (x + 2)(x + 4)$

47.

$x^2 - 7x + 10 = (x + m)x + n)$

Two integer factors of 10 whose sum is -7 are -5 and -2. Thus, m = -5 and n = -2.

So, $x^2 - 7x + 10 = (x - 5)(x - 2)$

49.

$x^2 - 6x - 27 = (x + m)(x + n)$

Two integer factors of -27 whose sum is -6 are 3 and -9. Thus, m = 3 and n = -9.

So, $x^2 - 6x - 27 = (x + 3)(x - 9)$

51.

$x^2 + 7x - 8 = (x + m)(x + n)$

Two integer factors of -8 whose sum is 7 are 8 and -1.

Thus, m = 8 and n = -1.

So, $x^2 + 7x - 8 = (x + 8)(x - 1)$

53.

$x^2 - 3x + 4 = (x + m)(x + n)$

There are no two integer factors whose product is 4 and sum is -3. The trinomial is prime.

55.

$-x^2 + 2x + 35 = -(x^2 - 2x - 35)$

$= -(x + m)(x + n)$

Two integers whose product is -35 and sum is -2 are 5 and -7. Thus, m = 5 and n = -7.

So, $-x^2 + 2x + 35 = -(x + 5)(x - 7)$

57.

$x^2 + 12x - 13 = (x + m)(x + n)$

Two integer factors of -13 whose sum is 12 are -1 and 13. Thus, m = -1 and n = 13.

So, $x^2 + 12x - 13 = (x - 1)(x + 13)$

59.

$3x^2 - 36x + 96 = 3(x^2 - 12x + 32)$

$= 3(x + m)(x + n)$

#59 continued

Two integer factors of 32 whose sum is -12 are -4 and -8.

Thus, m= -4 and n = -8.

So, $3x^2 - 36x + 96 = 3(x - 4)(x - 8)$

61.

$$4x^4 - 16x^3 - 48x^2 = 4x^2(x^2 - 4x - 12)$$
$$= 4x^2(x + m)(x + n)$$

Two integer factors of -12 whose sum is -4 are -6 and 2. Thus, m = -6 and n = 2.

So, $4x^4 - 16x^3 - 48x^2 = 4x^2(x - 6)(x + 2)$

63.

$$x^2 - 11xy + 24y^2 = (x + my)(x + ny)$$

Two factors of 24 whose sum is -11 are -8 and -3.

Thus, m = -8 and n = -3.

So, $x^2 - 11xy + 24y^2 = (x - 8y)(x - 3y)$

65.

$$2x^2 - 14xy - 60y^2 = 2(x^2 - 7xy - 30y^2)$$
$$= 2(x + my)(x + ny)$$

Two factors of -30 whose sum is -7 are -10 and 3.

Thus, m = -10 and n = 3.

So, $2x^2 - 14xy - 60y^2 = 2(x - 10y)(x + 3y)$

67.

$$5x^4y - 45x^3 y^2 + 100x^2y^3$$
$$= 5x^2y(x^2 - 9xy + 20y^2)$$
$$= 5x^2y(x + my)(x + ny)$$

Two factors of 20 whose sum is -9 are -4 and -5.

Thus, m = -4 and n = -5.

So, $5x^4y - 45x^3 y^2 + 100x^2y^3$
$$= 5x^2y(x - 4y)(x - 5y)$$

69.

$$2x^2 + 5x + 3 = (2x + h)(x + k)$$

<u>Possible Factor Combinations</u>

$h \cdot k = 3$

3 · 1	Both signs
1 · 3	are positive

The factor combination that gives the correct product is 3 and 1.

Thus, $(2x + 3)(x + 1) = 2x^2 + 5x + 3$

So, $2x^2 + 5x + 3 = (2x + 3)(x + 1)$

71.

$$3x^2 - 5x + 2 = (3x + h)(x + k)$$

#71 continued

<u>Possible Factor Combinations</u>

$h \cdot k = 2$

-1(-2)	-2(-1)	Both signs are negative

The factor combination that gives the correct product is -2 and 1.

Thus $(3x - 2)(x - 1) = 3x^2 - 5x + 2$

So, $3x^2 - 5x + 2 = (3x - 2)(x - 1)$

73.

$$4x^2 - 11x + 6 = (4x + h)(x + k)$$
$$\text{or}$$
$$= (2x + h)(2x + k)$$

<u>Possible Factor Combinations</u>

$h \cdot k = 6$

-6(-1)	-1 · (-6)	Both signs
-3(-2)	-2(-3)	are negative

The factor combination that gives the correct product is -3 and -2 with $(4x + h)(x + k)$

Thus, $(4x - 3)(x - 2) = 4x^2 - 11x + 6$

So $4x^2 - 11x + 6 = (4x - 3)(x - 2)$

75.

$$6x^2 - 11x - 2 = (6x + h)(x + k)$$
$$\text{or}$$
$$= (2x + h)(3x + k)$$

<u>Possible Factor Combinations</u>

$h \cdot k = -2$

-1 · 2	-2 · 1	Signs are
2(-1)	1(-2)	opposite

The factor combination that gives the correct product is 1 and -2 with $(6x + h)(x + k)$.

Thus, $(6x + 1)(x - 2) = 6x^2 - 11x - 2$.

So, $6x^2 - 11x - 2 = (6x + 1)(x - 2)$

77.

$$4x^2 + x - 2 = (4x + h)(x + k)$$
$$\text{or}$$
$$= (2x + h)(2x + k)$$

<u>Possible Factor Combinations</u>

$h \cdot k = -2$

1 · (-2)	2(-1)	Signs are
-2 · 1	-1 · 2	opposite

79.

$-4x^2 - 4x + 3 = -(4x^2 + 4x - 3)$
$ = -(4x\ h)(x + k)$
$ $ or
$ = -(2x + h)(2x + k)$

<u>Possible Factor Combinations</u>
$$h \cdot k = -3$$

1 · (-3)	3(-1)	Signs are
-3 · 1	-1 · 3	opposite

The factor combination that gives the correct product is 3 and -1.

Thus, $-(2x + 3)(2x - 1) = -(4x^2 + 4x - 3)$
$ = -4x^2 - 4x + 3$
So, $-4x^2 - 4x + 3 = -(2x + 3)(2x - 1)$

81.

$18x^2 + 18x - 20 = 2(9x^2 + 9x - 10)$
$ = 2(9x + h)(x + k)$
$ $ or
$ = 2(3x + h)(3x + k)$

<u>Possible Factor Combinations</u>
$$h \cdot k = -10$$

1 · (-10)	10(-1)	Signs are
-10 · 1	-1 · 10	opposite
2(-5)	5(-2)	
-5 · 2	-2 · 5	

The factor combination that gives the correct product is 5 and -2 with $2(3x + h)(3x + k)$.

Thus, $2(3x + 5)(3x - 2) = 2(9x^2 + 9x - 10)$
$ = 18x^2 + 18x - 20$

So, $18x^2 + 18x - 20 = 2(3x + 5)(3x - 2)$

83.

$6x^3 - 7x^2 - 10x = x(6x^2 - 7x - 10)$
$ = x(6x + h)(x + k)$
$ $ or
$ = x(2x + h)(3x + k)$

<u>Possible Factor Combinations</u>
$$h \cdot k = -10$$

1 · (-10)	10(-1)	Signs are
-10 · 1	-1 · 10	opposite
2(-5)	5(-2)	
-5 · 2	-2 · 5	

#83 continued

The factor combination that gives the correct product is 5 and -2 with $x(6x + h)(x + k)$.

Thus, $x(6x + 5)(x - 2) = x(6x^2 - 7x - 10)$
$ = 6x^3 - 7x^2 - 10x$
So, $6x^3 - 7x^2 - 10x = x(6x + 5)(x - 2)$

85.

$6x^3 + 3x^2 + 9x = 3x(2x^2 + x + 3)$
$ = 3x(3x + h)(x + k)$

<u>Possible Factor Combinations</u>
$$h \cdot k = 3$$

3 · 1	1 · 3	Both signs
		are positive

There are no combination of factors that will give the correct product
Thus, $6x^3 + 3x^2 + 9x = 3x(2x^2 + x + 3)$

87.

$7x^2 + 23xy + 6y^2 = (7x + hy)(x + ky)$

<u>Possible Factor Combinations</u>
$$h \cdot k = 6$$

6 · 1	2 · 3	Both signs
1 · 6	3 · 2	are positive

The factor combination that gives the correct product is 2 and 3.

Thus, $(7x + 2y)(x + 3y) = 7x^2 + 23xy + 6y^2$
So, $7x^2 + 23xy + 6y^2 = (7x + 2y)(x + 3y)$

89.

$16x^2 + 88xy + 40y^2 = 8(2x^2 + 11xy + 5y^2)$
$ = 8(2x + hy)(x + 5y)$

<u>Possible Factor Combinations</u>
$$h \cdot k = 5$$

5 · 1	1 · 5	Both signs
		are positive

The factor combination that gives the correct product is 1 and 5.
Thus, $8(2x + y)(x + 5y) = 8(2x^2 + 11xy + 5y^2)$
$ = 16x^2 + 88xy + 40y^2$
So, $16x^2 + 88xy + 40y^2 = 8(2x + y)(x + 5y)$

91.

$-12x^5 y + 22x^4 y^2 - 6x^3 y^3$
$= -2x^3 y(6x^2 - 11xy + 3y^2)$

#91 continued

$= -2x^3 y(6x + hy)(x + ky)$

or

$= -2x^3 y(2x + hy)(3x + ky)$

<u>Possible Factor Combinations</u>

$h \cdot k = 3$

-1(-3)   -3(-1)   Both signs
                           are negative

The factor combination that gives the correct product is -3 and -1 with

$-2x^3 y(2x + hy)(3x + ky)$

Thus, $-2x^3 y(2x - 3y)(3x - y)$

$\qquad = -2x^3 y(6x^2 - 11xy + 3y^2)$

$\qquad = -12x^5 y + 22x^4 y^2 - 6x^3 y^3$

So, $-12x^5 y + 22x^4 y^2 - 6x^3 y^3$

$\qquad = -2x^3 y(2x - 3y)(3x - y)$

93.

$6xy + 21x + 10y + 35$

$= (6xy + 21x) + (10y + 35)$	Group
$= 3x(2y + 7) + 5(2y + 7)$	Factor
$= (3x + 5)(2y + 7)$	Factor

95.

$8xy + 20x - 14y - 35$

$= (8xy + 20x) + (-14y - 35)$	Group
$= 4x(2y + 5) - 7(2y + 5)$	Factor
$= (4x - 7)(2y + 5)$	Factor

97.

$6x^3 + 14x^2 + 3x + 7$

$= (6x^3 + 14x^2) + (3x + 7)$	Group
$= 2x^2(3x + 7) + 1 \cdot (3x + 7)$	Factor
$= (2x^2 + 1)(3x + 7)$	Factor

99.

$12xy + 18x + 2ay + 3a$

$= (12xy + 18x) + (2ay + 3a)$	Group
$= 6x(2y + 3) + a(2y + 3)$	Factor
$= (6x + a)(2y + 3)$	Factor

101.

$5x^3 + 2x^2 - 45x - 18$

$= (5x^3 + 2x^2) + (-45x - 18)$	Group
$= x^2(5x + 2) - 9(5x + 2)$	Factor
$= (x^2 - 9)(5x + 2)$	Factor
$= (x + 3)(x - 3)(5x + 2)$	Factor

103.

$6x^2 y + 24x^2 + 27xy + 108x$

$= 3x(2xy + 8x + 9y + 36)$	Factor
$= 3x(2xy + 9y + 8x + 36)$	Commutative property
$= 3x((2xy + 9y) + (8x + 36))$	Group
$= 3x(y(2x + 9) + 4(2x + 9))$	Factor
$= 3x(y + 4)(2x + 9)$	Factor

105.

$6x^2 + 31x + 40$

<u>Product</u>

$6 \cdot 40 = 240$

<u>Factors</u>	<u>Sum</u>
$16 \cdot 15 = 240$	$16 + 15 = 31$

Replace $31x$ with $16x + 15x$.

So, $6x^2 + 31x + 40$

$= 6x^2 + 16x + 15x + 40$

$= (6x^2 + 16x) + (15x + 40)$

$= 2x(3x + 8) + 5(3x + 8)$

$= (2x + 5)(3x + 8)$

107.

$10x^2 - x - 2$

<u>Product</u>

$10(-2) = -20$

<u>Factors</u>	<u>Sum</u>
$4(-5) = -20$	$4 + (-5) = -1$

Replace $-x$ with $4x - 5x$

So, $10x^2 - x - 2 \quad = 10x^2 + 4x - 5x - 2$

$\qquad\qquad\qquad = (10x^2 + 4x) + (-5x - 2)$

$\qquad\qquad\qquad = 2x(5x + 2) - 1(5x + 2)$

$\qquad\qquad\qquad = (2x - 1)(5x + 2)$

109.

$6x^2 - 17xy - 14y^2$

<u>Product</u>

$6(-14) = -84$

<u>Factors</u>	<u>Sum</u>
$4(-21) = -84$	$4 - 21 = -17$

Replace $-17xy$ with $4xy - 21xy$.

So, $6x^2 + 4xy - 21xy - 14y^2$

$= (6x^2 + 4xy) + (-21xy - 14y^2)$

$= 2x(3x + 2y) - 7y(3x + 2y)$

$= (2x - 7y)(3x + 2y)$

111.

$6x^2 + 11xy - 30y^2$

#111 continued

Product
$6(-30) = -180$

Factors	Sum
$-9 \cdot 20 = -180$	$-9 + 20 = 11$

Replace $11xy$ with $-9xy + 20xy$.
So,
$6x^2 + 11xy - 30y^2$

$$= 6x^2 - 9xy + 20xy - 30y^2$$
$$= (6x^2 - 9xy) + (20xy - 30y^2)$$
$$= 3x(2x - 3y) + 10y(2x - 3y)$$
$$= (3x + 10y)(2x - 3y)$$

113.
$x^3 - 125$ $= (x)^3 - (5)^3$
$= (x - 5)(x^2 + 5 \cdot x + 5^2)$
$= (x - 5)(x^2 + 5x + 25)$

115.
$8m^3 + 27$ $= (2m)^3 + (3)^3$
$= (2m + 3)((2m)^2 - 2m \cdot 3 + 3^2)$
$= (2m + 3)(4m^2 - 6m + 9)$

117.
$4y^3 - 4$ $= 4(y^3 - 1)$
$= 4(y - 1)(y^2 + y + 1)$

119.
$54x^4y^3 + 2xy^6$
$= 2xy^3(27x^3 + y^3)$
$= 2xy^3(3x + y)(9x^2 - 3xy + y^2)$

121.
$64x^6 - y^3$
$= (4x^2)^3 - y^3$
$= (4x^2 - y)((4x^2)^2 + 4x^2 \cdot y + y^2)$
$= (4x^2 - y)(16x^4 + 4x^2y + y^2)$

123
$(x - 2)(x + 5) = 0$

$x - 2 = 0$ or $x + 5 = 0$
$x = 2$ $x = -5$

A check verifies that the solution set
is {2, -5}.

125.

$(2x + 5)(2x - 3)(x - 7) = 0$

$2x + 5 = 0$ or $2x - 3 = 0$ or $x - 7 = 0$
$2x = -5$ $2x = 3$ $x = 7$
$x = -\dfrac{5}{2}$ $x = \dfrac{3}{2}$

#125 continued

A check verifies that the solution set
is $\left\{-\dfrac{5}{2}, \dfrac{3}{2}, 7\right\}$.

127.
$4x^2 - 49 = 0$
$(2x - 7)(2x + 7) = 0$

$2x - 7 = 0$ or $2x + 7 = 0$
$2x = 7$ $2x = -7$
$x = \dfrac{7}{2}$ $x = -\dfrac{7}{2}$

A check verifies that the solution set
is $\left\{\dfrac{7}{2}, -\dfrac{7}{2}\right\}$.

129.
$x^2 - 13x + 40 = 0$
$(x - 5)(x - 8) = 0$

$x - 5 = 0$ or $x - 8 = 0$
$x = 5$ $x = 8$

A check verifies that the solution set is {5, 8}.

131.
$2x^2 + 5x - 10 = x^2 + 2x + 8$
$x^2 + 3x - 18 = 0$
$(x - 3)(x + 6) = 0$

$x - 3 = 0$ or $x + 6 = 0$
$x = 3$ $x = -6$

A check verifies that the solution set is
{-6, 3}.

133.
$3x^2 - 11x + 14 = (2x + 1)(x - 2)$
$3x^2 - 11x + 14 = 2x^2 - 3x - 2$
$x^2 - 8x + 16 = 0$
$(x - 4)(x - 4) = 0$

$x - 4 = 0$ or $x - 4 = 0$
$x = 4$ $x = 4$

A check verifies that the solution set is {4}.

135.
$(5x + 4)(x - 5) = -24$
$5x^2 - 21x - 20 = -24$
$5x^2 - 21x + 4 = 0$
$(5x - 1)(x - 4) = 0$

#135 continued

$$5x - 1 = 0 \quad \text{or} \quad x - 4 = 0$$
$$5x = 1 \qquad\qquad x = 4$$
$$x = \frac{1}{5}$$

A check verifies that the solution set is $\left\{\frac{1}{5}, 4\right\}$.

137.

$$(3x + 10)(3x - 2) = (x + 3)(3x + 10)$$
$$9x^2 + 24x - 20 = 3x^2 + 19x + 30$$
$$6x^2 + 5x - 50 = 0$$
$$(3x + 10)(2x - 5) = 0$$

$$3x + 10 = 0 \quad \text{or} \qquad 2x - 5 = 0$$
$$3x = -10 \qquad\qquad 2x = 5$$
$$x = -\frac{10}{3} \qquad\qquad x = \frac{5}{2}$$

A check verifies that the solution set is $\left\{-\frac{10}{3}, \frac{5}{2}\right\}$.

139.

$$4x^3 + 10x^2 - 50x = 0$$
$$2x(2x^2 + 5x - 25) = 0$$
$$2x(2x - 5)(x + 5) = 0$$

$$2x = 0 \quad \text{or} \quad 2x - 5 = 0 \quad \text{or} \quad x + 5 = 0$$
$$x = 0 \qquad\qquad 2x = 5 \qquad\qquad x = -5$$
$$x = \frac{5}{2}$$

A check verifies that the solution set is $\left\{-5, 0, \frac{5}{2}\right\}$.

141.

$$(2x + 5)(x^2 + x - 12) = 0$$
$$(2x + 5)(x - 3)(x + 4) = 0$$

$$2x + 5 = 0 \quad \text{or} \quad x - 3 = 0 \quad \text{or} \quad x + 4 = 0$$
$$2x = -5 \qquad\qquad x = 3 \qquad\qquad x = -4$$
$$x = -\frac{5}{2}$$

A check verifies that the solution set is $\left\{-4, -\frac{5}{2}, 3\right\}$.

143.

Let x = 1st even integer.

Let x + 2 = 2nd even integer.

#143 continued

The product is x(x + 2).

The larger integer plus 10 is (x + 2) + 10.

The equation is

$$x(x + 2) = (x + 2) + 10$$
$$x^2 + 2x = x + 12$$
$$x^2 + x - 12 = 0$$
$$(x - 3)(x + 4) = 0$$

$$x - 3 = 0 \quad \text{or} \quad x + 4 = 0$$
$$x = 3 \qquad\qquad x = -4$$

The two consecutive even integers are -4 and -2. Note: 3 is rejected since it is an odd integer.

145.

Let x = 1st consecutive integer.

Let x + 1 = 2nd consecutive integer.

The sum of the squares of each integer is $x^2 + (x + 1)^2$.

The larger integer plus 28 is (x + 1) + 28.

The equation is

$$x^2 + (x + 1)^2 = (x + 1) + 28$$
$$x^2 + x^2 + 2x + 1 = x + 29$$
$$2x^2 + x - 28 = 0$$
$$(2x - 7)(x + 4) = 0$$

$$2x - 7 = 0 \quad \text{or} \quad x + 4 = 0$$
$$2x = 7 \qquad\qquad x = -4$$
$$x = \frac{7}{2}$$

The two integers are -4 and -3.

Note: $\frac{7}{2}$ is not an integer.

147.

Let x = width measurement.

Let 4x = length measurement.

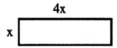

The area formula for a rectangle is $A = L \cdot W$.

The equation is

$$x(4x) = 36$$
$$4x^2 = 36$$
$$x^2 = 9$$
$$x^2 - 9 = 0$$
$$(x + 3)(x - 3) = 0$$

#147 continued

$x + 3 = 0$ or $x - 3 = 0$
$x = -3$ $x = 3$

There are two solutions, -3 and 3. Since measurement is not negative, -3 is rejected. The dimensions are 3 ft. by 12 ft.

149.

Let x = height measurement.

Let $2x - 4$ = base measurement.

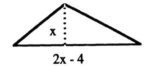

The area formula for a triangle is $A = \frac{1}{2}bh$.

The equation is

$$\frac{1}{2}(2x - 4)x = 35$$

$$\frac{1}{2}x(2x - 4) = 35$$

$$x^2 - 2x - 35 = 0$$

$$(x + 5)(x - 7) = 0$$

$x + 5 = 0$ or $x - 7 = 0$
$x = -5$ $x = 7$

There are two solutions, -5 and 7. Since measurement is not negative, -5 is rejected. The height is 7 m. and base is 10 m.
Note: $2x - 4 = 2 \cdot 7 - 4 = 14 - 4 = 10$.

151.

Let x = shorter leg measurement.

Let $2x + 4$ = longer leg measurement.

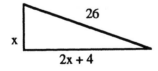

The Pythagorean formula is $a^2 + b^2 = c^2$.

The equation is
$$x^2 + (2x + 4)^2 = 26^2$$
$$x^2 + 4x^2 + 16x + 16 = 676$$
$$5x^2 + 16x - 660 = 0$$
$$(5x + 66)(x - 10) = 0$$

$5x + 66 = 0$ or $x - 10 = 0$
$5x = -66$ $x = 10$
$x = -\dfrac{66}{5}$

#151 continued

Measure is positive, so $-\dfrac{66}{5}$ is rejected.

The two legs measure 10 ft. and 24 ft.
Note: $2x + 4 = 2 \cdot 10 + 4 = 20 + 4 = 24$.

153.
$h = -16t^2 + 224t$

Solve for t when $h = 768$.

$$768 = -16t^2 + 224t$$
$$0 = -16t^2 + 224t - 768$$
$$0 = 16t^2 - 224t + 768$$
$$0 = t^2 - 14t + 48$$
$$0 = (t - 6)(t - 8)$$

$t - 6 = 0$ or $t - 8 = 0$
$t = 6$ $t = 8$

The object will be 768 feet above the ground at 6 seconds on its way up, and at 8 seconds on its way down.

Chapter 4 Test Solutions

1.
$4x^2 - 32x + 55$

Product
$4 \cdot 55 = 220$

Factors	Sum
$-10(-22) = 22$	$-10 - 22 = -32$

Replace $-32x$ with $-10x - 22x$.

So, $4x^2 - 32x + 55$

$= 4x^2 - 10x - 22x + 55$
$= (4x^2 - 10x) + (-22x + 55)$
$= 2x(2x - 5) - 11(2x - 5)$
$= (2x - 11)(2x - 5)$

3.
$9m^2 + 66m + 121$

Product
$9 \cdot 121 = 1089$

Factors	Sum
$33(33) = 1089$	$33 + 33 = 66$

Replace $66m$ with $33m + 33m$.

So, $9m^2 + 66m + 121$
$= 9m^2 + 33m + 33m + 121$
$= (9m^2 + 33m) + (33m + 121)$
$= 3m(3m + 11) + 11(3m + 11)$
$= (3m + 11)(3m + 11)$ or $(3m + 11)^2$

5.

$x^2 - 2x - 24$

Factors	Sum
$4(-6) = -24$	$4 + (-6) = -2$

So, $x^2 - 2x - 24 = (x + 4)(x - 6)$

7.

$32x^2 - 2 = 2(16x^2 - 1) = 2(4x + 1)(4x - 1)$

9.

$3x^2 - 7x - 26$

Product
$$3(-26) = -78$$

Factors	Sum
$6(-13) = -78$	$6 + (-13) = -7$

Replace $-7x$ with $6x - 13x$.

So, $3x^2 - 7x - 26$
$= 3x^2 + 6x - 13x - 26$
$= (3x^2 + 6x) + (-13x - 26)$
$= 3x(x + 2) - 13(x + 2)$
$= (3x - 13)(x + 2)$

11.

$8x^4 y^5 - 36x^3 y^6 + 20x^2 y^5$
$= 4x^2 y^5 (2x^2 - 9xy + 5)$

13.

$9y^2 - 49 = (3y + 7)(3y - 7)$

15.

$27x^4 + 27x^3 - 12x^2 = 3x^2(9x^2 + 9x - 4)$
$= 3x^2(9x^2 + 12x - 3x - 4)$
$= 3x^2[3x(3x + 4) - (3x + 4)]$
$= 3x^2(3x - 1)(3x + 4)$

17.

$y^2 - 14y + 33$

Factors	Sum
$-3(-11) = 33$	$-3 + (-11) = -14$

So, $y^2 - 14y + 33 = (y - 3)(y - 11)$

19.

$4x^2 y - 10x^2 + 20xy - 50x$
$= 2x(2xy - 5x + 10y - 25)$
$= 2x[(2xy - 5x) + (10y - 25)]$
$= 2x[x(2y - 5) + 5(2y - 5)]$
$= 2x(x + 5)(2y - 5)$

21.

$$3x^2 - 13x + 4 = 0$$
$$3x^2 - 12x - x + 4 = 0$$
$$3x(x - 4) - (x - 4) = 0$$
$$(3x - 1)(x - 4) = 0$$

$3x - 1 = 0 \quad$ or $\quad x - 4 = 0$
$x = \dfrac{1}{3} \quad$ or $\quad x = 4$

Solution set: $\left\{ \dfrac{1}{3}, 4 \right\}$

23.

$$4x^3 - 18x^2 + 18x = 0$$
$$2x(2x^2 - 9x + 9) = 0$$
$$2x(2x^2 - 3x - 6x + 9) = 0$$
$$2x[x(2x - 3) - 3(2x - 3)] = 0$$
$$2x(x - 3)(2x - 3) = 0$$

$2x = 0 \quad$ or $\quad x - 3 = 0 \quad$ or $\quad 2x - 3 = 0$
$x = 0 \quad$ or $\quad x = 3 \quad$ or $\quad x = \dfrac{3}{2}$

Solution set: $\left\{ 0, 3, \dfrac{3}{2} \right\}$

25.

Let x = measurement of the base.
Then $2x - 6$ = measurement of the height.

Thus, $A = \dfrac{1}{2} bh$.

$18 = \dfrac{1}{2}(x)(2x - 6)$

$18 = x^2 - 3x$
$0 = x^2 - 3x - 18$
$0 = (x + 3)(x - 6)$

$(x + 3) = 0 \quad$ or $\quad (x - 6) = 0$
$x = -3 \quad$ or $\quad x = 6$

Answer: $b = 6$ ft., $h = 6$ ft.

Measurement is never negative.

> **Test Your Memory**

1.

$$\dfrac{5^2 - 2^4}{3^2 + 6^2} = \dfrac{25 - 16}{9 + 36} = \dfrac{9}{45} = \dfrac{1}{5}$$

3.

$$4^{-2} - 4^0 = \dfrac{1}{4^2} - 1 = \dfrac{1}{16} - \dfrac{16}{16} = -\dfrac{15}{16}$$

5.

$$\frac{5}{14} - \frac{6}{7} + \frac{1}{4} = \frac{10 - 24 + 7}{28} \qquad \text{LCD} = 28$$

$$= -\frac{7}{28} = -\frac{1}{4}$$

7.

$$(5 \cdot 2^3)^2 = (5 \cdot 8)^2 = (40)^2 = 1600$$

9.

$$4 + 2(x + 3) = 3(2x - 1) + 9x$$
$$4 + 2x + 6 = 6x - 3 + 9x$$
$$2x + 10 = 15x - 3$$
$$2x - 2x + 10 = 15x - 2x - 3$$
$$10 = 13x - 3$$
$$10 + 3 = 13x - 3 + 3$$
$$13 = 13x$$
$$1 = x$$

Solution set: $\{ 1 \}$

11.

$$4(x + 3) - 7(x + 2) = 3(x + 5)$$
$$4x + 12 - 7x - 14 = 3x + 15$$
$$-3x - 2 = 3x + 15$$
$$-3x + 3x - 2 = 3x + 3x + 15$$
$$-2 = 6x + 15$$
$$-2 - 15 = 6x + 15 - 15$$
$$-17 = 6x$$
$$-\frac{17}{6} = x \qquad \text{Divide by 6}$$

Solution set: $\left\{ -\frac{17}{6} \right\}$

13.

$$x^2 - 6x + 5 = 0$$
$$(x - 5)(x - 1) = 0 \quad x - 5 = 0 \quad \text{or} \quad x - 1 = 0$$
$$x = 5 \qquad\qquad x = 1$$

Solution set: $\{ 5, 1 \}$

15.

$$(2x + 1)(x - 3) = 30$$
$$2x^2 - 5x - 3 = 30$$
$$2x^2 - 5x - 33 = 0$$
$$(x + 3)(2x - 11) = 0$$
$$x + 3 = 0 \quad \text{or} \quad 2x - 11 = 0$$
$$x = -3 \quad \text{or} \quad x = \frac{11}{2}$$

Solution set: $\left\{ -3, \frac{11}{2} \right\}$

17.

$$2(4x + 1) \le 5(2x - 1)$$
$$8x + 2 \le 10x - 5$$
$$8x - 8x + 2 \le 10x - 8x - 5$$
$$2 \le 2x - 5$$
$$2 + 5 \le 2x - 5 + 5$$
$$7 \le 2x$$
$$\frac{7}{2} \le x \text{ or } x \ge \frac{7}{2}$$

The graph is found in the answer section of the main text book.

19.

$$-8 \le 3x + 1 \le -2$$
$$-9 \le 3x \le -3 \qquad \text{Subtract 1}$$
$$-3 \le x \le -1 \qquad \text{Divide by 3}$$

The graph is found in the answer section of the main text book.

21.

$$\frac{12x^8}{20x^4} = \frac{12}{20} \cdot \frac{x^8}{x^4} = \frac{3}{5} \cdot \frac{1}{x^{4-(-8)}} = \frac{3}{5} \cdot \frac{1}{x^{12}} = \frac{3}{5x^{12}}$$

23.

$$\left(\frac{x^{-3}y^2}{x^4 y^{-3}} \right)^{-1} = \frac{x^3 y^{-2}}{x^{-4} y^3} = \frac{x^3}{x^{-4}} \cdot \frac{y^{-2}}{y^3}$$

$$= \frac{x^{3-(-4)}}{1} \cdot \frac{1}{y^{3-(-2)}} = \frac{x^7}{y^5}$$

25.

$$3(5x^2 - 4) - 5(x^2 - 4x + 3)$$
$$= 15x^2 - 12 - 5x^2 + 20x - 15$$
$$= 15x^2 - 5x^2 + 20x - 12 - 15$$
$$= 10x^2 + 20x - 27$$

27.

$$-5x^3 y^2 (2x^2 + 8xy - 3xy^3)$$
$$= -10x^5 y^2 - 40x^4 y^3 + 15x^4 y^5$$

29.

$$\frac{9x^4 - 15x^2 y^2 + 6xy^3}{-12x^2 y}$$

#29 continued

$$= \frac{9x^4}{-12x^2 y} + \frac{-15x^2 y^2}{-12x^2 y} + \frac{6xy^3}{-12x^2 y}$$

$$= -\frac{3x^2}{4y} + \frac{5y}{4} - \frac{y^2}{2x}$$

31.

$(3x - 5y)(11x - 2y)$

    F   O   I   L

$= 3x \cdot 11x + 3x(-2y) - 5y \cdot 11x - 5y(-2y)$

$= 33x^2 - 6xy - 55xy + 10y^2$

$= 33x^2 - 61xy + 10y^2$

33.

$6x^2 + 7xy + 2y^2$

<u>Product</u>

$(6 \cdot 2) = 12$

<u>Factors</u>	<u>Sum</u>
$3(4) = 12$	$3 + 4 = 7$

Replace $7xy$ with $3xy + 4xy$.

So,

$$6x^2 + 7xy + 2y^2 = 6x^2 + 3xy + 4xy + 2y^2$$
$$= 3x(2x + y) + 2y(2x + y)$$
$$= (3x + 2y)(2x + y)$$

35.

$x^2 - 8$ (Prime)

37.

$3x^3 + 3x^2 - 60x = 3x(x^2 + x - 20)$
$$= 3x(x + 5)(x - 4)$$

39.

$49k^2 - 16 = (7k + 4)(7k - 4)$

41.

$x^2 - 14x + 40$

<u>Factors</u>	<u>Sum</u>
$-4(-10) = 40$	$-4 + (-10) = -14$

So, $x^2 - 14x + 40 = (x - 4)(x - 10)$

43.

$$\frac{216}{9\frac{3}{5}} = \frac{216}{\frac{48}{5}} = \frac{216}{1} + \frac{48}{5} = \frac{216}{1} \cdot \frac{5}{48}$$

#43 continued

$$= \frac{9}{1} \cdot \frac{5}{2} \quad \text{Cancel}$$

$$= \frac{45}{2} \text{ or } 22.5$$

Answer: 22.5 mph

45.

Let x be amount invested in first account at 6%.

Then $4500 - x$ is the amount invested in second account.

Let $0.06x$ be the interest gained on the first account.

Let $0.09(4500 - x)$ be the interest gained on the second account.

$$\begin{aligned} \text{Thus,} \quad 0.06x + 0.09(4500 - x) &= 360 \\ 0.06x + 405 - 0.09x &= 360 \\ -0.03x &= -45 \\ x &= 1500 \\ (4500 - x) &= 3000 \end{aligned}$$

Answer: 6%, \$1,500; 9%, \$3,000

47.

Let x = one number.

Then $9 - x$ = the other number.

Also, x^2 and $(9 - x)^2$ are their squares

$$\begin{aligned} \text{Thus,} \quad x^2 + (9 - x)^2 &= 45 \\ x^2 + 81 - 18x + x^2 &= 45 \\ 2x^2 - 18x + 81 &= 45 \\ 2x^2 - 18x + 36 &= 0 \quad \text{Subtract 45} \\ x^2 - 9x + 18 &= 0 \quad \text{Divide by 2} \\ (x - 3)(x - 6) &= 0 \\ x = 3 \quad \text{or} \quad x &= 6 \end{aligned}$$

Answer: The numbers are 3 and 6.

49.

Let x = measurement of the width.

Then $2x - 3$ is the measurement of the length.

$$\begin{aligned} \text{Thus,} \quad A &= \ell \cdot w \\ 54 &= (2x - 3) \cdot x \\ 54 &= 2x^2 - 3x \\ 0 &= 2x^2 - 3x - 54 \\ 0 &= (x - 6)(2x + 9) \quad x = 6 \text{ or } x = -\frac{9}{2} \end{aligned}$$

Answer: 6 by 9 ft. Measurement is not negative.

CHAPTER 4 STUDY GUIDE

Completely factor the following polynomials.

1. $6x^2 - 49x + 30$

2. $2m^4 - 6mn + 2m$

3. $4x^2 - 52x + 169$

4. $x^2 - 36$

5. $x^2 + 2x - 15$

6. $6x^2 + 54$

7. $27x^2 - 3$

8. $x^2 - 18x + 81$

9. $10x^2 - 19x - 15$

10. $15x^5 y^4 + 10x^4 y^5 - 35x^3 y^4$

11. $x^2 + 18x + 45$

12. $30x^5 - 35x^4 - 15x^3$

13. $9x^2 + 18xy - 7y^2$

14. $y^2 - 15y + 56$

15. $3x^2 + 14x - 5$

16. $3x^4 - 21x^3 + 30x^2$

17. $12x^2 - 27x + 28x - 63$

Find the solutions of the following equations; also check your solutions.

18. $15x^2 + 22x - 5 = 0$

19. $12x^2 + 3x - 5 = 5(2x + 1)$

Solve the following problem.

20. The sum of the squares of three consecutive numbers is 110. Find the integers.

The worked-out solutions begin on the next page.

Self-Test Solutions

1.

$6x^2 - 49x + 30$

Product

$6(30) = 180$

Factors	Sum
$-45(-4) = 180$	$-45 + (-4) = -49$

Replace $-49x$ with $-45x - 4x$.
Thus, $6x^2 - 49x + 30$
$= 6x^2 - 45x - 4x + 30$
$= 3x(2x - 15) - 2(2x - 15)$
$= (3x - 2)(2x - 15)$

2.

$2m^4 - 6mn + 2m = 2m(m^3 - 3n + 1)$

3.

$4x^2 - 52x + 169 = (2x - 13)(2x - 13)$
$= (2x - 13)^2$

4.

$x^2 - 36 = (x + 6)(x - 6)$

5.

$x^2 + 2x - 15$

Factors	Sum
$-3(5) = -15$	$-3 + 5 = 2$

So, $x^2 + 2x - 15 = (x - 3)(x + 5)$

6.

$6x^2 + 54 = 6(x^2 + 9)$

7.

$27x^2 - 3 = 3(9x^2 - 1) = 3(3x + 1)(3x - 1)$

8.

$x^2 - 18x + 81 = (x - 9)(x - 9) = (x - 9)^2$

9.

$10x^2 - 19x - 15$

Product

$10(-15) = -150$

Factors	Sum
$-25(6) = -150$	$-25 + 6 = -19$

Replace $-19x$ with $-25x + 6x$.
So, $10x^2 - 19x - 15$
$= 10x^2 - 25x + 6x - 15$
$= 5x(2x - 5) + 3(2x - 5)$
$= (5x + 3)(2x - 5)$

10.

$15x^5y^4 + 10x^4y^5 - 35x^3y^4$
$= 5x^3y^4(3x^2 + 2xy - 7)$

11.

$x^2 + 18x + 45$

Factors	Sum
$3(15) = 45$	$3 + 15 = 18$

So, $x^2 + 18x + 45 = (x + 3)(x + 15)$

12.

$30x^5 - 35x^4 - 15x^3 = 5x^3(6x^2 - 7x - 3)$
Now we next factor $6x^2 - 7x - 3$.

Product

$6(-3) = -18$

Factors	Sum
$-9(2) = -18$	$-9 + 2 = -7$

Replace $-7x$ with $-9x + 2x$.
So,
$5x^3(6x^2 - 7x - 3) = 5x^3(6x^2 - 9x + 2x - 3)$
$= 5x^3[3x(2x - 3) + (2x - 3)]$
$= 5x^3(3x + 1)(2x - 3)$

13.

$9x^2 + 18xy - 7y^2$

Product

$9(-7) = -63$

Factors	Sum
$21(-3) = -63$	$21 + (-3) = 18$

Replace $18xy$ with $21xy - 3xy$.

So, $9x^2 + 18xy - 7y^2$
$= 9x^2 + 21xy - 3xy - 7y^2$
$= 3x(3x + 7y) - y(3x + 7y)$
$= (3x - y)(3x + 7y)$

14.

$y^2 - 15y + 56$

Factors	Sum
$-8(-7) = 56$	$-7 + (-8) = -15$

So, $y^2 - 15y + 56 = (y - 7)(y - 8)$

15.

$3x^2 + 14x - 5$

Product

$3(-5) = -15$

Factors	Sum
$15(-1) = -15$	$15 + (-1) = 14$

#15 continued

Replace 14x with 15x - x.

So, $3x^2 + 14x - 5 = 3x^2 + 15x - x - 5$

$ = 3x(x + 5) - (x + 5)$

$ = (3x - 1)(x + 5)$

16.

$3x^4 - 21x^3 + 30x^2 = 3x^2(x^2 - 7x + 10)$

Next, we factor $(x^2 - 7x + 10)$.

Factors	Sum
$-5(-2) = 10$	$-5 + (-2) = -7$

So, $3x^2(x^2 - 7x + 10) = 3x^2(x - 5)(x - 2)$

17.

$12x^2 - 27x + 28x - 63$

$= (3x + 7)(4x - 9)$

18.

$15x^2 + 22x - 5 = 0$

$(5x - 1)(3x + 5) = 0$

So, $\quad 5x - 1 = 0 \quad$ or $\quad 3x + 5 = 0$

$ x = \dfrac{1}{5} \quad$ or $\quad x = -\dfrac{5}{3}$

Solution set: $\left\{\dfrac{1}{5}, -\dfrac{5}{3}\right\}$.

19.

$12x^2 + 3x - 5 = 5(2x + 1)$

$12x^2 + 3x - 5 = 10x + 5$

$12x^2 - 7x - 10 = 0$

$(3x + 2)(4x - 5) = 0$

So, $3x + 2 = 0 \quad$ or $\quad 4x - 5 = 0$

$ x = -\dfrac{2}{3} \quad$ or $\quad x = \dfrac{5}{4}$

Solution set: $\left\{-\dfrac{2}{3}, \dfrac{5}{4}\right\}$.

20.

Let x, x + 1, and x + 2 be three consecutive integers.

Then, $\quad x^2 + (x + 1)^2 + (x + 2)^2 = 110$

$x^2 + x^2 + 2x + 1 + x^2 + 4x + 4 = 110$

$3x^2 + 6x - 105 = 0$

$x^2 + 2x - 35 = 0$

$(x - 5)(x + 7) = 0$

So, $\quad x = 5$ or $x = -7$

There are two sets of answers: 5, 6, 7 and -7, -6, -5.

CHAPTER 5   RATIONAL   EXPRESSIONS

Solutions   to   Text   Exercises

1.

$$\frac{16x^2y^4}{24x^2y^4} = \frac{2\cdot 8\cdot x^2\cdot y^4}{3\cdot 8\cdot x^2\cdot y^4} \quad \text{Factor}$$

$$= \frac{2\cdot \cancel{8}\cdot \cancel{x^2}\cdot \cancel{y^4}}{3\cdot \cancel{8}\cdot \cancel{x^2}\cdot \cancel{y^4}} \quad \text{Divide out}$$

$$= \frac{2}{3}$$

3.

$$\frac{27x^4y}{72xy^5} = \frac{3\cdot 9x^4y}{8\cdot 9xy^5} \quad \text{Factor}$$

$$= \frac{3\cdot 9x^{4-1}}{8\cdot 9y^{5-1}} \quad \text{Divide out}$$

$$= \frac{3x^3}{8y^4}$$

5.

$$\frac{6x^3y^2}{7x^5y^4} = \frac{6}{7x^{5-3}y^{4-2}} \quad \text{Divide out}$$

$$= \frac{6}{7x^2y^2}$$

7.

$$\frac{8xy^5}{2y^2} = \frac{2\cdot 4xy^5}{2y^2} \quad \text{Factor}$$

$$= \frac{\cancel{2}\cdot 4xy^{5-2}}{\cancel{2}\cdot 1} \quad \text{Divide out}$$

$$= 4xy^3$$

9.

$$\frac{6x^2y^2}{12x^3y^5} = \frac{6\cdot x^2y^2}{2\cdot 6x^3y^5}$$

$$= \frac{\cancel{6}\cdot 1}{2\cdot \cancel{6}x^{3-2}y^{5-2}} \quad \text{Divide out}$$

$$= \frac{1}{2xy^3}$$

11.

$$\frac{18x^2y^4}{12xy^2} = \frac{6\cdot 3x^2y^4}{6\cdot 2xy^2} \quad \text{Factor}$$

$$= \frac{\cancel{6}\cdot 3x^{2-1}y^{4-2}}{\cancel{6}\cdot 2} \quad \text{Divide out}$$

$$= \frac{3xy^2}{2}$$

13.

$$\frac{14x^3y^4}{18y^8} = \frac{2\cdot 7\cdot x^3\cdot y^4}{2\cdot 9\cdot y^8}$$

$$= \frac{\cancel{2}\cdot 7\cdot x^3}{\cancel{2}\cdot 9y^{8-4}}$$

$$= \frac{7x^3}{9y^4}$$

15.

$$\frac{3x^3}{4y^2} \quad \text{Reduced}$$

17.

$$\frac{6x^3y - 24x^2y}{9x^2y^2 + 27xy^2} = \frac{6x^2y(x-4)}{9xy^2(x+3)}$$

$$= \frac{2 \cdot 3x^2y(x-4)}{3 \cdot 3xy^2(x+3)}$$

$$= \frac{2 \cdot 3x^{2-1}(x-4)}{3 \cdot 3y^{2-1}(x+3)}$$

$$= \frac{2x(x-4)}{3y(x+3)}$$

19.

$$\frac{12x^4y^3 + 24x^3y^3}{32x^2y^5} = \frac{12x^3y^3(x+2)}{32x^2y^5}$$

$$= \frac{3 \cdot 4x^3y^3(x+2)}{4 \cdot 8x^2y^5}$$

$$= \frac{3 \cdot 4x^{3-2}(x+2)}{4 \cdot 8y^{5-3}}$$

$$= \frac{3x(x+2)}{8y^2}$$

21.

$$\frac{90x^4y^2 - 45x^3y^3}{9x^2y^2} = \frac{45x^3y^2(2x-y)}{9x^2y^2}$$

$$= \frac{5 \cdot 9x^3y^2(2x-y)}{9x^2y^2}$$

$$= \frac{5 \cdot 9x^{3-2}y^{2-2}(2x-y)}{9 \cdot 1}$$

$$= \frac{5xy^0(2x-y)}{1}$$

$$= 5x(2x-y)$$

23.

$$\frac{4x^2y}{8x^4y^2 + 32x^3y^2} = \frac{4x^2y}{8x^3y^2(x+4)}$$

$$= \frac{4 \cdot 1}{4 \cdot 2x^{3-2}y^{2-1}(x+4)}$$

$$= \frac{1}{2xy(x+4)}$$

25.

$$\frac{12x^2 - 4xy}{3y^3 - 9xy^2} = \frac{4x(3x-y)}{3y^2(y-3x)}$$

$$= \frac{4x(3x-y)}{3y^2(-1)(-y+3x)} \qquad \text{Factor out a } -1$$

$$= \frac{4x(3x-y)}{-3y^2(3x-y)} \qquad \begin{array}{l}\text{Commutative} \\ \text{property}\end{array}$$

$$= -\frac{4x}{3y^2}$$

27.

$$\frac{6x^3y^3 - 24x^2y^4}{18x^4y^5 + 36x^3y^6} = \frac{6x^2y^3(x-4y)}{18x^3y^5(x+2y)}$$

$$= \frac{6x^2y^3(x-4y)}{3 \cdot 6x^3y^5(x+2y)}$$

$$= \frac{6(x-4y)}{3 \cdot 6x^{3-2}y^{5-3}(x+2y)}$$

$$= \frac{x-4y}{3xy^2(x+2y)}$$

29.

$$\frac{x^2 + 5x + 6}{x^2 + 6x + 8} = \frac{(x+2)(x+3)}{(x+4)(x+2)} \qquad \text{Factor}$$

$$= \frac{(x+2)(x+3)}{(x+4)(x+2)}$$

$$= \frac{x+3}{x+4}$$

31.

$$\frac{4x^2 - 4x - 15}{6x^2 - 11x - 10} = \frac{(2x+3)(2x-5)}{(2x-5)(3x+2)} \qquad \text{Factor}$$

$$= \frac{(2x+3)\cancel{(2x-5)}}{\cancel{(2x-5)}(3x+2)}$$

$$= \frac{2x+3}{3x+2}$$

33.

$$\frac{x^2 - 1}{2x^2 - x - 3} = \frac{(x+1)(x-1)}{(x+1)(2x-3)}$$

$$= \frac{\cancel{(x+1)}(x-1)}{\cancel{(x+1)}(2x-3)}$$

$$= \frac{(x-1)}{(2x-3)}$$

35.

$$\frac{x^2 - 9}{2x^2 - 11x + 15} = \frac{(x+3)(x-3)}{(x-3)(2x-5)}$$

$$= \frac{(x+3)\cancel{(x-3)}}{\cancel{(x-3)}(2x-5)}$$

$$= \frac{x+3}{2x-5}$$

37.

$$\frac{3x^2 - 11x + 8}{3x^2 - 14x + 16} = \frac{(x-1)(3x-8)}{(x-2)(3x-8)}$$

$$= \frac{(x-1)\cancel{(3x-8)}}{(x-2)\cancel{(3x-8)}}$$

$$= \frac{x-1}{x-2}$$

39.

$$\frac{x^2 - x - 12}{x^2 - x - 2} = \frac{(x+3)(x-4)}{(x-2)(x+1)}$$

The rational expression is reduced.

41.

$$\frac{16 - x^2}{2x^2 - 8x} = \frac{-\left(x^2 - 16\right)}{2x^2 - 8x}$$

$$= \frac{-(x+4)(x-4)}{2x(x-4)}$$

$$= \frac{-(x+4)\cancel{(x-4)}}{2x\cancel{(x-4)}}$$

$$= -\frac{x+4}{2x}$$

43.

$$\frac{x^2 + 8x + 16}{x^2 + 6x + 8} = \frac{(x+4)(x+4)}{(x+2)(x+4)}$$

$$= \frac{(x+4)\cancel{(x+4)}}{(x+2)\cancel{(x+4)}}$$

$$= \frac{x+4}{x+2}$$

45.

$$\frac{6x^2}{6x^3 - 30x^2} = \frac{6x^2}{6x^2(x-5)}$$

$$= \frac{\cancel{6x^2} \cdot 1}{\cancel{6x^2}(x-5)}$$

$$= \frac{1}{x-5}$$

47.

$$\frac{x^2 + 3x - 10}{2x^2 + 10x} = \frac{(x+5)(x-2)}{2x(x+5)}$$

$$= \frac{\cancel{(x+5)}(x-2)}{2x\cancel{(x+5)}}$$

$$= \frac{x-2}{2x}$$

49.

$$\frac{2x-7}{7-2x} = \frac{2x-7}{-(-7+2x)}$$

$$= \frac{2x-7}{-(2x-7)}$$

$$= \frac{(2x-7)}{-(2x-7)}$$

$$= -1$$

51.

$$\frac{4x^2+12x+9}{2x^2-5x-12} = \frac{(2x+3)(2x+3)}{(2x+3)(x-4)}$$

$$= \frac{(2x+3)(2x+3)}{(2x+3)(x-4)}$$

$$= \frac{2x+3}{x-4}$$

53.

$$\frac{2-x}{x^2+6x-16} = \frac{2-x}{(x-2)(x+8)}$$

$$= \frac{-(x-2)}{(x-2)(x+8)}$$

$$= \frac{-1\cdot(x-2)}{(x-2)(x+8)}$$

$$= -\frac{1}{x+8}$$

55.

$$\frac{x^2-11x+30}{6-x} = \frac{(x-6)(x-5)}{-(x-6)}$$

$$= \frac{(x-6)(x-5)}{-1\cdot(x-6)}$$

$$= \frac{x-5}{-1}$$

$$= -(x-5) \text{ or } -x+5$$

57.

$$\frac{12x^2-8x}{4x} = \frac{4x(3x-2)}{4x}$$

$$= \frac{4x(3x-2)}{4x\cdot1}$$

$$= 3x-2$$

59.

$$\frac{5x^3y-25x^2y+20xy}{x^2-5x+4} = \frac{5xy(x^2-5x+4)}{x^2-5x+4}$$

$$= \frac{5xy(x^2-5x+4)}{(x^2-5x+4)\cdot1}$$

$$= 5xy$$

61.

$$\frac{12x-3y}{y-4x} = \frac{3(4x-y)}{-(4x-y)}$$

$$= \frac{3(4x-y)}{-1\cdot(4x-y)}$$

$$= -3$$

63.

$$\frac{x^2+3x+2}{3x^4+9x^3+6x^2} = \frac{x^2+3x+2}{3x^2(x^2+3x+2)}$$

$$= \frac{(x^2+3x+2)\cdot1}{3x^2(x^2+3x+2)}$$

$$= \frac{1}{3x^2}$$

65.

$$\frac{6x^3 + 9x^2}{4x^2 + 8x + 3} = \frac{3x^2(2x + 3)}{(2x + 3)(2x + 1)}$$

$$= \frac{3x^2\cancel{(2x + 3)}}{\cancel{(2x + 3)}(2x + 1)}$$

$$= \frac{3x^2}{2x + 1}$$

67.

$$\frac{15x^4 + 105x^3 + 180x^2}{5x^5 + 20x^4 + 15x^3} = \frac{15x^2\left(x^2 + 7x + 12\right)}{5x^3\left(x^2 + 4x + 3\right)}$$

$$= \frac{3 \cdot 5x^2(x + 3)(x + 4)}{5x^3(x + 3)(x + 1)}$$

$$= \frac{3 \cdot \cancel{5}\cancel{(x + 3)}(x + 4)}{\cancel{5}x^{3-2}\cancel{(x + 3)}(x + 1)}$$

$$= \frac{3(x + 4)}{x(x + 1)}$$

69.

$$\frac{6x^5 - 54x^4 + 120x^3}{3x^5 - 6x^4 - 45x^3} = \frac{6x^3\left(x^2 - 9x + 20\right)}{3x^3\left(x^2 - 2x - 15\right)}$$

$$= \frac{2 \cdot 3x^3(x - 4)(x - 5)}{3x^3(x + 3)(x - 5)}$$

$$= \frac{2 \cdot \cancel{3}\cancel{x^3}(x - 4)\cancel{(x - 5)}}{\cancel{3} \cdot \cancel{x^3}(x + 3)\cancel{(x - 5)}}$$

$$= \frac{2(x - 4)}{x + 3}$$

71.

$$\frac{2x^3y - 4x^2y - 48xy}{8x^4y^3 + 48x^3y^3 + 64x^2y^3}$$

$$= \frac{2xy\left(x^2 - 2x - 24\right)}{8x^2y^3\left(x^2 + 6x + 8\right)}$$

#71. continued.

$$= \frac{2xy(x + 4)(x - 6)}{2 \cdot 4 \cdot x^2y^3(x + 2)(x + 4)}$$

$$= \frac{\cancel{2}(x + 4)(x - 6)}{\cancel{2} \cdot 4x^{2-1}y^{3-1}(x + 2)\cancel{(x + 4)}}$$

$$= \frac{x - 6}{4xy^2(x + 2)}$$

73.

$$\frac{10x^3 - 40x}{6x^4 - 30x^3 + 36x^2} = \frac{10x\left(x^2 - 4\right)}{6x^2\left(x^2 - 5x + 6\right)}$$

$$= \frac{10x(x + 2)(x - 2)}{6x^2(x - 2)(x - 3)}$$

$$= \frac{\cancel{2} \cdot 5(x + 2)\cancel{(x - 2)}}{\cancel{2} \cdot 3x^{2-1}\cancel{(x - 2)}(x - 3)}$$

$$= \frac{5(x + 2)}{3x(x - 3)}$$

75.

$$\frac{108x^2 - 12x^4}{9x^3 - 36x^2 + 27x} = \frac{12x^2\left(9 - x^2\right)}{9x\left(x^2 - 4x + 3\right)}$$

$$= \frac{12x^2(3 + x)(3 - x)}{9x(x - 3)(x - 1)}$$

$$= \frac{\cancel{3} \cdot 4 \cdot x^{2-1}(3 + x)(-1)\cancel{(x - 3)}}{\cancel{3} \cdot 3\cancel{(x - 3)}(x - 1)}$$

$$= \frac{-4x(3 + x)}{3(x - 1)}$$

$$= -\frac{4x(x + 3)}{3(x - 1)}$$

77.

$$\frac{2x^4 + 6x^3 + 4x^2}{6x^3y + 3x^2y - 3xy} = \frac{2x^2\left(x^2 + 3x + 2\right)}{3xy\left(2x^2 + x - 1\right)}$$

$$= \frac{2x^2(x+2)(x+1)}{3xy(2x-1)(x+1)}$$

$$= \frac{2x^{2-1}(x+2)\cancel{(x+1)}}{3y(2x-1)\cancel{(x+1)}}$$

$$= \frac{2x(x+2)}{3y(2x-1)}$$

79.

$$\frac{ax + 2a + bx + 2b}{ax + 6a + bx + 6b} = \frac{a(x+2) + b(x+2)}{ax + bx + 6a + 6b}$$

$$= \frac{(a+b)(x+2)}{x(a+b) + 6(a+b)}$$

$$= \frac{\cancel{(a+b)}(x+2)}{\cancel{(a+b)}(x+6)}$$

$$= \frac{x+2}{x+6}$$

Exercises 5.2

1.

$$\frac{4x^3}{5x^2} \cdot \frac{3x^2}{2x^2} = \frac{4 \cdot 3x^5}{5 \cdot 2x^4}$$

$$= \frac{\cancel{2} \cdot 2 \cdot 3 \cdot x^{5-4}}{5 \cdot \cancel{2}}$$

$$= \frac{6x}{5}$$

3.

$$\frac{14x}{6x^3} \div \frac{2x}{3x^2} = \frac{14x}{6x^3} \cdot \frac{3x^2}{2x}$$

$$= \frac{14 \cdot 3 \cdot x^3}{6 \cdot 2 \cdot x^4}$$

$$= \frac{\cancel{2} \cdot 7 \cdot \cancel{3}}{\cancel{2} \cdot \cancel{3} \cdot 2x^{4-3}}$$

$$= \frac{7}{2x}$$

5.

$$\frac{5x^5}{2y^4} \cdot \frac{2y^2}{3x^4} = \frac{5 \cdot 2 \cdot x^5 \cdot y^2}{2 \cdot 3 \cdot x^4 \cdot y^4}$$

$$= \frac{5 \cdot \cancel{2} x^{5-4}}{\cancel{2} \cdot 3y^{4-2}}$$

$$= \frac{5x}{3y^2}$$

7.

$$\frac{8x^2y^3}{27xy^4} \div \frac{2y^2}{3xy} = \frac{8x^2y^3}{27xy^4} \cdot \frac{3xy}{2y^2}$$

$$= \frac{8 \cdot 3 \cdot x^3 \cdot y^4}{27 \cdot 2 \cdot x \cdot y^6}$$

$$= \frac{\cancel{2} \cdot 4 \cdot \cancel{3} \cdot x^{3-1}}{\cancel{3} \cdot 9 \cdot \cancel{2} \cdot y^{6-4}}$$

$$= \frac{4x^2}{9y^2}$$

9.

$$\frac{3x^2y^2}{6xy} \div \frac{6x^3y^2}{9x^2y} = \frac{3x^2y^2}{6xy} \cdot \frac{9x^2y}{6x^3y^2}$$

$$= \frac{3 \cdot 9 \cdot x^4 \cdot y^3}{6 \cdot 6 \cdot x^4 \cdot y^3}$$

$$= \frac{\cancel{3} \cdot \cancel{9} \cdot 3 \cdot x^{4-4}}{2 \cdot \cancel{3} \cdot 2 \cdot \cancel{3} \cdot y^{3-3}}$$

$$= \frac{3 \cdot x^0}{4 \cdot y^0} = \frac{3 \cdot 1}{4 \cdot 1} = \frac{3}{4}$$

11.

$$\frac{6xy^2}{5x} \cdot \frac{2x^2y^2}{6y^3} = \frac{6 \cdot 2 \cdot x^3 \cdot y^4}{5 \cdot 6 \cdot x \cdot y^3}$$

$$= \frac{\cancel{6} \cdot 2 \cdot x^{3-1} \cdot y^{4-3}}{5 \cdot \cancel{6}}$$

$$= \frac{2x^2y}{5}$$

13.

$$\frac{2xy^2}{5x^2y^3} \div \frac{6xy}{18x} = \frac{2xy^2}{5x^2y^3} \cdot \frac{18x}{6xy}$$

$$= \frac{2 \cdot 18 \cdot x^2 \cdot y^2}{5 \cdot 6 \cdot x^3 \cdot y^4}$$

$$= \frac{2 \cdot 3 \cdot \cancel{6}}{5 \cdot \cancel{6} \cdot x^{3-2}y^{4-2}}$$

$$= \frac{6}{5xy^2}$$

15.

$$\frac{3x^2}{27xy^3} \cdot \frac{3y^3}{2xy^2} = \frac{3 \cdot 3x^2y^3}{27 \cdot 2x^2y^5}$$

$$= \frac{\cancel{3} \cdot \cancel{3}x^{2-2}}{3 \cdot \cancel{3} \cdot \cancel{3} \cdot 2y^{5-3}}$$

$$= \frac{1}{6y^2}$$

17.

$$\frac{7}{2x+6} \div \frac{14x-28}{8x+24} = \frac{7}{2(x+3)} \cdot \frac{8(x+3)}{14(x-2)}$$

$$= \frac{\cancel{7}}{\cancel{2}(\cancel{x+3})} \cdot \frac{\cancel{2} \cdot \cancel{2} \cdot 2(\cancel{x+3})}{\cancel{2} \cdot \cancel{7}(x-2)}$$

$$= \frac{2}{(x-2)}$$

19.

$$\frac{x-3}{3x^2+9x} \cdot \frac{6x^3+6x^2}{5x-15} = \frac{x-3}{3x(x+3)} \cdot \frac{6x^2(x+1)}{5(x-3)}$$

$$= \frac{(\cancel{x-3})}{\cancel{3}(x+3)} \cdot \frac{2 \cdot \cancel{3} \cdot x^{2-1}(x+1)}{5(\cancel{x-3})}$$

$$= \frac{2x(x+1)}{5(x+3)} \quad \text{or} \quad \frac{2x^2+2x}{5x+15}$$

21.

$$\frac{6x^3y^2}{10x-20} \cdot \frac{2x^2-4x}{9xy^5} = \frac{6x^3y^2}{10(x-2)} \cdot \frac{2x(x-2)}{9xy^5}$$

$$= \frac{6 \cdot 2x^4y^2(x-2)}{10 \cdot 9xy^5(x-2)}$$

$$= \frac{\cancel{2} \cdot \cancel{3} \cdot 2x^{4-1}(\cancel{x-2})}{\cancel{2} \cdot 5 \cdot \cancel{3} \cdot 3y^{5-2}(\cancel{x-2})}$$

$$= \frac{2x^3}{15y^3}$$

23.

$$\frac{16x^2y^4}{6x^2+9x} \div \frac{4xy^2}{18x^3+27x^2}$$

$$= \frac{16x^2y^4}{3x(2x+3)} \cdot \frac{9x^2(2x+3)}{4xy^2} \qquad \text{Invert and factor}$$

$$= \frac{16 \cdot 9 \cdot x^4y^4(2x+3)}{3 \cdot 4x^2y^2(2x+3)}$$

#23. continued.

$$= \frac{\cancel{4} \cdot 4 \cdot \cancel{3} \cdot 3 \cdot x^{4-2} y^{4-2} \cancel{(2x+3)}}{\cancel{3} \cdot \cancel{4} \cancel{(2x+3)}}$$

$$= \frac{12x^2 y^2}{1}$$

$$= 12x^2 y^2$$

25.

$$\frac{x+3}{4x^2} \cdot \frac{2x^5}{2x^2 + x - 15}$$

$$= \frac{(x+3)}{2 \cdot 2 \cdot x^2} \cdot \frac{2x^5}{(2x-5)(x+3)}$$

$$= \frac{\cancel{(x+3)} \cdot \cancel{2} \cdot x^{5-2}}{\cancel{2} \cdot 2 \cdot (2x-5)\cancel{(x+3)}}$$

$$= \frac{x^3}{2(2x-5)}$$

27.

$$\frac{x^2 - 1}{2x^4 - 8x^3} \div \frac{2x^2 - x - 3}{3x^2 - 12x}$$

$$= \frac{(x+1)(x-1)}{2x^3(x-4)} \cdot \frac{3x(x-4)}{(2x-3)(x+1)} \qquad \text{Invert and factor}$$

$$= \frac{3\cancel{(x+1)}(x-1)\cancel{(x-4)}}{2x^{3-1}\cancel{(x-4)}(2x-3)\cancel{(x+1)}}$$

$$= \frac{3(x-1)}{2x^2(2x-3)}$$

29.

$$\frac{x^2 + 4x - 12}{18x^4} \div \frac{3x^4 - 6x^3}{6x^5}$$

$$= \frac{(x+6)(x-2)}{2 \cdot 3 \cdot 3x^4} \cdot \frac{2 \cdot 3x^5}{3x^3(x-2)} \qquad \text{Invert and factor}$$

$$= \frac{2 \cdot 3 \cdot x^5(x+6)(x-2)}{2 \cdot 3 \cdot 3 \cdot 3x^7(x-2)} \qquad \text{Multiply}$$

$$= \frac{\cancel{2} \cdot \cancel{3} \cdot (x+6)\cancel{(x-2)}}{\cancel{2} \cdot \cancel{3} \cdot 3 \cdot 3 \cdot x^{7-5}\cancel{(x-2)}}$$

$$= \frac{x+6}{9x^2}$$

31.

$$\frac{x^2 - 4x - 12}{3x^2} \cdot \frac{12x^6}{x^2 - 7x + 6}$$

$$= \frac{(x+2)(x-6)}{3x^2} \cdot \frac{3 \cdot 4x^6}{(x-6)(x-1)}$$

$$= \frac{\cancel{3} \cdot 4 \cdot x^{6-2}(x+2)\cancel{(x-6)}}{\cancel{3}\cancel{(x-6)}(x-1)}$$

$$= \frac{4x^4(x+2)}{x-1}$$

33.

$$\frac{x^2 - x - 12}{x^2 - 3x + 2} \cdot \frac{x^2 - 4x + 4}{x^2 + x - 6}$$

$$= \frac{(x+3)(x-4)}{(x-2)(x-1)} \cdot \frac{\cancel{(x-2)}\cancel{(x-2)}}{\cancel{(x+3)}\cancel{(x-2)}}$$

$$= \frac{x-4}{x-1}$$

35.

$$\frac{x^2 + 9x + 20}{x^2 + 7x + 12} \div \frac{x^2 - x - 30}{x^2 + x - 6}$$

$$= \frac{\cancel{(x+4)}(x+5)}{\cancel{(x+3)}\cancel{(x+4)}} \cdot \frac{\cancel{(x+3)}(x-2)}{\cancel{(x+5)}(x-6)} \qquad \text{Invert and factor}$$

$$= \frac{x-2}{x-6}$$

37.

$$\frac{4x^2 + 4x - 3}{6x^2 - 13x + 6} \div \frac{4x^2 + 16x + 15}{4x^2 + 4x - 15}$$

$$= \frac{\cancel{(2x+3)}(2x-1)}{\cancel{(2x-3)}(3x-2)} \cdot \frac{\cancel{(2x-3)}\cancel{(2x+5)}}{\cancel{(2x+3)}\cancel{(2x+5)}} \qquad \text{Invert and factor.}$$

$$= \frac{2x-1}{3x-2}$$

39.

$$\frac{2x^2 - x - 15}{x^2 - 3x + 2} \div \frac{4x^2 + 8x - 5}{3x^2 - 7x + 2}$$

$$= \frac{(x-3)(2x+5)}{(x-2)(x-1)} \cdot \frac{(3x-1)(x-2)}{(2x+5)(2x-1)} \quad \text{Invert and factor.}$$

$$= \frac{(x-3)(3x-1)}{(x-1)(2x-1)}$$

41.

$$\frac{16 - x^2}{x^2 + x - 2} \div \frac{x^2 - 2x - 8}{x^2 + 4x + 4}$$

$$= \frac{(4-x)(4+x)}{(x+2)(x-1)} \cdot \frac{(x+2)(x+2)}{(x-4)(x+2)} \quad \text{Invert and factor.}$$

$$= \frac{-(x-4)(x+4)}{(x+2)(x-1)} \cdot \frac{(x+2)(x+2)}{(x-4)(x+2)}$$

$$= \frac{-(x+4)}{x-1} \quad \text{or} \quad \frac{-x-4}{x-1}$$

43.

$$\frac{x-1}{2x^2 + 13x + 6} \div \frac{x^2 + x - 2}{2x + 1}$$

$$= \frac{(x-1)}{(2x+1)(x+6)} \cdot \frac{(2x+1)}{(x+2)(x-1)} \quad \text{Invert and factor.}$$

$$= \frac{1}{(x+6)(x+2)}$$

45.

$$\frac{2x^2 + x - 15}{x - 6} \cdot \frac{x^2 - 8x + 12}{2x - 5}$$

$$= \frac{(2x-5)(x+3)}{x-6} \cdot \frac{(x-6)(x-2)}{(2x-5)}$$

$$= (x+3)(x-2)$$

$$= x^2 + x - 6$$

47.

$$\frac{x^2 - 2x - 15}{x^2 + x - 2} \cdot \frac{x^2 - 3x + 2}{x^2 + 2x - 8}$$

$$= \frac{(x+3)(x-5)}{(x+2)(x-1)} \cdot \frac{(x-2)(x-1)}{(x-2)(x+4)}$$

$$= \frac{(x+3)(x-5)}{(x+2)(x+4)}$$

49.

$$\frac{2x^2 + 3x - 5}{3x^2 - 2x - 1} \cdot \frac{3x^2 - 17x + 10}{2x^2 - 5x - 25}$$

$$= \frac{(2x+5)(x-1)}{(3x+1)(x-1)} \cdot \frac{(3x-2)(x-5)}{(2x+5)(x-5)}$$

$$= \frac{3x-2}{3x+1}$$

51.

$$\frac{x^2 + 3x - 10}{x + 4} \div \frac{x + 5}{x^2 + 10x + 24}$$

$$= \frac{(x+5)(x-2)}{(x+4)} \cdot \frac{(x+6)(x+4)}{(x+5)} \quad \text{Invert and factor.}$$

$$= (x-2)(x+6)$$

$$= x^2 + 4x - 12$$

53.

$$\frac{2x^2 + 9x - 5}{3x^2 + 16x + 5} \div \frac{2x^2 - 5x + 2}{3x^2 - 5x - 2}$$

$$= \frac{(x+5)(2x-1)}{(x+5)(3x+1)} \cdot \frac{(3x+1)(x-2)}{(2x-1)(x-2)} \quad \text{Invert and factor.}$$

$$= 1$$

55.

$$\frac{2x^2 - 5x - 3}{2x^2 - x - 15} \cdot \frac{2x^2 + 17x + 30}{4x^2 - 1}$$

$$= \frac{(2x+1)(x-3)}{(2x+5)(x-3)} \cdot \frac{(2x+5)(x+6)}{(2x+1)(2x-1)}$$

$$= \frac{x+6}{2x-1}$$

57.

$$\frac{1-x^2}{2x^2 + 13x + 20} \cdot \frac{6x^2 + 17x + 5}{3x^2 - 2x - 1}$$

$$= \frac{(1-x)(1+x)}{(2x+5)(x+4)} \cdot \frac{(2x+5)(3x+1)}{(3x+1)(x-1)}$$

$$= \frac{-(x-1)(x+1)}{(2x+5)(x+4)} \cdot \frac{(2x+5)(3x+1)}{(3x+1)(x-1)}$$

$$= \frac{-(x+1)}{x+4}$$

59.

$$\frac{x^2 - x - 6}{5x^2 + 30x + 40} \cdot \frac{3x^2 + 6x - 24}{x^2 - 5x + 6}$$

$$= \frac{(x-3)(x+2)}{5(x+2)(x+4)} \cdot \frac{3(x-2)(x+4)}{(x-3)(x-2)}$$

$$= \frac{3}{5}$$

61.

$$\frac{x+3}{1} \cdot \frac{x^2 + 3x - 10}{x^2 + 2x - 3} = \frac{(x+3)}{1} \cdot \frac{(x+5)(x-2)}{(x+3)(x-1)}$$

$$= \frac{(x+5)(x-2)}{x-1}$$

63.

$$\frac{2x-3}{1} \cdot \frac{x^2 - 11x + 28}{2x^2 - 11x + 12}$$

$$= \frac{(2x-3)}{1} \cdot \frac{(x-7)(x-4)}{(2x-3)(x-4)}$$

$$= x - 7$$

65.

$$\frac{x^2 + 4x + 4}{3x^2 + 14x - 5} \div \frac{(x+2)}{1}$$

$$= \frac{(x+2)(x+2)}{(3x-1)(x+5)} \cdot \frac{1}{(x+2)}$$

$$= \frac{x+2}{(3x-1)(x+5)}$$

67.

$$\frac{(x-6)}{1} \div \frac{36 - x^2}{2x^2 + 5x + 3} \qquad \text{Invert and factor}$$

$$= \frac{(x-6)}{1} \cdot \frac{(2x+3)(x+1)}{(6-x)(6+x)}$$

$$= \frac{(x-6)}{1} \cdot \frac{(2x+3)(x+1)}{-(x-6)(x+6)}$$

$$= -\frac{(2x+3)(x+1)}{x+6}$$

69.

$$\frac{12x^2 - 30x}{3x^2 - 3x - 6} \div \frac{(2x-5)}{1}$$

$$= \frac{6x(2x-5)}{3(x+1)(x-2)} \cdot \frac{1}{(2x-5)}$$

$$= \frac{2 \cdot 3x(2x-5)}{3(x+1)(x-2)} \cdot \frac{1}{(2x-5)}$$

$$= \frac{2x}{(x+1)(x-2)}$$

71.

$$\frac{(4x+3)}{1} \cdot \frac{9x^2 - 36x}{12x^4 + 9x^3} = \frac{(4x+3)}{1} \cdot \frac{9x(x-4)}{3x^3(4x+3)}$$

$$= \frac{(4x+3)}{1} \cdot \frac{3 \cdot 3 \cdot (x-4)}{3 \cdot x^{3-1}(4x+3)}$$

$$= \frac{3(x-4)}{x^2}$$

73.

$$\frac{4x^3 + 12x^2}{x^2 - 3x - 10} \div \frac{8x^3 + 22x^2 - 6x}{4x^2 + 7x - 2}$$

$$= \frac{4x^2(x+3)}{(x+2)(x-5)} \cdot \frac{(x+2)(4x-1)}{2x(x+3)(4x-1)} \quad \text{Invert and factor.}$$

$$= \frac{2 \cdot 2 \cdot x^{2-1} \cancel{(x+3)} \cancel{(x+2)} \cancel{(4x-1)}}{\cancel{(x+2)}(x-5) \cdot 2 \cdot \cancel{(x+3)} \cancel{(4x-1)}}$$

$$= \frac{2x}{x-5}$$

75.

$$\frac{6x^4 + 12x^3 + 6x^2}{5x^2 - 5} \div \frac{3x^4 + 9x^3 + 6x^2}{x^2 + x - 2}$$

$$= \frac{6x^2(x+1)(x+1)}{5(x+1)(x-1)} \cdot \frac{(x+2)(x-1)}{3x^2(x+2)(x+1)} \quad \text{Invert and factor.}$$

$$= \frac{2 \cdot 3 \cdot x^{2-2} \cancel{(x+1)} \cancel{(x+1)} \cancel{(x+2)} \cancel{(x-1)}}{5 \cancel{(x+1)} \cancel{(x-1)} \cdot 3 \cdot \cancel{(x+2)} \cancel{(x+1)}}$$

$$= \frac{2x^0}{5} = \frac{2}{5}$$

77.

$$\frac{6x^4 + 12x^3 - 18x^2}{2x^3 - 2x^2 - 24x} \cdot \frac{2x^2 - 7x - 4}{8x^3 - 4x^2 - 4x}$$

$$= \frac{6x^2(x+3)(x-1)}{2x(x+3)(x-4)} \cdot \frac{(2x+1)(x-4)}{4x(2x+1)(x-1)}$$

$$= \frac{2 \cdot 3 \cdot x^{2-2} \cancel{(x+3)} \cancel{(x-1)} \cancel{(2x+1)} \cancel{(x-4)}}{2 \cdot 4 \cancel{(x+3)} \cancel{(x-4)} \cancel{(2x+1)} \cancel{(x-1)}}$$

$$= \frac{3x^0}{4} = \frac{3}{4}$$

79.

$$\frac{2x^5 + 10x^4 + 8x^3}{3x^2 + 11x - 4} \cdot \frac{3x^2 + 5x - 2}{6x^3 - 24x}$$

$$= \frac{2x^3(x+4)(x+1)}{(3x-1)(x+4)} \cdot \frac{(3x-1)(x+2)}{6x(x+2)(x-2)}$$

$$= \frac{2 \cdot x^{3-1} \cancel{(x+4)}(x+1) \cancel{(3x-1)} \cancel{(x+2)}}{2 \cdot 3 \cancel{(3x-1)} \cancel{(x+4)} \cancel{(x+2)}(x-2)}$$

$$= \frac{x^2(x+1)}{3(x-2)}$$

81.

$$\frac{4x^2 + 8x}{x^2 - 8x - 9} \cdot \frac{x+7}{2x^2 - 6x} \cdot \frac{x^2 - 2x - 3}{3x^2 + 6x}$$

$$= \frac{2 \cdot 2x \cancel{(x+2)}}{\cancel{(x+1)}(x-9)} \cdot \frac{(x+7)}{2x \cancel{(x-3)}} \cdot \frac{\cancel{(x-3)} \cancel{(x+1)}}{3x \cancel{(x+2)}}$$

$$= \frac{2(x+7)}{3x(x-9)}$$

83.

$$\frac{x^2 + 5x + 6}{3x^2 - 11x - 4} \cdot \frac{x^2 - 2x + 1}{2x^2 - 9x + 10} \cdot \frac{6x^2 - 13x - 5}{x^2 + 2x - 3}$$

$$= \frac{(x+2) \cancel{(x+3)}}{\cancel{(3x+1)}(x-4)} \cdot \frac{\cancel{(x-1)}(x-1)}{\cancel{(2x-5)}(x-2)} \cdot \frac{\cancel{(2x-5)} \cancel{(3x+1)}}{\cancel{(x+3)} \cancel{(x-1)}}$$

$$= \frac{(x+2)(x-1)}{(x-4)(x-2)}$$

85.

$$\frac{x+3}{x^2 + x - 20} \div \frac{2x^2 - 5x + 2}{x-4} \cdot \frac{2x^2 + 9x - 5}{2x^2 + 7x + 3}$$

$$= \frac{\cancel{(x+3)}}{\cancel{(x-4)} \cancel{(x+5)}} \cdot \frac{\cancel{(x-4)}}{\cancel{(2x-1)}(x-2)} \cdot \frac{\cancel{(2x-1)} \cancel{(x+5)}}{(2x+1) \cancel{(x+3)}}$$

$$= \frac{1}{(x-2)(2x+1)}$$

87.

$$\frac{6x^2+24x}{x^2-2x-15} \div \frac{2x^2+3x-2}{2x^2+4x} \div \frac{8x^2+32x}{2x^2+5x-3}$$

$$= \frac{6x(x+4)}{(x+3)(x-5)} \cdot \frac{2x(x+2)}{(2x-1)(x+2)} \cdot \frac{(2x-1)(x+3)}{8x(x+4)}$$

$$= \frac{2 \cdot 3x(x+4) \cdot 2 \cdot x(x+2)(2x-1)(x+3)}{(x+3)(x-5)(2x-1)(x+2) \cdot 2 \cdot 2 \cdot 2 \cdot x(x+4)}$$

$$= \frac{3x}{2(x-5)}$$

89.

$$\frac{2ax+2ay+bx+by}{8x^2-24x} \div \frac{3ax+3ay+2bx+2by}{4x^2-12x}$$

$$= \frac{(2a+b)(x+y)}{2 \cdot 4x(x-3)} \cdot \frac{4x(x-3)}{(3a+2b)(x+y)}$$

$$= \frac{2a+b}{2(3a+2b)}$$

Exercises 5.3

1.

$$\frac{5x}{3} + \frac{2x}{3} = \frac{5x+2x}{3}$$

$$= \frac{7x}{3}$$

3.

$$\frac{7x}{4} - \frac{4x}{4} = \frac{7x-4x}{4}$$

$$= \frac{3x}{4}$$

5.

$$\frac{y}{8} + \frac{3y}{8} = \frac{y+3y}{8}$$

$$= \frac{4y}{2 \cdot 4} = \frac{y}{2}$$

7.

$$\frac{2b}{6} - \frac{4b}{6} = \frac{2b-4b}{6}$$

$$= -\frac{2b}{2 \cdot 3}$$

$$= -\frac{b}{3}$$

9.

$$\frac{4}{7x} + \frac{2}{7x} = \frac{4+2}{7x}$$

$$= \frac{6}{7x}$$

11.

$$\frac{7y}{4x^2} - \frac{2y}{4x^2} = \frac{7y-2y}{4x^2}$$

$$= \frac{5y}{4x^2}$$

13.

$$\frac{8}{3xy} - \frac{2}{3xy} = \frac{8-2}{3xy}$$

$$= \frac{6}{3xy}$$

$$= \frac{2 \cdot 3}{3xy}$$

$$= \frac{2}{xy}$$

15.

$$\frac{4x}{6xy^2} + \frac{2x}{6xy^2} = \frac{4x+2x}{6xy^2}$$

$$= \frac{6x}{6xy^2}$$

$$= \frac{1}{y^2}$$

17.

$$\frac{3x+5}{8}+\frac{2x+1}{8}=\frac{(3x+5)+(2x+1)}{8}$$

$$=\frac{5x+6}{8}$$

19.

$$\frac{7x+5}{2y}+\frac{3x+1}{2y}=\frac{(7x+5)+(3x+1)}{2y}$$

$$=\frac{10x+6}{2y}$$

$$=\frac{2(5x+3)}{2y}$$

$$=\frac{5x+3}{y}$$

21.

$$\frac{2x+3}{6}-\frac{x}{6}=\frac{2x+3-x}{6}$$

$$=\frac{x+3}{6}$$

23.

$$\frac{5x-2}{3y}-\frac{2x+3}{3y}=\frac{(5x-2)-(2x+3)}{3y}$$

$$=\frac{5x-2-2x-3}{3y}$$

$$=\frac{3x-5}{3y}$$

25.

$$\frac{2x+3}{2y}-\frac{2x-3}{2y}=\frac{(2x+3)-(2x-3)}{2y}$$

$$=\frac{2x+3-2x+3}{2y}$$

$$=\frac{6}{2y}$$

$$=\frac{2\cdot3}{2y}=\frac{3}{y}$$

27.

$$\frac{4x}{3x-2y}+\frac{8x}{3x-2y}=\frac{4x+8x}{3x-2y}$$

$$=\frac{12x}{3x-2y}$$

29.

$$\frac{2x}{x+3}+\frac{6}{x+3}=\frac{2x+6}{x+3}=\frac{2(x+3)}{x+3}=2$$

31.

$$\frac{4x}{2x-5}-\frac{2x}{2x-5}=\frac{4x-2x}{2x-5}=\frac{2x}{2x-5}$$

33.

$$\frac{2x}{3x+2}-\frac{x+5}{3x+2}=\frac{2x-(x+5)}{3x+2}$$

$$=\frac{2x-x-5}{3x+2}$$

$$=\frac{x-5}{3x+2}$$

35.

$$\frac{2x^2+17x}{x+6}+\frac{x^2-6}{x+6}=\frac{\left(2x^2+17x\right)+\left(x^2-6\right)}{x+6}$$

$$=\frac{3x^2+17x-6}{x+6}$$

$$=\frac{(3x-1)(x+6)}{(x+6)}$$

$$=3x-1$$

37.

$$\frac{4x-2}{3x-4}-\frac{2x-7}{3x-4}=\frac{(4x-2)-(2x-7)}{3x-4}$$

$$=\frac{4x-2-2x+7}{3x-4}$$

$$=\frac{2x+5}{3x-4}$$

39.

$$\frac{3x+13}{x^2-16}+\frac{2x+7}{x^2-16}=\frac{(3x+13)+(2x+7)}{x^2-16}$$

$$=\frac{5x+20}{x^2-16}$$

$$=\frac{5(x+4)}{(x-4)(x+4)}$$

$$=\frac{5}{x-4}$$

41.

$$\frac{2x+5}{x^2-4}-\frac{x+3}{x^2-4}=\frac{(2x+5)-(x+3)}{x^2-4}$$

$$=\frac{2x+5-x-3}{x^2-4}$$

$$=\frac{(x+2)}{(x-2)(x+2)}$$

$$=\frac{1}{x-2}$$

43.

$$\frac{7x-2}{6x+3}+\frac{3x+7}{6x+3}=\frac{(7x-2)+(3x+7)}{6x+3}$$

$$=\frac{10x+5}{6x+3}$$

$$=\frac{5(2x+1)}{3(2x+1)}$$

$$=\frac{5}{3}$$

45.

$$\frac{7x-3}{21x-28}-\frac{x+5}{21x-28}=\frac{(7x-3)-(x+5)}{21x-28}$$

$$=\frac{7x-3-x-5}{21x-28}$$

$$=\frac{6x-8}{21x-28}$$

$$=\frac{2(3x-4)}{7(3x-4)}$$

$$=\frac{2}{7}$$

47.

$$\frac{2x+5}{x^2-x-7}+\frac{x-12}{x^2-x-7}=\frac{(2x+5)+(x-12)}{x^2-x-7}$$

$$=\frac{3x-7}{x^2-x-7}$$

49.

$$\frac{2x^2+1}{x^2-x-12}-\frac{x^2-15}{x^2-x-12}=\frac{\left(2x^2+1\right)-\left(x^2-15\right)}{x^2-x-12}$$

$$=\frac{2x^2+1-x^2+15}{x^2-x-12}$$

$$=\frac{x^2+16}{x^2-x-12}$$

51.

$$\frac{5x^2-7}{2x^2+13x+15}-\frac{x^2+2}{2x^2+13x+15}$$

$$=\frac{\left(5x^2-7\right)-\left(x^2+2\right)}{2x^2+13x+15}$$

$$=\frac{5x^2-7-x^2-2}{2x^2+13x+15}$$

$$=\frac{4x^2-9}{2x^2+13x+15}$$

$$=\frac{(2x+3)(2x-3)}{(2x+3)(x+5)}$$

$$=\frac{2x-3}{x+5}$$

53.

$$\frac{x-5}{2x^2-5x+3}+\frac{x+2}{2x^2-5x+3}=\frac{(x-5)+(x+2)}{2x^2-5x+3}$$

$$=\frac{2x-3}{2x^2-5x+3}$$

$$=\frac{\cancel{(2x-3)}}{\cancel{(2x-3)}(x-1)}$$

$$=\frac{1}{x-1}$$

55.

$$\frac{2x^2+3x-31}{x^2-2x-15}+\frac{x^2-9x-14}{x^2-2x-15}$$

$$=\frac{\left(2x^2+3x-31\right)+\left(x^2-9x-14\right)}{x^2-2x-15}$$

$$=\frac{3x^2-6x-45}{x^2-2x-15}$$

$$=\frac{3\cancel{\left(x^2-2x-15\right)}}{\cancel{\left(x^2-2x-15\right)}}$$

$$=3$$

57.

$$\frac{5x^2-2x+8}{x^2-4x-32}-\frac{3x^2-2x+14}{x^2-4x-32}$$

$$=\frac{\left(5x^2-2x+8\right)-\left(3x^2-2x+14\right)}{x^2-4x-32}$$

$$=\frac{5x^2-2x+8-3x^2+2x-14}{x^2-4x-32}$$

$$=\frac{2x^2-6}{x^2-4x-32}$$

$$=\frac{2\left(x^2-3\right)}{(x+4)(x-8)}$$

Factoring does not give any common factors,

so the difference is $\dfrac{2x^2-6}{x^2-4x-32}$.

59.

$$\frac{x^2+7x-10}{x^2-5x-6}+\frac{x^2-2x-2}{x^2-5x-6}$$

$$=\frac{\left(x^2+7x-10\right)+\left(x^2-2x-2\right)}{x^2-5x-6}$$

$$=\frac{2x^2+5x-12}{x^2-5x-6}$$

$$=\frac{(2x-3)(x+4)}{(x-6)(x+1)}$$

Factoring does not give any common

factors, so the sum is $\dfrac{2x^2+5x-12}{x^2-5x-6}$.

61.

$$\frac{3x^2-4x-8}{x^2+3x-18}-\frac{x^2-3x+7}{x^2+3x-18}$$

$$=\frac{\left(3x^2-4x-8\right)-\left(x^2-3x+7\right)}{x^2+3x-18}$$

$$=\frac{3x^2-4x-8-x^2+3x-7}{x^2+3x-18}$$

$$=\frac{2x^2-x-15}{x^2+3x-18}$$

$$=\frac{(2x+5)\cancel{(x-3)}}{\cancel{(x-3)}(x+6)}$$

$$=\frac{2x+5}{x+6}$$

63.

$$\frac{8}{x-1} + \frac{5}{1-x} = \frac{8}{x-1} + \frac{5}{(1-x)} \cdot \frac{-1}{-1}$$

$$= \frac{8}{x-1} + \frac{-5}{-1+x}$$

$$= \frac{8}{x-1} + \frac{-5}{x-1}$$

$$= \frac{8-5}{x-1} = \frac{3}{x-1}$$

65.

$$\frac{4}{2x-3} - \frac{5}{3-2x} = \frac{4}{2x-3} - \frac{5}{(3-2x)} \cdot \frac{-1}{-1}$$

$$= \frac{4}{2x-3} - \frac{-5}{-3+2x}$$

$$= \frac{4}{2x-3} - \frac{-5}{2x-3}$$

$$= \frac{4-(-5)}{2x-3}$$

$$= \frac{4+5}{2x-3} = \frac{9}{2x-3}$$

67.

$$\frac{4}{5x-4} + \frac{6}{4-5x} = \frac{4}{5x-4} + \frac{6}{(4-5x)} \cdot \frac{-1}{-1}$$

$$= \frac{4}{5x-4} + \frac{-6}{-4+5x}$$

$$= \frac{4}{5x-4} + \frac{-6}{5x-4}$$

$$= \frac{4-6}{5x-4}$$

$$= \frac{-2}{5x-4}$$

69.

$$\frac{3x+4}{5x-2} + \frac{x-9}{2-5x} = \frac{3x+4}{5x-2} + \frac{(x-9)}{(2-5x)} \cdot \frac{-1}{-1}$$

$$= \frac{3x+4}{5x-2} + \frac{-x+9}{-2+5x}$$

$$= \frac{(3x+4)+(-x+9)}{5x-2}$$

$$= \frac{2x+13}{5x-2}$$

71.

$$\frac{x-8}{2x-3} - \frac{2x+5}{3-2x} = \frac{x-8}{2x-3} - \frac{(2x+5)}{(3-2x)} \cdot \frac{-1}{-1}$$

$$= \frac{x-8}{2x-3} - \frac{-2x-5}{-3+2x}$$

$$= \frac{(x-8)-(-2x-5)}{2x-3}$$

$$= \frac{x-8+2x+5}{2x-3}$$

$$= \frac{3x-3}{2x-3}$$

73.

$$\frac{6x+9}{2x-7} + \frac{8x+2}{7-2x} = \frac{6x+9}{2x-7} + \frac{(8x+2)}{(7-2x)} \cdot \frac{-1}{-1}$$

$$= \frac{6x+9}{2x-7} + \frac{-8x-2}{-7+2x}$$

$$= \frac{(6x+9)+(-8x-2)}{2x-7}$$

$$= \frac{-2x+7}{2x-7}$$

$$= \frac{-\cancel{(2x-7)}}{\cancel{(2x-7)}}$$

$$= -1$$

75.

$$\frac{5x^2-2}{9x^2-9}+\frac{x^2+2}{9-9x^2}=\frac{5x^2-2}{9x^2-9}+\frac{\left(x^2+2\right)}{\left(9-9x^2\right)}\cdot\frac{-1}{-1}$$

$$=\frac{5x^2-2}{9x^2-9}+\frac{-x^2-2}{-9+9x^2}$$

$$=\frac{\left(5x^2-2\right)+\left(-x^2-2\right)}{9x^2-9}$$

$$=\frac{4x^2-4}{9x^2-9}$$

$$=\frac{4\left(x^2-1\right)}{9\left(x^2-1\right)}=\frac{4}{9}$$

77.

$$\frac{3x+7}{x^2-9}-\frac{x+5}{9-x^2}=\frac{3x+7}{x^2-9}-\frac{(x+5)}{\left(9-x^2\right)}\cdot\frac{-1}{-1}$$

$$=\frac{3x+7}{x^2-9}-\frac{-x-5}{-9+x^2}$$

$$=\frac{(3x+7)-(-x-5)}{x^2-9}$$

$$=\frac{3x+7+x+5}{x^2-9}$$

$$=\frac{4x+12}{x^2-9}$$

$$=\frac{4(x+3)}{(x+3)(x-3)}$$

$$=\frac{4}{x-3}$$

79.

$$\frac{15y}{14x}+\frac{11y}{14x}-\frac{5y}{14x}=\frac{15y+11y-5y}{14x}$$

$$=\frac{21y}{14x}$$

$$=\frac{3\cdot7y}{2\cdot7x}$$

$$=\frac{3y}{2x}$$

81.

$$\frac{5x+8}{4x+4}+\frac{3x+5}{4x+4}+\frac{x-4}{4x+4}$$

$$=\frac{(5x+8)+(3x+5)+(x-4)}{4x+4}$$

$$=\frac{9x+9}{4x+4}$$

$$=\frac{9(x+1)}{4(x+1)}$$

$$=\frac{9}{4}$$

83.

$$\frac{x+3}{2x-5}-\frac{6x+5}{2x-5}+\frac{x-1}{2x-5}$$

$$=\frac{(x+3)-(6x+5)+(x-1)}{2x-5}$$

$$=\frac{x+3-6x-5+x-1}{2x-5}$$

$$=\frac{-4x-3}{2x-5}$$

85.

$$\frac{4x^2-x}{x+4}-\frac{x^2-10x}{x+4}-\frac{2x+20}{x+4}$$

$$=\frac{\left(4x^2-x\right)-\left(x^2-10x\right)-(2x+20)}{x+4}$$

$$=\frac{4x^2-x-x^2+10x-2x-20}{x+4}$$

$$=\frac{3x^2+7x-20}{x+4}$$

$$=\frac{\cancel{(x+4)}(3x-5)}{\cancel{(x+4)}}$$

$$=3x-5$$

87.

$$\frac{7x+3}{x-2}+\frac{x-5}{2-x}-\frac{3x+4}{x-2}$$

$$=\frac{7x+3}{x-2}+\frac{(x-5)}{(2-x)}\cdot\frac{-1}{-1}-\frac{3x+4}{x-2}$$

$$=\frac{7x+3}{x-2}+\frac{-x+5}{x-2}-\frac{3x+4}{x-2}$$

$$=\frac{(7x+3)+(-x+5)-(3x+4)}{x-2}$$

$$=\frac{7x+3-x+5-3x-4}{x-2}$$

$$=\frac{3x+4}{x-2}$$

89.

$$\frac{4x+1}{3x^2-16x-12}-\frac{2x-5}{3x^2-16x-12}+\frac{x-4}{3x^2-16x-12}$$

$$=\frac{(4x+1)-(2x-5)+(x-4)}{3x^2-16x-12}$$

$$=\frac{4x+1-2x+5+x-4}{3x^2-16x-12}$$

$$=\frac{3x+2}{3x^2-16x-12}$$

$$=\frac{\cancel{(3x+2)}}{\cancel{(3x+2)}(x-6)}=\frac{1}{x-6}$$

1.

$$\frac{3x}{5}+\frac{2x}{7}\qquad\text{LCD}=5\cdot7\text{ or }35$$

$$=\frac{3x}{5}\cdot\frac{7}{7}+\frac{2x}{7}\cdot\frac{5}{5}$$

$$=\frac{21x}{35}+\frac{10x}{35}$$

$$=\frac{21x+10x}{35}=\frac{31x}{35}$$

3.

$$\frac{4x}{3}-\frac{2x}{9}\qquad\text{LCD}=9$$

$$=\frac{4x}{3}\cdot\frac{3}{3}-\frac{2x}{9}$$

$$=\frac{12x}{9}-\frac{2x}{9}$$

$$=\frac{10x}{9}$$

5.

$$\frac{x}{2}+\frac{3x}{10}\qquad\text{LCD}=10$$

$$=\frac{x}{2}\cdot\frac{5}{5}+\frac{3x}{10}$$

$$=\frac{5x}{10}+\frac{3x}{10}$$

$$=\frac{8x}{10}=\frac{\cancel{2}\cdot4x}{\cancel{2}\cdot5}=\frac{4x}{5}$$

7.

$$\frac{7x}{30}-\frac{13x}{18}$$

$$=\frac{7x}{2\cdot3\cdot5}-\frac{13x}{2\cdot3\cdot3}\qquad\text{LCD}=2\cdot3\cdot3\cdot5$$

$$=90$$

$$=\frac{7x}{2\cdot3\cdot5}\cdot\frac{3}{3}-\frac{13x}{2\cdot3\cdot3}\cdot\frac{5}{5}$$

$$=\frac{21x}{90}-\frac{65x}{90}$$

$$=-\frac{44x}{90}=-\frac{\cancel{2}\cdot2\cdot11x}{3\cdot3\cdot\cancel{2}\cdot5}=-\frac{22x}{45}$$

9.

$$\frac{2}{3x} + \frac{5}{2x^2}$$

$$\text{LCD} = 3 \cdot 2 \cdot x^2$$
$$= 6x^2$$

$$= \frac{2}{3x} \cdot \frac{2x}{2x} + \frac{5}{2x^2} \cdot \frac{3}{3}$$

$$= \frac{4x}{6x^2} + \frac{15}{6x^2}$$

$$= \frac{4x + 15}{6x^2}$$

11.

$$\frac{3}{2x} - \frac{1}{6x}$$

$$= \frac{3}{2x} - \frac{1}{2 \cdot 3x}$$

$$\text{LCD} = 2 \cdot 3 \cdot x$$
$$= 6x$$

$$= \frac{3}{2x} \cdot \frac{3}{3} - \frac{1}{2 \cdot 3x}$$

$$= \frac{9}{6x} - \frac{1}{6x}$$

$$= \frac{9 - 1}{6x}$$

$$= \frac{8}{6x} = \frac{\cancel{2} \cdot 4}{\cancel{2} \cdot 3x} = \frac{4}{3x}$$

13.

$$\frac{3y}{10x} + \frac{y}{5x}$$

$$\text{LCD} = 10x$$

$$= \frac{3y}{10x} + \frac{y}{5x} \cdot \frac{2}{2}$$

$$= \frac{3y}{10x} + \frac{2y}{10x}$$

$$= \frac{3y + 2y}{10x}$$

$$= \frac{5y}{10x} = \frac{\cancel{5}y}{2 \cdot \cancel{5}x} = \frac{y}{2x}$$

15.

$$\frac{2y}{5x} - \frac{4y}{7x^2}$$

$$\text{LCD} = 5 \cdot 7 \cdot x^2$$
$$= 35x^2$$

$$= \frac{2y}{5x} \cdot \frac{7x}{7x} - \frac{4y}{7x^2} \cdot \frac{5}{5}$$

$$= \frac{14xy}{35x^2} - \frac{20y}{35x^2}$$

$$= \frac{14xy - 20y}{35x^2}$$

17.

$$\frac{2}{x} + \frac{4}{y}$$

$$\text{LCD} = xy$$

$$= \frac{2}{x} \cdot \frac{y}{y} + \frac{4}{y} \cdot \frac{x}{x}$$

$$= \frac{2y}{xy} + \frac{4x}{xy}$$

$$= \frac{2y + 4x}{xy}$$

19.

$$\frac{2y}{3x} - \frac{3x}{2y}$$

$$\text{LCD} = 2 \cdot 3 \cdot x \cdot y$$
$$= 6xy$$

$$= \frac{2y}{3x} \cdot \frac{2y}{2y} - \frac{3x}{2y} \cdot \frac{3x}{3x}$$

$$= \frac{4y^2}{6xy} - \frac{9x^2}{6xy}$$

$$= \frac{4y^2 - 9x^2}{6xy}$$

21.

$$\frac{5}{x^2y} + \frac{6}{xy^3} \qquad \text{LCD} = x^2y^3$$

$$= \frac{5}{x^2y} \cdot \frac{y^2}{y^2} + \frac{6}{xy^3} \cdot \frac{x}{x}$$

$$= \frac{5y^2}{x^2y^3} + \frac{6x}{x^2y^3}$$

$$= \frac{5y^2 + 6x}{x^2y^3}$$

23.

$$\frac{1}{3xy} - \frac{5}{9y} \qquad \text{LCD} = 9xy$$

$$= \frac{1}{3xy} \cdot \frac{3}{3} - \frac{5}{9y} \cdot \frac{x}{x}$$

$$= \frac{3}{9xy} - \frac{5x}{9xy}$$

$$= \frac{3 - 5x}{9xy}$$

25.

$$\frac{1}{x+3} + \frac{4}{2x+1} \qquad \text{LCD} = (x+3)(2x+1)$$

$$= \frac{1}{x+3} \cdot \frac{2x+1}{2x+1} + \frac{4}{2x+1} \cdot \frac{x+3}{x+3}$$

$$= \frac{2x+1}{(x+3)(2x+1)} + \frac{4(x+3)}{(x+3)(2x+1)}$$

$$= \frac{2x+1+4x+12}{(x+3)(2x+1)}$$

$$= \frac{6x+13}{(x+3)(2x+1)}$$

27.

$$\frac{5}{2x-1} - \frac{2}{x-2} \qquad \text{LCD} = (2x-1)(x-2)$$

$$= \frac{5}{2x-1} \cdot \frac{x-2}{x-2} - \frac{2}{x-2} \cdot \frac{2x-1}{2x-1}$$

$$= \frac{5(x-2)}{(2x-1)(x-2)} - \frac{2(2x-1)}{(2x-1)(x-2)}$$

$$= \frac{5(x-2) - 2(2x-1)}{(2x-1)(x-2)}$$

#27. continued.

$$= \frac{5x - 10 - 4x + 2}{(2x-1)(x-2)}$$

$$= \frac{x-8}{(2x-1)(x-2)}$$

29.

$$\frac{2}{x-5} - \frac{7}{x+3} \qquad \text{LCD} = (x-5)(x+3)$$

$$= \frac{2}{x-5} \cdot \frac{x+3}{x+3} - \frac{7}{x+3} \cdot \frac{x-5}{x-5}$$

$$= \frac{2(x+3)}{(x-5)(x+3)} - \frac{7(x-5)}{(x-5)(x+3)}$$

$$= \frac{2(x+3) - 7(x-5)}{(x-5)(x+3)}$$

$$= \frac{2x+6-7x+35}{(x-5)(x+3)}$$

$$= \frac{-5x+41}{(x-5)(x+3)}$$

31.

$$\frac{6x}{x+2} + \frac{3}{x-4} \qquad \text{LCD} = (x+2)(x-4)$$

$$= \frac{6x}{x+2} \cdot \frac{x-4}{x-4} + \frac{3}{x-4} \cdot \frac{x+2}{x+2}$$

$$= \frac{6x(x-4)}{(x+2)(x-4)} + \frac{3(x+2)}{(x+2)(x-4)}$$

$$= \frac{6x^2 - 24x + 3x + 6}{(x+2)(x-4)}$$

$$= \frac{6x^2 - 21x + 6}{(x+2)(x-4)}$$

33.

$$\frac{x}{x-4} - \frac{2}{x-1} \qquad \text{LCD} = (x\text{-}4)(x\text{-}1)$$

$$= \frac{x}{x-4} \cdot \frac{x-1}{x-1} - \frac{2}{x-1} \cdot \frac{x-4}{x-4}$$

$$= \frac{x(x-1)}{(x\text{-}4)(x\text{-}1)} - \frac{2(x-4)}{(x\text{-}1)(x\text{-}4)}$$

$$= \frac{x(x-1) - 2(x-4)}{(x-4)(x-1)}$$

$$= \frac{x^2 - x - 2x + 8}{(x-4)(x-1)}$$

$$= \frac{x^2 - 3x + 8}{(x-4)(x-1)}$$

35.

$$\frac{2x}{x-3} - \frac{x}{2x+1} \qquad \text{LCD} = (x\text{-}3)(2x+1)$$

$$= \frac{2x}{x-3} \cdot \frac{2x+1}{2x+1} - \frac{x}{2x+1} \cdot \frac{x-3}{x-3}$$

$$= \frac{2x(2x+1)}{(x-3)(2x+1)} - \frac{x(x-3)}{(2x+1)(x-3)}$$

$$= \frac{2x(2x+1) - x(x-3)}{(x-3)(2x+1)}$$

$$= \frac{4x^2 + 2x - x^2 + 3x}{(x-3)(2x+1)}$$

$$= \frac{3x^2 + 5x}{(x-3)(2x+1)}$$

37.

$$\frac{5x}{x+1} + \frac{2}{3x} \qquad \text{LCD} = 3x(x+1)$$

$$= \frac{5x}{x+1} \cdot \frac{3x}{3x} + \frac{2}{3x} \cdot \frac{x+1}{x+1}$$

$$= \frac{15x^2}{3x(x+1)} + \frac{2(x+1)}{3x(x+1)}$$

$$= \frac{15x^2 + 2x + 2}{3x(x+1)}$$

39.

$$\frac{3}{2x-1} - \frac{5}{2x} \qquad \text{LCD} = 2x(2x\text{-}1)$$

$$= \frac{3}{2x-1} \cdot \frac{2x}{2x} - \frac{5}{2x} \cdot \frac{2x-1}{2x-1}$$

$$= \frac{6x}{2x(2x-1)} - \frac{5(2x-1)}{2x(2x-1)}$$

$$= \frac{6x - 5(2x-1)}{2x(2x-1)}$$

$$= \frac{6x - 10x + 5}{2x(2x-1)}$$

$$= \frac{-4x + 5}{2x(2x-1)}$$

41.

$$\frac{2x}{4x-1} - \frac{2}{5x} \qquad \text{LCD} = 5x(4x\text{-}1)$$

$$= \frac{2x}{4x-1} \cdot \frac{5x}{5x} - \frac{2}{5x} \cdot \frac{4x-1}{4x-1}$$

$$= \frac{10x^2}{5x(4x-1)} - \frac{2(4x-1)}{5x(4x-1)}$$

$$= \frac{10x^2 - 2(4x-1)}{5x(4x-1)}$$

$$= \frac{10x^2 - 8x + 2}{5x(4x-1)}$$

43.

$$\frac{x}{4x+3} + \frac{9}{8x} \qquad \text{LCD} = 8x(4x+3)$$

$$= \frac{x}{4x+3} \cdot \frac{8x}{8x} + \frac{9}{8x} \cdot \frac{4x+3}{4x+3}$$

$$= \frac{8x^2}{8x(4x+3)} + \frac{9(4x+3)}{8x(4x+3)}$$

$$= \frac{8x^2 + 9(4x+3)}{8x(4x+3)}$$

$$= \frac{8x^2 + 36x + 27}{8x(4x+3)}$$

45.

$$\frac{2x+3}{x-5}+\frac{x+2}{x-2} \qquad \text{LCD} = (x-5)(x-2)$$

$$= \frac{2x+3}{x-5}\cdot\frac{(x-2)}{(x-2)}+\frac{x+2}{x-2}\cdot\frac{x-5}{x-5}$$

$$= \frac{(2x+3)(x-2)}{(x-5)(x-2)}+\frac{(x+2)(x-5)}{(x-2)(x-5)}$$

$$= \frac{2x^2-x-6+x^2-3x-10}{(x-5)(x-2)}$$

$$= \frac{3x^2-4x-16}{(x-5)(x-2)}$$

47.

$$\frac{x-2}{x-4}-\frac{x-5}{2x-3} \qquad \text{LCD} = (x-4)(2x-3)$$

$$= \frac{x-2}{x-4}\cdot\frac{2x-3}{2x-3}-\frac{x-5}{2x-3}\cdot\frac{x-4}{x-4}$$

$$= \frac{(x-2)(2x-3)}{(x-4)(2x-3)}-\frac{(x-5)(x-4)}{(2x-3)(x-4)}$$

$$= \frac{(x-2)(2x-3)-(x-5)(x-4)}{(x-4)(2x-3)}$$

$$= \frac{2x^2-7x+6-\left(x^2-9x+20\right)}{(x-4)(2x-3)}$$

$$= \frac{2x^2-7x+6-x^2+9x-20}{(x-4)(2x-3)}$$

$$= \frac{x^2+2x-14}{(x-4)(2x-3)}$$

49.

$$\frac{2x+1}{3x-5}-\frac{x-1}{x-2} \qquad \text{LCD} = (3x-5)(x-2)$$

$$= \frac{2x+1}{3x-5}\cdot\frac{x-2}{x-2}-\frac{x-1}{x-2}\cdot\frac{3x-5}{3x-5}$$

$$= \frac{(2x+1)(x-2)}{(3x-5)(x-2)}-\frac{(x-1)(3x-5)}{(x-2)(3x-5)}$$

$$= \frac{(2x+1)(x-2)-(x-1)(3x-5)}{(3x-5)(x-2)}$$

$$= \frac{2x^2-3x-2-\left(3x^2-8x+5\right)}{(3x-5)(x-2)}$$

#49. continued.

$$= \frac{2x^2-3x-2-3x^2+8x-5}{(3x-5)(x-2)}$$

$$= \frac{-x^2+5x-7}{(3x-5)(x-2)}$$

51.

$$\frac{3x-1}{2x+3}+\frac{2x-5}{3x-7} \qquad \text{LCD} = (2x+3)(3x-7)$$

$$= \frac{3x-1}{2x+3}\cdot\frac{3x-7}{3x-7}+\frac{2x-5}{3x-7}\cdot\frac{2x+3}{2x+3}$$

$$= \frac{(3x-1)(3x-7)}{(2x+3)(3x-7)}+\frac{(2x-5)(2x+3)}{(3x-7)(2x+3)}$$

$$= \frac{9x^2-24x+7+4x^2-4x-15}{(2x+3)(3x-7)}$$

$$= \frac{13x^2-28x-8}{(2x+3)(3x-7)}$$

53.

$$\frac{5x}{x^2-x-6}+\frac{2x}{x^2+6x+8}$$

$$= \frac{5x}{(x+2)(x-3)}+\frac{2x}{(x+2)(x+4)}$$

$$\text{LCD} = (x+2)(x-3)(x+4)$$

$$= \frac{5x}{(x+2)(x-3)}\cdot\frac{x+4}{x+4}+\frac{2x}{(x+2)(x+4)}\cdot\frac{x-3}{x-3}$$

$$= \frac{5x(x+4)}{(x+2)(x-3)(x+4)}+\frac{2x(x-3)}{(x+2)(x+4)(x-3)}$$

$$= \frac{5x^2+20x+2x^2-6x}{(x+2)(x-3)(x+4)}$$

$$= \frac{7x^2+14x}{(x+2)(x-3)(x+4)}$$

$$= \frac{7x\cancel{(x+2)}}{\cancel{(x+2)}(x-3)(x+4)}$$

$$= \frac{7x}{(x-3)(x+4)}$$

55.

$$\frac{4}{3x^2+12x} - \frac{1}{x^2+5x+4}$$

$$= \frac{4}{3x(x+4)} - \frac{1}{(x+4)(x+1)}$$

$$\text{LCD} = 3x(x+4)(x+1)$$

$$= \frac{4}{3x(x+4)} \cdot \frac{x+1}{x+1} - \frac{1}{(x+4)(x+1)} \cdot \frac{3x}{3x}$$

$$= \frac{4(x+1)}{3x(x+4)(x+1)} - \frac{3x}{(x+4)(x+1)3x}$$

$$= \frac{4x+4-3x}{3x(x+4)(x+1)}$$

$$= \frac{\cancel{(x+4)}}{3x\cancel{(x+4)}(x+1)}$$

$$= \frac{1}{3x(x+1)}$$

57.

$$\frac{3x}{x^2+x-2} + \frac{2}{x^2-4x+3}$$

$$= \frac{3x}{(x+2)(x-1)} + \frac{2}{(x-3)(x-1)}$$

$$\text{LCD} = (x+2)(x-1)(x-3)$$

$$= \frac{3x}{(x+2)(x-1)} \cdot \frac{x-3}{x-3} + \frac{2}{(x-3)(x-1)} \cdot \frac{x+2}{x+2}$$

$$= \frac{3x(x-3)}{(x+2)(x-1)(x-3)} + \frac{2(x+2)}{(x-3)(x-1)(x+2)}$$

$$= \frac{3x^2-9x+2x+4}{(x+2)(x-1)(x-3)}$$

$$= \frac{3x^2-7x+4}{(x+2)(x-1)(x-3)}$$

$$= \frac{(3x-4)\cancel{(x-1)}}{(x+2)\cancel{(x-1)}(x-3)}$$

$$= \frac{3x-4}{(x+2)(x-3)}$$

59.

$$\frac{3}{x^2-2x+1} - \frac{1}{x^2+4x-5}$$

$$= \frac{3}{(x-1)(x-1)} - \frac{1}{(x+5)(x-1)}$$

$$\text{LCD} = (x-1)^2(x+5)$$

$$= \frac{3}{(x-1)^2} \cdot \frac{x+5}{x+5} - \frac{1}{(x+5)(x-1)} \cdot \frac{x-1}{x-1}$$

$$= \frac{3(x+5)}{(x-1)^2(x+5)} - \frac{x-1}{(x+5)(x-1)^2}$$

$$= \frac{3x+15-(x-1)}{(x-1)^2(x+5)}$$

$$= \frac{3x+15-x+1}{(x-1)^2(x+5)}$$

$$= \frac{2x+16}{(x-1)^2(x+5)}$$

61.

$$\frac{x-4}{2x^2+8x} + \frac{7}{x^2+x-12}$$

$$= \frac{x-4}{2x(x+4)} + \frac{7}{(x-3)(x+4)}$$

$$\text{LCD} = 2x(x+4)(x-3)$$

$$= \frac{x-4}{2x(x+4)} \cdot \frac{x-3}{x-3} + \frac{7}{(x-3)(x+4)} \cdot \frac{2x}{2x}$$

$$= \frac{(x-4)(x-3)}{2x(x+4)(x-3)} + \frac{7 \cdot 2x}{(x-3)(x+4)2x}$$

$$= \frac{x^2-7x+12+14x}{2x(x+4)(x-3)}$$

$$= \frac{x^2+7x+12}{2x(x+4)(x-3)}$$

$$= \frac{(x+3)\cancel{(x+4)}}{2x\cancel{(x+4)}(x-3)}$$

$$= \frac{x+3}{2x(x-3)}$$

63.

$$\frac{x-4}{x^2-2x-8}-\frac{7}{x^2+8x+12}$$

$$=\frac{x-4}{(x-4)(x+2)}-\frac{7}{(x+2)(x+6)}$$

$$\text{LCD}=(x-4)(x+2)(x+6)$$

$$=\frac{x-4}{(x-4)(x+2)}\cdot\frac{x+6}{x+6}-\frac{7}{(x+2)(x+6)}\cdot\frac{x-4}{x-4}$$

$$=\frac{(x-4)(x+6)-7(x-4)}{(x-4)(x+2)(x+6)}$$

$$=\frac{x^2+2x-24-7x+28}{(x-4)(x+2)(x+6)}$$

$$=\frac{x^2-5x+4}{(x-4)(x+2)(x+6)}$$

$$=\frac{(x-4)(x-1)}{(x-4)(x+2)(x+6)}$$

$$=\frac{x-1}{(x+2)(x+6)}$$

65.

$$\frac{x+1}{x^2+8x+16}+\frac{3}{x^2-16}$$

$$=\frac{x+1}{(x+4)(x+4)}+\frac{3}{(x-4)(x+4)}$$

$$\text{LCD}=(x+4)^2(x-4)$$

$$=\frac{x+1}{(x+4)^2}\cdot\frac{x-4}{x-4}+\frac{3}{(x-4)(x+4)}\cdot\frac{x+4}{x+4}$$

$$=\frac{(x+1)(x-4)+3(x+4)}{(x+4)^2(x-4)}$$

$$=\frac{x^2-3x-4+3x+12}{(x+4)^2(x-4)}$$

$$=\frac{x^2+8}{(x+4)^2(x-4)}$$

67.

$$\frac{x+7}{x^2+2x-3}-\frac{6}{x^2+x-2}$$

$$=\frac{x+7}{(x+3)(x-1)}-\frac{6}{(x+2)(x-1)}$$

$$\text{LCD}=(x+3)(x-1)(x+2)$$

$$=\frac{x+7}{(x+3)(x-1)}\cdot\frac{x+2}{x+2}-\frac{6}{(x+2)(x-1)}\cdot\frac{x+3}{x+3}$$

$$=\frac{(x+7)(x+2)-6(x+3)}{(x+3)(x-1)(x+2)}$$

$$=\frac{x^2+9x+14-6x-18}{(x+3)(x-1)(x+2)}$$

$$=\frac{x^2+3x-4}{(x+3)(x-1)(x+2)}$$

$$=\frac{(x+4)(x-1)}{(x+3)(x-1)(x+2)}$$

$$=\frac{x+4}{(x+3)(x+2)}$$

69.

$$\frac{x+1}{x^2+6x+8}+\frac{x-4}{x^2-3x-10}$$

$$=\frac{x+1}{(x+4)(x+2)}+\frac{x-4}{(x-5)(x+2)}$$

$$\text{LCD}=(x+4)(x+2)(x-5)$$

$$=\frac{x+1}{(x+4)(x+2)}\cdot\frac{x-5}{x-5}+\frac{x-4}{(x-5)(x+2)}\cdot\frac{x+4}{x+4}$$

$$=\frac{(x+1)(x-5)+(x-4)(x+4)}{(x+4)(x+2)(x-5)}$$

$$=\frac{x^2-4x-5+x^2-16}{(x+4)(x+2)(x-5)}$$

$$=\frac{2x^2-4x-21}{(x+4)(x+2)(x-5)}$$

71.

$$\frac{x-3}{x^2+2x-15} - \frac{x-2}{x^2+3x-10}$$

$$= \frac{(x-3)}{(x-3)(x+5)} - \frac{(x-2)}{(x-2)(x+5)}$$

$$= \frac{1}{x+5} - \frac{1}{x+5}$$

$$= 0$$

73.

$$\frac{x-10}{x^2-2x-8} + \frac{x+12}{x^2-x-6}$$

$$= \frac{x-10}{(x-4)(x+2)} + \frac{x+12}{(x+2)(x-3)}$$

$$\text{LCD} = (x-4)(x+2)(x-3)$$

$$= \frac{x-10}{(x-4)(x+2)} \cdot \frac{x-3}{x-3} + \frac{x+12}{(x+2)(x-3)} \cdot \frac{x-4}{x-4}$$

$$= \frac{(x-10)(x-3) + (x+12)(x-4)}{(x-4)(x+2)(x-3)}$$

$$\frac{x^2-13x+30 + x^2+8x-48}{(x-4)(x+2)(x-3)}$$

$$= \frac{2x^2-5x-18}{(x-4)(x+2)(x-3)}$$

$$= \frac{(2x-9)(x+2)}{(x-4)(x+2)(x-3)}$$

$$= \frac{2x-9}{(x-4)(x-3)}$$

75.

$$\frac{2x-8}{x^2-2x-24} - \frac{x-4}{x^2+3x-4}$$

$$= \frac{2x-8}{(x+4)(x-6)} - \frac{x-4}{(x+4)(x-1)}$$

$$\text{LCD} = (x+4)(x-6)(x-1)$$

$$= \frac{2x-8}{(x+4)(x-6)} \cdot \frac{x-1}{x-1} - \frac{x-4}{(x+4)(x-1)} \cdot \frac{x-6}{x-6}$$

$$= \frac{(2x-8)(x-1) - (x-4)(x-6)}{(x+4)(x-6)(x-1)}$$

#75. continued.

$$\frac{2x^2-10x+8 - \left(x^2-10x+24\right)}{(x+4)(x-6)(x-1)}$$

$$= \frac{2x^2-10x+8 - x^2+10x-24}{(x+4)(x-6)(x-1)}$$

$$= \frac{x^2-16}{(x+4)(x-6)(x-1)}$$

$$= \frac{(x+4)(x-4)}{(x+4)(x-6)(x-1)}$$

$$= \frac{x-4}{(x-6)(x-1)}$$

77.

$$\frac{2x}{x-2} + \frac{5}{x+2} - \frac{8x}{x^2-4}$$

$$= \frac{2x}{x-2} + \frac{5}{x+2} - \frac{8x}{(x-2)(x+2)}$$

$$\text{LCD} = (x-2)(x+2)$$
$$= x^2-4$$

$$= \frac{2x}{x-2} \cdot \frac{x+2}{x+2} + \frac{5}{x+2} \cdot \frac{x-2}{x-2} - \frac{8x}{x^2-4}$$

$$= \frac{2x(x+2) + 5(x-2) - 8x}{x^2-4}$$

$$= \frac{2x^2+4x+5x-10-8x}{x^2-4}$$

$$= \frac{2x^2+x-10}{x^2-4}$$

$$= \frac{(2x+5)(x-2)}{(x+2)(x-2)}$$

$$= \frac{2x+5}{x+2}$$

79.

$$\frac{3}{x-6} + \frac{x}{2x+2} - \frac{21}{x^2-5x-6}$$

$$= \frac{3}{x-6} + \frac{x}{2(x+1)} - \frac{21}{(x-6)(x+1)}$$

$$\qquad\qquad LCD = 2(x-6)(x+1)$$

$$= \frac{3}{x-6} \cdot \frac{2(x+1)}{2(x+1)} + \frac{x}{2(x+1)} \cdot \frac{x-6}{x-6} - \frac{21}{(x-6)(x+1)} \cdot \frac{2}{2}$$

$$= \frac{6(x+1) + x(x-6) - 21 \cdot 2}{2(x-6)(x+1)}$$

$$= \frac{6x+6+x^2-6x-42}{2(x-6)(x+1)}$$

$$= \frac{x^2-36}{2(x-6)(x+1)}$$

$$= \frac{(x+6)(x-6)}{2(x-6)(x+1)}$$

$$= \frac{x+6}{2(x+1)}$$

Exercises 5.5

1.

$$\frac{\frac{3}{5}}{\frac{7}{2}} = \frac{3}{5} + \frac{7}{2} = \frac{3}{5} \cdot \frac{2}{7} = \frac{6}{35} \qquad \text{By method 1}$$

3.

$$\frac{\frac{2}{3}}{\frac{8}{9}} = \frac{2}{3} + \frac{8}{9} = \frac{2}{3} \cdot \frac{9}{8} = \frac{2 \cdot 3 \cdot 3}{3 \cdot 2 \cdot 4} = \frac{3}{4} \qquad \text{By method 1}$$

5.

$$\frac{\frac{12}{9}}{\frac{2}{3}} = \frac{12}{9} + \frac{2}{3} = \frac{12}{9} \cdot \frac{3}{2} = \frac{2 \cdot 2 \cdot 3 \cdot 3}{3 \cdot 3 \cdot 2}$$

$$\qquad\qquad = 2 \qquad \text{By method 1}$$

7.

$$\frac{\frac{x^3}{y^2}}{\frac{x^4}{y^6}} = \frac{x^3}{y^2} + \frac{x^4}{y^6} = \frac{x^3}{y^2} \cdot \frac{y^6}{x^4}$$

$$= \frac{y^{6-2}}{x^{4-3}}$$

$$= \frac{y^4}{x} \qquad \text{By method 1}$$

9.

$$\frac{\frac{2x^4}{9y^3}}{\frac{8x^2}{3y}} = \frac{2x^4}{9y^3} + \frac{8x^2}{3y} = \frac{2x^4}{9y^3} \cdot \frac{3y}{8x^2}$$

$$= \frac{2 \cdot 3 \cdot x^4 \cdot y}{3 \cdot 3 \cdot 2 \cdot 4x^2y^3}$$

$$= \frac{2 \cdot 3 x^{4-2}}{3 \cdot 3 \cdot 2 \cdot 4 y^{3-1}}$$

$$= \frac{x^2}{12y^2} \qquad \text{By method 1}$$

11.

$$\frac{\frac{5x^2}{3y}}{\frac{7y^6}{2x^3}} = \frac{5x^2}{3y} + \frac{7y^6}{2x^3} = \frac{5x^2}{3y} \cdot \frac{2x^3}{7y^6} = \frac{10x^5}{21y^7}$$

13.

$$\frac{\frac{4x^2}{3y^3}}{2} = \frac{4x^2}{3y^3} + \frac{2}{1} = \frac{4x^2}{3y^3} \cdot \frac{1}{2} = \frac{2 \cdot 2x^2}{2 \cdot 3y^3} = \frac{2x^2}{3y^3}$$

15.

$$\frac{\frac{5x^4}{9y^3}}{x^2} = \frac{5x^4}{9y^3} + \frac{x^2}{1} = \frac{5x^4}{9y^3} \cdot \frac{1}{x^2} = \frac{5x^{4-2}}{9y^3} = \frac{5x^2}{9y^3}$$

17.

$$\frac{\dfrac{9x^4}{10}}{3x^3} = \frac{9x^4}{10} \div \frac{3x^3}{1} = \frac{9x^4}{10} \cdot \frac{1}{3x^3} = \frac{3 \cdot 3x^{4-3}}{10 \cdot 3} = \frac{3x}{10}$$

19.

$$\frac{\dfrac{2x^2}{3y^3}}{4x^3} = \frac{2x^2}{3y^3} \div \frac{4x^3}{1} = \frac{2x^2}{3y^3} \cdot \frac{1}{4x^3}$$

$$= \frac{\cancel{2}}{3 \cdot \cancel{2} \cdot 2x^{3-2}y^3}$$

$$= \frac{1}{6xy^3} \qquad \text{By method 1}$$

21.

$$\frac{\dfrac{6x^3}{5y}}{\dfrac{1}{10y^2}} = \frac{6x^3}{5y} \div \frac{1}{10y^2} = \frac{6x^3}{5y} \cdot \frac{10y^2}{1}$$

$$= \frac{6 \cdot 2 \cdot \cancel{5}x^3y^{2-1}}{\cancel{5}}$$

$$= 12x^3y \qquad \text{By method 1}$$

23.

$$\frac{\dfrac{1}{x-2}}{\dfrac{1}{x^2-4}} = \frac{1}{x-2} \div \frac{1}{x^2-4}$$

$$= \frac{1}{x-2} \cdot \frac{x^2-4}{1}$$

$$= \frac{1}{\cancel{(x-2)}} \cdot \frac{\cancel{(x-2)}(x+2)}{1}$$

$$= x+2 \qquad \text{By method 1}$$

25.

$$\frac{\dfrac{5}{x^2+x-2}}{\dfrac{10}{x^2+5x+6}} = \frac{5}{x^2+x-2} \div \frac{10}{x^2+5x+6}$$

$$= \frac{5}{x^2+x-2} \cdot \frac{x^2+5x+6}{10}$$

$$= \frac{\cancel{5}(x+2)(x+3)}{2 \cdot \cancel{5}(x+2)(x-1)}$$

$$= \frac{x+3}{2(x-1)}$$

27.

$$\frac{\dfrac{2}{x^2+x-6}}{\dfrac{6}{x-2}} = \frac{2}{x^2+x-6} \div \frac{6}{x-2}$$

$$= \frac{2}{x^2+x-6} \cdot \frac{x-2}{6}$$

$$= \frac{\cancel{2}\cancel{(x-2)}}{\cancel{2} \cdot 3\cancel{(x-2)}(x+3)}$$

$$= \frac{1}{3(x+3)} \qquad \text{By method 1}$$

29.

$$\frac{\dfrac{x^2-9}{7}}{x+3} = \frac{x^2-9}{7} \div \frac{x+3}{1}$$

$$= \frac{x^2-9}{7} \cdot \frac{1}{x+3}$$

$$= \frac{\cancel{(x+3)}(x-3)}{7\cancel{(x+3)}}$$

$$= \frac{x-3}{7}$$

31.

$$\frac{\dfrac{4}{2x+3}}{12} = \frac{4}{2x+3} + \frac{12}{1}$$

$$= \frac{4}{2x+3} \cdot \frac{1}{12}$$

$$= \frac{\cancel{4}}{3 \cdot \cancel{4}(2x+3)}$$

$$= \frac{1}{3(2x+3)}$$

33.

$$\frac{\dfrac{6x^3}{3x+1}}{8x} = \frac{6x^3}{3x+1} + \frac{8x}{1}$$

$$= \frac{6x^3}{3x+1} \cdot \frac{1}{8x}$$

$$= \frac{\cancel{2} \cdot 3x^{3-1}}{\cancel{2} \cdot 4(3x+1)}$$

$$= \frac{3x^2}{4(3x+1)}$$

35.

$$\frac{\dfrac{12x^2}{x+5}}{\dfrac{15x^4}{x-1}} = \frac{12x^2}{x+5} + \frac{15x^4}{x-1}$$

$$= \frac{12x^2}{x+5} \cdot \frac{x-1}{15x^4}$$

$$= \frac{\cancel{3} \cdot 4(x-1)}{\cancel{3} \cdot 5 \cdot x^{4-2}(x+5)}$$

$$= \frac{4(x-1)}{5x^2(x+5)} \qquad \text{By method 1}$$

37.

$$\frac{\dfrac{2x^2-2x}{x+3}}{\dfrac{5x^3-5x^2}{x-5}} = \frac{2x^2-2x}{x+3} + \frac{5x^3-5x^2}{x-5}$$

$$= \frac{2x^2-2x}{x+3} \cdot \frac{x-5}{5x^3-5x^2}$$

$$= \frac{2x(x-1)}{(x+3)} \cdot \frac{(x-5)}{5x^2(x-1)}$$

$$= \frac{2\cancel{(x-1)}(x-5)}{5x^{2-1}(x+3)\cancel{(x-1)}}$$

$$= \frac{2(x-5)}{5x(x+3)} \qquad \text{By method 1}$$

39.

$$\frac{\dfrac{5x^2+10x}{2x-6}}{\dfrac{2x^2+4x}{3x-9}} = \frac{5x^2+10x}{2x-6} + \frac{2x^2+4x}{3x-9}$$

$$= \frac{5\cancel{x}\cancel{(x+2)}}{2\cancel{(x-3)}} \cdot \frac{3\cancel{(x-3)}}{2\cancel{x}\cancel{(x+2)}}$$

$$= \frac{15}{4} \qquad \text{By method 1}$$

41.

$$\frac{\dfrac{4x}{3x+9}}{\dfrac{8x^2-12x}{x+3}} = \frac{4x}{3x+9} + \frac{8x^2-12x}{x+3}$$

$$= \frac{\cancel{4x}}{3\cancel{(x+3)}} \cdot \frac{\cancel{(x+3)}}{\cancel{4x}(2x-3)} = \frac{1}{3(2x-3)}$$

43.

$$\frac{\dfrac{1}{2}+\dfrac{1}{4}}{\dfrac{1}{3}+\dfrac{1}{6}}=\frac{\dfrac{2}{4}+\dfrac{1}{4}}{\dfrac{2}{6}+\dfrac{1}{6}}$$

Add numerator and denominator.

$$=\frac{\dfrac{3}{4}}{\dfrac{3}{6}}$$

$$=\frac{3}{4}\div\frac{3}{6}$$

$$=\frac{3}{4}\cdot\frac{6}{3}$$

$$=\frac{3\cdot2\cdot3}{2\cdot2\cdot3}=\frac{3}{2}$$

45.

$$\frac{\dfrac{2}{3}-\dfrac{1}{2}}{\dfrac{3}{4}-\dfrac{1}{6}}=\frac{\dfrac{4}{6}-\dfrac{3}{6}}{\dfrac{9}{12}-\dfrac{2}{12}}$$

Subtract numerator and denominator.

$$=\frac{\dfrac{1}{6}}{\dfrac{7}{12}}$$

$$=\frac{1}{6}\cdot\frac{12}{7}$$

Invert and multiply.

$$=\frac{2\cdot6}{6\cdot7}=\frac{2}{7}$$

By method 1

47.

$$\frac{\dfrac{5}{4}-\dfrac{2}{3}}{\dfrac{5}{6}+\dfrac{1}{2}}=\frac{\dfrac{15}{12}-\dfrac{8}{12}}{\dfrac{5}{6}+\dfrac{3}{6}}=\frac{\dfrac{7}{12}}{\dfrac{8}{6}}=\frac{\dfrac{7}{12}}{\dfrac{4}{3}}$$

$$=\frac{7}{12}\cdot\frac{3}{4}$$

Invert and multiply

$$=\frac{7}{3\cdot4}\cdot\frac{3}{4}$$

$$=\frac{7}{16}$$

By method 1

49.

$$\frac{3+\dfrac{3}{2x}}{x+\dfrac{1}{2}}$$

LCD = 2x

Multiply by $\dfrac{2x}{2x}$.

$$=\frac{\left(3+\dfrac{3}{2x}\right)}{\left(x+\dfrac{1}{2}\right)}\cdot\frac{2x}{2x}$$

$$=\frac{\left(3+\dfrac{3}{2x}\right)\cdot2x}{\left(x+\dfrac{1}{2}\right)\cdot2x}$$

$$=\frac{6x+3}{2x^2+x}$$

$$=\frac{3(2x+1)}{x(2x+1)}\qquad\text{or}\qquad\frac{3}{x}$$

51.

$$\frac{2-\dfrac{3}{x}}{\dfrac{10}{3}-\dfrac{5}{x}}$$

LCD = 3x

Multiply by $\dfrac{3x}{3x}$.

$$=\frac{\left(2-\dfrac{3}{x}\right)}{\left(\dfrac{10}{3}-\dfrac{5}{x}\right)}\cdot\frac{3x}{3x}$$

$$=\frac{\left(2-\dfrac{3}{x}\right)3x}{\left(\dfrac{10}{3}-\dfrac{5}{x}\right)3x}$$

$$=\frac{6x-9}{10x-15}$$

$$=\frac{3(2x-3)}{5(2x-3)}=\frac{3}{5}$$

By method 2

53.

$$\dfrac{\dfrac{9}{4}-\dfrac{1}{x^2}}{\dfrac{3}{2}+\dfrac{1}{x}} \qquad \text{LCD} = 4x^2$$

Multiply by $\dfrac{4x^2}{4x^2}$.

$$=\dfrac{\left(\dfrac{9}{4}-\dfrac{1}{x^2}\right)}{\left(\dfrac{3}{2}+\dfrac{1}{x}\right)}\cdot\dfrac{4x^2}{4x^2}$$

$$=\dfrac{\left(\dfrac{9}{4}-\dfrac{1}{x^2}\right)4x^2}{\left(\dfrac{3}{2}+\dfrac{1}{x}\right)4x^2}$$

$$=\dfrac{9x^2-4}{6x^2+4x}$$

$$=\dfrac{(3x+2)(3x-2)}{2x(3x+2)}$$

$$=\dfrac{3x-2}{2x} \qquad \text{By method 2}$$

55.

$$\dfrac{\dfrac{6}{x}+\dfrac{3}{2}}{9} \qquad \text{LCD} = 2x$$

Multiply by $\dfrac{2x}{2x}$.

$$=\dfrac{\left(\dfrac{6}{x}+\dfrac{3}{2}\right)}{9}\cdot\dfrac{2x}{2x}$$

$$=\dfrac{\left(\dfrac{6}{x}+\dfrac{3}{2}\right)2x}{9\cdot 2x}$$

$$=\dfrac{12+3x}{18x}$$

$$=\dfrac{3(4+x)}{3\cdot 6x}$$

$$=\dfrac{4+x}{6x}\ \text{ or }\ \dfrac{x+4}{6x} \qquad \text{By method 2}$$

57.

$$\dfrac{\dfrac{1}{x}+\dfrac{1}{2}}{x+2} \qquad \text{LCD} = 2x$$

Multiply by $\dfrac{2x}{2x}$

$$=\dfrac{\left(\dfrac{1}{x}+\dfrac{1}{2}\right)}{(x+2)}\cdot\dfrac{2x}{2x}$$

$$=\dfrac{\left(\dfrac{1}{x}+\dfrac{1}{2}\right)2x}{(x+2)2x}$$

$$=\dfrac{2+x}{2x(x+2)}$$

$$=\dfrac{(x+2)}{2x(x+2)}=\dfrac{1}{2x}$$

59.

$$\dfrac{x-5}{\dfrac{1}{25}-\dfrac{1}{x^2}} \qquad \text{LCD} = 25x^2$$

Multiply by $\dfrac{25x^2}{25x^2}$.

$$=\dfrac{(x-5)}{\left(\dfrac{1}{25}-\dfrac{1}{x^2}\right)}\cdot\dfrac{25x^2}{25x^2}$$

$$=\dfrac{(x-5)25x^2}{\left(\dfrac{1}{25}-\dfrac{1}{x^2}\right)25x^2}$$

$$=\dfrac{25x^2(x-5)}{\left(x^2-25\right)}$$

$$=\dfrac{25x^2(x-5)}{(x-5)(x+5)}=\dfrac{25x^2}{x+5}$$

61.

$$\frac{2x+1}{\frac{1}{4}-\frac{3}{x}}$$

LCD = 4x

Multiply by $\frac{4x}{4x}$.

$$=\frac{(2x+1)}{\left(\frac{1}{4}-\frac{3}{x}\right)}\cdot\frac{4x}{4x}$$

$$=\frac{(2x+1)4x}{\left(\frac{1}{4}-\frac{3}{x}\right)4x}$$

$$=\frac{4x(2x+1)}{x-12}$$

By method 2

63.

$$\frac{1+\frac{1}{x}-\frac{20}{x^2}}{1+\frac{4}{x}-\frac{5}{x^2}}$$

LCD = x^2

Multiply by $\frac{x^2}{x^2}$.

$$=\frac{\left(1+\frac{1}{x}-\frac{20}{x^2}\right)}{\left(1+\frac{4}{x}-\frac{5}{x^2}\right)}\cdot\frac{x^2}{x^2}$$

$$=\frac{\left(1+\frac{1}{x}-\frac{20}{x^2}\right)x^2}{\left(1+\frac{4}{x}-\frac{5}{x^2}\right)x^2}$$

$$=\frac{x^2+x-20}{x^2+4x-5}$$

$$=\frac{(x+5)(x-4)}{(x+5)(x-1)}$$

$$=\frac{x-4}{x-1}$$

65.

$$\frac{1+\frac{5}{x}-\frac{14}{x^2}}{1+\frac{3}{x}-\frac{10}{x^2}}$$

LCD = x^2

Multiply by $\frac{x^2}{x^2}$.

$$=\frac{\left(1+\frac{5}{x}-\frac{14}{x^2}\right)x^2}{\left(1+\frac{3}{x}-\frac{10}{x^2}\right)x^2}$$

$$=\frac{x^2+5x-14}{x^2+3x-10}$$

$$=\frac{(x+7)(x-2)}{(x+5)(x-2)}$$

$$=\frac{x+7}{x+5}$$

67.

$$\frac{1-\frac{4}{x}+\frac{3}{x^2}}{1-\frac{6}{x}+\frac{5}{x^2}}$$

LCD = x^2

Multiply by $\frac{x^2}{x^2}$.

$$=\frac{\left(1-\frac{4}{x}+\frac{3}{x^2}\right)x^2}{\left(1-\frac{6}{x}+\frac{5}{x^2}\right)x^2}$$

$$=\frac{x^2-4x+3}{x^2-6x+5}$$

$$=\frac{(x-3)(x-1)}{(x-5)(x-1)}$$

$$=\frac{x-3}{x-5}$$

69.

$$\frac{3+\dfrac{1}{x+2}}{\dfrac{2}{x+2}+\dfrac{1}{4}}$$

LCD $= 4(x+2)$

Multiply by $\dfrac{4(x+2)}{4(x+2)}$

$$=\frac{\left(3+\dfrac{1}{x+2}\right)\cdot 4(x+2)}{\left(\dfrac{2}{x+2}+\dfrac{1}{4}\right)\cdot 4(x+2)}$$

$$=\frac{12(x+2)+4}{8+(x+2)}$$

$$=\frac{12x+24+4}{8+x+2}$$

$$=\frac{12x+28}{x+10}$$

73.

$$\frac{\dfrac{1}{3}+\dfrac{1}{x-4}}{\dfrac{2}{3}+\dfrac{4}{x-4}}$$

LCD $= 3(x-4)$

Multiply by $\dfrac{3(x-4)}{3(x-4)}$

$$=\frac{\left(\dfrac{1}{3}+\dfrac{1}{x-4}\right)\cdot 3(x-4)}{\left(\dfrac{2}{3}+\dfrac{4}{x-4}\right)\cdot 3(x-4)}$$

$$=\frac{x-4+3}{2(x-4)+12}$$

$$=\frac{x-1}{2x-8+12}$$

$$=\frac{x-1}{2x+4}$$

71.

$$\frac{5-\dfrac{3}{x-1}}{\dfrac{3}{x-1}+\dfrac{2}{3}}$$

LCD $= 3(x-1)$

Multiply by $\dfrac{3(x-1)}{3(x-1)}$

$$=\frac{\left(5-\dfrac{3}{x-1}\right)\cdot 3(x-1)}{\left(\dfrac{3}{x-1}+\dfrac{2}{3}\right)\cdot 3(x-1)}$$

$$=\frac{15(x-1)-9}{9+2(x-1)}$$

$$=\frac{15x-15-9}{9+2x-2}$$

$$=\frac{15x-24}{2x+7}$$

75.

$$\frac{\dfrac{3}{x}-\dfrac{1}{x+2}}{\dfrac{5}{x}-\dfrac{4}{x+2}}$$

LCD $= x(x+2)$

Multiply by $\dfrac{x(x+2)}{x(x+2)}$

$$=\frac{\left(\dfrac{3}{x}-\dfrac{1}{x+2}\right)\cdot x(x+2)}{\left(\dfrac{5}{x}-\dfrac{4}{x+2}\right)\cdot x(x+2)}$$

$$=\frac{3(x+2)-x}{5(x+2)-4x}$$

$$=\frac{3x+6-x}{5x+10-4x}$$

$$=\frac{2x+6}{x+10}$$

Exercises 5.6

1.

$$\frac{2x}{3} + \frac{x}{4} = \frac{11}{12} \qquad LCD = 12$$

$$\frac{12}{1}\left(\frac{2x}{3} + \frac{x}{4}\right) = \frac{11}{12} \cdot \frac{12}{1}$$

$$\frac{12}{1} \cdot \frac{2x}{3} + \frac{12}{1} \cdot \frac{x}{4} = \frac{11}{12} \cdot \frac{12}{1}$$

$$8x + 3x = 11$$

$$11x = 11$$

$$x = 1$$

Solution set: $\{1\}$

3.

$$\frac{5x}{2} - \frac{2x}{3} = \frac{-11}{18} \qquad LCD = 18$$

$$\frac{18}{1}\left(\frac{5x}{2} - \frac{2x}{3}\right) = \frac{-11}{18} \cdot \frac{18}{1}$$

$$\frac{18}{1} \cdot \frac{5x}{2} - \frac{18}{1} \cdot \frac{2x}{3} = \frac{-11}{18} \cdot \frac{18}{1}$$

$$45x - 12x = -11$$

$$33x = -11$$

$$x = -\frac{1}{3}$$

Solution set: $\left\{-\frac{1}{3}\right\}$

5.

$$\frac{x}{10} - 1 = \frac{x}{15} \qquad LCD = 30$$

$$\frac{30}{1}\left(\frac{x}{10} - 1\right) = \frac{x}{15} \cdot \frac{30}{1}$$

$$\frac{30}{1} \cdot \frac{x}{10} - \frac{30}{1} \cdot 1 = \frac{x}{15} \cdot \frac{30}{1}$$

$$3x - 30 = 2x$$

$$3x - 2x = 30$$

$$x = 30$$

Solution set: $\{30\}$

7.

$$\frac{5x}{6} + \frac{1}{2} = \frac{5x}{4} \qquad LCD = 12$$

$$\frac{12}{1}\left(\frac{5x}{6} + \frac{1}{2}\right) = \frac{5x}{4} \cdot \frac{12}{1}$$

$$\frac{12}{1} \cdot \frac{5x}{6} + \frac{12}{1} \cdot \frac{1}{2} = \frac{5x}{4} \cdot \frac{12}{1}$$

$$10x + 6 = 15x$$

$$6 = 15x - 10x$$

$$6 = 5x$$

$$\frac{6}{5} = x$$

Solution set: $\left\{\frac{6}{5}\right\}$

9.

$$\frac{x}{2} - \frac{4}{3} = \frac{2x - 5}{3} \qquad LCD = 6$$

$$\frac{6}{1}\left(\frac{x}{2} - \frac{4}{3}\right) = \frac{6}{1}\left(\frac{2x - 5}{3}\right)$$

$$\frac{6}{1} \cdot \frac{x}{2} - \frac{6}{1} \cdot \frac{4}{3} = \frac{6(2x - 5)}{3}$$

$$3x - 8 = 2(2x - 5)$$

$$3x - 8 = 4x - 10$$

$$-8 + 10 = 4x - 3x$$

$$2 = x$$

Solution set: $\{2\}$

11.
$$\frac{2x}{5} + 3 = \frac{4x+1}{2} \qquad \text{LCD} = 10$$

$$\frac{10}{1}\left(\frac{2x}{5} + 3\right) = \frac{10}{1}\left(\frac{4x+1}{2}\right)$$

$$\frac{10}{1} \cdot \frac{2x}{5} + \frac{10}{1} \cdot 3 = \frac{10(4x+1)}{2}$$

$$4x + 30 = 5(4x+1)$$

$$4x + 30 = 20x + 5$$

$$30 - 5 = 20x - 4x$$

$$25 = 16x$$

$$\frac{25}{16} = x$$

Solution set: $\left\{\dfrac{25}{16}\right\}$

15.
$$\frac{x+1}{4} + \frac{2x-5}{2} = \frac{x-3}{8} \qquad \text{LCD} = 8$$

$$\frac{8}{1}\left(\frac{x+1}{4} + \frac{2x-5}{2}\right) = \frac{8}{1}\left(\frac{x-3}{8}\right)$$

$$\frac{8}{1}\left(\frac{x+1}{4}\right) + \frac{8}{1}\left(\frac{2x-5}{2}\right) = \frac{8}{1}\left(\frac{x-3}{8}\right)$$

$$2(x+1) + 4(2x-5) = x - 3$$

$$2x + 2 + 8x - 20 = x - 3$$

$$10x - 18 = x - 3$$

$$10x - x = -3 + 18$$

$$9x = 15$$

$$x = \frac{15}{9} \ \text{ or } \ \frac{5}{3}$$

Solution set: $\left\{\dfrac{5}{3}\right\}$

13.
$$\frac{x-1}{2} - \frac{3x+1}{5} = \frac{7}{10} \qquad \text{LCD} = 10$$

$$\frac{10}{1}\left(\frac{x-1}{2} - \frac{3x+1}{5}\right) = \frac{7}{10} \cdot \frac{10}{1}$$

$$\frac{10}{1}\left(\frac{x-1}{2}\right) - \frac{10}{1}\left(\frac{3x+1}{5}\right) = 7$$

$$5(x-1) - 2(3x+1) = 7$$

$$5x - 5 - 6x - 2 = 7$$

$$-x - 7 = 7$$

$$-x = 7 + 7$$

$$-x = 14$$

$$x = -14$$

Solution set: $\{-14\}$

17.
$$\frac{4x}{3} = \frac{1}{2} \quad \text{LCD} = 6$$

$$\frac{6}{1}\left(\frac{4x}{3}\right) = \frac{6}{1}\left(\frac{1}{2}\right)$$

$$8x = 3$$

$$x = \frac{3}{8}$$

Solution set: $\left\{\dfrac{3}{8}\right\}$

19.

$$\frac{x-1}{2} = \frac{3x}{4} \qquad \text{LCD} = 4$$

$$\frac{4}{1}\left(\frac{x-1}{2}\right) = \frac{4}{1}\left(\frac{3x}{4}\right)$$

$$2(x-1) = 3x$$

$$2x - 2 = 3x$$

$$-2 = 3x - 2x$$

$$-2 = x$$

Solution set: $\{-2\}$

21.

$$\frac{x+1}{6} = \frac{2x-3}{4} \qquad \text{LCD} = 12$$

$$\frac{12}{1}\left(\frac{x+1}{6}\right) = \frac{12}{1}\left(\frac{2x-3}{4}\right)$$

$$2(x+1) = 3(2x-3)$$

$$2x + 2 = 6x - 9$$

$$2 + 9 = 6x - 2x$$

$$11 = 4x$$

$$\frac{11}{4} = x$$

Solution set: $\left\{\frac{11}{4}\right\}$

23.

$$\frac{x-3}{3} = \frac{3x+1}{4} \qquad \text{LCD} = 12$$

$$\frac{12}{1}\left(\frac{x-3}{3}\right) = \frac{12}{1}\left(\frac{3x+1}{4}\right)$$

$$4(x-3) = 3(3x+1)$$

$$4x - 12 = 9x + 3$$

$$-12 - 3 = 9x - 4x$$

$$-15 = 5x$$

$$-3 = x$$

Solution set: $\{-3\}$

25.

$$\frac{3}{2x} + \frac{1}{x} = \frac{3}{4} \qquad \text{LCD} = 4x$$

$$\frac{4x}{1}\left(\frac{3}{2x} + \frac{1}{x}\right) = \frac{4x}{1} \cdot \frac{3}{4}$$

$$\frac{4x}{1} \cdot \frac{3}{2x} + \frac{4x}{1} \cdot \frac{1}{x} = \frac{4x}{1} \cdot \frac{3}{4}$$

$$6 + 4 = 3x$$

$$10 = 3x$$

$$\frac{10}{3} = x$$

Solution set: $\left\{\frac{10}{3}\right\}$

27.

$$\frac{4}{3x} - 1 = \frac{1}{2x} \qquad \text{LCD} = 6x$$

$$\frac{6x}{1}\left(\frac{4}{3x} - 1\right) = \frac{6x}{1} \cdot \frac{1}{2x}$$

$$\frac{6x}{1} \cdot \frac{4}{3x} - \frac{6x}{1} \cdot 1 = \frac{6x}{1} \cdot \frac{1}{2x}$$

$$8 - 6x = 3$$

$$8 - 3 = 6x$$

$$5 = 6x$$

$$\frac{5}{6} = x$$

Solution set: $\left\{\frac{5}{6}\right\}$

29.
$$\frac{x}{2} + \frac{1}{x} = \frac{9}{4} \qquad \text{LCD} = 4x$$

$$\frac{4x}{1}\left(\frac{x}{2} + \frac{1}{x}\right) = \frac{4x}{1} \cdot \frac{9}{4}$$

$$\frac{4x}{1} \cdot \frac{x}{2} + \frac{4x}{1} \cdot \frac{1}{x} = \frac{4x}{1} \cdot \frac{9}{4}$$

$$2x^2 + 4 = 9x$$

$$2x^2 - 9x + 4 = 0$$

$$(2x - 1)(x - 4) = 0$$

$$2x - 1 = 0 \quad \text{or} \quad x - 4 = 0$$

$$2x = 1 \qquad\qquad x = 4$$

$$x = \frac{1}{2}$$

Solution set: $\left\{\frac{1}{2}, 4\right\}$

31.
$$\frac{x}{8} - \frac{5}{8x} = -\frac{1}{2} \qquad \text{LCD} = 8x$$

$$\frac{8x}{1}\left(\frac{x}{8} - \frac{5}{8x}\right) = \frac{8x}{1}\left(-\frac{1}{2}\right)$$

$$\frac{8x}{1} \cdot \frac{x}{8} - \frac{8x}{1} \cdot \frac{5}{8x} = \frac{8x}{1}\left(-\frac{1}{2}\right)$$

$$x^2 - 5 = -4x$$

$$x^2 + 4x - 5 = 0$$

$$(x + 5)(x - 1) = 0$$

$$x + 5 = 0 \quad \text{or} \quad x - 1 = 0$$

$$x = -5 \qquad\qquad x = 1$$

Solution set: $\{-5, 1\}$

33.
$$\frac{5}{2x - 1} = \frac{4}{3x} \qquad \text{LCD} = 3x(2x - 1)$$

$$\frac{3x(2x - 1)}{1} \cdot \frac{5}{2x - 1} = \frac{3x(2x - 1)}{1} \cdot \frac{4}{3x}$$

$$\frac{15x(2x - 1)}{(2x - 1)} = \frac{4 \cdot 3x(2x - 1)}{3x}$$

$$15x = 4(2x - 1)$$

$$15x = 8x - 4$$

$$7x = -4$$

$$x = -\frac{4}{7}$$

Solution set: $\left\{-\frac{4}{7}\right\}$

35.
$$\frac{x}{x + 7} = \frac{5}{x + 3} \qquad \text{LCD} = (x + 7)(x + 3)$$

$$\frac{(x + 7)(x + 3)}{1} \cdot \frac{x}{(x + 7)} = \frac{(x + 7)(x + 3)}{1} \cdot \frac{5}{(x + 3)}$$

$$x(x + 3) = 5(x + 7)$$

$$x^2 + 3x = 5x + 35$$

$$x^2 + 3x - 5x - 35 = 0$$

$$x^2 - 2x - 35 = 0$$

$$(x - 7)(x + 5) = 0$$

$$x - 7 = 0 \quad \text{or} \quad x + 5 = 0$$

$$x = 7 \qquad\qquad x = -5$$

Solution set: $\{7, -5\}$

37.

$$\frac{x+1}{x+2} = \frac{x-5}{x-3} \quad LCD = (x+2)(x-3)$$

$$\frac{(x+2)(x-3)}{1} \cdot \frac{(x+1)}{(x+2)} = \frac{(x+2)(x-3)}{1} \cdot \frac{(x-5)}{(x-3)}$$

$$(x-3)(x+1) = (x+2)(x-5)$$

$$x^2 - 2x - 3 = x^2 - 3x - 10$$

$$x^2 - x^2 - 2x + 3x = -10 + 3$$

$$x = -7$$

Solution set: {-7}

39.

$$\frac{x+9}{x+12} = \frac{4x+3}{3x+4} \quad LCD = (x+12)(3x+4)$$

$$\frac{(x+12)(3x+4)}{1} \cdot \frac{(x+9)}{(x+12)}$$

$$= \frac{(x+12)(3x+4)}{1} \cdot \frac{(4x+3)}{(3x+4)}$$

$$(3x+4)(x+9) = (x+12)(4x+3)$$

$$3x^2 + 31x + 36 = 4x^2 + 51x + 36$$

$$0 = 4x^2 - 3x^2 + 51x - 31x + 36 - 36$$

$$0 = x^2 + 20x$$

$$0 = x(x+20)$$

$$x = 0 \quad \text{or} \quad x + 20 = 0$$

$$x = -20$$

Solution set: {0, -20}

41.

$$\frac{x}{x-2} - 3 = \frac{2}{x-2} \quad LCD = x - 2$$

$$\frac{x-2}{1}\left(\frac{x}{x-2} - 3\right) = \frac{x-2}{1} \cdot \frac{2}{x-2}$$

$$\frac{(x-2)}{1} \cdot \frac{x}{(x-2)} - \frac{x-2}{1} \cdot 3 = \frac{(x-2)}{1} \cdot \frac{2}{(x-2)}$$

$$x - 3(x-2) = 2$$

$$x - 3x + 6 = 2$$

$$-2x = 2 - 6$$

$$-2x = -4$$

$$x = 2$$

Solution set: ∅
Note: x ≠ 2

43.

$$\frac{x}{x-3} + 5 = \frac{21}{x-3} \quad LCD = x - 3$$

$$\frac{x-3}{1}\left(\frac{x}{x-3} + 5\right) = \frac{x-3}{1} \cdot \frac{21}{x-3}$$

$$\frac{(x-3)}{1} \cdot \frac{x}{(x-3)} + \frac{x-3}{1} \cdot 5 = \frac{(x-3)}{1} \cdot \frac{21}{(x-3)}$$

$$x + 5(x-3) = 21$$

$$x + 5x - 15 = 21$$

$$6x = 21 + 15$$

$$6x = 36$$

$$x = 6$$

Solution set: {6}

45.

$$\frac{3}{x+4}+\frac{5}{x-2}=\frac{16}{x^2+2x-8}$$

$$\frac{3}{x+4}+\frac{5}{x-2}=\frac{16}{(x+4)(x-2)}$$

$$LCD=(x+4)(x-2)$$

$$\frac{(x+4)(x-2)}{1}\left(\frac{3}{(x+4)}+\frac{5}{(x-2)}\right)=$$

$$\frac{(x+4)(x-2)}{1}\left(\frac{16}{(x+4)(x-2)}\right)$$

$$3(x-2)+5(x+4)=16 \quad \text{Distributive property}$$
$$\text{and cancel}$$

$$3x-6+5x+20=16$$
$$8x=16+6-20$$
$$8x=2$$
$$x=\frac{2}{8}\ \text{or}\ \frac{1}{4}$$

Solution set: $\left\{\frac{1}{4}\right\}$

47.

$$\frac{5}{x+2}-\frac{3}{x+1}=\frac{9}{x^2+3x+2}$$

$$\frac{5}{x+2}-\frac{3}{x+1}=\frac{9}{(x+2)(x+1)}$$

$$LCD=(x+2)(x+1)$$

$$\frac{(x+2)(x+1)}{1}\left(\frac{5}{x+2}-\frac{3}{x+1}\right)$$

$$=\frac{(x+2)(x+1)}{1}\frac{9}{(x+2)(x+1)}$$

$$5(x+1)-3(x+2)=9 \quad \text{Distributive}$$
$$\text{property and cancel}$$

$$5x+5-3x-6=9$$
$$2x=9-5+6$$
$$2x=10$$
$$x=5$$

Solution set: $\{5\}$

49.

$$\frac{4}{5x}+\frac{1}{x-2}=\frac{1}{5x^2-10x}$$

$$\frac{4}{5x}+\frac{1}{x-2}=\frac{1}{5x(x-2)} \quad LCD=5x(x-2)$$

$$\frac{5x(x-2)}{1}\left(\frac{4}{5x}+\frac{1}{x-2}\right)=\frac{5x(x-2)}{1}\cdot\frac{1}{5x(x-2)}$$

$$4(x-2)+5x=1 \quad \text{Distributive property and cancel}$$

$$4x-8+5x=1$$
$$9x=1+8$$
$$9x=9$$
$$x=1$$

Solution set: $\{1\}$

51.

$$\frac{3}{2x}+\frac{4}{5x+1}=\frac{3}{10x^2+2x}$$

$$\frac{3}{2x}+\frac{4}{5x+1}=\frac{3}{2x(5x+1)} \quad LCD=2x(5x+1)$$

$$\frac{2x(5x+1)}{1}\left(\frac{3}{2x}+\frac{4}{5x+1}\right)=\frac{2x(5x+1)}{1}\cdot\frac{3}{2x(5x+1)}$$

$$3(5x+1)+4(2x)=3 \quad \text{Distributive property and cancel}$$

$$15x+3+8x=3$$
$$23x=3-3$$
$$23x=0$$
$$x=0$$

Solution set: $\varnothing$

Note: $x\neq0\ \text{or}\ -\frac{1}{5}$

53.

$$\frac{1}{x-2}-\frac{1}{x+2}=\frac{4}{5} \quad LCD=5(x-2)(x+2)$$

$$\frac{5(x+2)(x-2)}{1}\left(\frac{1}{x-2}-\frac{1}{x+2}\right)=\frac{5(x+2)(x-2)}{1}\cdot\frac{4}{5}$$

#53 continued

$$5(x+2)-5(x-2)=4(x+2)(x-2) \quad \text{Distributive property and cancel}$$

$$5x+10-5x+10=4(x^2-4)$$
$$20=4x^2-16$$
$$0=4x^2-16-20$$
$$0=4x^2-36$$
$$0=x^2-9$$
$$0=(x+3)(x-3)$$
$$x-3=0 \quad \text{or} \quad x+3=0$$
$$x=3 \qquad x=-3$$

Solution set: $\{3, -3\}$

55.

$$\frac{x}{x+4}+\frac{3}{x-2}=\frac{18}{x^2+2x-8}$$
$$\frac{x}{x+4}+\frac{3}{x-2}=\frac{18}{(x+4)(x-2)}$$
$$\text{LCD}=(x+4)(x-2)$$

$$\frac{(x+4)(x-2)}{1}\left(\frac{x}{x+4}+\frac{3}{x-2}\right)$$
$$=\frac{(x+4)(x-2)}{1}\left(\frac{18}{(x+4)(x-2)}\right)$$

$$x(x-2)+3(x+4)=18 \quad \text{Distributive property and cancel}$$

$$x^2-2x+3x+12=18$$
$$x^2+x-6=0$$
$$(x-2)(x+3)=0$$
$$x=2 \quad \text{or} \quad x=-3$$

Solution set: $\{-3\}$

Note: $x \ne -4$ or 2

57.

$$\frac{x}{x-2}-\frac{2}{x-3}=\frac{-2}{x^2-5x+6}$$
$$\frac{x}{x-2}-\frac{2}{x-3}=\frac{-2}{(x-2)(x-3)}$$
$$\text{LCD}=(x-2)(x-3)$$

#57 continued

$$\frac{(x-2)(x-3)}{1}\left(\frac{x}{x-2}-\frac{2}{x-3}\right)$$
$$=\frac{(x-2)(x-3)}{1}\left(\frac{-2}{(x-2)(x-3)}\right)$$

$$x(x-3)-2(x-2)=-2$$

Distributive property and cancel

$$x^2-3x-2x+4=-2$$
$$x^2-5x+6=0$$
$$(x-2)(x-3)=0$$
$$x-2=0 \quad \text{or} \quad x-3=0$$
$$x=2 \quad \text{or} \quad x=3$$

Solution set: $\varnothing$

Note: $x \ne 2$ or 3

59.

$$\frac{2x}{x+3}+\frac{17}{x-4}=\frac{56}{x^2-x-12}$$
$$\frac{2x}{x+3}+\frac{17}{x-4}=\frac{56}{(x+3)(x-4)}$$
$$\text{LCD}=(x+3)(x-4)$$

$$\frac{(x+3)(x-4)}{1}\left(\frac{2x}{x+3}+\frac{17}{x-4}\right)$$
$$=\frac{(x+3)(x-4)}{1}\left(\frac{56}{(x+3)(x-4)}\right)$$

$$2x(x-4)+17(x+3)=56$$

Distributive property and cancel

$$2x^2-8x+17x+51=56$$
$$2x^2+9x-5=0$$
$$(2x-1)(x+5)=0$$
$$2x-1=0 \quad \text{or} \quad x+5=0$$
$$x=\frac{1}{2} \qquad x=-5$$

Solution set: $\left\{\frac{1}{2}, -5\right\}$

61.

$$\frac{2x}{x-5} - \frac{x}{x+3} = \frac{-24}{x^2-2x-15}$$

$$\frac{2x}{x-5} - \frac{x}{x+3} = \frac{-24}{(x-5)(x+3)}$$

$$LCD = (x-5)(x+3)$$

$$\frac{(x-5)(x+3)}{1}\left(\frac{2x}{x-5} - \frac{x}{x+3}\right)$$

$$= \frac{(x-5)(x+3)}{1}\left(\frac{-24}{(x-5)(x+3)}\right)$$

$$2x(x+3) - x(x-5) = -24$$

Distributive property and cancel

$$2x^2 + 6x - x^2 + 5x = -24$$

$$x^2 + 11x + 24 = 0$$

$$(x+3)(x+8) = 0$$

$$x+3 = 0 \quad \text{or} \quad x+8 = 0$$

$$x = -3 \qquad\qquad x = -8$$

Solution set: $\{-8\}$

Note: $x \neq -3$ or 5

Exercises 5.7

1. - 17.
The answers do not require worked out solutions.

19.

$$\frac{2\text{ feet}}{18\text{ inches}} = \frac{24\text{ inches}}{18\text{ inches}}$$

$$= \frac{4}{3} \qquad \text{Reduce}$$

21.

$$\frac{28\text{ hours}}{2\text{ days}} = \frac{28\text{ hours}}{48\text{ hours}}$$

$$= \frac{7}{12} \qquad \text{Reduce}$$

23.

$$\frac{x}{6} = \frac{2}{3}$$

$$3x = 12 \quad \text{Equating the cross products}$$

$$x = 4$$

Solution set: $\{4\}$

25.

$$\frac{2x+1}{5} = \frac{7}{2}$$

$$2(2x+1) = 5\cdot 7 \quad \text{Equating the cross products}$$

$$4x + 2 = 35$$

$$4x = 35 - 2$$

$$4x = 33$$

$$x = \frac{33}{4}$$

Solution set: $\left\{\frac{33}{4}\right\}$

27.

$$\frac{x-4}{5} = \frac{2x}{9}$$

$$9(x-4) = 5(2x) \quad \text{Equating the cross products}$$

$$9x - 36 = 10x$$

$$-36 = 10x - 9x$$

$$-36 = x$$

Solution set: $\{-36\}$

29.

$$\frac{3x-1}{5} = \frac{2x-3}{6}$$

$$6(3x-1) = 5(2x-3) \quad \text{Equating the cross products}$$

$$18x - 6 = 10x - 15$$

$$18x - 10x = -15 + 6$$

$$8x = -9$$

$$x = -\frac{9}{8}$$

Solution set: $\left\{-\frac{9}{8}\right\}$

31.

$$\frac{2}{x} = \frac{4}{5}$$

$4x = 10$ Equating the cross products

$$x = \frac{10}{4} \quad \text{or} \quad \frac{5}{2}$$

Solution set: $\left\{\frac{5}{2}\right\}$

33.

$$\frac{1}{2x-1} = \frac{5}{8}$$

$5(2x - 1) = 8 \cdot 1$ Equating the cross products

$10x - 5 = 8$

$10x = 8 + 5$

$10x = 13$

$$x = \frac{13}{10}$$

Solution set: $\left\{\frac{13}{10}\right\}$

35.

$$\frac{2}{3x+2} = \frac{4}{3x}$$

$4(3x + 2) = 2(3x)$ Equating the cross products

$12x + 8 = 6x$

$12x - 6x = -8$

$6x = -8$

$$x = \frac{-8}{6} \quad \text{or} \quad -\frac{4}{3}$$

Solution set: $\left\{-\frac{4}{3}\right\}$

37.

$$\frac{5}{x+3} = \frac{4}{2x+1}$$

$5(2x + 1) = 4(x + 3)$ Equating the cross products

$10x + 5 = 4x + 12$

$10x - 4x = 12 - 5$

$6x = 7$

$$x = \frac{7}{6}$$

Solution set: $\left\{\frac{7}{6}\right\}$

39.

$$\frac{4}{x} = \frac{2x+5}{3}$$

$4 \cdot 3 = x(2x + 5)$ Equating the cross products

$12 = 2x^2 + 5x$

$0 = 2x^2 + 5x - 12$

$0 = (2x - 3)(x + 4)$

$2x - 3 = 0 \quad \text{or} \quad x + 4 = 0$

$2x = 3 \qquad\qquad x = -4$

$$x = \frac{3}{2}$$

Solution set: $\left\{\frac{3}{2}, \ -4\right\}$

41.

$$\frac{x}{3} = \frac{-2}{x-5}$$

$x(x-5) = 3(-2)$ Equating the cross products

$x^2 - 5x = -6$

$x^2 - 5x + 6 = 0$

$(x-3)(x-2) = 0$

$x - 3 = 0$ or $x - 2 = 0$

$x = 3$ $x = 2$

Solution set: $\{2, \ 3\}$

43.

$$\frac{3x+2}{4x} = \frac{5}{3}$$

$3(3x+2) = 5(4x)$ Equating the cross products

$9x + 6 = 20x$

$6 = 20x - 9x$

$6 = 11x$

$$\frac{6}{11} = x$$

Solution set: $\left\{\dfrac{6}{11}\right\}$

45.

$$\frac{x+1}{x-3} = \frac{x-2}{x+4}$$

$(x+1)(x+4) = (x-2)(x-3)$

                 Equating the cross products

$x^2 + 5x + 4 = x^2 - 5x + 6$

$x^2 - x^2 + 5x + 5x = 6 - 4$

$10x = 2$

$x = \dfrac{2}{10}$ or $\dfrac{1}{5}$

Solution set: $\left\{\dfrac{1}{5}\right\}$

47.

$$\frac{2x+1}{x-3} = \frac{x}{x-3}$$

$(2x+1)(x-3) = x(x-3)$ Equating the

                       cross products

$2x^2 - 5x - 3 = x^2 - 3x$

$2x^2 - x^2 - 5x + 3x - 3 = 0$

$x^2 - 2x - 3 = 0$

$(x-3)(x+1) = 0$

$x - 3 = 0$ or $x + 1 = 0$

$x = 3$ $x = -1$

Solution set: $\{-1\}$

Note: $x \neq 3$

49.

$$\frac{2x+3}{x+2} = \frac{6}{x+5}$$

$(2x+3)(x+5) = 6(x+2)$ Equating the

                         cross products

$2x^2 + 13x + 15 = 6x + 12$

$2x^2 + 13x - 6x + 15 - 12 = 0$

$2x^2 + 7x + 3 = 0$

$(2x+1)(x+3) = 0$

$2x + 1 = 0$ or $x + 3 = 0$

$x = -\dfrac{1}{2}$ $x = -3$

Solution set: $\left\{-\dfrac{1}{2}, \ -3\right\}$

51.

$$\frac{3}{x+2} = \frac{5}{x+2}$$

$3(x+2) = 5(x+2)$ Equating the cross products

$3x + 6 = 5x + 10$

$6 - 10 = 5x - 3x$

$-4 = 2x$

$-2 = x$

Solution set: $\varnothing$

Note: $x \neq -2$

53.

$$\frac{2x+3}{x+3} = \frac{x-1}{x-3}$$

$(2x+3)(x-3) = (x-1)(x+3)$ Equating the cross products

$$2x^2 - 3x - 9 = x^2 + 2x - 3$$

$$2x^2 - x^2 - 3x - 2x - 9 + 3 = 0$$

$$x^2 - 5x - 6 = 0$$
$$(x-6)(x+1) = 0$$
$$x - 6 = 0 \quad \text{or} \quad x + 1 = 0$$
$$x = 6 \qquad\qquad x = -1$$

Solution set: $\{6, -1\}$

55.
Let x = cost of 8 apples
Then,

$$\frac{5}{2} = \frac{8}{x}$$
$$5x = 16$$
$$x = \frac{16}{5} \text{ or } 3.2$$

The cost is $3.20

Note:
$$\begin{array}{r} 3.2 \\ 5\overline{)16.0} \\ \underline{15} \\ 10 \\ \underline{10} \end{array}$$

57.
Let x = miles in 10k
Then,

$$\frac{1.6}{1} = \frac{10}{x}$$
$$1.6x = 10$$
$$x = \frac{10}{1.6} \text{ or } 6.25$$

#57. continued.

There are 6.25 miles in 10k.

Note: $\dfrac{10}{1.6} = \dfrac{100}{16}$

$$\begin{array}{r} 6.25 \\ 16\overline{)100.00} \\ \underline{96} \\ 40 \\ \underline{32} \\ 80 \\ \underline{80} \end{array}$$

59.
Let x = glasses of water
Then,

$$\frac{9}{160} = \frac{x}{200}$$
$$160x = 1800$$
$$x = \frac{1800}{160}$$
$$x = \frac{45}{4}, \text{ or } 11.25$$

Sharky must drink 11.25 glasses of water each day.

61.
Let x = cups of milk
Then,

$$\frac{2}{6} = \frac{x}{26}$$
$$6x = 52$$
$$x = \frac{52}{6} \text{ or } 8\frac{2}{3}$$

$8\frac{2}{3}$ cups of milk

63.
Let x = gallons of gasoline

Then,

$$\frac{2.5}{45} = \frac{x}{108}$$
$$45x = 2.5 \cdot 108$$
$$45x = 270$$
$$x = 6$$

6 gallons of gasoline

65.
Let x = pounds of fertilizer

$$\frac{5}{120} = \frac{x}{300}$$
$$1500 = 120x$$
$$\frac{1500}{120} = x$$
$$12.5 = x$$

12.5 pounds of fertilizer are required.

Exercises 5.8

1.
Let x = the number to be added.

The numerator is $5 + x$ and the denominator is $17 + x$.

The equation is

$$\frac{5+x}{17+x} = \frac{2}{5}$$
$$5(5+x) = 2(17+x)$$
$$25 + 5x = 34 + 2x$$
$$5x - 2x = 34 - 25$$
$$3x = 9$$
$$x = 3$$

3 must be added.

3.
Let x = the number to be subtracted.

The numerator is 12 - x and the denominator is 19 - x.

The equation is

$$\frac{12-x}{19-x} = \frac{1}{2}$$
$$2(12 - x) = 19 - x$$
$$24 - 2x = 19 - x$$
$$24 - 19 = -x + 2x$$
$$5 = x$$

5 must be subtracted.

5.
Let x = the number in the numerator.

The denominator is $x + 4$.

The original fraction is $\dfrac{x}{x+4}$.

Subtract 1 from the numerator and add 1 to the denominator gives $\dfrac{x-1}{(x+4)+1}$.

The equation is

$$\frac{x-1}{x+5} = \frac{2}{3}$$
$$2(x+5) = 3(x-1)$$
$$2x + 10 = 3x - 3$$
$$10 + 3 = 3x - 2x$$
$$13 = x$$

The original fraction is $\dfrac{x}{x+4}$ or $\dfrac{13}{13+4} = \dfrac{13}{17}$.

7.
Let x = one of the numbers.

The other number is $x + 8$.

The reciprocals are $\dfrac{1}{x}$ and $\dfrac{1}{x+8}$.

The equation is $\dfrac{1}{x} + \dfrac{1}{x+8} = \dfrac{1}{3}$ LCD $= 3x(x+8)$

$$3x(x+8)\left(\frac{1}{x} + \frac{1}{x+8}\right) = 3x(x+8) \cdot \frac{1}{3}$$
$$3(x+8) + 3x = x(x+8)$$
$$3x + 24 + 3x = x^2 + 8x$$

#7. continued.

$$6x + 24 = x^2 + 8x$$
$$0 = x^2 + 2x - 24$$
$$0 = (x + 6)(x - 4)$$
$$x + 6 = 0 \text{ or } x - 4 = 0$$
$$x = -6 \qquad x = 4$$
$$x + 8 = 2 \qquad x + 8 = 12$$

The numbers are - 6 and 2, or 4 and 12.

9.

Let x = one of the numbers.

The other number is 10 - x.

The reciprocals are $\dfrac{1}{x}$ and $\dfrac{1}{10 - x}$.

The equation is $\dfrac{1}{x} + \dfrac{1}{10 - x} = \dfrac{5}{8}$ LCD = 8x(10 - x)

$$8x(10 - x)\left(\frac{1}{x} + \frac{1}{10 - x}\right) = 8x(10 - x) \cdot \frac{5}{8}$$
$$8(10 - x) + 8x = 50x - 5x^2$$
$$80 - 8x + 8x = 50x - 5x^2$$
$$5x^2 - 50x + 80 = 0$$
$$x^2 - 10x + 16 = 0$$
$$(x - 2)(x - 8) = 0$$
$$x - 2 = 0 \text{ or } x - 8 = 0$$
$$x = 2 \qquad x = 8$$
$$10 - x = 8 \qquad 10 - x = 2$$

The numbers are 8 and 2.

11.

Let x = one of the numbers.

The other number is x + 2.

The reciprocals are $\dfrac{1}{x}$ and $\dfrac{1}{x + 2}$.

#11. continued.

The equation is $\dfrac{1}{x} + \dfrac{1}{x + 2} = \dfrac{12}{5}$ LCD = 5x(x + 2)

$$5x(x + 2)\left(\frac{1}{x} + \frac{1}{x + 2}\right) = 5x(x + 2) \cdot \frac{12}{5}$$
$$5x + 10 + 5x = 12x^2 + 24x$$
$$10x + 10 = 12x^2 + 24x$$
$$0 = 12x^2 + 14x - 10$$
$$0 = 6x^2 + 7x - 5$$
$$0 = (3x + 5)(2x - 1)$$
$$3x + 5 = 0 \text{ or } 2x - 1 = 0$$
$$x = -\frac{5}{3} \qquad x = \frac{1}{2}$$
$$x + 2 = \frac{1}{3} \qquad x + 2 = \frac{5}{2}$$

The numbers are $-\dfrac{5}{3}$ and $\dfrac{1}{3}$, or $\dfrac{1}{2}$ and $\dfrac{5}{2}$.

13.

	Terry	Gerry	Both
Part hour	$\dfrac{1}{4}$	$\dfrac{1}{2}$	$\dfrac{1}{x}$

The equation is

$$\frac{1}{4} + \frac{1}{2} = \frac{1}{x} \qquad \text{LCD} = 4x$$
$$4x\left(\frac{1}{4} + \frac{1}{2}\right) = 4x \cdot \frac{1}{x}$$

$x + 2x = 4$ Distributive property
                  and cancel

$$3x = 4$$
$$x = \frac{4}{3} = 1\frac{1}{3}$$

$1\dfrac{1}{3}$ hr. or 1 hr. 20 min.

15.

	Kathy	Melissa	Both
$\dfrac{\text{Part}}{\text{minute}}$	$\dfrac{1}{40}$	$\dfrac{1}{60}$	$\dfrac{1}{x}$

The equation is

$$\frac{1}{40}+\frac{1}{60}=\frac{1}{x} \qquad LCD = 120x$$

$$120x\left(\frac{1}{40}+\frac{1}{60}\right)=120x\cdot\frac{1}{x}$$

$$3x+2x=120 \qquad \text{Distributive property}$$
$$\text{and cancel}$$

$$5x=120$$

$$x=24$$

24 minutes.

17.

	Butch	Bart	Both
$\dfrac{\text{Part}}{\text{hours}}$	$\dfrac{1}{2}$	$\dfrac{1}{x}$	$\dfrac{1}{\frac{6}{5}}$

Note: 1 hr. 12 min. $=1\dfrac{12}{60}=1\dfrac{1}{5}=\dfrac{6}{5}$ hr.

The equation is

$$\frac{1}{2}+\frac{1}{x}=\frac{1}{\frac{6}{5}}$$

$$\frac{1}{2}+\frac{1}{x}=\frac{5}{6} \qquad LCD = 6x$$

$$6x\left(\frac{1}{2}+\frac{1}{x}\right)=6x\cdot\frac{5}{6}$$

$$3x+6=5x \qquad \text{Distributive property}$$
$$\text{and cancel}$$

$$6=2$$

$$3=x$$

3 hours

19.

Let x = number of minutes to drain for drain I (the faster drain).

Let $2x$ = number of minutes to drain for drain II (the slower drain).

	Drain I	Drain II	Both
$\dfrac{\text{Part}}{\text{minute}}$	$\dfrac{1}{x}$	$\dfrac{1}{2x}$	$\dfrac{1}{4}$

The equation is $\dfrac{1}{x}+\dfrac{1}{2x}=\dfrac{1}{4} \qquad LCD = 4x$

$$4x\left(\frac{1}{x}+\frac{1}{2x}\right)=4x\cdot\frac{1}{4}$$

$$4+2=x$$

$$6=x$$

$$12=2x$$

6 minutes and 12 minutes

21.

Let x = number of minutes for the pile of rocks to be broken down.

	Butch	Bart	Both
$\dfrac{\text{Part}}{\text{hour}}$	$\dfrac{1}{2}$	$\dfrac{1}{3}$	$\dfrac{1}{x}$

The equation is $\dfrac{1}{2}-\dfrac{1}{3}=\dfrac{1}{x} \qquad LCD = 6x$

$$6x\left(\frac{1}{2}-\frac{1}{3}\right)=6x\cdot\frac{1}{x}$$

$$3x-2x=6$$

$$x=6$$

6 hours

23.
Let x = number of hours to fill pool
with drain open.

	Faucets open	Drain open	Both
$\dfrac{\text{Part}}{\text{hour}}$	$\dfrac{1}{4}$	$\dfrac{1}{10}$	$\dfrac{1}{x}$

The equation is $\dfrac{1}{4}-\dfrac{1}{10}=\dfrac{1}{x}$ LCD $=20x$

$$20x\left(\dfrac{1}{4}-\dfrac{1}{10}\right)=20x\cdot\dfrac{1}{x}$$
$$5x-2x=20$$
$$3x=20$$
$$x=\dfrac{20}{3}\ \text{ or }\ 6\dfrac{2}{3}$$

$6\dfrac{2}{3}$ hours or 6 hr. 40 min.

25.
Let x = speed of boat in still water.
Let x + 1 = speed of boat down bayou.
Let x - 1 = speed of boat up bayou.

	D	R	T
Down Bayou	10	$x+1$	$\dfrac{10}{x+1}$
Up Bayou	6	$x-1$	$\dfrac{6}{x-1}$

Note: since $T=\dfrac{D}{R}$, then $T_D=\dfrac{10}{x+1}$ and $T_U=\dfrac{6}{x-1}$.

Since time is the same, the equation is

$$\dfrac{10}{x+1}=\dfrac{6}{x-1}$$
$$10(x-1)=6(x+1)$$
$$10x-10=6x+6$$
$$10x-6x=6+10$$
$$4x=16$$
$$x=4$$

The speed is 4 mph.

27.
Let x = speed of the current.
Let 45 + x = speed of the boat down river.
Let 45 - x = speed of the boat up river.

	D	R	T
Down River	17	$45+x$	$\dfrac{17}{45+x}$
Up River	13	$45-x$	$\dfrac{13}{45-x}$

Note: Since $T=\dfrac{D}{R}$,

then $T_D=\dfrac{17}{45+x}$ and $T_U=\dfrac{13}{45-x}$.

Since time is the same, the equation is

$$\dfrac{17}{45+x}=\dfrac{13}{45-x}$$
$$17(45-x)=13(45+x)$$
$$765-17x=585+13x$$
$$765-585=13x+17x$$
$$180=30x$$
$$6=x$$

The speed of the current is 6 mph

29.

Let x = speed of the boat in still water.

Let x - 4 = speed of the boat up river.

Let x + 4 = speed of the boat down river.

	D $=$ R $\cdot$ T
Up River	48	$x-4$	$\dfrac{48}{x-4}$
Down River	48	$x+4$	$\dfrac{48}{x+4}$

Note: 1. Since $T = \dfrac{D}{R}$, then $T_U = \dfrac{48}{x-4}$

and $T_D = \dfrac{48}{x+4}$.

2. The total trip time is 5 hours.

The equation is $\dfrac{48}{x-4} + \dfrac{48}{x+4} = 5$ LCD $= (x-4)(x+4)$

$$(x-4)(x+4)\left(\dfrac{48}{x-4} + \dfrac{48}{x+4}\right) = (x-4)(x+4)\cdot 5$$

$$48(x+4) + 48(x-4) = \left(x^2 - 16\right)5$$

$$48x + 192 + 48x - 192 = 5x^2 - 80$$

$$5x^2 - 96x + 80 = 0$$

$$(5x + 4)(x - 20) = 0$$

$$5x + 4 = 0 \ \text{ or } \ x - 20 = 0$$

$$x = -\dfrac{4}{5} \qquad x = 20$$

The speed of the boat in still water is 20 mph.

31.

Let x = Distance between Maria's home and San Antonio.

	D $=$ R $\cdot$ T
Jet	x	250	$\dfrac{x}{250}$
Train	x	62.5	$\dfrac{x}{62.5}$

Note: 1. Since $T = \dfrac{D}{R}$,

then $T_J = \dfrac{x}{250}$ and $T_T = \dfrac{x}{62.5}$

2. The total trip time is 4 hours.

The equation is

$$\dfrac{x}{250} + \dfrac{x}{62.5} = 4$$

$$\dfrac{x}{250} + \dfrac{10x}{625} = 4$$

$$\dfrac{x}{2\cdot 5^3} + \dfrac{10x}{5^4} = 4 \qquad \text{LCD} = 5^4 \cdot 2$$

$$5^4 \cdot 2\left(\dfrac{x}{2\cdot 5^3} + \dfrac{10x}{5^4}\right) = 5^4 \cdot 2 \cdot 4$$

$$5x + 20x = 5000$$

$$25x = 5000$$

$$x = 200$$

The distance is 200 miles

33.

Let x = speed of plane in still air.

Let x + 15 = speed of plane with the wind.

Let x - 15 = speed of plane against the wind.

	D $=$ R $\cdot$ T		
With Wind	210	x + 15	$\dfrac{210}{x+15}$
Against Wind	165	x - 15	$\dfrac{165}{x-15}$

Note: 1. Since $T = \dfrac{D}{R}$,

then $T_W = \dfrac{210}{x+15}$ and $T_A = \dfrac{165}{x-15}$

2. The total flight time is 3 hours.

The equation is

$$\frac{210}{x+15} + \frac{165}{x-15} = 3$$

$$LCD = (x+15)(x-15)$$

$$(x+15)(x-15)\left(\frac{210}{x+15} + \frac{165}{x-15}\right) = (x+15)(x-15)\cdot 3$$

$$210(x-15) + 165(x+15) = \left(x^2 - 225\right)\cdot 3$$

$$210x - 3150 + 165x + 2475 = 3x^2 - 675$$

$$3x^2 - 375x = 0$$

$$3x(x-125) = 0$$

$$3x = 0 \text{ or } x - 125 = 0$$

$$x = 0 \qquad x = 125$$

The speed of the plane is 125 mph.

35.

Let x = Bruce's speed from the stables.
Let x-20 = Bruce's speed back to the stables.

	D $=$ R $\cdot$ T		
Out	$10\frac{1}{2}$	x	$\dfrac{10\frac{1}{2}}{x}$
Back	$10\frac{1}{2}$	x-20	$\dfrac{10\frac{1}{2}}{x-20}$

Note: 1. Since $T = \dfrac{D}{R}$, then $T_O = \dfrac{10\frac{1}{2}}{x}$.

2. The total riding time is 1 hour.

The equation is

$$\frac{10\frac{1}{2}}{x} + \frac{10\frac{1}{2}}{x-20} = 1$$

$$\frac{\frac{21}{2}}{x} + \frac{\frac{21}{2}}{x-20} = 1$$

$$\frac{21}{x} + \frac{21}{(x-20)} = 2 \qquad\qquad \text{Multiply by 2}$$

$$LCD = x(x-20)$$

$$x(x-20)\left(\frac{21}{x} + \frac{21}{(x-20)}\right) = x(x-20)\cdot 2$$

$$21(x-20) + 21x = 2x(x-20)$$

$$21x - 420 + 21x = 2x^2 - 40x$$

$$42x - 420 = 2x^2 - 40x$$

$$2x^2 - 82x + 420 = 0$$

$$x^2 - 41x + 210 = 0$$

$$(x-35)(x-6) = 0$$

$$x - 35 = 0 \text{ or } x - 6 = 0$$

$$x = 35 \qquad x = 6$$

If x = 35, then x - 20 = 15.

But if x = 6, then x - 20 = -14.

Thus, 6 is not a solution.

Bruce's speeds are 35 mph and 15 mph.

Review Exercises

1.

$$\frac{27x^3y}{18x^2y^4} = \frac{3 \cdot 3 \cdot 3x^{3-2}}{3 \cdot 3 \cdot 2y^{4-1}}$$

$$= \frac{3x}{2y^3}$$

3.

$$\frac{3x^4 - 6x^3y^2}{24x^9 + 12x^9y^3} = \frac{3x^3(x - 2y^2)}{12x^9(2 + y^3)}$$

$$= \frac{3(x - 2y^2)}{3 \cdot 4x^{9-3}(2 + y^3)}$$

$$= \frac{x - 2y^2}{4x^6(2 + y^3)}$$

5.

$$\frac{2x - 6y}{3x^2y - x^3} = \frac{2(x - 3y)}{x^2(3y - x)}$$

$$= \frac{-2(3y - x)}{x^2(3y - x)}$$

$$= -\frac{2}{x^2}$$

7.

$$\frac{2x^3 + 8x^2}{12x^2 + 18x} = \frac{2x^2(x + 4)}{6x(2x + 3)}$$

$$= \frac{2x^{2-1}(x + 4)}{2 \cdot 3(2x + 3)}$$

$$= \frac{x(x + 4)}{3(2x + 3)}$$

9.

$$\frac{3x^2 + 5x - 2}{x^2 - 5x - 14} = \frac{(3x - 1)(x + 2)}{(x + 2)(x - 7)}$$

$$= \frac{3x - 1}{x - 7}$$

11.

$$\frac{3 - x}{3x^2 - 5x - 12} = \frac{3 - x}{(x - 3)(3x + 4)}$$

$$= \frac{-(x - 3)}{(x - 3)(3x + 4)}$$

$$= -\frac{1}{3x + 4}$$

13.

$$\frac{6x^3 - 9x^2y - 6xy^2}{6x^2 + 5xy + y^2} = \frac{3x(2x^2 - 3xy - 2y^2)}{16x^2 + 5xy + y^2}$$

$$= \frac{3x(2x + y)(x - 2y)}{(2x + y)(3x + y)}$$

$$= \frac{3x(x - 2y)}{3x + y}$$

15.

$$\frac{6x^5 - 22x^4 + 12x^3}{12x^3 + 16x^2 - 16x} = \frac{2x^3(3x^2 - 11x + 6)}{4x(3x^2 + 4x - 4)}$$

$$= \frac{2x^{3-1}(3x - 2)(x - 3)}{2 \cdot 2 \cdot (3x - 2)(x + 2)}$$

$$= \frac{x^2(x - 3)}{2(x + 2)}$$

17.

$$\frac{8x^3}{5x} \cdot \frac{10x^3}{12x^4} = \frac{4 \cdot 2 \cdot 2 \cdot 5x^6}{5 \cdot 4 \cdot 3 \cdot x^5}$$

$$= \frac{\cancel{4} \cdot 2 \cdot 2 \cdot \cancel{5}x^{6-5}}{\cancel{5} \cdot \cancel{4} \cdot 3}$$

$$= \frac{4x}{3}$$

19.

$$\frac{11y^4}{9x^2} \div \frac{2y^8}{3x^6} = \frac{11y^4}{9x^2} \cdot \frac{3x^6}{2y^8}$$

$$= \frac{11 \cdot 3 \cdot x^6 \cdot y^4}{3 \cdot 3 \cdot 2 \cdot x^2 \cdot y^8}$$

$$= \frac{11 \cdot \cancel{3} \cdot x^{6-2}}{\cancel{3} \cdot 3 \cdot 2 \cdot y^{8-4}}$$

$$= \frac{11x^4}{6y^4}$$

21.

$$\frac{5x-10}{3x-4} \cdot \frac{8-6x}{x-2} = \frac{5(x-2)}{(3x-4)} \cdot \frac{2(4-3x)}{(x-2)}$$

$$= \frac{5\cancel{(x-2)}}{\cancel{(3x-4)}} \cdot \frac{-2\cancel{(3x-4)}}{\cancel{(x-2)}}$$

$$= \frac{-10}{1} \text{ or } -10$$

23.

$$\frac{3x^2-9xy}{6x^2-6y^2} \div \frac{x^3-3x^2y}{xy+y^2}$$

$$= \frac{3x(x-3y)}{6(x^2-y^2)} \cdot \frac{y(x+y)}{x^2(x-3y)}$$

$$= \frac{\cancel{3}\cancel{(x-3y)}}{2 \cdot \cancel{3}\cancel{(x+y)}(x-y)} \cdot \frac{y\cancel{(x+y)}}{x^{2-1}\cancel{(x-3y)}}$$

$$= \frac{y}{2x(x-y)}$$

25.

$$\frac{x^2+10x+21}{x^2-4x+4} \cdot \frac{x^2-4}{x^2+9x+14}$$

$$= \frac{\cancel{(x+7)}(x+3)}{(x-2)\cancel{(x-2)}} \cdot \frac{\cancel{(x+2)}\cancel{(x-2)}}{\cancel{(x+7)}\cancel{(x+2)}}$$

$$= \frac{x+3}{x-2}$$

27.

$$\frac{2x^2-7xy+3y^2}{2x^2+xy-y^2} \div \frac{3y^2-4yx+x^2}{x^2-xy-2y^2}$$

$$= \frac{(2x-y)(x-3y)}{(2x-y)(x+y)} \cdot \frac{(x-2y)(x+y)}{(3y-x)(y-x)}$$

$$= \frac{\cancel{(x-3y)}(x-2y)}{-\cancel{(x-3y)}(y-x)}$$

$$= -\frac{x-2y}{y-x} \text{ or } \frac{x-2y}{x-y}$$

Note:
$$-\frac{x-2y}{y-x} = \frac{1}{-1} \cdot \frac{x-2y}{y-x}$$

$$= \frac{x-2y}{-1(y-x)}$$

$$= \frac{x-2y}{-y+x} \text{ or } \frac{x-2y}{x-y}$$

29.

$$\frac{2x^2+8x}{2x^2-4x-30} \cdot \frac{3x^2+7x-6}{3x^2-2x}$$

$$= \frac{2x(x+4)}{2(x^2-2x-15)} \cdot \frac{3x^2+7x-6}{x(3x-2)}$$

$$= \frac{\cancel{2}\cancel{x}(x+4)}{\cancel{2}(x+3)(x-5)} \cdot \frac{\cancel{(3x-2)}(x+3)}{\cancel{x}\cancel{(3x-2)}}$$

$$= \frac{x+4}{x-5}$$

31.

$$\frac{x^4 - 6x^3 + 9x^2}{x^2 - x - 6} \div \frac{3x^2 - 9x}{9x^2 + 24x + 12}$$

$$= \frac{x^2\left(x^2 - 6x + 9\right)}{x^2 - x - 6} \cdot \frac{3\left(3x^2 + 8x + 4\right)}{3x(x - 3)}$$

$$= \frac{x^2 \cdot 1 \cancel{(x-3)}\cancel{(x-3)}}{\cancel{(x+2)}\cancel{(x-3)}} \cdot \frac{\cancel{3}(3x+2)\cancel{(x+2)}}{\cancel{3}\cancel{(x-3)}}$$

$$= x(3x + 2)$$

33.

$$\frac{x^2 - 8x - 9}{14x^2 + 63x + 49} \cdot \frac{14x^2 + 49x}{x^3 - 5x^2 - 36x} \cdot \frac{2x + 8}{2x^2 - 4x - 6}$$

$$= \frac{x^2 - 8x - 9}{7\left(2x^2 + 9x + 7\right)} \cdot \frac{7x(2x + 7)}{x\left(x^2 - 5x - 36\right)} \cdot \frac{2(x + 4)}{2\left(x^2 - 2x - 3\right)}$$

$$= \frac{\cancel{(x-9)}\cancel{(x+1)}}{\cancel{7}\cancel{(2x+7)}\cancel{(x+1)}} \cdot \frac{\cancel{7}\cancel{x}\cancel{(2x+7)}}{\cancel{x}\cancel{(x+4)}\cancel{(x-9)}} \cdot \frac{\cancel{2}\cancel{(x+4)}}{\cancel{2}(x-3)(x+1)}$$

$$= \frac{1}{(x-3)(x+1)}$$

35.

$$\frac{x^2y^2 - 2xy^3 - 3y^4}{2x^2 - xy} \div \frac{3x^2y + 6xy^2 + 3y^3}{2x^2 + 7xy - 4y^2}$$

$$\qquad \div \frac{xy + 4y^2}{3x^3 + 33x^2y + 30xy^2}$$

$$= \frac{y^2\left(x^2 - 2xy - 3y^2\right)}{x(2x - y)} \div \frac{3y\left(x^2 + 2xy + y^2\right)}{2x^2 + 7xy - 4y^2}$$

$$\qquad \div \frac{y(x + 4y)}{3x\left(x^2 + 11xy + 10y^2\right)}$$

#35. continued.

$$= \frac{y^2(x - 3y)(x + y)}{x(2x - y)} \cdot \frac{(2x - y)(x + 4y)}{3y(x + y)(x + y)}$$

$$\qquad \cdot \frac{3x(x + 10y)(x + y)}{y(x + 4y)}$$

$$= \frac{\cancel{3xy^2}(x - 3y)(x + 10y)}{\cancel{3xy^2}}$$

$$= (x - 3y)(x + 10y)$$

37.

$$\frac{7x}{9} + \frac{5x}{9} = \frac{7x + 5x}{9} = \frac{12x}{9} = \frac{\cancel{3} \cdot 4x}{\cancel{3} \cdot 3} = \frac{4x}{3}$$

39.

$$\frac{4x + 7}{5y} - \frac{2x - 3}{5y} = \frac{4x + 7 - (2x - 3)}{5y}$$

$$= \frac{4x + 7 - 2x + 3}{5y} = \frac{2x + 10}{5y}$$

41.

$$\frac{6x - 5y}{2x - y} - \frac{4x - 4y}{2x - y} = \frac{6x - 5y - (4x - 4y)}{2x - y}$$

$$= \frac{6x - 5y - 4x + 4y}{2x - y}$$

$$= \frac{\cancel{2x - y}}{\cancel{2x - y}}$$

$$= 1$$

43.

$$\frac{3x + 5}{x^2 - x - 12} + \frac{x + 4}{x^2 - x - 12} = \frac{2x + 5 + x + 4}{x^2 - x - 12}$$

$$= \frac{3x + 9}{x^2 - x - 12}$$

$$= \frac{3\cancel{(x+3)}}{\cancel{(x+3)}(x - 4)}$$

$$= \frac{3}{x - 4}$$

45.

$$\frac{x^2-15}{2x-7}+\frac{x^2-x-6}{2x-7}=\frac{x^2-15+x^2-x-6}{2x-7}$$

$$=\frac{2x^2-x-21}{2x-7}$$

$$=\frac{(2x-7)(x+3)}{2x-7}$$

$$=x+3$$

47.

$$\frac{x^2+3x-8}{x^2-2x-15}-\frac{2x^2-3x+5}{x^2-2x-15}$$

$$=\frac{x^2+3x-8-\left(2x^2-3x+5\right)}{x^2-2x-15}$$

$$=\frac{x^2+3x-8-2x^2+3x-5}{x^2-2x-15}$$

$$=\frac{-x^2+6x-13}{x^2-2x-15}$$

49.

$$\frac{6x+1}{3x-4}-\frac{3x-13}{4-3x}=\frac{6x+1}{3x-4}-\frac{3x-13}{-(3x-4)}$$

$$=\frac{6x+1}{3x-4}+\frac{3x-13}{3x-4}$$

$$=\frac{6x+1+3x-13}{3x-4}$$

$$=\frac{9x-12}{3x-4}$$

$$=\frac{3(3x-4)}{3x-4}$$

$$=3$$

51.

$$\frac{x-4}{2x+9}-\frac{5x+2}{2x+9}+\frac{2x-3}{2x+9}=\frac{x-4-(5x+2)+2x-3}{2x+9}$$

$$=\frac{x-4-5x-2+2x-3}{2x+9}$$

$$=\frac{-2x-9}{2x+9}$$

$$=\frac{-(2x+9)}{2x+9}$$

$$=-1$$

53.

$$\frac{3}{5x}+\frac{9}{2x}\qquad\qquad LCD=10x$$

$$=\frac{3}{5x}\cdot\frac{2}{2}+\frac{9}{2x}\cdot\frac{5}{5}$$

$$=\frac{6}{10x}+\frac{45}{10x}=\frac{6+45}{10x}=\frac{51}{10x}$$

55.

$$\frac{3}{8x}-\frac{2}{x-3}\qquad\qquad LCD=8x(x-3)$$

$$=\frac{3}{8x}\cdot\frac{x-3}{x-3}-\frac{2}{x-3}\cdot\frac{8x}{8x}$$

$$=\frac{3(x-3)-2(8x)}{8x(x-3)}$$

$$=\frac{3x-9-16x}{8x(x-3)}$$

$$=\frac{-13x-9}{8x(x-3)}$$

57.

$$\frac{5}{2x+1} + \frac{2x}{x-3} \qquad \text{LCD} = (2x+1)(x-3)$$

$$= \frac{5}{(2x+1)} \cdot \frac{(x-3)}{(x-3)} + \frac{2x}{(x-3)} \cdot \frac{(2x+1)}{(2x+1)}$$

$$= \frac{5(x-3) + 2x(2x+1)}{(2x+1)(x-3)}$$

$$= \frac{5x - 15 + 4x^2 + 2x}{(2x+1)(x-3)}$$

$$= \frac{4x^2 + 7x - 15}{(2x+1)(x-3)}$$

59.

$$\frac{3x-4}{x+2} - \frac{3x+1}{x-2} \qquad \text{LCD} = (x+2)(x-2)$$

$$= \frac{3x-4}{x+2} \cdot \frac{x-2}{x-2} - \frac{3x+1}{x-2} \cdot \frac{x+2}{x+2}$$

$$= \frac{(3x-4)(x-2) - (3x+1)(x+2)}{(x+2)(x-2)}$$

$$= \frac{3x^2 - 10x + 8 - \left(3x^2 + 7x + 2\right)}{(x+2)(x-2)}$$

$$= \frac{3x^2 - 10x + 8 - 3x^2 - 7x - 2}{(x+2)(x-2)}$$

$$= \frac{-17x + 6}{x^2 - 4}$$

61.

$$\frac{3}{x^2 + 3x + 2} + \frac{2x+1}{2x^2 + 7x + 6}$$

$$= \frac{3}{(x+2)(x+1)} + \frac{2x+1}{(x+2)(2x+3)}$$

$$\text{LCD} = (x+2)(x+1)(2x+3)$$

$$= \frac{3}{(x+2)(x+1)} \cdot \frac{(2x+3)}{(2x+3)} + \frac{2x+1}{(x+2)(2x+3)} \cdot \frac{(x+1)}{(x+1)}$$

$$= \frac{3(2x+3) + (2x+1)(x+1)}{(x+2)(x+1)(2x+3)}$$

$$= \frac{6x + 9 + 2x^2 + 3x + 1}{(x+2)(x+1)(2x+3)}$$

#61. continued.

$$= \frac{2x^2 + 9x + 10}{(x+2)(x+1)(2x+3)}$$

$$= \frac{(2x+5)(x+2)}{(x+2)(x+1)(2x+3)}$$

$$= \frac{2x+5}{(x+1)(2x+3)}$$

63.

$$\frac{x+5}{x^2 - 4x + 3} - \frac{10 - 2x}{x^2 - 5x + 6}$$

$$= \frac{x+5}{(x-3)(x-1)} - \frac{10 - 2x}{(x-3)(x-2)}$$

$$\text{LCD} = (x-3)(x-1)(x-2)$$

$$= \frac{x+5}{(x-3)(x-1)} \cdot \frac{x-2}{x-2} - \frac{10-2x}{(x-3)(x-2)} \cdot \frac{x-1}{x-1}$$

$$= \frac{(x+5)(x-2) - (10-2x)(x-1)}{(x-3)(x-1)(x-2)}$$

$$= \frac{x^2 + 3x - 10 - \left(10x + 2x - 2x^2 - 10\right)}{(x-3)(x-1)(x-2)}$$

$$= \frac{x^2 + 3x - 10 - 10x - 2x + 2x^2 + 10}{(x-3)(x-1)(x-2)}$$

$$= \frac{3x^2 - 9x}{(x-3)(x-1)(x-2)}$$

$$= \frac{3x(x-3)}{(x-3)(x-1)(x-2)}$$

$$= \frac{3x}{(x-1)(x-2)}$$

65.

$$\frac{x+4}{2x^2-7x+5}-\frac{3}{x^2-2x+1}$$

$$=\frac{x+4}{(2x-5)(x-1)}-\frac{3}{(x-1)(x-1)}$$

$$\text{LCD}=(x-1)^2(2x-5)$$

$$=\frac{x+4}{(2x-5)(x-1)}\cdot\frac{x-1}{x-1}-\frac{3}{(x-1)(x-1)}\cdot\frac{2x-5}{2x-5}$$

$$=\frac{(x+4)(x-1)-3(2x-5)}{(2x-5)(x-1)^2}$$

$$=\frac{x^2+3x-4-6x+15}{(2x-5)(x-1)^2}$$

$$=\frac{x^2-3x+11}{(2x-5)(x-1)^2}$$

67.

$$\frac{1}{x-3}+\frac{x}{3x-12}-\frac{2x-7}{x^2-7x+12}$$

$$=\frac{1}{x-3}+\frac{x}{3(x-4)}-\frac{2x-7}{(x-3)(x-4)}$$

$$\text{LCD}=3(x-3)(x-4)$$

$$=\frac{1}{x-3}\cdot\frac{3(x-4)}{3(x-4)}+\frac{x}{3(x-4)}\cdot\frac{x-3}{x-3}$$

$$-\frac{2x-7}{(x-3)(x-4)}\cdot\frac{3}{3}$$

$$=\frac{3(x-4)+x(x-3)-(2x-7)\cdot3}{3(x-3)(x-4)}$$

$$=\frac{3x-12+x^2-3x-6x+21}{3(x-3)(x-4)}$$

$$=\frac{x^2-6x+9}{3(x-3)(x-4)}$$

$$=\frac{(x-3)(x-3)}{3(x-3)(x-4)}$$

$$=\frac{x-3}{3(x-4)}$$

69.

$$\frac{\dfrac{5x^3}{12y^2}}{\dfrac{15x}{8y^8}}=\frac{5x^3}{12y^2}+\frac{15x}{8y^8}=\frac{5x^3}{12y^2}\cdot\frac{8y^8}{15x}$$

$$=\frac{5x^{3-1}\cdot4\cdot2y^{8-2}}{3\cdot4\cdot3\cdot5}$$

$$=\frac{2x^2y^6}{9}$$

71.

$$\frac{\dfrac{2}{x-3}}{\dfrac{4}{x^2-9}}=\frac{2}{x-3}+\frac{4}{x^2-9}$$

$$=\frac{2}{x-3}\cdot\frac{(x-3)(x+3)}{2\cdot2}=\frac{x+3}{2}$$

73.

$$\frac{\dfrac{4x^2-1}{15}}{10x-5}=\frac{4x^2-1}{15}+(10x-5)$$

$$=\frac{(2x+1)(2x-1)}{15}\cdot\frac{1}{5(2x-1)}=\frac{2x+1}{75}$$

75.

$$\frac{\dfrac{7x^4-14x^3}{x+3}}{\dfrac{21x^3-42x^2}{x-5}}=\frac{7x^4-14x^3}{x+3}+\frac{21x^3-42x^2}{x-5}$$

$$=\frac{7x^3(x-2)}{x+3}\cdot\frac{x-5}{21x^2(x-2)}$$

$$=\frac{7\cdot x^{3-2}(x-2)(x-5)}{3\cdot7(x+3)(x-2)}$$

$$=\frac{x(x-5)}{3(x+3)}$$

77.

$$\frac{\dfrac{5}{6}-\dfrac{1}{3}}{\dfrac{5}{12}+\dfrac{1}{3}} \qquad \text{LCD}=12$$

$$=\frac{\left(\dfrac{5}{6}-\dfrac{1}{3}\right)}{\left(\dfrac{5}{12}+\dfrac{1}{3}\right)}\cdot\frac{12}{12}$$

$$=\frac{\left(\dfrac{5}{6}-\dfrac{1}{3}\right)\cdot 12}{\left(\dfrac{5}{12}+\dfrac{1}{3}\right)\cdot 12}$$

$$=\frac{10-4}{5+4} \qquad \text{Distributive property}$$

$$=\frac{6}{9} \quad \text{or} \quad \frac{2}{3}$$

79.

$$\frac{\dfrac{3}{5x}-\dfrac{5x}{3}}{x-\dfrac{3}{5}} \qquad \text{LCD}=15x$$

$$=\frac{\left(\dfrac{3}{5x}-\dfrac{5x}{3}\right)}{\left(x-\dfrac{3}{5}\right)}\cdot\frac{15x}{15x}$$

$$=\frac{\left(\dfrac{3}{5x}-\dfrac{5x}{3}\right)\cdot 15x}{\left(x-\dfrac{3}{5}\right)\cdot 15x}$$

$$=\frac{9-25x^2}{15x^2-9x} \qquad \text{Distributive property}$$

$$=\frac{-\left(25x^2-9\right)}{15x^2-9x}$$

$$=\frac{-\cancel{(5x-3)}(5x+3)}{3x\cancel{(5x-3)}}$$

$$=-\frac{5x+3}{3x}$$

81.

$$\frac{\dfrac{2x+7}{4}}{\dfrac{4}{7}-\dfrac{7}{x^2}} \qquad \text{LCD}=7x^2$$

$$=\frac{(2x+7)}{\left(\dfrac{4}{7}-\dfrac{7}{x^2}\right)}\cdot\frac{7x^2}{7x^2}$$

$$=\frac{(2x+7)\cdot 7x^2}{\left(\dfrac{4}{7}-\dfrac{7}{x^2}\right)\cdot 7x^2}$$

$$=\frac{(2x+7)\cdot 7x^2}{4x^2-49} \qquad \text{Distributive property}$$

$$=\frac{\cancel{(2x+7)}\cdot 7x^2}{\cancel{(2x+7)}(2x-7)}$$

$$=\frac{7x^2}{2x-7}$$

83.

$$\frac{\dfrac{1}{2}-\dfrac{1}{x+1}}{\dfrac{2}{x+1}-1} \qquad \text{LCD}=2(x+1)$$

$$=\frac{\left(\dfrac{1}{2}-\dfrac{1}{x+1}\right)}{\left(\dfrac{2}{x+1}-1\right)}\cdot\frac{2(x+1)}{2(x+1)}$$

$$=\frac{\left(\dfrac{1}{2}-\dfrac{1}{x+1}\right)\cdot 2(x+1)}{\left(\dfrac{2}{x+1}-1\right)\cdot 2(x+1)}$$

$$=\frac{x+1-2}{4-2(x+1)} \qquad \text{Distributive property}$$

$$=\frac{x-1}{4-2x-2}$$

$$=\frac{x-1}{-2x+2}$$

$$=\frac{\cancel{(x-1)}}{-2\cancel{(x-1)}}=-\frac{1}{2}$$

85.

$$\frac{3x}{5} - \frac{x}{2} = \frac{7}{10} \qquad LCD = 10$$

$$10\left(\frac{3x}{5} - \frac{x}{2}\right) = \frac{7}{10} \cdot 10$$

$$6x - 5x = 7 \qquad \text{Distributive property}$$

$$x = 7$$

Solution set: $\{7\}$

87.

$$\frac{3x+2}{6} - \frac{x-2}{4} = \frac{1}{3} \qquad LCD = 12$$

$$12\left(\frac{3x+2}{6} - \frac{x-2}{4}\right) = \frac{1}{3} \cdot 12$$

$$2(3x+2) - 3(x-2) = 4 \qquad \text{Distributive property}$$

$$6x + 4 - 3x + 6 = 4$$

$$3x + 10 = 4$$

$$3x = -6$$

$$x = -2$$

Solution set: $\{-2\}$

89.

$$\frac{3}{4x} + \frac{1}{x} = \frac{7}{12} \qquad LCD = 12x$$

$$12x\left(\frac{3}{4x} + \frac{1}{x}\right) = \frac{7}{12} \cdot 12x$$

$$9 + 12 = 7x \qquad \text{Distributive property}$$

$$21 = 7x$$

$$3 = x$$

Solution set: $\{3\}$

91.

$$\frac{x+4}{x-1} = \frac{6}{x-3}$$

$$6(x-1) = (x+4)(x-3) \qquad \text{Cross multiply}$$

$$6x - 6 = x^2 + x - 12$$

$$0 = x^2 - 5x - 6$$

$$0 = (x-6)(x+1)$$

$$x - 6 = 0 \quad \text{or} \quad x + 1 = 0$$

$$x = 6 \qquad\qquad x = -1$$

Solution set: $\{-1, \ 6\}$

93.

$$\frac{x}{x-3} - 2 = \frac{1}{x-3} \qquad LCD = x - 3$$

$$(x-3)\left(\frac{x}{x-3} - 2\right) = \frac{1}{x-3} \cdot (x-3)$$

$$x - 2(x-3) = 1 \qquad \text{Distributive property}$$

$$x - 2x + 6 = 1$$

$$-x = -5$$

$$x = 5$$

Solution set: $\{5\}$

95.

$$\frac{x}{x+1} - \frac{5}{x-3} = \frac{-20}{x^2 - 2x - 3}$$

$$\frac{x}{x+1} - \frac{5}{x-3} = \frac{-20}{(x+1)(x-3)} \qquad LCD = (x+1)(x-3)$$

$$(x+1)(x-3)\left(\frac{x}{x+1} - \frac{5}{x-3}\right)$$

$$= \left(\frac{-20}{(x+1)(x-3)}\right) \cdot (x+1)(x-3)$$

$$x(x-3) - 5(x+1) = -20 \qquad \text{Distributive property}$$

$$x^2 - 3x - 5x - 5 = -20$$

$$x^2 - 8x + 15 = 0$$

$$(x-3)(x-5) = 0$$

$$x - 3 = 0 \quad \text{or} \quad x - 5 = 0$$

$$x = 3 \qquad\qquad x = 5$$

Solution set: $\{5\}$

Note: $x \neq 3$

97.

$$\frac{16}{12} = \frac{4 \cdot 4}{4 \cdot 3} = \frac{4}{3}$$

99.

$$\frac{24 \text{ feet}}{30 \text{ feet}} = \frac{6 \cdot 4 \text{ feet}}{6 \cdot 5 \text{ feet}} = \frac{4}{5}$$

101.

$$\frac{2x+5}{7} = \frac{1}{3}$$

$$3(2x+5) = 1 \cdot 7$$

$$6x+15 = 7$$

$$6x = -8$$

$$x = -\frac{8}{6} \quad \text{or} \quad -\frac{4}{3}$$

Solution set: $\left\{-\frac{4}{3}\right\}$

103.

$$\frac{4}{3x-2} = \frac{2}{5}$$

$$4 \cdot 5 = 2(3x-2)$$

$$20 = 6x-4$$

$$24 = 6x$$

$$4 = x$$

Solution set: $\{4\}$

105.

$$\frac{-1}{2x-7} = \frac{x}{5}$$

$$-1 \cdot 5 = x(2x-7)$$

$$-5 = 2x^2 - 7x$$

$$0 = 2x^2 - 7x + 5$$

$$0 = (2x-5)(x-1)$$

$$2x-5 = 0 \quad \text{or} \quad x-1 = 0$$

$$x = \frac{5}{2} \qquad x = 1$$

Solution set: $\left\{1, \frac{5}{2}\right\}$

107.

$$\frac{x+3}{x+2} = \frac{x-6}{x-10}$$

$$(x+3)(x-10) = (x+2)(x-6)$$

$$x^2 - 7x - 30 = x^2 - 4x - 12$$

$$x^2 - x^2 - 30 + 12 = -4x + 7x$$

$$-18 = 3x$$

$$-6 = x$$

Solution set: $\{-6\}$

109.

$$\frac{3 \text{ pounds}}{\$4.77} = \frac{5 \text{ pounds}}{x}$$

$$\frac{3}{4.77} = \frac{5}{x}$$

$$3x = 5(4.77)$$

$$3x = 23.85$$

$$x = 7.95$$

The cost is $7.95.

111.

Let x = number to be added.

The resulting fraction is $\frac{2+x}{7+x}$.

The equation is $\frac{2+x}{7+x} = \frac{2}{3}$

$$3(2+x) = 2(7+x)$$

$$6+3x = 14+2x$$

$$3x-2x = 14-6$$

$$x = 8$$

The number is 8.

113.

Let x = number of hours for Robin to paint alone.

The equation is $\dfrac{1}{4}+\dfrac{1}{x}=\dfrac{1}{3}$ LCD = 12x

$12x\left(\dfrac{1}{4}+\dfrac{1}{x}\right)=\dfrac{1}{3}\cdot 12x$

$\qquad 3x+12=4x$

$\qquad\qquad 12=4x-3x$

$\qquad\qquad 12=x$

It takes Robin 12 hours.

115.

Let x = speed of canoe in still water.

Let x + 3 = speed of canoe down river.

Let x - 3 = speed of canoe up river.

	D =	R ·	T
Up	8	x - 3	$\dfrac{8}{x-3}$
Down	20	x + 3	$\dfrac{20}{x+3}$

Note: 1. Since $T=\dfrac{D}{R}$, then $T_U=\dfrac{8}{x-3}$,

and $T_D=\dfrac{20}{x+3}$.

2. Both times are the same.

The equation is $\dfrac{8}{x-3}=\dfrac{20}{x+3}$

$8(x+3)=20(x-3)$

$8x+24=20x-60$

$24+60=20x-8x$

$\qquad 84=12x$

$\qquad\ 7=x$

The speed of the canoe in still water is 7mph.

Chapter 5 Test Solutions

1.

$\dfrac{5x^2y-15x^3}{3x^2+2xy-y^2}=\dfrac{-5x^2(-y+3x)}{(3x-y)(x+y)}$

$\qquad\qquad =\dfrac{-5x^2(3x-y)}{(3x-y)(x+y)}=\dfrac{-5x^2}{x+y}$

3.

$\dfrac{3x+10}{x^2+11x+30}-\dfrac{2x+5}{x^2+11x+30}=\dfrac{3x+10-2x-5}{x^2+11x+30}$

$=\dfrac{x+5}{x^2+11x+30}=\dfrac{\cancel{(x+5)}}{\cancel{(x+5)}(x+6)}=\dfrac{1}{x+6}$

5.

$\dfrac{x+1}{2x^2+9x+10}+\dfrac{1}{x^2+5x+6}$

$\qquad\qquad\qquad$ LCD $=(2x+5)(x+2)(x+3)$

$=\dfrac{(x+3)}{(x+3)}\cdot\dfrac{x+1}{(2x+5)(x+2)}+\dfrac{1}{(x+2)(x+3)}\cdot\dfrac{(2x+5)}{(2x+5)}$

$=\dfrac{x^2+4x+3}{(2x+5)(x+2)(x+3)}+\dfrac{2x+5}{(2x+5)(x+2)(x+3)}$

$=\dfrac{x^2+6x+8}{(2x+5)(x+2)(x+3)}$

$=\dfrac{(x+4)\cancel{(x+2)}}{(2x+5)\cancel{(x+2)}(x+3)}=\dfrac{x+4}{(2x+5)(x+3)}$

7.

$\dfrac{1}{6x}-2=\dfrac{5}{2x}$ LCD = 6x

$6x\left(\dfrac{1}{6x}-2\right)=\left(\dfrac{5}{2x}\right)6x$

$\dfrac{6x}{6x}-12x=\dfrac{30x}{2x}$

$1-12x=15$

$\qquad -12x=14$

$\qquad\qquad x=-\dfrac{14}{12}\text{ or }-\dfrac{7}{6}$

Solution set: $\left\{-\dfrac{7}{6}\right\}$

9.

$\dfrac{42}{66}=\dfrac{6\cdot 7}{6\cdot 11}=\dfrac{7}{11}$

11.

$\dfrac{\text{Gallons}}{\text{Miles}}=\dfrac{5.5}{154}=\dfrac{x}{175}$

$175(5.5)=154x$ Cross multiply

$962.5=154x$

$6.25=x$

Answer: 6.25 gallons

Test Your Memory

1.

$$\frac{2}{3}+\frac{3}{4}-\frac{5}{6} \qquad LCD = 12$$

$$=\frac{8}{12}+\frac{9}{12}-\frac{10}{12}=\frac{8+9-10}{12}=\frac{7}{12}$$

3.

$$\left(\frac{1}{2}\cdot\frac{3}{5}\right)^2=\left(\frac{3}{10}\right)^2=\frac{9}{100}$$

5.

$$x^2-13x+12=(x-12)(x-1)$$

7.

$$8x^2+6xy-4x-3y=2x(4x+3y)-1\cdot(4x+3y)$$
$$=(2x-1)(4x+3y)$$

9.

$$-1\le 2x-3\le 13$$
$$-1+3\le 2x-3+3\le 13+3$$
$$2\le 2x\le 16$$
$$1\le x\le 8 \qquad Divide\ by\ 2$$

The graph is in the answer section of the main text.

11.

$$8-2(3x+1)=5(2-x)+7$$
$$8-6x-2=10-5x+7$$
$$6-6x=17-5x$$
$$6=17+x \qquad Add\ 6x$$
$$-11=x \qquad Subtract\ 17$$

Solution set: {-11}

13.

$$(3x+1)(x-2)=48$$
$$3x^2-5x-2=48$$
$$3x^2-5x-50=0 \qquad Subtract\ 48$$
$$(3x+10)(x-5)=0$$
$$3x+10=0 \quad or \quad x-5=0$$
$$x=-\frac{10}{3} \quad or \quad x=5$$

#13   continued

Solution set: $\left\{-\dfrac{10}{3}\ ,\ 5\right\}$

15.

$$\frac{2}{5x}-\frac{1}{2x}=\frac{3}{10} \qquad LCD=10x$$

$$10x\left(\frac{2}{5x}-\frac{1}{2x}\right)=\left(\frac{3}{10}\right)10x$$

$$\frac{20x}{5x}-\frac{10x}{2x}=\frac{30x}{10}$$

$$4-5=3x$$
$$-1=3x$$
$$-\frac{1}{3}=x$$

Solution set: $\left\{-\dfrac{1}{3}\right\}$

17.

$$\frac{5}{x}+\frac{3}{x+2}=\frac{1}{2x} \qquad LCD=2x(x+2)$$

$$2x(x+2)\left[\frac{5}{x}+\frac{3}{x+2}\right]=\left(\frac{1}{2x}\right)2x(x+2)$$

$$10(x+2)+6x=x+2 \qquad Multiply\ by\ 2x(x+2)$$
$$10x+20+6x=x+2$$
$$16x+20=x+2$$
$$15x+20=2 \qquad Subtract\ x$$
$$15x=-18 \qquad Subtract\ 20$$
$$x=-\frac{18}{15}or-\frac{6}{5} \qquad Divide\ by\ 15$$

Solution set: $\left\{-\dfrac{6}{5}\right\}$

19.

$$\frac{4}{x+1}=\frac{2}{3x-1}$$

$$4(3x-1)=2(x+1) \qquad Cross\ multiply$$
$$12x-4=2x+2$$
$$10x-4=2 \qquad Subtract\ 2x$$
$$10x=6 \qquad Add\ 4$$
$$x=\frac{6}{10}\ or\ \frac{3}{5} \qquad Divide\ by\ 10$$

Solution set: $\left\{\dfrac{3}{5}\right\}$

21.

$$\frac{x}{9} = \frac{-2}{x - 9}$$

$$x(x - 9) = -2(9) \qquad \text{Cross multiply}$$

$$x^2 - 9x = -18$$

$$x^2 - 9x + 18 = 0$$

$$(x - 3)(x - 6) = 0$$

$$x - 3 = 0 \quad \text{or} \quad x - 6 = 0$$

$$x = 3 \quad \text{or} \quad x = 6$$

Solution set: {3, 6}

23.

$$\left(\frac{x^{-2}y^{-3}}{x^{-1}y^{-8}}\right)^{-2} = \frac{x^4 y^6}{x^2 y^{16}}$$

$$= \frac{x^{4-2}}{y^{16-6}} = \frac{x^2}{y^{10}}$$

25.

$$\frac{3x^3 - 12x^2}{x^2 - 16} = \frac{3x^2(x - 4)}{(x - 4)(x + 4)} = \frac{3x^2}{x + 4}$$

27.

$$\frac{\frac{3}{x} + \frac{1}{2}}{\frac{2}{x} + \frac{1}{3}} = \frac{6x\left(\frac{3}{x} + \frac{1}{2}\right)}{6x\left(\frac{2}{x} + \frac{1}{3}\right)} \qquad \text{LCD} = 6x$$

$$= \frac{18 + 3x}{12 + 2x} = \frac{3(6 + x)}{2(6 + x)} = \frac{3}{2}$$

29.

$$\frac{\frac{1}{x} + \frac{2}{x + 1}}{\frac{1}{x} + 3} = \frac{x(x + 1)\left(\frac{1}{x} + \frac{2}{x + 1}\right)}{x(x + 1)\left(\frac{1}{x} + 3\right)}$$

$$\text{LCD} = x(x + 1)$$

$$= \frac{x + 1 + 2x}{x + 1 + 3x(x + 1)}$$

$$= \frac{3x + 1}{x + 1 + 3x^2 + 3x}$$

$$= \frac{3x + 1}{3x^2 + 4x + 1}$$

#29 continued

$$= \frac{(3x + 1)}{(3x + 1)(x + 1)} = \frac{1}{x + 1}$$

31.

$$-2x^3 y(5x^2 + 3xy - 4y^4)$$

$$= -10x^5 y - 6x^4 y^2 + 8x^3 y^5$$

33.

$$(8x^3 - 24x^2 + 24x - 32) \div (2x - 5)$$

$$\begin{array}{r} 4x^2 - 2x + 7 \\ 2x - 5 \enclose{longdiv}{8x^3 - 24x^2 + 24x - 32} \\ \underline{-8x^3 \pm 20x^2} \\ -4x^2 + 24x \\ \underline{\pm 4x^2 \mp 10x} \\ 14x - 32 \\ \underline{-14x \pm 35} \\ 3 \end{array}$$

Answer: $4x^2 - 2x + 7 + \dfrac{3}{2x - 5}$

35.

$$\frac{x^3 - 4x^2}{2x^2 - 7x + 6} \cdot \frac{2x - 3}{5x^2 + 10x}$$

$$= \frac{x \cdot x(x - 4)}{(2x - 3)(x - 2)} \cdot \frac{(2x - 3)}{5x(x + 2)}$$

$$= \frac{x(x - 4)}{5(x - 2)(x + 2)}$$

37.

$$\frac{9x^2 - 3x}{3x^2 - 7x + 2} + \frac{12x + 9}{4x^2 - 5x - 6}$$

$$= \frac{3x(3x - 1)}{(3x - 1)(x - 2)} + \frac{3(4x + 3)}{(4x + 3)(x - 2)} \qquad \text{Factor}$$

$$= \frac{3x(3x - 1)}{(3x - 1)(x - 2)} \cdot \frac{(4x + 3)(x - 2)}{3(4x + 3)} = x$$

39.

$$\frac{2x^2 - 5x}{x^2 - 9} + \frac{4x - 21}{x^2 - 9} = \frac{2x^2 - 5x + 4x - 21}{x^2 - 9}$$

$$= \frac{2x^2 - x - 21}{x^2 - 9} = \frac{(2x - 7)(x + 3)}{(x + 3)(x - 3)} = \frac{2x - 7}{x - 3}$$

41.

$$\frac{7}{x + 3} - \frac{5}{x - 4}$$

$$= \left(\frac{x - 4}{x - 4}\right)\left(\frac{7}{x + 3}\right) - \left(\frac{5}{x - 4}\right)\left(\frac{x + 3}{x + 3}\right)$$

$$= \frac{7(x - 4)}{(x + 3)(x - 4)} - \frac{5(x + 3)}{(x + 3)(x - 4)}$$

$$= \frac{7x - 28 - 5(x + 3)}{(x + 3)(x - 4)}$$

$$= \frac{7x - 28 - 5x - 15}{(x + 3)(x - 4)} = \frac{2x - 43}{(x + 3)(x - 4)}$$

43.

Let x = the number of quarters.

Then 4x + 1 = the number of nickels.

Let 25x = value of quarters in cents
and 5(4x + 1) = value of nickels in cents.

So,

$$25x + 5(4x + 1) = 140$$
$$25x + 20x + 5 = 140$$
$$45x = 135$$
$$x = 3$$
$$4x + 1 = 13$$
$$4(3) + 1 = 13$$

Answer: 3 quarters and 13 nickels

45.

Let x = measurement of the width.

Then 2x - 3 = measurement of length.

So, P = 2W + 2L.

$$30 = 2x + 2(2x - 3)$$
$$30 = 2x + 4x - 6$$
$$30 = 6x - 6$$
$$36 = 6x \qquad \text{Add 6}$$
$$6 = x \qquad \text{Divide by 6}$$
$$9 = 2x - 3$$

Answer: 6 x 9 mm

47.

$$\frac{\text{Number of Eggs}}{\text{Time}} = \frac{15}{25} = \frac{x}{60} \qquad 1 \text{ hr.} = 60 \text{ min.}$$

$$15(60) = 25x$$
$$900 = 25x$$
$$36 = x$$

Answer: 36 eggs

49.

Let x be the number of minutes father takes alone.

Then 3x is the number of minutes son takes alone.

So,

$$\frac{1}{x} + \frac{1}{3x} = \frac{1}{36} \qquad LCD = 36x$$

$$36x\left(\frac{1}{x} + \frac{1}{3x}\right) = \frac{1}{36}(36x)$$

$$36 + 12 = x$$
$$48 = x$$

Answer: 48 minutes

CHAPTER 5 STUDY GUIDE

Self Test Exercises

1. Reduce to lowest terms.

$$\frac{3x^2y - 6x^3}{2x^2 - 3xy + y^2}$$

2. Simplify the complex fraction.

$$\frac{\dfrac{2}{x}}{2 - \dfrac{1}{x}}$$

Perform the indicated operations, and reduce the answers to lowest terms.

3. $\dfrac{2x - 13}{x^2 - 10x + 21} - \dfrac{x - 10}{x^2 - 10x + 21}$

4. $\dfrac{x^2 - x - 2}{x^2 + 5x - 14} \cdot \dfrac{2x^2 + 13x - 7}{3x^2 + x - 2}$

5. $\dfrac{x + 1}{2x^2 - 9x - 5} + \dfrac{1 - x}{x^2 - 4x - 5}$

6. $\dfrac{x^2 + 7x + 12}{x^2 - x - 20} + \dfrac{x^2 + 6x + 9}{x^2 - 10x + 25}$

Solve for x in the following equations.

7. $\dfrac{3}{x} - \dfrac{1}{6} = \dfrac{5}{2x}$

8. $\dfrac{1}{x^2 - x - 30} - \dfrac{2}{x^2 - 11x + 30} = \dfrac{-3}{x^2 - 25}$

9. Write the following ratio as a fraction in lowest terms.

   36 feet to 88 feet

10. Solve for x in the following proportion.

$$\frac{4}{m - 8} = \frac{4 - m}{8 - m}$$

11. Use a proportion to find the solution of the following problem.

   A car uses 7.5 gallons of gasoline on a 150 mile trip. How many gallons of gasoline will it use on a 275 mile trip?

12. Solve the following problem.

   It takes Sarah 4 hours to paint a certain area of a fence. It takes Sid 5 hours to do the same job. Working together, how long would it take them to do the same paint job?

The worked-out solutions begin on the next page.

1.

$$\frac{3x^2y - 6x^3}{2x^2 - 3xy + y^2}$$

$$= \frac{-3x^2(-y + 2x)}{(2x - y)(x - y)}$$

$$= \frac{-3x^2(2x - y)}{(2x - y)(x - y)} = \frac{-3x^2}{x - y}$$

2.

$$\frac{\frac{2}{x}}{2 - \frac{1}{x}} = \frac{x \cdot \frac{2}{x}}{x\left(2 - \frac{1}{x}\right)} = \frac{2}{2x - 1}$$

3.

$$\frac{2x - 13}{x^2 - 10x + 21} - \frac{x - 10}{x^2 - 10x + 21}$$

$$= \frac{2x - 13 - x + 10}{x^2 - 10x + 21}$$

$$= \frac{(x - 3)}{(x - 7)(x - 3)} = \frac{1}{x - 7}$$

4.

$$\frac{x^2 - x - 2}{x^2 + 5x - 14} \cdot \frac{2x^2 + 13x - 7}{3x^2 + x - 2}$$

$$= \frac{(x - 2)(x + 1)}{(x + 7)(x - 2)} \cdot \frac{(2x - 1)(x + 7)}{(3x - 2)(x + 1)} = \frac{2x - 1}{3x - 2}$$

5.

$$\frac{x + 1}{2x^2 - 9x - 5} + \frac{1 - x}{x^2 - 4x - 5}$$

$$= \frac{x + 1}{(2x + 1)(x - 5)} + \frac{x - 1}{(x - 5)(x + 1)}$$

$$\text{LCD} = (2x + 1)(x - 5)(x + 1)$$

$$= \frac{x + 1}{x + 1} \cdot \frac{x + 1}{(2x + 1)(x - 5)}$$

$$+ \frac{x - 1}{(x - 5)(x + 1)} \cdot \frac{2x + 1}{2x + 1}$$

$$= \frac{x^2 + 2x + 1 + 2x^2 - x - 1}{(2x + 1)(x - 5)(x + 1)}$$

$$= \frac{3x^2 + x}{(2x + 1)(x - 5)(x + 1)}$$

6.

$$\frac{x^2 + 7x + 12}{x^2 - x - 20} \div \frac{x^2 + 6x + 9}{x^2 - 10x + 25}$$

$$= \frac{x^2 + 7x + 12}{x^2 - x - 20} \cdot \frac{x^2 - 10x + 25}{x^2 + 6x + 9}$$

$$= \frac{(x + 4)(x + 3)}{(x - 5)(x + 4)} \cdot \frac{(x - 5)(x - 5)}{(x + 3)(x + 3)} = \frac{x - 5}{x + 3}$$

7.

$$\frac{3}{x} - \frac{1}{6} = \frac{5}{2x} \qquad \text{LCD} = 6x$$

$$6x\left(\frac{3}{x} - \frac{1}{6}\right) = \frac{5}{2x} \cdot 6x$$

$$18 - x = 15$$

$$-x = -3$$

$$x = 3$$

Solution set: {3}

8.

$$\frac{1}{x^2 - x - 30} - \frac{2}{x^2 - 11x + 30} = \frac{-3}{x^2 - 25}$$

$$\frac{1}{(x + 5)(x - 6)} - \frac{2}{(x - 5)(x - 6)}$$

$$= \frac{-3}{(x + 5)(x - 5)}$$

$$\text{LCD} = (x + 5)(x - 5)(x - 6)$$

$$(x + 5)(x - 5)(x - 6)\left(\frac{1}{(x + 5)(x - 6)} - \frac{2}{(x - 5)(x - 6)}\right)$$

$$= \frac{-3}{(x + 5)(x - 5)}(x + 5)(x - 5)(x - 6)$$

$$x - 5 - 2(x + 5) = -3(x - 6)$$

$$x - 5 - 2x - 10 = -3x + 18$$

$$-x - 15 = -3x + 18$$

$$2x - 15 = 18$$

$$2x = 33$$

$$x = \frac{33}{2}$$

Solution set: $\left\{\frac{33}{2}\right\}$

9.

$$\frac{36}{88} = \frac{4 \cdot 9}{4 \cdot 22} = \frac{9}{22}$$

10.

$$\frac{4}{m - 8} = \frac{4 - m}{8 - m}$$

$$4(8 - m) = (m - 8)(4 - m)$$

Cross multiply

$$32 - 4m = 4m - 32 - m^2 + 8m$$

$$32 - 4m = 12m - 32 - m^2$$

$$m^2 - 16m + 64 = 0$$

$$(m - 8)(m - 8) = 0$$

$$m - 8 = 0$$

$$m = 8$$

Solution set: $\emptyset$ Note: $m - 8 \neq 0$
So, $m \neq 8$.

11.

$$\frac{gallons}{miles} = \frac{7.5}{150} = \frac{x}{275}$$

$$7.5(275) = 150x$$

$$2062.5 = 150x$$

$$13.75 = x$$

Answer: 13.75 gallons

12.

Let x be the number of hours it would take for both Sarah and Sid.

$$\frac{1}{4} + \frac{1}{5} = \frac{1}{x} \qquad LCD = 20x$$

$$20x\left(\frac{1}{4} + \frac{1}{5}\right) = \frac{1}{x} \cdot 20x$$

$$5x + 4x = 20$$

$$9x = 20$$

$$x = \frac{20}{9} \text{ or } 2\frac{2}{9}$$

Answer: $2\frac{2}{9}$ hours, or approximately

2 hours, 13 minutes

CHAPTER 6   GRAPHS OF LINEAR EQUATIONS AND INEQUALITIES IN TWO VARIABLES

Solutions to Text Exercises

Exercises 6.1

1.
$x + 4y = 7$, $(3, 1)$
$3 + 4 \cdot 1 = 3 + 4 = 7$; yes

3.
$5x - y = 3$, $(2, 7)$
$5 \cdot 2 - 7 = 10 - 7 = 3$; yes

5.
$2x - 7y = 8$, $(5, 1)$
$2 \cdot 5 - 7 \cdot 1 = 10 - 7 = 3$; no

7.
$7x + 3y = 3$, $(6, -13)$
$7 \cdot 6 + 3(-13) = 42 - 39 = 3$; yes

9.
$12x - 5y + 9 = 0$, $(3, 9)$
$12 \cdot 3 - 5 \cdot 9 + 9 = 36 - 45 + 9 = 0$; yes

11.
$3x + 8y + 19 = 0$, $(1, -2)$
$3 \cdot 1 + 8(-2) + 19 = 3 - 16 + 19 = 6$; no

13.
$5x - 2y = 7$, $\left(2, \dfrac{3}{2}\right)$

$5 \cdot 2 - 2\left(\dfrac{3}{2}\right) = 10 - 3 = 7$; yes

15.
$\dfrac{7}{2}x - 4y = 2$, $(-4, -2)$

$\dfrac{7}{2}(-4) - 4(-2) = -14 + 8 = -6$; no

17.
$\dfrac{3}{4}x + \dfrac{5}{6}y = 2$, $(-2, 3)$
$\dfrac{3}{4}(-2) + \dfrac{5}{6} \cdot 3 = -\dfrac{3}{2} + \dfrac{5}{2} = \dfrac{2}{2} = 1$; no

19.
$2.4x - 3.5y = 5$, $(5, 2)$
$2.4(5) - 3.5(2) = 12 - 7 = 5$; yes

21.
$x = 3$, $(3, 7)$
$3 = 3$; yes

23.
$y - 2 = 0$, $(8, -2)$
$-2 - 2 = -4$; no

25.
$2x - 5y = 10$

$(-5, ?)$, $2(-5) - 5y = 10$
$\quad\quad -10 - 5y = 10$
$\quad\quad\quad - 5y = 10 + 10$
$\quad\quad\quad - 5y = 20$
$\quad\quad\quad\quad y = -4$, $(-5, -4)$

$(0, ?)$, $2 \cdot 0 - 5y = 10$
$\quad\quad\quad - 5y = 10$
$\quad\quad\quad\quad y = -2$, $(0, -2)$

$(?, 0)$, $2x - 5 \cdot 0 = 10$
$\quad\quad\quad 2x = 10$
$\quad\quad\quad\quad x = 5$, $(5, 0)$

$(?, 2)$, $2x - 5 \cdot 2 = 10$
$\quad\quad\quad 2x - 10 = 10$
$\quad\quad\quad\quad 2x = 20$
$\quad\quad\quad\quad\quad x = 10$, $(10, 2)$

27.
$7x + 4y = 14$

$(-2, ?)$, $7(-2) + 4y = 14$
$\quad\quad\quad -14 + 4y = 14$

#27 continued

$$4y = 28$$
$$y = 7, (-2, 7)$$

$(0,?),$ $\overline{7(0) + 4y = 14}$
$$4y = 14$$
$$y = \frac{7}{2}, \left(0, \frac{7}{2}\right)$$

$(?,0),$ $\overline{7x + 4 \cdot 0 = 14}$
$$7x = 14$$
$$x = 2, (2, 0)$$

$(?,-7),$ $\overline{7x + 4(-7) = 14}$
$$7x - 28 = 14$$
$$7x = 42$$
$$x = 6, (6, -7)$$

29.
$5x + 3y = -8$

$(2,?),$ $5 \cdot 2 + 3y = -8$
$$10 + 3y = -8$$
$$3y = -18$$
$$y = -6, (2, -6)$$

$(-1,?),$ $\overline{5(-1) + 3y = -8}$
$$-5 + 3y = -8$$
$$3y = -3$$
$$y = -1, (-1, -1)$$

$(?,4),$ $\overline{5x + 3 \cdot 4 = -8}$
$$5x + 12 = -8$$
$$5x = -20$$
$$x = -4, (-4, 4)$$

$(?, -11),$ $\overline{5x + 3(-11) = -8}$
$$5x - 33 = -8$$
$$5x = 25$$
$$x = 5, (5, -11)$$

31.
$6x - y = -10$

$(1,?),$ $6 \cdot 1 - y = -10$
$$6 - y = -10$$
$$-y = -16$$
$$y = 16, (1,16)$$

#31 continued

$(-2,?),$ $6(-2) - y = -10$
$$-12 - y = -10$$
$$-y = 2$$
$$y = -2, (-2, -2)$$

$(?,-8),$ $\overline{6x - (-8) = -10}$
$$6x + 8 = -10$$
$$6x = -18$$
$$x = -3, (-3, -8)$$

$(?,4),$ $\overline{6x - 4 = -10}$
$$6x = -6$$
$$x = -1, (-1, 4)$$

33.
$x + 3y = 5$

$(-1,?),$ $-1 + 3y = 5$
$$3y = 6$$
$$y = 2, (-1, 2)$$

$(11,?),$ $\overline{11 + 3y = 5}$
$$3y = -6$$
$$y = -2, (11, -2)$$

$(?,4),$ $\overline{x + 3 \cdot 4 = 5}$
$$x + 12 = 5$$
$$x = -7, (-7, 4)$$

$(?,-1),$ $\overline{x + 3(-1) = 5}$
$$x - 3 = 5$$
$$x = 8, (8, -1)$$

35.
$x + y + 6 = 0$

$\left(-\frac{5}{2}, ?\right),$ $-\frac{5}{2} + y + 6 = 0$
$$y = \frac{5}{2} - 6$$
$$y = -\frac{7}{2} \quad \left(-\frac{5}{2}, -\frac{7}{2}\right)$$

$\left(\frac{4}{3}, ?\right),$ $\overline{\frac{4}{3} + y + 6 = 0}$
$$y = -\frac{4}{3} - 6$$

#35 continued

$$y = -\frac{22}{3} \quad \left(\frac{4}{3}, -\frac{22}{3}\right)$$

$(?,2), \quad \overline{x + 2 + 6 = 0}$
$$x + 8 = 0$$
$$x = -8, \quad (-8,2)$$

$\left(?, -\frac{7}{4}\right) \quad \overline{x - \frac{7}{4} + 6 = 0}$

$$x = \frac{7}{4} - 6$$

$$x = -\frac{17}{4}, \left(-\frac{17}{4}, -\frac{7}{4}\right)$$

37.
$2x - 8y - 5 = 0$

$\left(-\frac{3}{2}, ?\right) \quad 2\left(-\frac{3}{2}\right) - 8y - 5 = 0$
$$-3 - 8y - 5 = 0$$
$$-8y - 8 = 0$$
$$-8y = 8$$
$$y = -1, \quad \left(-\frac{3}{2}, -1\right)$$

$(4,?), \quad \overline{2 \cdot 4 - 8y - 5 = 0}$
$$8 - 8y - 5 = 0$$
$$-8y + 3 = 0$$
$$-8y = -3$$
$$y = \frac{3}{8}, \left(4, \frac{3}{8}\right)$$

$\left(? \ -\frac{1}{2}\right) \quad \overline{2x - 8\left(-\frac{1}{2}\right) - 5 = 0}$

$$2x + 4 - 5 = 0$$
$$2x - 1 = 0$$
$$2x = 1$$
$$x = \frac{1}{2}, \left(\frac{1}{2}, -\frac{1}{2}\right)$$

$\left(? \ \frac{1}{3}\right) \quad \overline{2x - 8\left(\frac{1}{3}\right) - 5 = 0}$

$$2x - \frac{8}{3} - 5 = 0$$

#37 continued

$$2x = \frac{8}{3} + 5$$
$$2x = \frac{23}{3}$$
$$x = \frac{23}{6}, \left(\frac{23}{6}, \frac{1}{3}\right)$$

39.
$\frac{1}{2}x + y = 5$

$(4,?), \quad \frac{1}{2} \cdot 4 + y = 5$
$$2 + y = 5$$
$$y = 3, (4,3)$$

$(-6,?), \quad \overline{\frac{1}{2}(-6) + y = 5}$
$$-3 + y = 5$$
$$y = 8, (-6,8)$$

$(?,4), \quad \overline{\frac{1}{2}x + 4 = 5}$
$$\frac{1}{2}x = 1$$
$$x = 2, \quad (2,4)$$

$(?,10), \quad \overline{\frac{1}{2}x + 10 = 5}$
$$\frac{1}{2}x = -5$$
$$x = -10, (-10,10)$$

41.
$3x + \frac{3}{4}y = 9$

$\left(\frac{3}{2}, ?\right) \quad 3\left(\frac{3}{2}\right) + \frac{3}{4}y = 9$
$$\frac{9}{2} + \frac{3}{4}y = 9$$
$$\frac{3}{4}y = 9 - \frac{9}{2}$$
$$\frac{3}{4}y = \frac{9}{2}$$
$$y = \frac{4}{3} \cdot \frac{9}{2}$$

#41 continued

$$y = 6, \left(\frac{3}{2}, 6\right)$$

$(-1,?)$, $3(-1) + \frac{3}{4}y = 9$

$$-3 + \frac{3}{4}y = 9$$

$$\frac{3}{4}y = 9 + 3$$

$$\frac{3}{4}y = 12$$

$$y = \frac{4}{3} \cdot \frac{12}{1}$$

$$y = 16, (-1, 16)$$

$(?, 8)$, $3x + \frac{3}{4} \cdot 8 = 9$

$$3x + 6 = 9$$

$$3x = 9 - 6$$

$$3x = 3$$

$$x = 1, (1, 8)$$

$\left(?, -\frac{8}{5}\right)$, $3x + \frac{3}{4}\left(-\frac{8}{5}\right) = 9$

$$3x - \frac{6}{5} = 9$$

$$3x = 9 + \frac{6}{5}$$

$$3x = \frac{51}{5}$$

$$x = \frac{1}{3} \cdot \frac{51}{5}$$

$$x = \frac{17}{5}, \left(\frac{17}{5}, -\frac{8}{5}\right)$$

43.
$2.7x - 4y = 5.4$

$(2,?)$, $2.7 \cdot 2 - 4y = 5.4$
$$5.4 - 4y = 5.4$$
$$-4y = 5.4 - 5.4$$
$$-4y = 0$$
$$y = 0, (2, 0)$$

$(10,?)$, $\overline{2.7 \cdot 10 - 4y = 5.4}$
$$27 - 4y = 5.4$$
$$-4y = 5.4 - 27$$

#43 continued

$$-4y = -21.6$$
$$y = 5.4, (10, 5.4)$$

$(?, -1.35)$, $2.7x - 4(-1.35) = 5.4$
$$2.7x + 5.40 = 5.4$$
$$2.7x = 0$$
$$x = 0, (0, -1.35)$$

$(?, 6.75)$, $2.7x - 4(6.75) = 5.4$
$$2.7x - 27.00 = 5.4$$
$$2.7x = 5.4 + 27$$
$$2.7x = 32.4$$
$$x = 12, (12, 6.75)$$

---

45. - 47.
The answers do not require worked-out
solutions.

49.
$3x + y = 8;$ $\quad\quad x = -2, 0, 3$
$y = -3x + 8$

x	Work
-2	$y = -3(-2) + 8 = 6 + 8 = 14$
	Solution: (-2, 14)
0	$y = -3(0) + 8 = 0 + 8 = 8$
	Solution: (0, 8)
3	$y = -3 \cdot 3 + 8 = -9 + 8 = -1$
	Solution: (3, -1)

51.
$5x - y = -4;$ $\quad x = -4, 0, 1$
$-y = -5x - 4$
$y = 5x + 4$

x	Work
-4	$y = 5(-4) + 4 = -20 + 4 = -16$
	Solution: (-4, -16)
0	$y = 5(0) + 4 = 4$
	Solution: (0, 4)
1	$y = 5 \cdot 1 + 4 = 5 + 4 = 9$
	Solution: (1, 9)

53.
$2x + 5y = 10;$ $\quad\quad x = 5, 0, -10$
$5y = -2x + 10$
$$y = \frac{-2}{5}x + 2$$

#53 continued

x	Work
5	$y = -\frac{2}{5} \cdot 5 + 2 = -2 + 2 = 0$
	Solution: (5,0)
0	$y = -\frac{2}{5} \cdot 0 + 2 = 0 + 2 = 2$
	Solution: (0,2)
-10	$y = -\frac{2}{5}(-10) + 2 = 4 + 2 = 6$
	Solution: (-10,6)

55.

$-6x + 3y = 5; \qquad x = -\frac{1}{2}, 2, \frac{1}{6}$

$3y = 6x + 5$

$y = 2x + \frac{5}{3}$

x	Work
$-\frac{1}{2}$	$y = 2\left(-\frac{1}{2}\right) + \frac{5}{3} = -1 + \frac{5}{3} = \frac{2}{3}$
	Solution: $\left(-\frac{1}{2}, \frac{2}{3}\right)$
2	$y = 2 \cdot 2 + \frac{5}{3} = 4 + \frac{5}{3} = \frac{17}{3}$
	Solution: $\left(2, \frac{17}{3}\right)$
$\frac{1}{6}$	$y = 2\left(\frac{1}{6}\right) + \frac{5}{3} = \frac{1}{3} + \frac{5}{3} = \frac{6}{3} = 2$
	Solution: $\left(\frac{1}{6}, 2\right)$

57.

$4x - 7y + 7 = 0; \qquad x = -3, 7, 14$

$-7y = -4x - 7$

$y = \frac{4}{7}x + 1$

x	Work
-3	$y = \frac{4}{7}(-3) + 1 = \frac{-12}{7} + 1 = \frac{-5}{7}$
	Solution: $\left(-3, -\frac{5}{7}\right)$
7	$y = \frac{4}{7} \cdot 7 + 1 = 4 + 1 = 5$
	Solution: (7,5)
14	$y = \frac{4}{7} \cdot 14 + 1 = 8 + 1 = 9$

#57 continued

Solution: (14,9)

59.

$\frac{2}{3}x - \frac{1}{2}y = 5; \qquad x = 6, \frac{3}{4}, -3$

$-\frac{1}{2}y = -\frac{2}{3}x + 5$

$y = \frac{4}{3}x - 10$

x	Work
6	$y = \frac{4}{3} \cdot 6 - 10 = 8 - 10 = -2$
	Solution: (6,-2)
$\frac{3}{4}$	$y = \frac{4}{3} \cdot \frac{3}{4} - 10 = 1 - 10 = -9$
	Solution: $\left(\frac{3}{4}, -9\right)$
-3	$y = \frac{4}{3}(-3) - 10 = -4 - 10 = -14$
	Solution: (-3,-14)

61.

$3y + 2 = 0; \qquad x = 5, 0, -2$

$3y = -2$

$y = -\frac{2}{3}$

x can be any real number, but y is always $-\frac{2}{3}$.

Solutions: $\left(5, -\frac{2}{3}\right), \left(0, -\frac{2}{3}\right), \left(-2, -\frac{2}{3}\right)$

63.

$x + 5y = 2; \qquad y = 1, 0, -1$

$x = -5y + 2$

y	Work
1	$x = -5 \cdot 1 + 2 = -5 + 2 = -3$
	Solution: (-3,1)
0	$x = -5 \cdot 0 + 2 = 2$
	Solution: (2,0)
-1	$x = -5(-1) + 2 = 5 + 2 = 7$
	Solution: (7,-1)

65.

$-x + 4y = 3; \qquad y = -2, 0, 3$

$-x = -4y + 3$

$x = 4y - 3$

#65 continued

y	Work
-2	$x = 4(-2) - 3 = -8 - 3 = -11$
	Solution: $(-11,-2)$
0	$x = 4 \cdot 0 - 3 = -3$
	Solution: $(-3,0)$
3	$x = 4 \cdot 3 - 3 = 12 - 3 = 9$
	Solution: $(9,3)$

67.

$3x + 4y = 12;$ $y = 3,0,-6$

$3x = -4y + 12$

$x = -\frac{4}{3}y + 4$

y	Work
3	$x = -\frac{4}{3} \cdot 3 + 4 = -4 + 4 = 0$
	Solution: $(0,3)$
0	$x = -\frac{4}{3} \cdot 0 + 4 = 4$
	Solution: $(4,0)$
-6	$x = -\frac{4}{3}(-6) + 4 = 8 + 4 = 12$
	Solution: $(12,-6)$

69.

$2x - 8y = 5;$ $y = \frac{1}{8}, 1, -\frac{3}{4}$

$2x = 8y + 5$

$x = 4y + \frac{5}{2}$

y	Work
$\frac{1}{8}$	$x = 4 \cdot \frac{1}{8} + \frac{5}{2} = \frac{1}{2} + \frac{5}{2} = \frac{6}{2} = 3$
	Solution: $\left(3, \frac{1}{8}\right)$
1	$x = 4 \cdot 1 + \frac{5}{2} = 4 + \frac{5}{2} = \frac{13}{2}$
	Solution: $\left(\frac{13}{2}, 1\right)$
$-\frac{3}{4}$	$x = 4\left(-\frac{3}{4}\right) + \frac{5}{2} = -3 + \frac{5}{2} = -\frac{1}{2}$
	Solution: $\left(-\frac{1}{2}, -\frac{3}{4}\right)$

71.

$-4x + 9y + 4 = 0;$ $y = -1,-4,4$

#71 continued

$-4x = -9y - 4$

$x = \frac{9}{4}y + 1$

y	Work
-1	$x = \frac{9}{4}(-1) + 1 = -\frac{9}{4} + 1 = \frac{-5}{4}$
	Solution: $\left(-\frac{5}{4}, -1\right)$
-4	$x = \frac{9}{4}(-4) + 1 = -9 + 1 = -8$
	Solution: $(-8,-4)$
4	$x = \frac{9}{4} \cdot 4 + 1 = 9 + 1 = 10$
	Solution: $(10,4)$

73.

$-\frac{1}{4}x + \frac{3}{2}y = 2;$ $y = \frac{2}{3}, 2,-1$

$-\frac{1}{4}x = -\frac{3}{2}y + 2$

$x = 6y - 8$

y	Work
$\frac{2}{3}$	$x = 6 \cdot \frac{2}{3} - 8 = 4 - 8 = -4$
	Solution: $\left(-4, \frac{2}{3}\right)$
2	$x = 6 \cdot 2 - 8 = 12 - 8 = 4$
	Solution: $(4,2)$
-1	$x = 6(-1) - 8 = -6 - 8 = -14$
	Solution: $(-14,-1)$

75.

$3x - 4 = 0;$ $y = 8,0,-\frac{7}{2}$

$x = \frac{4}{3}$

y can be any real number, but x is always $\frac{4}{3}$.

Solution: $\left(\frac{4}{3}, 8\right)\left(\frac{4}{3}, 0\right)\left(\frac{4}{3}, \frac{-7}{2}\right)$

77.

$2L + 2W = 2200$

Let $W = 430$

$2L + 2(430) = 2200$

$2L + 860 = 2200$

$2L = 2200 - 860$

#77 continued

$$2L = 1340$$
$$L = 670$$
The length is 670 ft.

Let L = 850
$$2(850) + 2W = 2200$$
$$1700 + 2W = 2200$$
$$2W = 2200 - 1700$$
$$2W = 500$$
$$W = 250$$
The width is 250 ft.

79.
$$2L + 3W = 2500$$
Let W = 350
$$2L + 3(350) = 2500$$
$$2L + 1050 = 2500$$
$$2L = 2500 - 1050$$
$$2L = 1450$$
$$L = 725$$

The length is 725 ft.

$$2L + 3W = 2500$$
Let L = 830
$$2(830) + 3W = 2500$$
$$1660 + 3W = 2500$$
$$3W = 2500 - 1660$$
$$3W = 840$$
$$W = 280$$
The width is 280 ft.

81.
$$5F - 9C = 160$$
$$5(32) - 9C = 160$$
$$160 - 9C = 160$$
$$-9C = 160 - 160$$
$$-9C = 0$$
$$C = 0$$
The equivalent is 0°
$$5F - 9C = 160$$
$$5F - 9(100) = 160$$
$$5F - 900 = 160$$
$$5F = 160 + 900$$
$$5F = 1060$$
$$F = 212$$
The equivalent is 212°

1. - 13.
The answers do not require worked-out solutions.

1. - 27.
The graphs are in the answer section of the main text.

29.
$$x + y = 3$$
For the y-intercept let x = 0.
$$0 + y = 3$$
$$y = 3 ; (0,3)$$
For the x-intercept let y = 0.
$$x + 0 = 3$$
$$x = 3; (3,0)$$

31.
$$x - y = -6$$
For the y-intercept let x = 0.
$$0 - y = -6$$
$$y = 6; (0,6)$$
For the x-intercept let y = 0.
$$x - 0 = -6$$
$$x = -6; (-6,0)$$

33.
$$x + 4y = -4$$
For the y-intercept let x = 0.
$$0 + 4y = -4$$
$$y = -1; (0,-1)$$
For the x-intercept let y = 0.
$$x + 4 \cdot 0 = -4$$
$$x = -4; (-4,0)$$

35.
$$6x + y = 3$$
For the y-intercept let x = 0.
$$6 \cdot 0 + y = 3$$
$$y = 3; (0,3)$$
For the x-intercept let y = 0.
$$6x + 0 = 3$$
$$6x = 3$$

#35 continued

$$x = \frac{1}{2}; \left(\frac{1}{2}, 0\right)$$

37.

$3x + 4y = 12$

For the y-intercept let x = 0.

$3 \cdot 0 + 4y = 12$

$\qquad 4y = 12$

$\qquad y = 3; \ (0, 3)$

For the x-intercept let y = 0.

$3x + 4 \cdot 0 = 12$

$\qquad 3x = 12$

$\qquad x = 4; \ (4, 0)$

39.

$2x - 7y = 7$

For the y-intercept let x = 0.

$2 \cdot 0 - 7y = 7$

$\qquad -7y = 7$

$\qquad y = -1; \ (0, -1)$

For the x-intercept let y = 0.

$2x - 7 \cdot 0 = 7$

$\qquad 2x = 7$

$$x = \frac{7}{2}; \left(\frac{7}{2}, 0\right)$$

41.

$-5x + 9y = 12$

For the y-intercept let x = 0.

$-5 \cdot 0 + 9y = 12$

$\qquad 9y = 12$

$$y = \frac{4}{3}; \left(0, \frac{4}{3}\right)$$

For the x-intercept let y = 0.

$-5x + 9 \cdot 0 = 12$

$\qquad -5x = 12$

$$x = -\frac{12}{5}; \left(-\frac{12}{5}, 0\right)$$

43.

$y = \frac{2}{3}x + 4$

For the y-intercept let x = 0.

$$y = \frac{2}{3} \cdot 0 + 4$$

$\qquad y = 4; \ (0, 4)$

For the x-intercept let y = 0.

#43 continued

$$0 = \frac{2}{3}x + 4$$

$$-4 = \frac{2}{3}x$$

$-6 = x; \ (-6, 0)$

45.

$$\frac{2}{5}x - y = \frac{3}{2}$$

For the y-intercept let x = 0.

$$\frac{2}{5} \cdot 0 - y = \frac{3}{2}$$

$$y = -\frac{3}{2}; \left(0, -\frac{3}{2}\right)$$

For the x-intercept let y = 0.

$$\frac{2}{5}x - 0 = \frac{3}{2}$$

$$\frac{2}{5}x = \frac{3}{2}$$

$$x = \frac{15}{4}; \left(\frac{15}{4}, 0\right)$$

47.

$$\frac{1}{2}x + \frac{1}{3}y + 2 = 0$$

For the y-intercept let x = 0.

$$\frac{1}{2} \cdot 0 + \frac{1}{3}y = -2$$

$$\frac{1}{3}y = -2$$

$\qquad y = -6; \ (0, -6)$

For the x-intercept let y = 0.

$$\frac{1}{2}x + \frac{1}{3} \cdot 0 + 2 = 0$$

$$\frac{1}{2}x = -2$$

$\qquad x = -4; \ (-4, 0)$

49.

$0.6x + 2.7y = -1.8$

For the y-intercept let x = 0.

$0.6 \cdot 0 + 2.7y = -1.8$

$\qquad 2.7y = -1.8$

$$y = -\frac{1.8}{2.7}$$

$$y = -\frac{2}{3}; \left(0, -\frac{2}{3}\right)$$

#49 continued

For the x-intercept let $y = 0$.
$0.6x + 2.7 \cdot 0 = -1.8$
$\qquad 0.6x = -1.8$
$\qquad\qquad x = -3; \quad (-3,0)$

51.
$3x - 5y = 0$
For the y-intercept let $x = 0$.
$3 \cdot 0 - 5y = 0$
$\qquad -5y = 0$
$\qquad\quad y = 0; \quad (0,0)$
Note: $(0,0)$ is both the x and y intercept!
For the x-intercept let $y = 0$.
$3x - 5 \cdot 0 = 0$
$\qquad 3x = 0$
$\qquad\; x = 0; \quad (0,0)$

53.
$4x + 3y = 0$
For the y-intercept let $x = 0$.
$4 \cdot 0 + 3y = 0$
$\qquad\quad y = 0; \quad (0,0)$
For the x-intercept let $y = 0$.
$4x + 3 \cdot 0 = 0$
$\qquad\quad x = 0; \quad (0,0)$

55.
$y = x; \; y = 0; \; x = 0$

57.
$3x + y = 4$
$\quad y = -3x + 4$

x	y
0	4
1	1
-1	7

Check these points on the graph in the answers.

59.
$4x - y = -6$
$\quad -y = -4x - 6$
$\quad\; y = 4x + 6$

x	y
0	6
-1	2
-2	-2

#59 continued

Check these points on the graph in the answers.

61.
$-2x + 5y = 10$
$\qquad 5y = 2x + 10$
$\qquad\; y = \dfrac{2}{5}x + 2$

x	y
0	2
-5	0
5	4

Check these points on the graph in the answers.

63.
$6x - 4y = 3$
$\quad -4y = -6x + 3$
$\qquad y = \dfrac{3}{2}x - \dfrac{3}{4}$

x	y
0	$-\dfrac{3}{4}$
$\dfrac{1}{2}$	0
2	$\dfrac{9}{4}$

Check these points on the graph in the answers

65.
$5x + 4y + 8 = 0$
$\qquad 4y = -5x - 8$
$\qquad\; y = -\dfrac{5}{4}x - 2$

x	y
0	-2
4	-7
-4	3

Check these points on the graph in the answers.

67.
$\dfrac{7x - 3y}{5} = 3$
$\quad 7x - 3y = 15$

263

#67 continued

$$-3y = -7x + 15$$

$$y = \frac{7}{3}x - 5$$

x	y
0	-5
3	2
1	$-\frac{8}{3}$

Check these points on the graph in the
answers.

69.

$$\frac{3}{5}x + \frac{5}{2}y = \frac{7}{4}$$

$$\frac{5}{2}y = -\frac{3}{5}x + \frac{7}{4}$$

$$\frac{2}{5} \cdot \frac{5}{2}y = \frac{2}{5}\left(-\frac{3}{5}x + \frac{7}{4}\right)$$

$$y = -\frac{6}{25}x + \frac{7}{10}$$

x	y
0	$\frac{7}{10}$
5	$-\frac{1}{2}$
-5	$\frac{19}{10}$

Check these points on the graph in the answer
section of the main text.

73. - 79.
The graphs are in the answer section of the
main text.

Exercises 6.4

1.
(5, 3), (2, 1)

$$m = \frac{y_2 - y_1}{x_2 - x_1} = \frac{1 - 3}{2 - 5} = \frac{-2}{-3} = \frac{2}{3}$$

3.
(-4, 2), (-2, 6)

$$m = \frac{y_2 - y_1}{x_2 - x_1} = \frac{6 - 2}{-2 - (-4)} = \frac{4}{-2 + 4} = \frac{4}{2} = 2$$

5.
(2,-3), (-1, -7)

$$m = \frac{y_2 - y_1}{x_2 - x_1} = \frac{-7 - (-3)}{-1 - 2} = \frac{-7 + 3}{-3}$$

$$= \frac{-4}{-3} = \frac{4}{3}$$

7.
(4,-3), (2,-3)

$$m = \frac{y_2 - y_1}{x_2 - x_1} = \frac{-3 - (-3)}{2 - 4} = \frac{-3 + 3}{-2} = \frac{0}{2} = 0$$

9.
(1, 4), (1, -2)

$$m = \frac{y_2 - y_1}{x_2 - x_1} = \frac{-2 - 4}{1 - 1} = \frac{-6}{0}$$

Undefined

11.

$$\left(\frac{2}{3}, -1\right), \left(\frac{1}{6}, \frac{3}{2}\right)$$

$$m = \frac{y_2 - y_1}{x_2 - x_1} = \frac{\frac{3}{2} - (-1)}{\frac{1}{6} - \frac{2}{3}} = \frac{\frac{3}{2} + 1}{\frac{1}{6} - \frac{2}{3}}$$

$$= \frac{9 + 6}{1 - 4}$$

$$= \frac{15}{-3} = -5$$

13.
(2.3, 4), (-0.5, 0.5)

$$m = \frac{y_2 - y_1}{x_2 - x_1} = \frac{0.5 - 4}{-0.5 - 2.3} = \frac{-3.5}{-2.8} = 1.25$$

15.
(0, 3), (-5, 0)

$$m = \frac{y_2 - y_1}{x_2 - x_1} = \frac{0 - 3}{-5 - 0} = \frac{-3}{-5} = \frac{3}{5}$$

17.
4x - 3y = 7
Two ordered pair solutions are (1, -1) and
(-2, -5).

#17 continued

$$m = \frac{y_2 - y_1}{x_2 - x_1} = \frac{-5 - (-1)}{-2 - 1} = \frac{-5 + 1}{-3}$$
$$= \frac{-4}{-3} = \frac{4}{3}$$

19.
$7x + 5y = 10$
Two ordered pair solutions are (0,2) and (5,-5).
$$m = \frac{y_2 - y_1}{x_2 - x_1} = \frac{2 - (-5)}{0 - 5} = \frac{2 + 5}{-5} = -\frac{7}{5}$$

21.
$2x - 9y = -5$
Two ordered pair solutions are (2,1) and (-7,-1).
$$m = \frac{y_2 - y_1}{x_2 - x_1} = \frac{1 - (-1)}{2 - (-7)} = \frac{1 + 1}{9} = \frac{2}{9}$$

23.
$x + \frac{2}{3}y = -4$
Two ordered pair solutions are (0,-6) and (2,-9).
$$m = \frac{y_2 - y_1}{x_2 - x_1} = \frac{-6 - (-9)}{0 - 2} = \frac{-6 + 9}{0 - 2} = -\frac{3}{2}$$

25.
$y = 2x - 5$
Two ordered pair solutions are (0,-5) and (1,-3).
$$m = \frac{y_2 - y_1}{x_2 - x_1} = \frac{-3 - (-5)}{1 - 0} = \frac{-3 + 5}{1} = 2$$

27.
$y = -\frac{3}{4}x + 5$
Two ordered pair solutions are (0,5) and (4,2).
$$m = \frac{y_2 - y_1}{x_2 - x_1} = \frac{2 - 5}{4 - 0} = -\frac{3}{4}$$

29.
$y = \frac{4x - 3}{7}$

#29 continued

Two ordered pair solutions are (-1, -1) and (-8, -5).
$$m = \frac{y_2 - y_1}{x_2 - x_1} = \frac{-5 - (-1)}{-8 - (-1)} = \frac{-5 + 1}{-8 + 1} = \frac{4}{7}$$

31.
$x = \frac{3y + 6}{4}$
Two ordered pair solutions are (0,-2), and (3 ,2).
$$m = \frac{y_2 - y_1}{x_2 - x_1} = \frac{2 - (-2)}{3 - 0} = \frac{2 + 2}{3} = \frac{4}{3}$$

33.
$y = -4$
Two ordered pair solutions are (0,-4) and (1,-4).
$$m = \frac{y_2 - y_1}{x_2 - x_1} = \frac{-4 - (-4)}{1 - 0} = \frac{0}{1} = 0$$

35.
$2x - 3 = 0$
Two ordered pair solutions are
$\left(\frac{3}{2}, 0\right)$ and $\left(\frac{3}{2}, 1\right)$.
$$m = \frac{y_2 - y_1}{x_2 - x_1} = \frac{1 - 0}{\frac{3}{2} - \frac{3}{2}} = \frac{1}{0}$$

Undefined

---

37. - 59.
The answers do not require worked-out solutions.

Exercises 6.5

1.
$y = 3x + 2$
$y = mx + b$
$m = 3, b = 2$

3.
$4x + 5y = 10$
$\qquad 5y = -4x + 10$

#3 continued

$$y = -\frac{4}{5}x + 2$$
$$y = mx + b$$
$$m = -\frac{4}{5}, b = 2$$

5.
$$7x - 3y = 5$$
$$-3y = -7x + 5$$
$$y = \frac{7}{3}x - \frac{5}{3}$$
$$y = mx + b$$
$$m = \frac{7}{3}, b = -\frac{5}{3}$$

7.
$$x + 2y = -3$$
$$2y = -x - 3$$
$$y = -\frac{1}{2}x - \frac{3}{2}$$
$$y = mx + b$$
$$m = -\frac{1}{2}, b = -\frac{3}{2}$$

9.
$$2x - y + 7 = 0$$
$$-y = -2x - 7$$
$$y = 2x + 7$$
$$y = mx + b$$
$$m = 2, b = 7$$

11.
$$3y + 4 = 0$$
$$3y = 0x - 4$$
$$y = 0x - \frac{4}{3}$$
$$y = mx + b$$
$$m = 0, b = -\frac{4}{3}$$

13.
$$2x - 7 = 0$$
m is undefined

15.
$$y = mx + b$$
$$y = \frac{2}{3}x + 4$$

17.
$$m = -\frac{1}{2}, (3,-1)$$
$$y - y_1 = m(x - x_1)$$
$$y + 1 = -\frac{1}{2}(x - 3)$$
$$y + 1 = -\frac{1}{2}x + \frac{3}{2}$$
$$y = -\frac{1}{2}x + \frac{3}{2} - 1$$
$$y = -\frac{1}{2}x + \frac{1}{2}$$

19.
$$m = -3, (-1,-4)$$
$$y - y_1 = m(x - x_1)$$
$$y + 4 = -3(x + 1)$$
$$y + 4 = -3x - 3$$
$$y = -3x - 7$$

21.
$$m = \frac{3}{4}, (4,2)$$
$$y - y_1 = m(x - x_1)$$
$$y - 2 = \frac{3}{4}(x - 4)$$
$$y - 2 = \frac{3}{4}x - 3$$
$$y = \frac{3}{4}x - 1$$

23.
$$m = 0, (-5,6)$$
$$y - y_1 = m(x - x_1)$$
$$y - 6 = 0(x + 5)$$
$$y - 6 = 0x + 0$$
$$y = 0x + 6$$

25.
$$m = 2, (-3,0)$$
$$y - y_1 = m(x - x_1)$$
$$y - 0 = 2(x + 3)$$
$$y = 2x + 6$$

27.
Through (-5,2) and (-1,-2)

Step 1: $m = \dfrac{-2-2}{-1+5} = \dfrac{-4}{4} = -1$

Step 2: Use m = -1 and (-1, -2).

$y - y_1 = m(x - x_1)$
$y + 2 = -1(x + 1)$
$y + 2 = -x - 1$
$y = -x - 3$

29.
Through (5,1) and (2,3)

Step 1: $m = \dfrac{3-1}{2-5} = \dfrac{2}{-3}$

Step 2: Use $m = -\dfrac{2}{3}$ and (2,3).

$y - y_1 = m(x - x_1)$
$y - 3 = -\dfrac{2}{3}(x - 2)$
$y - 3 = -\dfrac{2}{3}x + \dfrac{4}{3}$
$y = -\dfrac{2}{3}x + \dfrac{4}{3} + 3$
$y = -\dfrac{2}{3}x + \dfrac{13}{3}$

31.
Through (2,4) and (-4,4)

Step 1: $m = \dfrac{4-4}{-4-2} = \dfrac{0}{-6} = 0$

Step 2: use m = 0 and (2,4).

$y - y_1 = m(x - x_1)$
$y - 4 = 0(x - 2)$
$y = 0x + 4$

33.
Through (4,0) and (0,3)

Step 1: $m = \dfrac{3-0}{0-4} = -\dfrac{3}{4}$

Step 2: use $m = -\dfrac{3}{4}$ and (4,0).

$y - y_1 = m(x - x_1)$
$y - 0 = -\dfrac{3}{4}(x - 4)$
$y = -\dfrac{3}{4}x + 3$

35.
$m = \dfrac{3}{5}$, (-1,2)

$y - y_1 = m(x - x_1)$
$y - 2 = \dfrac{3}{5}(x + 1)$
$y - 2 = \dfrac{3}{5}x + \dfrac{3}{5}$
$y = \dfrac{3}{5}x + \dfrac{3}{5} + 2$
$y = \dfrac{3}{5}x + \dfrac{13}{5}$
$5y = 3x + 13$
$-3x + 5y = 13$
$3x - 5y = -13$

37.
$m = -\dfrac{7}{2}$, (1,-5)

$y - y_1 = m(x - x_1)$
$y + 5 = -\dfrac{7}{2}(x - 1)$
$y + 5 = -\dfrac{7}{2}x + \dfrac{7}{2}$
$y = -\dfrac{7}{2}x + \dfrac{7}{2} - 5$
$y = -\dfrac{7}{2}x - \dfrac{3}{2}$
$2y = -7x - 3$
$7x + 2y = -3$

39.
m = 4, (-3,-7)

$y - y_1 = m(x - x_1)$
$y + 7 = 4(x + 3)$
$y + 7 = 4x + 12$
$y = 4x + 5$
$-4x + y = 5$
$4x - y = -5$

41.
m = 0, (6,3)

$y - y_1 = m(x - x_1)$
$y - 3 = 0(x - 6)$
$y - 3 = 0x$
$0x + y = 3$

43.
m is undefined, (4, -1)
The line is vertical passing through
(4, -1).
The equation is $x + 0y = 4$

45.
Through (-5, 8) and (-2, 2)
Step 1: $m = \dfrac{2 - 8}{-2 + 5} = \dfrac{-6}{3} = -2$
Step 2: Use m = -2 and (-2, 2).
$y - y_1 = m(x - x_1)$
$y - 2 = -2(x + 2)$
$y - 2 = -2x - 4$
$\quad\quad y = -2x - 2$
$2x + y = -2$

47.
Through (3, -6) and (5, 1)
Step 1: $m = \dfrac{1 + 6}{5 - 3} = \dfrac{7}{2}$
Step 2: Use $m = \dfrac{7}{2}$ and (5, 1).
$y - y_1 = m(x - x_1)$
$y - 1 = \dfrac{7}{2}(x - 5)$
$y - 1 = \dfrac{7}{2}x - \dfrac{35}{2}$
$\quad\quad y = \dfrac{7}{2}x - \dfrac{35}{2} + 1$
$\quad\quad y = \dfrac{7}{2}x - \dfrac{33}{2}$
$2y = 7x - 33$
$-7x + 2y = -33$
$7x - 2y = 33$

49.
Through (-4, 3) and (0, 3)
Step 1: $m = \dfrac{3 - 3}{-4 - 0} = \dfrac{0}{-4} = 0$
Step 2: Use m = 0 and (0, 3),
$y - y_1 = m(x - x_1)$
$y - 3 = 0(x - 0)$
$y - 3 = 0x$
$0x + y = 3$

51.
Through (-3, 2) and (-3, -6)
Step 1: $m = \dfrac{-6 - 2}{-3 + 3} = \dfrac{-8}{0}$
$\quad\quad$ m is undefined
The line is vertical passing through
(-3, 2) and (-3, -6).
The equation is $x + 0y = -3$

53.
Through (3, 0) and (0, -3)
Step 1: $m = \dfrac{-3 - 0}{0 - 3} = \dfrac{-3}{-3} = 1$
Step 2: Use m = 1 and (3, 0),
$y - y_1 = m(x - x_1)$
$y - 0 = 1(x - 3)$
$\quad\quad y = x - 3$
$-x + y = -3$
$x - y = 3$

55.
(1) $y = 3x - 4$ (2) $y = 3 - 4x$
$\quad\quad m = 3$ $\quad\quad\quad y = -4x + 3$
$\quad\quad\quad\quad\quad\quad\quad\quad m = -4$
Neither

57.
(1) $2x - 4y = 5$ $\quad\quad$ (2) $2x + y = -3$
$\quad\quad -4y = -2x + 5$ $\quad\quad\quad y = -2x - 3$
$\quad\quad\quad y = \dfrac{1}{2}x - \dfrac{5}{4}$ $\quad\quad\quad m = -2$
$\quad\quad\quad m = \dfrac{1}{2}$

Perpendicular: The slopes are negative
reciprocals.

59.
(1) $x - 3y = 4$ $\quad\quad$ (2) $-2x + 6y = 7$
$\quad\quad -3y = -x + 4$ $\quad\quad\quad 6y = 2x + 7$
$\quad\quad\quad y = \dfrac{1}{3}x - \dfrac{4}{3}$ $\quad\quad\quad y = \dfrac{1}{3}x + \dfrac{7}{6}$
$\quad\quad\quad m = \dfrac{1}{3}$ $\quad\quad\quad\quad m = \dfrac{1}{3}$

Parallel: The slopes are equal.

61.
(1) $5x + 3y = -2$ $\quad\quad$ (2) $-\dfrac{5}{3}x - y = 4$

#61 continued

$$3y = -5x - 2 \qquad -y = \frac{5}{3}x + 4$$

$$y = -\frac{5}{3}x - \frac{2}{3} \qquad y = -\frac{5}{3}x - 4$$

$$m = -\frac{5}{3} \qquad m = -\frac{5}{3}$$

Parallel: The slopes are equal.

63.

(1) $4x - 5y + 8 = 0$

$$-5y = -4x + 8$$

$$y = \frac{4}{5}x - \frac{8}{5}$$

$$m = \frac{4}{5}$$

(2) $2x - 10y - 3 = 0$

$$-10y = -2x + 3$$

$$y = \frac{1}{5}x - \frac{3}{10}$$

$$m = \frac{1}{5}$$

Neither

65.

(1) $3x + 5 = 0 \qquad$ (2) $4y - 7 = 0$

$\qquad 3x = -5 \qquad\qquad\quad 4y = 7$

$$x = -\frac{5}{3} \qquad\qquad y = \frac{7}{4}$$

m is undefined $\qquad m = \frac{7}{4}$

Perpendicular: Since one line is vertical and the other horizontal, the lines are perpendicular.

67.

$(-3 , 7); \; y = 5x + 4$

$m = 5$: parallel $\qquad m = -\frac{1}{5}$: perpendicular

Parallel $\qquad\qquad\qquad$ Perpendicular

$y - 7 = 5(x + 3) \qquad y - 7 = -\frac{1}{5}(x + 3)$

$y - 7 = 5x + 15 \qquad y - 7 = -\frac{1}{5}x - \frac{3}{5}$

$\qquad y = 5x + 22 \qquad\qquad y = -\frac{1}{5}x - \frac{3}{5} + 7$

$$\qquad\qquad\qquad\qquad\qquad y = -\frac{1}{5}x - \frac{3}{5} + \frac{35}{5}$$

#67 continued

$$y = -\frac{1}{5}x + \frac{32}{5}$$

69.

$(-1 , -2); \; 2x + 3y = 7$

$2x + 3y = 7$

$3y = -2x + 7$

$$y = -\frac{2}{3}x + \frac{7}{3}$$

$m = -\frac{2}{3}$: parallel $\qquad m = \frac{3}{2}$: perpendicular

Parallel $\qquad\qquad\qquad$ Perpendicular

$y + 2 = -\frac{2}{3}(x + 1) \qquad y + 2 = \frac{3}{2}(x + 1)$

$y + 2 = -\frac{2}{3}x - \frac{2}{3} \qquad y + 2 = \frac{3}{2}x + \frac{3}{2}$

$y = -\frac{2}{3}x - \frac{2}{3} - 2 \qquad y = \frac{3}{2}x + \frac{3}{2} - 2$

$y = -\frac{2}{3}x - \frac{2}{3} - \frac{6}{3} \qquad y = \frac{3}{2}x + \frac{3}{2} - \frac{4}{2}$

$y = -\frac{2}{3}x - \frac{8}{3} \qquad\quad y = \frac{3}{2}x - \frac{1}{2}$

71.

The exercise does not require a worked-out solution.

73. - 77.

The graphs are in the answer section of the main text.

Exercises 6.6

1. - 31.

The answers do not require worked-out solutions.

Review Exercises

1.

$3x - y = 8, \; (2,-2)$

$3 \cdot 2 -(-2) = 6 + 2 = 8$; yes

3.

$y = 2x - 6, \left(\dfrac{3}{2}, 3\right)$

$3 = 2\left(\dfrac{3}{2}\right) - 6 = 3 - 6 = -3;$ no

5.
$2.5x + 3.2 y = 4, \ (8, -5)$
$2.5 \cdot 8 + 3.2(-5) = 20.0 - 16.0 = 4;$ yes

7.
$6x - 5y = 10$

$(5, ?), \ 6 \cdot 5 - 5y = 10$
$\qquad 30 - 5y = 10$
$\qquad -5y = -20$
$\qquad y = 4, \ (5, 4)$

$(0, ?), \ 6 \cdot 0 - 5y = 10$
$\qquad -5y = 10$
$\qquad y = -2, \ (0, -2)$

$(?, 0) \quad 6x - 5 \cdot 0 = 10$
$\qquad 6x = 10$
$\qquad x = \dfrac{5}{3}, \left(\dfrac{5}{3}, 0\right)$

$(?, -8), \ 6x - 5(-8) = 10$
$\qquad 6x + 40 = 10$
$\qquad 6x = -30$
$\qquad x = -5, \ (-5, -8)$

9.
$3x - 4y + 9 = 0$

$(1, ?), \ 3 \cdot 1 - 4y + 9 = 0$
$\qquad -4y + 12 = 0$
$\qquad -4y = -12$
$\qquad y = 3, \ (1, 3)$

$\left(\dfrac{7}{3}, ?\right), \ 3\left(\dfrac{7}{3}\right) - 4y + 9 = 0$
$\qquad 7 - 4y + 9 = 0$
$\qquad -4y + 16 = 0$
$\qquad -4y = -16$
$\qquad y = 4, \left(\dfrac{7}{3}, 4\right)$

#9 continued

$(?, -6), \ 3x - 4(-6) + 9 = 0$
$\qquad 3x + 24 + 9 = 0$
$\qquad 3x + 33 = 0$
$\qquad 3x = -33$
$\qquad x = -11, \ (-11, -6)$

$\left(?, \dfrac{3}{4}\right), \ 3x - 4\left(\dfrac{3}{4}\right) + 9 = 0$
$\qquad 3x - 3 + 9 = 0$
$\qquad 3x = -6$
$\qquad x = -2, \left(-2, \dfrac{3}{4}\right)$

11.
$1.3x - 2y = 2.6$

$(4, ?), \ 1.3 \cdot 4 - 2y = 2.6$
$\qquad 5.2 - 2y = 2.6$
$\qquad -2y = -2.6$
$\qquad y = 1.3, \ (4, 1.3)$

$(-10, ?), \ 1.3(-10) - 2y = 2.6$
$\qquad -13 - 2y = 2.6$
$\qquad -2y = 15.6$
$\qquad y = -7.8, \ (-10, -7.8)$

$(?, 0), \ 1.3x - 2 \cdot 0 = 2.6$
$\qquad 1.3x = 2.6$
$\qquad x = 2, \ (2, 0)$

$(?, 0.65), \ 1.3x - 2(0.65) = 2.6$
$\qquad 1.3x - 1.30 = 2.6$
$\qquad 1.3x = 3.9$
$\qquad x = 3, \ (3, 0.65)$

13.
$4x - y = 7; \qquad\qquad x = -2, 0, 5$
$\qquad -y = -4x + 7$
$\qquad y = 4x - 7$

x	Work
-2	$y = 4(-2) - 7 = -8 - 7 = -15$
	Solution: (-2, -15)
0	$y = 4 \cdot 0 - 7 = -7$
	Solution: (0, -7)
5	$y = 4 \cdot 5 - 7 = 20 - 7 = 13$

#13 continued

Solution: (5, 13)

15.

$\frac{1}{4}x + \frac{2}{3}y = 6;$ 　　　　$x = 8, -4, \frac{8}{3}$

$\frac{2}{3}y = -\frac{1}{4}x + 6$

$y = -\frac{3}{8}x + 9$

x	Work
8	$y = -\frac{3}{8} \cdot 8 + 9 = -3 + 9 = 6$ Solution: (8, 6)
-4	$y = -\frac{3}{8}(-4) + 9 = \frac{3}{2} + 9 = \frac{21}{2}$ Solution: $\left(-4, \frac{21}{2}\right)$
$\frac{8}{3}$	$y = -\frac{3}{8} \cdot \frac{8}{3} + 9 = -1 + 9 = 8$ Solution: $\left(\frac{8}{3}, 8\right)$

17.

$3x - 5y = 9;$ 　　　　$y = -3, -2, 1$

$3x = 5y + 9$

$x = \frac{5}{3}y + 3$

y	Work
-3	$x = \frac{5}{3}(-3) + 3 = -5 + 3 = -2$ Solution: (-2, -3)
-2	$x = \frac{5}{3}(-2) + 3 = -\frac{10}{3} + 3 = -\frac{1}{3}$ Solution: $\left(-\frac{1}{3}, -2\right)$
1	$x = \frac{5}{3} \cdot 1 + 3 = \frac{5}{3} + 3 = \frac{14}{3}$ Solution: $\left(\frac{14}{3}, 1\right)$

19.

$P = 2W + 2L$

$72 = 2W + 2L$

Let W = 6

$72 = 2 \cdot 6 + 2L$

$72 = 12 + 2L$

$60 = 2L$

$30 = L$

The length is 30 ft.

Let L = 26

$72 = 2W + 2 \cdot 26$

$72 = 2W + 52$

$20 + 2W$

$10 = W$

The width is 10 ft.

---

21. - 31.

The answers do not require worked-out solutions.

33.

$-x + y = -5$

For the y-intercept let x = 0.

$0 + y = -5$

$y = -5; \quad (0, -5)$

For the x-intercept let y = 0.

$-x + 0 = -5$

$-x = -5$

$x = 5; \quad (5, 0)$

35.

$2x - 6y = -15$

For the y-intercept let x = 0.

$2 \cdot 0 - 6y = -15$

$y = \frac{5}{2}; \quad \left(0, \frac{5}{2}\right)$

For the x-intercept let y = 0.

$2x - 6 \cdot 0 = -15$

$2x = -15$

$x = -\frac{15}{2}; \quad \left(-\frac{15}{2}, 0\right)$

37.

$\frac{5}{6}x + \frac{3}{4}y = 5$

For the y-intercept let x = 0.

$\frac{5}{6} \cdot 0 + \frac{3}{4}y = 5$

$\frac{3}{4}y = 5$

$y = \frac{20}{3}; \left(0 \; \frac{20}{3}\right)$

For the x-intercept let y = 0.

$\frac{5}{6}x + \frac{3}{4} \cdot 0 = 5$

$\frac{5}{6}x = 5$

$x = 6; \; (6,0)$

39.

$y = -\frac{5}{8}x + 2$

For the y-intercept let x = 0.

$y = -\frac{5}{8} \cdot 0 + 2$

$y = 2; \; (0,2)$

For the x-intercept let y = 0.

$0 = -\frac{5}{8}x + 2$

$-2 = -\frac{5}{8}x$

$\frac{16}{5} = x; \left(\frac{16}{5} \; 0\right)$

41.

$-4x + y = -3$

$y = 4x - 3$

x	y
0	-3
1	1
2	5

Check these points on the graph in the answers.

43.

$9x + 2y + 6 = 0$

$2y = -9x - 6$

#43 continued

$y = -\frac{9}{2}x - 3$

x	y
0	-3
-2	6

Check these points on the graph in the answers.

45.

$(4, -5), (2, -1)$

$m = \frac{y_2 - y_1}{x_2 - x_1} = \frac{-1 -(-5)}{2 - 4} = \frac{-1 + 5}{-2}$

$= \frac{4}{-2} = -2$

47.

$\left(\frac{1}{4}, -2\right), \left(\frac{7}{4}, 1\right)$

$m = \frac{y_2 - y_1}{x_2 - x_1} = \frac{1 -(-2)}{\frac{7}{4} - \frac{1}{4}} = \frac{3}{\frac{6}{4}} = 3 \cdot \left(\frac{2}{3}\right) = 2$

49.

$8x + 5y = 2$

Two ordered pair solutions are
$(4, -6), (-6, 10)$.

$m = \frac{y_2 - y_1}{x_2 - x_1} = \frac{-6 - 10}{4 -(-6)} = \frac{-16}{4 + 6}$

$= \frac{-16}{10} = -\frac{8}{5}$

51.

$x + \frac{7}{4}y = -7$

Two ordered pair solutions are $(7, -8)$ and $(0, -4)$.

$m = \frac{y_2 - y_1}{x_2 - x_1} = \frac{-8 -(-4)}{7 - 0} = \frac{-8 + 4}{7} = -\frac{4}{7}$

53.

$x - 3 = 0$

The slope is undefined.

55. - 59.
The answers do not require worked-out solutions.

61.

$y = \dfrac{2}{5} x - 2$

$y = mx + b$

$m = \dfrac{2}{5}, \ b = -2$

63.

$2x + 4y - 7 = 0$

$4y = -2x + 7$

$y = -\dfrac{1}{2} x + \dfrac{7}{4}$

$y = mx + b$

$m = -\dfrac{1}{2}, \ b = \dfrac{7}{4}$

65.

$m = -\dfrac{1}{3}, \ (-4, 2)$

$y - y_1 = m(x - x_1)$

$y - 2 = -\dfrac{1}{3} (x + 4)$

$y - 2 = -\dfrac{1}{3} x - \dfrac{4}{3}$

$y = -\dfrac{1}{3} x - \dfrac{4}{3} + 2$

$y = -\dfrac{1}{3} x + \dfrac{2}{3}$

67.

$m = 0, \ (3, -6)$

$y - y_1 = m(x - x_1)$

$y + 6 = 0(x - 3)$

$y = 0x - 6$

69.
Through $(-1, 5)$ and $(2, 6)$

Step 1: $m = \dfrac{5 - 6}{-1 - 2} = \dfrac{1}{3}$

Step 2: Use $m = \dfrac{1}{3}$ and $(2, 6)$.

$y - y_1 = m(x - x_1)$

$y - 6 = \dfrac{1}{3} (x - 2)$

#69 continued

$y - 6 = \dfrac{1}{3} x - \dfrac{2}{3}$

$y = \dfrac{1}{3} x - \dfrac{2}{3} + 6$

$y = \dfrac{1}{3} x + \dfrac{16}{3}$

71.

$m = \dfrac{6}{5}, \quad (-5, -2)$

$y - y_1 = m(x - x_1)$

$y + 2 = \dfrac{6}{5} (x + 5)$

$y + 2 = \dfrac{6}{5} x + 6$

$y = \dfrac{6}{5} x + 4$

$5y = 6x + 20$

$-6x + 5y = 20$

$6x - 5y = -20$

73.

$m = 0, \ (9, 3)$

$y - y_1 = m(x - x_1)$

$y - 3 = 0(x - 9)$

$y - 3 = 0x$

$0x + y = 3$

75.
Through $(-3, 4)$ and $(6, -2)$

Step 1: $m = \dfrac{-2 - 4}{6 + 3} = \dfrac{-6}{9} = -\dfrac{2}{3}$

Step 2: Use $m = -\dfrac{2}{3}$ and $(-3, 4)$.

$y - y_1 = m(x - x_1)$

$y - 4 = -\dfrac{2}{3} (x + 3)$

$y - 4 = -\dfrac{2}{3} x - 2$

$y = -\dfrac{2}{3} x + 2$

$3y = -2x + 6$

$2x + 3y = 6$

77.

(1) $3x - 4y = 9$

$\quad -4y = -3x + 9$

$\quad y = \dfrac{3}{4}x - \dfrac{9}{4}$

$\quad m = \dfrac{3}{4}$

(2) $-\dfrac{3}{8}x + \dfrac{1}{2}y = 5$

$\quad \dfrac{1}{2}y = \dfrac{3}{8}x + 5$

$\quad y = \dfrac{3}{4}x + 10$

$\quad m = \dfrac{3}{4}$

Parallel: The slopes are equal.

79.

(1) $6x + 8y = 7$

$\quad 8y = -6x + 7$

$\quad y = -\dfrac{3}{4}x + \dfrac{7}{8}$

$\quad m = -\dfrac{3}{4}$

(2) $4x - 3y = -3$

$\quad -3y = -4x - 3$

$\quad y = \dfrac{4}{3}x + 1$

$\quad m = \dfrac{4}{3}$

Perpendicular: The slopes are negative reciprocals.

81. - 87.
The answers do not require worked-out solutions.

Chapter 6 Test Solutions

1.

The exercise does not require a worked out solution.

3-5.

The graphs are in the answer section of the main text.

7.

$(-3, 5), (-1, -5)$

#7 continued

$m = \dfrac{-5 - 5}{-1 + 3} = \dfrac{-10}{2} = -5$

9.

$7x - 3y = 9$

$\quad -3y = -7x + 9$

$\quad y = \dfrac{7}{3}x - 3$

$\quad m = \dfrac{7}{3}$

11.

The graph is in the answer section of the main text.

13.

This is a vertiacal line passing through $x = -6$.

15.

$(4, -5), (1, -5)$

This is a horizontal line passing through $y = -5$.

17.

The graph is in the answer section of the main text.

Test Your Memory

1—7.

The graphs are in the answer section of the main text.

9.

$5(x - 2) + 6(x + 1) = 4(x - 1)$

$\quad 5x - 10 + 6x + 6 = 4x - 4$

$\quad\quad\quad 11x - 4 = 4x - 4$

$\quad\quad\quad\quad 7x - 4 = -4 \quad\quad$ Subtract 4x

$\quad\quad\quad\quad\quad 7x = 0 \quad\quad$ Add 4

$\quad\quad\quad\quad\quad x = 0 \quad\quad$ Divide by 7

Solution set: {0}

11.

$$(2x + 5)(x + 1) = 2$$
$$2x^2 + 7x + 5 = 2$$
$$2x^2 + 7x + 3 = 0 \qquad \text{Subtract 2}$$
$$(2x + 1)(x + 3) = 0$$
$$2x + 1 = 0 \ \text{or} \ x + 3 = 0$$
$$x = -\frac{1}{2} \ \text{or} \ x = -3$$

Solution set: $\left\{-\dfrac{1}{2}, -3\right\}$

13.

$$\frac{5}{3} = \frac{3x}{x - 2}$$
$$5(x - 2) = 9x \qquad \text{Cross multiply}$$
$$5x - 10 = 9x$$
$$-10 = 4x \qquad \text{Subtract 5x}$$
$$-\frac{10}{4} = x \qquad \text{Divide by 4}$$
$$-\frac{5}{2} = x$$

Solution set: $\left\{-\dfrac{5}{2}\right\}$

15.

$$\frac{5}{x + 1} - \frac{3}{x} = \frac{1}{3x} \qquad \text{LCD} = 3x(x + 1)$$
$$3x(x + 1)\left(\frac{5}{x + 1} - \frac{3}{x}\right) = \frac{1}{3x}[3x(x + 1)]$$
$$15x - 9(x + 1) = x + 1$$
$$15x - 9x - 9 = x + 1$$
$$6x - 9 = x + 1$$
$$5x - 9 = 1 \qquad \text{Subtract x}$$
$$5x = 10 \qquad \text{Add 9}$$
$$x = \frac{10}{5} \qquad \text{Divide by 5}$$
$$x = 2$$

Solution set: {2}

17.

$$\frac{x - 2}{x + 3} - \frac{1}{x + 2} = \frac{5}{x^2 + 5x + 6}$$
$$\frac{x - 2}{x + 3} - \frac{1}{x + 2} = \frac{5}{(x + 3)(x + 2)}$$
$$\text{LCD} = (x + 3)(x + 2)$$

#17 continued

$$(x + 3)(x + 2)\left(\frac{x - 2}{x + 3} - \frac{1}{x + 2}\right)$$
$$= \frac{5}{(x + 3)(x + 2)}(x + 3)(x + 2)$$
$$(x + 2)(x - 2) - (x + 3) = 5$$
$$x^2 - 4 - x - 3 = 5$$
$$x^2 - x - 7 = 5$$
$$x^2 - x - 12 = 0$$
$$(x + 3)(x - 4) = 0$$
$$x + 3 = 0 \ \text{or} \ x - 4 = 0$$
$$x = -3 \ \text{or} \ x = 4$$

Solution set: {4}
(-3 is an extraneous solution.)

19.

$$\left(\frac{x^3 y^{-4}}{x^5 y^{-1}}\right)^{-3} = \frac{x^{-9} y^{12}}{x^{-15} y^3}$$
$$= \frac{x^{-9 + 15} y^{12 - 3}}{1} = x^6 y^9$$

21.

$$\frac{16x^2 + 24xy + 9y^2}{16x^2 - 9y^2}$$
$$= \frac{(4x + 3y)(4x + 3y)}{(4x + 3y)(4x - 3y)} = \frac{(4x + 3y)}{(4x - 3y)}$$

23.

$$\frac{\dfrac{1}{4} - \dfrac{1}{2x}}{\dfrac{1}{4} - \dfrac{1}{x^2}} \qquad \text{LCD} = 4x^2$$
$$= \frac{4x^2\left(\dfrac{1}{4} - \dfrac{1}{2x}\right)}{4x^2\left(\dfrac{1}{4} - \dfrac{1}{x^2}\right)} = \frac{x^2 - 2x}{x^2 - 4} = \frac{x(x - 2)}{(x + 2)(x - 2)} = \frac{x}{x + 2}$$

25.

$$6x^2 y^3(2x^4 - 7xy^2 - 3y^3)$$
$$= 12x^6 y^3 - 42x^3 y^5 - 18x^2 y^6$$

27.

$$3x - 2 \overline{\smash{\big)}\, 12x^3 - 11x^2 - 10x + 6} \quad \frac{4x^2 - x - 4}{}$$

$$\underline{-12x^3 \pm 8x^2} \qquad \text{Change}$$

$$-3x^2 - 10x$$

$$\underline{\pm 3x^2 \mp 2x} \qquad \text{Change}$$

$$-12x + 6$$

$$\underline{\pm 12x \mp 8} \qquad \text{Change}$$

$$-2$$

Answer: $4x^2 - x - 4 - \dfrac{2}{3x-2}$

29.

$$\frac{2x^2 - 3x}{x^2 + 2x + 1} + \frac{2x - 3}{x^2 + 2x + 1}$$

$$= \frac{2x^2 - 3x + 2x - 3}{x^2 + 2x + 1} = \frac{2x^2 - x - 3}{x^2 + 2x + 1}$$

$$= \frac{(2x - 3)(x + 1)}{(x + 1)(x + 1)} = \frac{2x - 3}{x + 1}$$

31.

$$\frac{3}{4xy} - \frac{9}{4x^2} \qquad \text{LCD} = 4x^2y$$

$$= \frac{x}{x} \cdot \frac{3}{4xy} - \frac{9}{4x^2} \cdot \frac{y}{y}$$

$$= \frac{3x}{4x^2y} - \frac{9y}{4x^2y} = \frac{3x-9y}{4x^2y}$$

33.

$$\frac{5}{x^2 + 4x + 4} - \frac{1}{x^2 - x - 6}$$

$$= \frac{5}{(x + 2)(x + 2)} - \frac{1}{(x - 3)(x + 2)}$$

$$\qquad\qquad \text{LCD} = (x + 2)^2(x - 3)$$

$$= \frac{x-3}{x-3} \cdot \frac{5}{(x+2)^2} - \frac{1}{(x-3)(x+2)} \cdot \frac{x+2}{x+2}$$

$$= \frac{5(x - 3) - (x + 2)}{(x + 2)^2(x - 3)}$$

$$= \frac{5x - 15 - x - 2}{(x + 2)^2(x - 3)} = \frac{4x - 17}{(x + 2)^2(x - 3)}$$

35.

$(-3, 8), (7, -4)$

$$m = \frac{-4 - 8}{7 + 3} = \frac{-12}{10} = -\frac{6}{5}$$

37.

$$2x - 5y = 8$$

$$-5y = -2x + 8$$

$$y = \frac{2}{5}x - \frac{8}{5}$$

$$m = \frac{2}{5}$$

39.

$$m = \frac{3}{4}, \ (-1, 2)$$

$$y - 2 = \frac{3}{4}(x + 1)$$

$$4y - 8 = 3(x + 1) \qquad \text{Multiply by 4}$$

$$4y - 8 = (3x + 3)$$

$$4y = 3x + 11 \qquad \text{Add 8}$$

$$-11 = 3x - 4y \qquad \text{Subtract 11 and 4y}$$

$$3x - 4y = -11$$

41.

$(7, -4), (-5, 4)$

$$m = \frac{4 + 4}{-5 - 7} = \frac{8}{-12} = -\frac{2}{3}$$

$$y - 4 = -\frac{2}{3}(x + 5)$$

$$3y - 12 = -2(x + 5) \qquad \text{Multiply by 3}$$

$$3y - 12 = -2x - 10$$

$$2x + 3y = 2 \qquad \text{Add 3x and 12}$$

43.

$$3x - 2y = -8 \qquad\qquad 2x + 3y = -3$$

$$-2y = -3x - 8 \qquad\qquad 3y = -2x - 3$$

$$y = \frac{3}{2}x + 4 \qquad\qquad y = -\frac{2}{3}x - 1$$

$$m = \frac{3}{2} \qquad\qquad m = -\frac{2}{3}$$

They are perpendicular.

45.

Let x = the number of dimes.

Then $18 - x$ = the number of quarters.

#45 continued

Thus,

$$10x + 25(18 - x) = 270$$
$$10x + 450 - 25x = 270$$
$$-15x = -180$$
$$x = 12$$
$$18 - x = 6$$

Answer: 12 dimes, 6 quarters

47.

Let x = measurement of the width.

Then 3x - 4 = measurement of the length.

Thus, $A = W \cdot L$
$$32 = x(3x - 4)$$
$$32 = 3x^2 - 4x$$
$$0 = 3x^2 - 4x - 32$$
$$0 = (3x + 8)(x - 4)$$
$$x = 4 \text{ or } \cancel{x = -\frac{8}{3}}$$
$$3x - 4 = 8$$

Answer: 4 x 8 feet

49.

$$\frac{1}{15} - \frac{1}{20} = \frac{1}{x} \qquad LCD = 60x$$

$$60x\left(\frac{1}{15} - \frac{1}{20}\right) = \frac{1}{x} \cdot 60x$$

$$4x - 3x = 60$$

$$x = 60$$

Answer: 60 hr

CHAPTER 6 STUDY GUIDE

Self-Test Exercises

1. Find solutions for the equation $5x - 4y = -3$ using the following values for x or y.

$$(5,), (, -1), (0,), (, 0)$$

2. Solve the equation $4x + 7y = 14$ for y, then find solutions using $x = 0$, $x = 7$, $x = -7$.

Graph the following equations.

3. $x = 2$

4. $5x - 7y = -35$

5. $y = 4$

6. $3x + 7y = -21$, using intercepts.

Find the slopes of the lines passing through the following pairs of points.

7. $(-1, 2), (3, -4)$

8. $(5, 7), (-2, 5)$

Find the slopes of the lines whose equations are given.

9. $3x - 7y = 14$

10. $x + 5 = 0$

11. Graph the line with $m = \dfrac{-3}{2}$, passing through $(-1, 2)$.

Find the equations of the lines satisfying the given information. Write your answers in standard form.

12. $m = -\dfrac{1}{2}$, passing through $(-4, 3)$.

13. m is 0, passing through $(-3, 5)$.

14. Passing through $(-1, -2)$ and $(-7, -4)$.

15. Passing through $(2, 5)$ and $(2, -7)$.

16. Determine whether the lines $5y = -4x + 10$ and $4y = 5x + 4$ are perpendicular, parallel, or neither.

Graph the following inequalities.

17. $3x + 2y \geq 6$

18. $5y - 2x < 10$

The worked-out solutions begin on the next page.

Self-Test Solutions	#2 continued

1.

For (5,): $5 \cdot 5 - 4y = -3$

$25 - 4y = -3$

$-4y = -3 - 25$

$-4y = -28$

$y = 7$

Answer: (5, 7).

For (, -1): $5x - 4(-1) = -3$

$5x + 4 = -3$

$5x = -7$

$x = -\frac{7}{5}$

Answer: $(-\frac{7}{5}, -1)$.

For (0,): $5 \cdot 0 - 4y = -3$

$-4y = -3$

$y = \frac{3}{4}$

Answer: $(0, \frac{3}{4})$.

For (, 0): $5x - 4 \cdot 0 = -3$

$5x = -3$

$x = -\frac{3}{5}$

Answer: $(-\frac{3}{5}, 0)$.

2.

$4x + 7y = 14$

$7y = -4x + 14$

$y = -\frac{4}{7}x + 2$

For x = 0: $y = -\frac{4}{7} \cdot 0 + 2$

$y = 2$

Answer: (0, 2).

For x = 7: $y = -\frac{4}{7} \cdot 7 + 2$

$y = -4 + 2$

$y = -2$

Answer: (7, -2).

#2 continued

For x = -7: $y = -\frac{4}{7}(-7) + 2$

$y = 4 + 2$

$y = 6$

Answer: (-7, 6).

3.

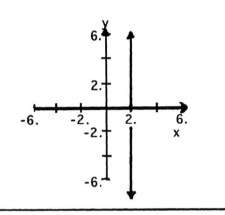

4.

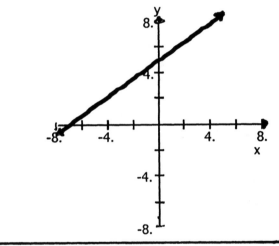

5.

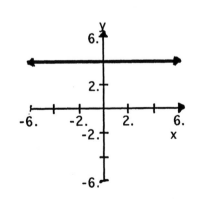

279

6.

$3x + 7y = -21$

For the y-intercept let x = 0.

$3 \cdot 0 + 7y = -21$

$y = -3;\quad (0, -3)$

For the x-intercept let y = 0.

$3x + 7 \cdot 0 = -21$

$3x = -21$

$x = -7;\quad (-7, 0)$

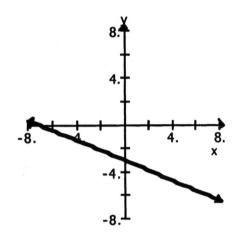

7.

$(-1, 2), (3, -4)$

$m = \dfrac{-4 - 2}{3 + 1} = \dfrac{-6}{4} = -\dfrac{3}{2}$

8.

$(5, 7), (-2, 5)$

$m = \dfrac{5 - 7}{-2 - 5} = \dfrac{-2}{-7} = \dfrac{2}{7}$

9.

$3x - 7y = 14$

$-7y = -3x + 14$

$y = \dfrac{3}{7}x - \dfrac{14}{7}$

$m = \dfrac{3}{7}$

10.

$x + 5 = 0$

$x = -5$

Slope is undefined.

11.

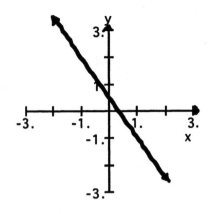

12.

$m = -\dfrac{1}{2},\ (-4,3)$

$y - y_1 = m(x - x_1)$

$y - 3 = -\dfrac{1}{2}(x + 4)$

$y - 3 = -\dfrac{1}{2}x - 2$

$y = -\dfrac{1}{2}x - 2 + 3$

$y = -\dfrac{1}{2}x + 1$

$\dfrac{1}{2}x + y = 1$

$x + 2y = 2$

13.

$y = 5$

14.

$(-1, -2), (-7, -4)$

$m = \dfrac{-4 + 2}{-7 + 1} = \dfrac{-2}{-6} = \dfrac{1}{3}$

$y - y_1 = m(x - x_1)$

$y + 2 = \dfrac{1}{3}(x + 1)$

$y + 2 = \dfrac{1}{3}x + \dfrac{1}{3}$

$y = \dfrac{1}{3}x + \dfrac{1}{3} - 2$

$y = \dfrac{1}{3}x - \dfrac{5}{3}$

$3y = x - 5$

$-x + 3y = -5$

$x - 3y = 5$

15.

(2, 5), (2, -7)

$m = \dfrac{-7 - 5}{2 - 2} = \dfrac{-12}{0}$

m is undefined, therefore, x = 2.

16.

$$5y = -4x + 10$$
$$y = -\frac{4x}{5} + 2$$
$$4y = 5x + 4$$
$$y = \frac{5}{4}x + 1$$

Since $-\dfrac{4}{5} \cdot \dfrac{5}{4} = -1$, the lines are perpendicular.

17.

$3x + 2y \geq 6$

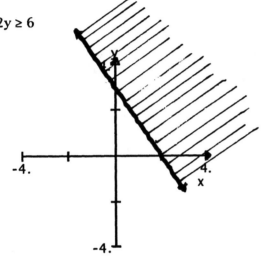

18.

$5y - 2x < 10$

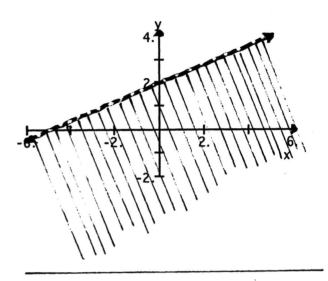

CHAPTER 7 SYSTEMS OF LINEAR EQUATIONS AND INEQUALITIES IN TWO VARIABLES

Solutions to Text Exercises

Exercises 7.1

1.
(1) $2x - y = 1$, (4,7)
   $2 \cdot 4 - 7 = 8 - 7 = 1$

(2) $x - y = -3$, (4,7)
   $4 - 7 = -3$
(4,7) is the solution to the system.

3.
(1) $4x + y = 0$, (-1,4)
   $4(-1) + 4 = -4 + 4 = 0$

(2) $3x - y = 7$, (-1,4)
   $3(-1) - 4 = -3 - 4 = -7 \neq 7$
(-1,4) is not the solution to the system.

5.
(1) $x - 5y = -23$, (-8,-3)
   $-8 - 5(-3) = -8 + 15 = 7 \neq -23$

(2) $x + y = -11$, (-8,-3)
   $-8 + (-3) = -11$
(-8,-3) is not the solution to the system.

7.
(1) $7x + 2y = -1$, (3,-11)
   $7 \cdot 3 + 2(-11) = 21 - 22 = -1$

(2) $2x - 3y = 39$, (3,-11)
   $2 \cdot 3 - 3(-11) = 6 + 33 = 39$
(3,-11) is the solution to the system.

9.
(1) $6x + y = 1$, $\left(\dfrac{1}{2}, -2\right)$

   $6 \cdot \dfrac{1}{2} - 2 = 3 - 2 = 1$

(2) $10x + 3y = -1$, $\left(\dfrac{1}{2}, -2\right)$

   $10 \cdot \dfrac{1}{2} + 3(-2) = 5 - 6 = -1$

#9 continued

$\left(\dfrac{1}{2}, -2\right)$ is the solution to the system.

11.

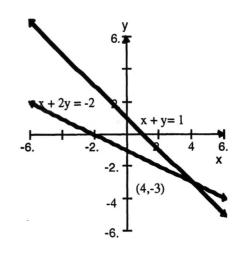

13.

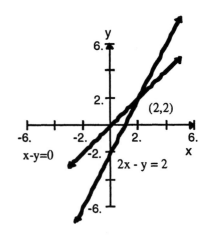

15.

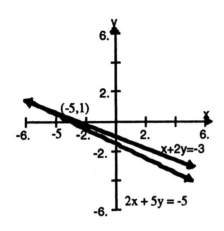

(-5,1)

x+2y=-3

2x + 5y = -5

21.

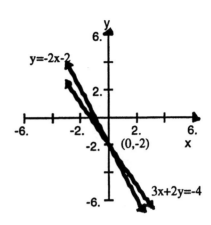

y=-2x-2

(0,-2)

3x+2y=-4

17.

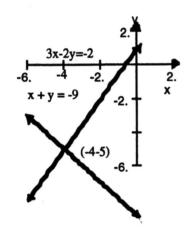

3x-2y=-2

x + y = -9

(-4-5)

23.

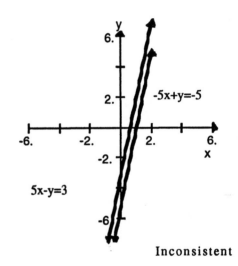

-5x+y=-5

5x-y=3

Inconsistent

19.

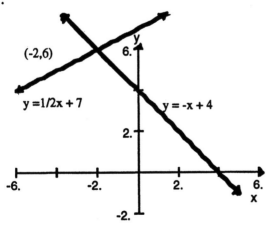

(-2,6)

y =1/2x + 7

y = -x + 4

25.

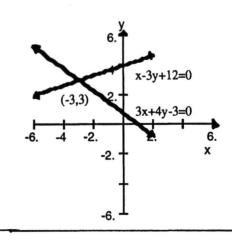

x-3y+12=0

(-3,3)

3x+4y-3=0

27.

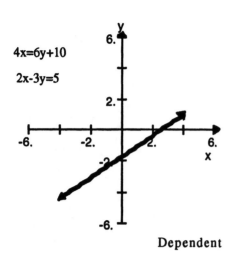

$4x=6y+10$

$2x-3y=5$

Dependent

33.

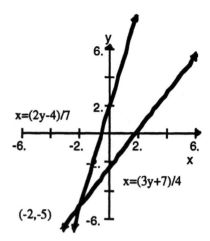

$x=(2y-4)/7$

$x=(3y+7)/4$

$(-2,-5)$

29.

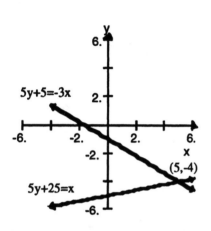

$5y+5=-3x$

$5y+25=x$

$(5,-4)$

35.

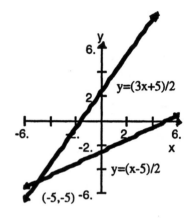

$y=(3x+5)/2$

$y=(x-5)/2$

$(-5,-5)$

31.

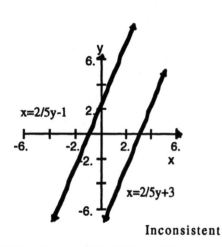

$x=2/5y-1$

$x=2/5y+3$

Inconsistent

37.

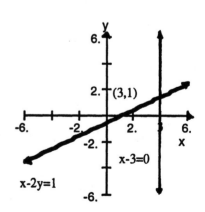

$(3,1)$

$x-3=0$

$x-2y=1$

284

39.

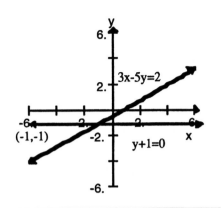

3x-5y=2

y+1=0

(-1,-1)

41.
a)

(1) $2x - 3y = 3$, $\left(\dfrac{1}{2}, -\dfrac{2}{3}\right)$

$2\left(\dfrac{1}{2}\right) - 3\left(-\dfrac{2}{3}\right) = 1 + 2 = 3$

(2) $\dfrac{2}{3}x + 5y = -3$, $\left(\dfrac{1}{2}, -\dfrac{2}{3}\right)$

$\dfrac{2}{3}\left(\dfrac{1}{2}\right) + 5\left(-\dfrac{2}{3}\right) = \dfrac{1}{3} - \dfrac{10}{3} = -3$

$\left(\dfrac{1}{2}, -\dfrac{2}{3}\right)$ is the solution.

b) The system is not dependent since it has only one solution.

43.
a)

(1) $4x + 12y = -9$, $\left(\dfrac{3}{4}, -1\right)$

$4\left(\dfrac{3}{4}\right) + 12(-1) = 3 - 12 = -9$

(2) $x + 3y + \dfrac{9}{4} = 0$, $\left(\dfrac{3}{4}, -1\right)$

$\dfrac{3}{4} + 3(-1) + \dfrac{9}{4} = \dfrac{3}{4} - \dfrac{12}{4} + \dfrac{9}{4} = 0$

(1) $4x + 12y = -9$, $\left(-\dfrac{3}{2}, -\dfrac{1}{4}\right)$

$4\left(-\dfrac{3}{2}\right) + 12\left(-\dfrac{1}{4}\right) = -6 - 3 = -9$

#43 continued

(2) $x + 3y + \dfrac{9}{4} = 0$, $\left(-\dfrac{3}{2}, -\dfrac{1}{4}\right)$

$-\dfrac{3}{2} + 3\left(-\dfrac{1}{4}\right) + \dfrac{9}{4} = -\dfrac{6}{4} - \dfrac{3}{4} + \dfrac{9}{4} = 0$

b) There is more than one solution, so the system is dependent.

Exercises 7.2

1.
(1) $x = 4$
(2) $3x - 2y = 2$

Substitute 4 for x in (2).
$3 \cdot 4 - 2y = 2$
$12 - 2y = 2$
$-2y = -10$
$y = 5$

Solution set: $\{(4,5)\}$

3.
(1) $x + 4y = 4$
(2) $y = \dfrac{3}{2}$

Substitute $\dfrac{3}{2}$ for y in (1).

$x + 4\left(\dfrac{3}{2}\right) = 4$
$x + 6 = 4$
$x = -2$

Solution set: $\left\{\left(-2, \dfrac{3}{2}\right)\right\}$

5.
(1) $x = 7y + 1$
(2) $2x - 5y = -7$

Substitute $7y + 1$ for x in (2).
$2(7y + 1) - 5y = -7$
$14y + 2 - 5y = -7$
$2 + 9y = -7$
$9y = -9$
$y = -1$

#5 continued

Substitute -1 for y in (1).

$x = 7(-1) + 1$

$x = -7 + 1$

$x = -6$

Solution set: $\{(-6,-1)\}$

7.

(1) $3x - 2y = 5$

(2) $x = \dfrac{2}{3}y + 3$

Substitute $\dfrac{2}{3}y + 3$ for x in (1).

$3\left(\dfrac{2}{3}y + 3\right) - 2y = 5$

$2y + 9 - 2y = 5$

$9 \neq 5$

No solution.
Inconsistant

9.

(1) $y = 2 - 4x$

(2) $9x + 2y = 2$

Substitute $2 - 4x$ for y in (2).

$9x + 2(2 - 4x) = 2$

$9x + 4 - 8x = 2$

$x + 4 = 2$

$x = -2$

Substitute -2 for x in (1).

$y = 2 - 4(-2)$

$y = 2 + 8$

$y = 10$

Solution set: $\{(-2,10)\}$

11.

(1) $5x + 4y = 1$

(2) $y = -\dfrac{3}{2}x + 1$

Substitute $-\dfrac{3}{2}x + 1$ for y in (1).

#11. continued.

$5x + 4\left(-\dfrac{3}{2}x + 1\right) = 1$

$5x - 6x + 4 = 1$

$-x + 4 = 1$

$-x = -3$

$x = 3$

Substitute 3 for x in (2).

$y = -\dfrac{3}{2} \cdot 3 + 1$

$y = -\dfrac{9}{2} + 1$

$y = -\dfrac{7}{2}$

Solution set: $\left\{\left(3, -\dfrac{7}{2}\right)\right\}$

13.

(1) $x + y = -1$

(2) $5x - y = 5$

Solve (1) for y.

(3) $y = -x - 1$

Substitute $-x - 1$ for y in (2).

$5x - (-x - 1) = 5$

$5x + x + 1 = 5$

$6x = 4$

$x = \dfrac{2}{3}$

Substitute $\dfrac{2}{3}$ for x in (3).

$y = -\dfrac{2}{3} - 1$

$y = -\dfrac{5}{3}$

Solution set: $\left\{\left(\dfrac{2}{3}, -\dfrac{5}{3}\right)\right\}$

15.

(1) $6x + 5y = 23$

(2) $x + y = 5$

Solve (1) for y.

(3) $y = -x + 5$

#15. continued.

Substitute $-x + 5$ for y in (1).

$6x + 5(-x + 5) = 23$

$\qquad 6x - 5x + 25 = 23$

$\qquad\qquad\qquad x = -2$

Substitute -2 for x in (3).

$y = -(-2) + 5$

$y = 7$

Solution set: $\{(-2, 7)\}$

17.

(1) $x - 3y = 0$

(2) $2x - y = 5$

Solve (1) for x.

(3) $x = 3y$

Substitute 3y for x in (2).

$2(3y) - y = 5$

$\qquad 6y - y = 5$

$\qquad\quad 5y = 5$

$\qquad\quad\ y = 1$

Substitute 1 for y in (3).

$x = 3 \cdot 1$

$x = 3$

Solution set: $\{(3, 1)\}$

19.

(1) $y - 4x = 0$

(2) $2x - 3y = \dfrac{5}{2}$

Solve (1) for y.

(3) $y = 4x$

Substitute 4x for y in (2).

$2x - 3(4x) = \dfrac{5}{2}$

$\quad 2x - 12x = \dfrac{5}{2}$

$\qquad -10x = \dfrac{5}{2}$

$\qquad\qquad x = -\dfrac{1}{4}$

#19. continued.

Substitute $-\dfrac{1}{4}$ for x in (3).

$y = 4\left(-\dfrac{1}{4}\right)$

$y = -1$

Solution set: $\left\{\left(-\dfrac{1}{4}, -1\right)\right\}$

21.

(1) $x - y = 5$

(2) $3x - 7y = 3$

Solve (1) for x.

(3) $x = y + 5$

Substitute y + 5 for x in (2).

$3(y + 5) - 7y = 3$

$\ 3y + 15 - 7y = 3$

$\qquad\qquad -4y = -12$

$\qquad\qquad\quad y = 3$

Substitute 3 for y in (3).

$x = 3 + 5$

$x = 8$

Solution set: $\{(8, 3)\}$

23.

(1) $x - 6y = 3$

(2) $2x + 9y = 13$

Solve (1) for x.

(3) $x = 6y + 3$

Substitute 6y + 3 for x in (2).

$2(6y + 3) + 9y = 13$

$\quad 12y + 6 + 9y = 13$

$\qquad\qquad\quad 21y = 7$

$\qquad\qquad\qquad y = \dfrac{1}{3}$

Substitute $\dfrac{1}{3}$ for y in (3).

$x = 6 \cdot \dfrac{1}{3} + 3$

$x = 2 + 3$

$x = 5$

#23. continued.

Solution set: $\left\{\left(5, \dfrac{1}{3}\right)\right\}$

25.
(1) $y - 3x = 5$
(2) $6x - 2y = -10$

Solve (1) for y.
  $y = 3x + 5$

Substitute $3x + 5$ for y in (2).
$6x - 2(3x + 5) = -10$
$6x - 6x - 10 = -10$
    $0 = 0$

The system is dependent.

27.
(1) $2x - y = -1$
(2) $5x - 3y = 2$

Solve (1) for y.
    $-y = -2x - 1$
(3) $y = 2x + 1$

Substitute $2x + 1$ for y in (2).
$5x - 3(2x + 1) = 2$
    $5x - 6x - 3 = 2$
        $-x = 5$
         $x = -5$

Substitute -5 for x in (3).
    $y = 2(-5) + 1$
    $y = -10 + 1$
    $y = -9$

Solution set: $\{(-5,-9)\}$

29.
(1) $3x + 8y = -29$
(2) $5x - y = 9$

Solve (2) for y.
      $-y = -5x + 9$
 (3) $y = 5x - 9$

Substitute $5x - 9$ for y in (1).
$3x + 8(5x - 9) = -29$

#29 continued

$3x + 40x - 72 = -29$
      $43x = 43$
        $x = 1$

Substitute 1 for x in (3).
  $y = 5 \cdot 1 - 9$
  $y = -4$

Solution set: $\{(1,-4)\}$

31.
(1) $3x - 15y = 2$
(2) $3y - x = 2$

Solve (2) for x.
   $-x = -3y + 2$
(3) $x = 3y - 2$

Substitute $3y - 2$ for x in (1).
$3(3y - 2) - 15y = 2$
    $9y - 6 - 15y = 2$
          $-6y = 8$
            $y = -\dfrac{4}{3}$

Substitute $-\dfrac{4}{3}$ for y in (3).

$x = 3\left(-\dfrac{4}{3}\right) - 2$
$x = -4 - 2$
$x = -6$

Solution set: $\left\{\left(-6, -\dfrac{4}{3}\right)\right\}$

33.
(1) $-x + 8y = 1$
(2) $\dfrac{1}{2}x + 5y = 4$

Solve (1) for x
    $-x = -8y + 1$
(3) $x = 8y - 1$

Substitute $8y - 1$ for x in (2).
$\dfrac{1}{2}(8y - 1) + 5y = 4$

$4y - \dfrac{1}{2} + 5y = 4$

#33 continued

$$9y = 4 + \frac{1}{2}$$

$$9y = \frac{9}{2}$$

$$y = \frac{1}{2}$$

Substitute $\frac{1}{2}$ for y in (3).

$$x = 8\left(\frac{1}{2}\right) - 1$$

$$x = 4 - 1$$

$$x = 3$$

Solution set: $\left\{\left(3, \frac{1}{2}\right)\right\}$

35.
(1) 5x + 2y = 27
(2) 2x - 3y = 26

Solve (2) for x.

$$2x = 3y + 26$$

$$(3) \qquad x = \frac{3}{2}y + 13$$

Substitute $\frac{3}{2}y + 13$ for x in (1).

$$5\left(\frac{3}{2}y + 13\right) + 2y = 27$$

$$\frac{15}{2}y + 65 + 2y = 27$$

$$\frac{19}{2}y = -38$$

$$y = \frac{2}{19}(-38)$$

$$y = -4$$

Substitute -4 for y in (3).

$$x = \frac{3}{2}(-4) + 13$$

$$x = -6 + 13$$

$$x = 7$$

Solution set: {(7, -4)}

37.
(1) 3x - 10y = -14
(2) 2x - 25y = 9

Solve (2) for x.

$$2x = 25y + 9$$

$$(3) \qquad x = \frac{25}{2}y + \frac{9}{2}$$

Substitute $\frac{25}{2}y + \frac{9}{2}$ for x in (1).

$$3\left(\frac{25}{2}y + \frac{9}{2}\right) - 10y = -14$$

$$\frac{75}{2}y + \frac{27}{2} - 10y = -14$$

$$\frac{55}{2}y = -\frac{55}{2}$$

$$y = -1$$

Substitue -1 for y in (3).

$$x = \frac{25}{2}(-1) + \frac{9}{2}$$

$$x = \frac{-25}{2} + \frac{9}{2}$$

$$x = \frac{-16}{2} \text{ or } -8$$

Solution set: {(-8, -1)}

39.
(1) 4x - 10y = 9
(2) -2x + 5y = 3

Solve (2) for y.

$$5y = 2x + 3$$

$$(3) \qquad y = \frac{2}{5}x + \frac{3}{5}$$

Substitute $\frac{2}{5}x + \frac{3}{5}$ for y in (1).

$$4x - 10\left(\frac{2}{5}x + \frac{3}{5}\right) = 9$$

$$4x - 4x - 6 = 9$$

$$-6 \neq 9$$

Inconsistent.   No solution.

41.
(1) 2x + 7y = 11
(2) 13x - 2y = 5
Solve (1) for x.

$$2x = -7y + 11$$

#41 continued

Solve (1) for x.
$$2x = -7y + 11$$
(3) $$x = -\frac{7}{2}y + \frac{11}{2}$$

Substitute $-\frac{7}{2}y + \frac{11}{2}$ for x in (2).

$$13\left(-\frac{7}{2}y + \frac{11}{2}\right) - 2y = 5$$

$$-\frac{91}{2}y + \frac{143}{2} - 2y = 5$$

$$\frac{-95}{2}y = -\frac{133}{2}$$

$$y = \frac{-133}{2}\left(-\frac{2}{95}\right)$$

$$y = \frac{7}{5}$$

Substitute $\frac{7}{5}$ for y in (3).

$$x = -\frac{7}{2}\left(\frac{7}{5}\right) + \frac{11}{2}$$

$$x = \frac{-49}{10} + \frac{11}{2}$$

$$x = \frac{6}{10} \text{ or } \frac{3}{5}$$

Solution set: $\left\{\left(\frac{3}{5}, \frac{7}{5}\right)\right\}$

43.

(1) $\frac{3}{4}x - \frac{5}{2}y = 8$

(2) $\frac{7}{2}x + 2y = 10$

Multiply (1) by 4 to simplify.
(3) $3x - 10y = 32$

Solve (2) for y.
$$2y = -\frac{7}{2}x + 10$$
(4) $$y = -\frac{7}{4}x + 5$$

Substitute $-\frac{7}{4}x + 5$ for y in (3).

#43 continued

$$3x - 10\left(-\frac{7}{4}x + 5\right) = 32$$

$$3x + \frac{35}{2}x - 50 = 32$$

$$\frac{41}{2}x = 82$$

$$x = 4$$

Substitute 4 for x in (4).

$$y = -\frac{7}{4} \cdot 4 + 5$$

$$y = -2$$

Solution set: $\{(4,-2)\}$

45.

(1) $-\frac{5}{6}x - \frac{7}{3}y = 3$

(2) $\frac{1}{4}x + \frac{5}{2}y = \frac{9}{2}$

Multiply (1) by 6 and (2) by 4 to clear fractions.
(3) $-5x - 14y = 18$
(4) $x + 10y = 18$

Solve (4) for x.
(5) $x = -10y + 18$

Substitute $-10y + 18$ for x in (3).
$$-5(-10y + 18) - 14y = 18$$
$$50y - 90 - 14y = 18$$
$$36y = 108$$
$$y = 3$$

Substitute 3 for y in (5).
$$x = -10 \cdot 3 + 18$$
$$x = -30 + 18$$
$$x = -12$$

Solution set: $\{(-12,3)\}$

47.

(1) $\frac{2}{9}x + 16y = 3$

(2) $\frac{11}{15}x - \frac{3}{5}y = 1$

#47 continued

Multiply (2) by 15 to clear fractions.
(3) $11x - 9y = 15$

Solve (1) for y.
$$16y = -\frac{2}{9}x + 3$$
(4) $$y = -\frac{1}{72}x + \frac{3}{16}$$

Substitute $-\frac{1}{72}x + \frac{3}{16}$ for y in (3).

$$11x - 9\left(-\frac{1}{72}x + \frac{3}{16}\right) = 15$$
$$11x + \frac{1}{8}x - \frac{27}{16} = 15$$
$$\frac{89}{8}x = \frac{27}{16} + \frac{240}{16}$$
$$\frac{89}{8}x = \frac{267}{16}$$
$$x = \frac{3}{2}$$

Substitute $\frac{3}{2}$ for x in (4).

$$y = -\frac{1}{72} \cdot \frac{3}{2} + \frac{3}{16}$$
$$y = -\frac{1}{48} + \frac{3}{16}$$
$$y = \frac{1}{6}$$

Solution set: $\left\{\left(\frac{3}{2}, \frac{1}{6}\right)\right\}$

49.
(1) $2(5x + 3y) = 1$
(2) $\frac{1}{3}(2x + 6y) = 7$

Simplify equations (1) and (2).
(3) $10x + 6y = 1$
(4) $\frac{2}{3}x + 2y = 7$

Solve (4) for y.
$$2y = -\frac{2}{3}x + 7$$
(5) $$y = -\frac{1}{3}x + \frac{7}{2}$$

#49 continued

Substitute $-\frac{1}{3}x + \frac{7}{2}$ for y in (3).

$$10x + 6\left(-\frac{1}{3}x + \frac{7}{2}\right) = 1$$
$$10x - 2x + 21 = 1$$
$$8x = 1 - 21$$
$$8x = -20$$
$$x = -\frac{5}{2}$$

Substitute $-\frac{5}{2}$ for x in (5).

$$y = -\frac{1}{3}\left(-\frac{5}{2}\right) + \frac{7}{2}$$
$$y = \frac{5}{6} + \frac{7}{2}$$
$$y = \frac{26}{6} \text{ or } \frac{13}{3}$$

Solution set: $\left\{\left(-\frac{5}{2}, \frac{13}{3}\right)\right\}$

51.
(1) $\frac{1}{4}\left(\frac{3}{2}x + 2y\right) = -2$
(2) $\frac{1}{10}\left(\frac{5}{6}x + \frac{1}{3}y\right) = \frac{1}{3}$

Multiply (1) by 8 and (2) by 60 to clear fractions.
(3) $3x + 4y = -16$
(4) $5x + 2y = 20$

Solve (3) for y.
$$4y = -3x - 16$$
(5) $$y = -\frac{3}{4}x - 4$$

Substitute $-\frac{3}{4}x - 4$ for y in (4).

$$5x + 2\left(-\frac{3}{4}x - 4\right) = 20$$
$$5x - \frac{3}{2}x - 8 = 20$$
$$\frac{7}{2}x = 28$$
$$x = 8$$

#51 continued

Substitute 8 for x in (5).

$$y = -\frac{3}{4} \cdot 8 - 4$$

$$y = -6 - 4$$

$$y = -10$$

Solution set: $\{(8, -10)\}$

53.

(1) $6x - 2y + 5 = 3x + 9$

(2) $x - 4y - 5 = 9 - 4x$

Simplify (1) and (2).

(3) $3x - 2y = 4$

(4) $5x - 4y = 14$

Solve (3) for y.

$$-2y = -3x + 4$$

(5) $y = \frac{3}{2}x - 2$

Substitute $\frac{3}{2}x - 2$ for y in (4).

$$5x - 4\left(\frac{3}{2}x - 2\right) = 14$$

$$5x - 6x + 8 = 14$$

$$-x = 6$$

$$x = -6$$

Substitute - 6 for x in (5).

$$y = \frac{3}{2}(-6) - 2$$

$$y = -9 - 2$$

$$y = -11$$

Solution set: $\{(-6, -11)\}$

55.

(1) $4(x - 2y) - 3 = 2(1 - y)$

(2) $3(y - 3x) + 10 = 7(2 - x)$

Simplify (1) and (2).

(3) $4x - 6y = 5$

(4) $-2x + 3y = 4$

Solve (4) for x.

$$-2x = -3y + 4$$

#55 continued

(5) $x = \frac{3}{2}y - 2$

Substitute $\frac{3}{2}y - 2$ for x in (3).

$$4\left(\frac{3}{2}y - 2\right) - 6y = 5$$

$$6y - 8 - 6y = 5$$

$$-8 \neq 5$$

Inconsistent. No solution.

Exercises 7.3

1.

(1) $-x + 2y = -19$

(2) $\underline{x - 5y = \quad 40}$

$$-3y = 21$$

$$y = -7$$

Substitute - 7 for y in (2).

$$x - 5(-7) = 40$$

$$x + 35 = 40$$

$$x = 5$$

Solution set: $\{(5, -7)\}$

3.

(1) $x + y = 11$

(2) $\underline{3x - y = 25}$

$$4x = 36$$

$$x = 9$$

Substitute 9 for x in (1).

$$9 + y = 11$$

$$y = 2$$

Solution set: $\{(9, 2)\}$

5.

(1) $7x + y = 10$

(2) $\underline{\quad -y = 3x - 4}$

$$7x = 3x + 6$$

$$4x = 6$$

$$x = \frac{3}{2}$$

#5 continued

Substitute $\frac{3}{2}$ for x in (1).

$7\left(\frac{3}{2}\right) + y = 10$

$\frac{21}{2} + y = 10$

$y = 10 - \frac{21}{2}$

$y = -\frac{1}{2}$

Solution set: $\left\{\left(\frac{3}{2}, -\frac{1}{2}\right)\right\}$

7.
(1) $3x - 2y = 10$
(2) $\underline{2y = 6x - 12}$
$3x = 6x - 2$
$ 2 = 3x$
$ \frac{2}{3} = x$

Substitute $\frac{2}{3}$ for x in (2).

$2y = 6\left(\frac{2}{3}\right) - 12$

$2y = 4 - 12$
$2y = -8$
$y = -4$

Solution set: $\left\{\left(\frac{2}{3}, -4\right)\right\}$

9.
(1) $4x + 3y = 5$
(2) $\underline{-4x + y = 23}$
$ 4y = 28$
$ y = 7$

Substitute 7 for y in (1).
$4x + 3 \cdot 7 = 5$
$ 4x = 5 - 21$
$ 4x = -16$
$ x = -4$

Solution set: $\{(-4,7)\}$

11.
(1) $2x - 5y = -1$
(2) $\underline{3x + 5y = -14}$
$5x = -15$
$ x = -3$

Substitute -3 for x in (2).
$3(-3) + 5y = -14$
$ 5y = -14 + 9$
$ 5y = -5$
$ y = -1$

Solution set: $\{(-3,-1)\}$

13.
(1) $x + 4y = 23$
(2) $-2x + 3y = 42$

Multiply (1) by 2.
(3) $2x + 8y = 46$
(2) $\underline{-2x + 3y = 42}$
$ 11y = 88$
$ y = 8$

Substitute 8 for y in (1).
$x + 4 \cdot 8 = 23$
$ x = 23 - 32$
$ x = -9$

Solution set: $\{(-9,8)\}$

15.
(1) $3x - 9y = 8$
(2) $x - 3y = 5$

Multiply (2) by -3.
(3) $-3x + 9y = -15$
(1) $\underline{3x - 9y = 8}$
$ 0 \neq -7$

The system has no solution and is inconsistent.

17.
(1) $6x + y = 15$
(2) $7x - 5y = 36$

#17 continued

Multiply (1) by 5.
(3) $30x + 5y = 75$
(2) $\underline{7x - 5y = 36}$
$\quad 37x = 111$
$\qquad\qquad x = 3$

Substitute 3 for x in (1).
$6 \cdot 3 + y = 15$
$y = -3$

Solution set: $\{(3,-3)\}$

19.
(1) $4x - y = 3$
(2) $8x + 3y = -1$

Multiply (1) by 3.
(3) $12x - 3y = 9$
(2) $\underline{8x + 3y = -1}$
$\quad 20x = 8$
$\qquad\qquad x = \dfrac{2}{5}$

Substitute $\dfrac{2}{5}$ for x in (1).

$4 \cdot \dfrac{2}{5} - y = 3$

$\phantom{4 \cdot \dfrac{2}{5}} -y = 3 - \dfrac{8}{5}$

$\phantom{4 \cdot \dfrac{2}{5}} -y = \dfrac{7}{5}$

$\phantom{4 \cdot \dfrac{2}{5}} y = -\dfrac{7}{5}$

Solution set: $\left\{\left(\dfrac{2}{5}, -\dfrac{7}{5}\right)\right\}$

21.
(1) $3x + 8y = -46$
(2) $2x - 5y = 21$

To eliminate the x's, multiply (1) by 2 and (2) by -3.
(3) $6x + 16y = -92$
(4) $\underline{-6x + 15y = -63}$
$\quad 31y = -155$
$\quady = -5$

Substitute -5 for y in (1).

#21. continued.

$3x + 8(-5) = -46$
$ 3x = -46 + 40$
$ 3x = -6$
$ x = -2$

Solution set: $\{(-2,-5)\}$

23.
(1) $-2x + 3y = -38$
(2) $5x + 4y = 3$

To eliminate the x's, multiply (1) by 5 and (2) by 2.
(3) $-10x + 15y = -190$
(4) $\underline{10x + 8y = 6}$
$\qquad 23y = -184$
$\qquady = -8$

Substitute -8 for y in (2).
$5x + 4(-8) = 3$
$ 5x = 3 + 32$
$ 5x = 35$
$ x = 7$

Solution set: $\{(7,-8)\}$

25.
(1) $7x + 3y = -15$
(2) $9x + 2y = -10$

To eliminate the y's, multiply (1) by 2 and (2) by -3.
(3) $14x + 6y = -30$
$\underline{-27x - 6y = 30}$
$\quad -13x = 0$
$\qquad\quad x = 0$

Substitute 0 for x in (2).
$9 \cdot 0 + 2y = -10$
$y = -5$

Solution set: $\{(0,-5)\}$

27.
(1) $-5x + 10y = 25$
(2) $3x - 6y = -15$

#27. continued.

To eliminate the x's multiply (1) by 3 and (2) by 5.
(3) $-15x + 30y = 75$
       $\underline{15x - 30y = -75}$
                $0 = 0$

This system is dependent.

29.
(1) $14x + 19y = 15$
(2) $4x - 7y = -4$

To eliminate the x's, multiply (1) by 2 and (2) by -7.
(3) $28x + 38y = 30$
(4) $\underline{-28x + 49y = 28}$
         $87y = 58$
          $y = \dfrac{2}{3}$

Substitute $\dfrac{2}{3}$ for y in (2).

$4x - 7 \cdot \dfrac{2}{3} = -4$

$4x - \dfrac{14}{3} = -4$

$4x = -4 + \dfrac{14}{3}$

$4x = \dfrac{2}{3}$

$x = \dfrac{1}{6}$

Solutions set: $\left\{ \left(\dfrac{1}{6}, \dfrac{2}{3} \right) \right\}$

31.
(1) $\dfrac{3}{4} x + \dfrac{4}{3} y = -3$
(2) $\dfrac{5}{2} x + \dfrac{11}{3} y = -3$

Simplify by multiplying (1) by 12 and (2) by 6.
(3) $9x + 16y = -36$
(4) $15x + 22y = -18$

To eliminate the x's, multiply (3) by 5 and (4) by -3.

#31 continued

(5) $45x + 80y = -180$
(6) $\underline{-45x - 66y = \quad 54}$
         $14y = -126$
          $y = -9$

Substitute -9 for y in (1).

$\dfrac{3}{4} x + \dfrac{4}{3}(-9) = -3$

$\dfrac{3}{4} x - 12 = -3$

$\dfrac{3}{4} x = 9$

$x = 12$

Solution set: $\{(12,-9)\}$

33.
(1) $\dfrac{5}{6} x - \dfrac{4}{9} y = -4$
(2) $\dfrac{7}{10} x - \dfrac{4}{15} y = -4$

Simplify by multiplying (1) by 18 and (2) by 30.
(3) $15x - 8y = -72$
(4) $21x - 8y = -120$

To eliminate the y's, multiply (4) by -1.
(3) $15x - 8y = -72$
(5) $\underline{-21x + 8y = 120}$
       $-6x = 48$
        $x = -8$

Substitute -8 for x in (3).
$15(-8) - 8y = -72$
 $-120 - 8y = -72$
    $-8y = 48$
     $y = -6$

Solution set: $\{(-8,-6)\}$

35.
(1) $-\dfrac{1}{5} x + \dfrac{1}{3} y = \dfrac{16}{15}$
(2) $\dfrac{1}{9} x - \dfrac{1}{4} y = -\dfrac{11}{12}$

#35 continued

Simplify by multiplying (1) by 15 and (2) by 36.

(3) - 3x + 5y = 16
(4)   4x - 9y = - 33

To eliminate the x's, multiply (3) by 4 and (4) by 3.

(5) - 12x + 20y = 64
(6)   12x - 27y = -99

$$- 7y = - 35$$
$$y = 5$$

Substitute 5 for y in (4).

$$4x - 9 \cdot 5 = - 33$$
$$4x = - 33 + 45$$
$$4x = 12$$
$$x = 3$$

Solution set: {(3,5)}

37.

(1) $y = \frac{2}{9} x + \frac{13}{3}$

(2)   3x + 14y = - 31

Place (1) in standard form.

(1) $- \frac{2}{9} x + y = \frac{13}{3}$

(2)   3x + 14y = - 31

Simplify by multiplying (1) by 9.

(3)   - 2x + 9y = 39
(2)   3x + 14y = - 31

To eliminate the x's, multiply (3) by 3 and (2) by 2.

(4)   - 6x + 27y = 117
(5)   6x + 28y = - 62

$$55y = 55$$
$$y = 1$$

Substitute 1 for y in (2).

$$3x + 14 \cdot 1 = - 31$$
$$3x = - 31 - 14$$
$$3x = - 45$$
$$x = - 15$$

Solution set: {(-15,1)}

39.

(1) 10x - 7y = - 8

(2) $x = \frac{7}{10} y - \frac{4}{5}$

Place (2) in standard form.

(1) 10x - 7y = - 8

(2) $x - \frac{7}{10} y = -\frac{4}{5}$

Simplify by multiplying (2) by 10.

(1) 10x - 7y = - 8
      10x - 7y = - 8

$$0 = 0$$

Dependent

41.

(1) $x = \frac{11y + 15}{2}$

(2) $y = \frac{5x + 3}{14}$

Simplify by multiplying (1) by 2 and (2) by 14.

(3)   2x = 11y + 15
(4) 14y = 5x + 3

Place (3) and (4) in standard form.

(5)   2x - 11y = 15
(6) -5x + 14y = 3

To eliminate the x's, multiply (5) by 5 and (6) by 2.

(7)   10x - 55y = 75
(8) -10x + 28y = 6

$$- 27y = 81$$
$$y = - 3$$

Substitute - 3 for y in (1).

$$x = \frac{11(-3) + 15}{2}$$
$$x = \frac{- 33 + 15}{2}$$
$$x = \frac{-18}{2} \text{ or } -9$$

Solution set: {(-9,-3)}

43.

(1) $y = \dfrac{5 - 3x}{13}$

(2) $x = \dfrac{22 - 8y}{5}$

Simplify by multiplying (1) by 13 and (2) by 5.
(3) $13y = 5 - 3x$
(4) $5x = 22 - 8y$

Place (3) and (4) in standard form.
(5) $3x + 13y = 5$
(6) $5x + 8y = 22$

To eliminate the x's, multiply (5) by 5 and (6) by -3.
(7) $15x + 65y = 25$
(8) $\underline{-15x - 24y = -66}$
$\quad\quad\quad 41y = -41$
$\quad\quad\quad\quad y = -1$

Substitute -1 for y in (2)

$x = \dfrac{22 - 8(-1)}{5}$

$x = \dfrac{22 + 8}{5}$

$x = 6$

Solution set: $\{(6,-1)\}$

45.
(1) $5(x - 2y) + 2x = -3 - 4y$
(2) $7(x - 2y) + 3x = -6 - 5y$

Simplify (1) and (2) and place in standard form.
(3) $7x - 6y = -3$
(4) $10x - 9y = -6$

To eliminate the y's, multiply (3) by 3 and (4) by -2.
(5) $21x - 18y = -9$
(6) $\underline{-20x + 18y = 12}$
$\quad\quad\quad x \quad\quad = 3$

Substitute 3 for x in (3).
$7 \cdot 3 - 6y = -3$
$\quad 21 - 6y = -3$
$\quad\quad\; -6y = -24$
$\quad\quad\quad\; y = 4$

#45 continued

Solution set: $\{(3,4)\}$

47.
(1) $3(2x - 3y - 2) = 2(x - 3y + 4)$
(2) $5(3x - y - 3) = 3(3x + 2y + 2)$

Simplify (1) and (2) and place in standard form.
(3) $4x - 3y = 14$
(4) $6x - 11y = 21$

To eliminate the x's, multiply (3) by 3 and (4) by -2.
(5) $12x - 9y = 42$
(6) $\underline{-12x + 22y = -42}$
$\quad\quad\quad 13y = 0$
$\quad\quad\quad\; y = 0$

Substitute 0 for y in (3).
$4x - 3 \cdot 0 = 14$
$\quad\quad\; 4x = 14$
$\quad\quad\; x = \dfrac{7}{2}$

Solution set: $\left\{\left(\dfrac{7}{2},\; 0\right)\right\}$

49.

(1) $\dfrac{4}{5}\left(\dfrac{1}{8} x + \dfrac{1}{6} y\right) = \dfrac{2}{3}$

(2) $\dfrac{2}{3}\left(\dfrac{1}{4} x + \dfrac{1}{3} y\right) = \dfrac{1}{2}$

Simplify by multiplying (1) by 30 and (2) by 18.
(3) $3x + 4y = 20$
(4) $3x + 4y = 9$

Multiply (4) by -1.
$\quad\quad 3x + 4y = 20$
(5) $\quad\underline{-3x - 4y = -9}$
$\quad\quad\quad\; 0 \neq 11$

Inconsistent.   No solution

51.

(1) $\frac{2}{3}\left(\frac{1}{6}x + \frac{1}{4}y\right) = \frac{1}{2}$

(2) $\frac{1}{6}\left(\frac{3}{2}x + 2y\right) = \frac{1}{2}$

Simplify by multiplying (1) by 18 and (2) by 12.

(3) $2x + 3y = 9$

(4) $3x + 4y = 6$

To eliminate the x's, multiply (3) by 3 and (4) by -2.

(5) $6x + 9y = 27$

(6) $\underline{-6x - 8y = -12}$

$y = 15$

Substitute 15 for y in (3).

$2x + 3(15) = 9$

$2x = 9 - 45$

$2x = -36$

$x = -18$

Solution set: $\{(-18, 15)\}$

53.

(1) $\frac{3}{x} + \frac{5}{y} = \frac{1}{2}$

(2) $\frac{1}{x} - \frac{10}{y} = \frac{5}{2}$

Let $u = \frac{1}{x}$ amd $v = \frac{1}{y}$.

Then

(3) $3u + 5v = \frac{1}{2}$

(4) $u - 10v = \frac{5}{2}$

Solve for u and v.

Simplify by multiplying (3) by 2 and (4) by -6.

(5) $6u + 10v = 1$

(6) $\underline{-6u + 60v = -15}$

$70v = -14$

$v = -\frac{1}{5}$

Substitute $-\frac{1}{5}$ for v in (4).

#53 continued

$u - 10\left(-\frac{1}{5}\right) = \frac{5}{2}$

$u + 2 = \frac{5}{2}$

$u = \frac{5}{2} - 2$

$u = \frac{1}{2}$

Thus,

$u = \frac{1}{x} = \frac{1}{2}$

$x = 2$ Cross multiply.

$v = \frac{1}{y} = -\frac{1}{5}$

$y = -5$ Cross multiply.

Solution set: $\{(2, -5)\}$

Exercises 7.4

1.

Let x = first number

Let y = second number

The equations are

(1) $x + y = 20$

(2) $\quad x = 2y - 1$

Place (2) In standard form as equation (3) below.

(1) $x + y = 20$

(3) $x - 2y = -1$

The eliminate the x's, multiply (3) by -1.

(1) $\quad x + y = 20$

(4) $\underline{-x + 2y = 1}$

$3y = 21$

$y = 7$

Substitute 7 for 7 in (2).

$x = 2 \cdot 7 - 1$

$x = 13$

The numbers are 7 and 13.

3.
Let x = first number
Let y = second number

The equations are
(1) $x - y = 10$
(2) $x = 3y + 12$

Place (2) in standard form as equation (3) below.
(1) $x - y = 10$
(3) $x - 3y = 12$

To eliminate the x's, multiply (3) by -1.
(1) $x - y = 10$
(4) $\underline{- x + 3y = - 12}$
            $2y = - 2$
             $y = - 1$

Substitute - 1 for y in (3).
    $x = 3(-1) + 12$
    $x = 9$

The numbers are - 1 and 9.

5.
Let t = tens digit
Let u = units digit.

The equations are
(1) $t + u = 7$
(2) $10t + u = 5u$

Place (2) in standard form as equation (3) below.
(1) $t + u = 7$
(3) $10t - 4u = 0$

To eliminate the u's, multiply (1) by 4.
(4) $4t + 4u = 28$
(3) $\underline{10t - 4u = 0}$
            $14t = 28$
             $t = 2$

Substitute 2 for t in (1).
    $2 + u = 7$
        $u = 5$

The number is 25.

7.
Let x = numerator
Let y = denominator

The equations are
(1) $x - y = 3$
(2) $\dfrac{x + 1}{y} = \dfrac{3}{2}$

Place (2) in standard form as equation (3) below.
(1) $x - y = 3$
(3) $2x - 3y = - 2$

To eliminate the x's, multiply (1) by -2.
(4) $- 2x + 2y = - 6$
(3) $\underline{2x - 3y = - 2}$
            $- y = - 8$
             $y = 8$

Substitute 8 for y in (1).
    $x - 8 = 3$
        $x = 11$

The original fraction is $\dfrac{11}{8}$.

9.
Let x = numerator
Let y = denominator

The equations are
(1) $y - x = 3$
(2) $\dfrac{x + 4}{y + 4} = \dfrac{2}{3}$

Place (1) and (2) in standard form as equations (3) and (4) below.
(3) $- x + y = 3$
(4) $3x - 2y = - 4$

To eliminate the x's, multiply (1) by 3.
(5) $- 3x + 3y = 9$
(4) $\underline{3x - 2y = - 4}$
             $y = 5$

Substitute 5 for y in (1).
    $5 - x = 3$
      $- x = - 2$
        $x = 2$

#9 continued

The original fraction is $\frac{2}{5}$.

11.

Let x = numerator
Let y = denominator

The equatiions are
(1) $x - y = 2$
(2) $\frac{x - 3}{y + 1} = \frac{3}{4}$

Place (2) in standard form as equation (3) below.
(1) $x - y = 2$
(3) $4x - 3y = 15$

To eliminate the x's, multiply (1) by -4.
(4) $-4x + 4y = -8$
(3) $\underline{4x - 3y = 15}$
          $y = 7$

Substitute 7 for y in (1).
$x - 7 = 2$
   $x = 9$

The original fraction is $\frac{9}{7}$.

13.

Let x = number of nickels
Let y = number of dimes
Let 5x = value of the nickels
Let 10y = value of the dimes

The equations are
(1) $x + y = 42$
(2) $5x + 10y = 355$

To eliminate the x's, multiply (1) by -5.
(3) $-5x - 5y = -210$
(2) $\underline{5x + 10y = 355}$
          $5y = 145$
            $y = 29$

Substitute 29 for y in (1).
   $x + 29 = 42$
        $x = 13$

There are 13 nickels and 29 dimes.

15.

Let x = number of dimes
Let y = number of quarters
Let 10x = value of dimes
Let 25y = value of quarters

The equations are
(1) $y = 3x - 2$
(2) $10x + 25y = 1480$

Place (1) in standard form.
(3) $-3x + y = -2$
(2) $10x + 25y = 1480$

To eliminate the y's, multiply (3) by -25.
(4) $75x - 25y = 50$
(2) $\underline{10x + 25y = 1480}$
        $85x \qquad = 1530$
             $x = 18$

Substitute 18 for x in (1).
      $y = 3(18) - 2$
      $y = 52$

There are 18 dimes and 52 quarters

17.

Let x = number of 25¢ stamps
Let y = number of 3¢ stamps
Let 25x = value of the 25¢ stamps
Let 3y = value of the 3¢ stamps

The equations are
(1) $x + y = 27$
(2) $25x + 3y = 301$

To eliminate the y's, multiply (1) by -3.
(3) $-3x - 3y = -81$
(2) $\underline{25x + 3y = 301}$
        $22x \qquad = 220$
           $x = 10$

Substitute 10 for x in (1)
      $10 + y = 27$
           $y = 17$

She bought 10 - 25¢ stamps and 17 - 3¢ stamps.

19.

Let x = number of adult tickets
Let y = number of children's tickets
Let 8.75x = value of adult tickets sold
Let 5.50y = value of children's tickets sold

The equations are
(1)   x + y = 154
(2)   8.75x + 5.50 y = 1256.50

Simplify (2) by multiplying by 100.
(1)   x + y = 154
(3)   875x + 550y = 125,650

To eliminate the y's, multiply (1) by -550.
(4)   - 550x - 550y = - 84,700
(3) $\underline{875x + 550y = 125.650}$
       325x   = 40,950
                x = 126

Substitute 126 for x in (1).
       126 + y = 154
              y = 28

There were 126 adult and 28 children's tickets.

21.

Let x = amount invested at 8%
Let y = amount invested at 10%
Let .08x = amount of interest yield at 8%
Let .10y = amount of interest yield at 10%

The total amount invested was $6000 and the total year interest yield was $550.

The equations are
(1)   x + y = 6000
(2)   .08x + .10y = 550

Simplify (2) by multiplying by 100.
(1)   x + y = 6,000
(3)   8x + 10y = 55,000

To eliminate the x's, multiply (1) by - 8.
(4)   - 8x - 8y = - 48,000
(3) $\underline{8x + 10y = 55.000}$
             2y = 7,000
              y = 3,500

#21   continued.

Substitute 3500 for y in (1).
       x + 3500 = 6000
              x = 2500

Amber invested $2500 at 8% and $3500 at 10%.

23.

Let x = amount invested at 11%
Let y = amount invested at 5%
Let .11x = amount of interest yield at 11%
Let .05y = amount of interest yield at 5%

The difference between the two amounts invested was $1600 and the total year interest yield was $400.

The equations are
(1)   x - y = 1600
(2)   .11x + .05y = 400

Simplify (2) by multiplying by 100.
(1)   x - y = 1,600
(3)   11x + 5y = 40,000

To eliminate the y's, multiply (1) by 5.
(4)   5x - 5y = 8,000
(3) $\underline{11x + 5y = 40.000}$
       16x   = 48,000
              x = 3,000

Substitute 3,000 for x in (1).
       3000 - y = 1,600
            - y = - 1,400
              y = 1,400

Heather invested $3,000 at 11% and $1,400 at 5%.

25.

Let x = amount invested at 18%
Let y = amount invested at 15%
Let .18x = amount of interest yield at 18%
Let .15y = amount of interest yield at 15%

The total year interest yield was $2,475.

#25. continued.

The equations are
(1) $x = 2y + 1000$
(2) $.18x + .15y = 2,475$

Place (1) in standard form and simplify (2) by multiplying by 100.
(3) $x - 2y = 1,000$
(4) $18x + 15y = 247,500$

To eliminate the x's, multiply (2) by - 18.
(5) $-18x + 36y = -18,000$
    $\underline{18x + 15y = \ 247,500}$
    $51y = 229,500$
    $y = 4,500$

Substitute 4,500 for y in (1).
    $x = 2(4500) + 1000$
    $x = 10,000$

Joy invested $10,000 at 18% and $4,500 at 15%.

27.
Let x = amount of Sun Valley
Let y = amount of Death Valley

	Amount ·	Percent =	Fruit Juice
Sun Valley	x	.15	.15x
Death Valley	y	.03	.03y
Big Valley	24	.11	.11(24)

The equations are
(1) $x + y = 24$
(2) $.15x + .03y = .11(24)$

Simplify (2) by multiplying by 100.
(1) $x + y = 24$
(3) $15x + 3y = 264$

#27. continued.

To eliminate the y's, multiply (1) by -3.
(4) $-3x - 3y = -72$
(3) $\underline{15x + 3y = 264}$
    $12x \ \ \ \ = 192$
    $x = 16$

Substitute 16 for x in (1).
    $16 + y = 24$
    $y = 8$

Use 16 oz of Sun Valley and 8 oz of Death Valley.

29.
Let x = amount of Ethel's
Let y = amount of Ebenezer's

	Amount ·	Percent =	Alcohol
Ethel's	x	.25	.25x
Ebenezer's	y	1.00	1.00y
Tommy's	20	.40	.40(20)

The equations are
(1) $x + y = 20$
(2) $.25x + 1.00y = .40(20)$

Simplify (2) by multiplying by 100.
(1) $x + y = 20$
(3) $25x + 100y = 800$

To eliminate the x's, multiply by -25.
(4) $-25x - 25y = -500$
(3) $\underline{25x + 100y = 800}$
    $75y = 300$
    $y = 4$

Substitute 4 for y in (1).
    $x + 4 = 20$
    $x = 16$

Use 16 oz of Ethel's and 4 oz of Ebenezer's.

31.

Let x = cost of each rose bush
Let y = cost of each azalea

The equations are
(1)   2x + 5y = 39.75
(2)   3x + 2y = 30.75

To eliminate the x's, multiply (1) by 3 and (2) by -2.
(3)   6x + 15y = 119.25
       -6x - 4y = -61.50
              11y = 57.75
               y = 5.25

Substitute 5.25 for y in (1).
       2x + 5(5.25) = 39.75
       2x +   26.25 = 39.75
              2x = 13.50
               x = 6.75

Each rose bush cost $6.75 and each azalea cost $5.25.

33.

Let x = cost of each chocolate chip cookie
Let y = cost of each butter cookie

The equations are
(1)   5x + 8y = 4.11
(2)   7x + 4y = 3.81

To eliminate the y's, multiply (2) by -2.
(1)   5x + 8y =   4.11
       -14x - 8y = - 7.62
         - 9x = - 3.51
               x = 0.39

Substitute 0.39 for x in (1).
5(0.39) + 8y = 4.11
    1.95 + 8y = 4.11
           8y = 2.16
            y = 0.27

Each chocolate chip cookie cost $0.39 and each butter cookie cost $0.27.

Exercises 7.5

1. - 23.
The exercises do not require worked-out solutions.

25.
Since the horizontal boundary passes through $y = 3$ and the boundary is included, the linear inequality is $y \geq 3$.

The oblique boundary has intercepts $(-1,0)$ and $(0,1)$ with a slope of 1.
The boundary equation is
$$y = x + 1 \quad \text{or} \quad x - y = -1$$

Since $(0,4)$ satisfies the linear inequality, and the boundary is included, the linear inequality is $x - y \leq -1$.

27.
The oblique boundary closest to the origin passes through $(0, -1)$ and $(3, -5)$.

The slope is $\dfrac{-1 + 5}{0 - 3} = -\dfrac{4}{3}$.

The equation is
$$y + 1 = -\frac{4}{3}(x - 0)$$
$$y + 1 = -\frac{4}{3}x$$
$$3y + 3 = -4x$$
$$4x + 3y = -3$$

Since $(0,-2)$ satisfies the linear inequality, and the boundary is not included, the linear inequality is $4x + 3y < -3$.

#27 continued

The other oblique boundary passes through $\left(-\dfrac{9}{2} \quad 0\right)$ and (-2,5).

The slope is $\dfrac{5 - 0}{-2 + \dfrac{9}{2}} = 2$

The equation is
$$y - 5 = 2(x + 2)$$
$$y - 5 = 2x + 4$$
$$2x - y = -9$$

Since (0,-2) satisfies the linear inequality and the boundary is not included, the linear inequality is $2x - y > -9$.

29.
The exercise does not require a worked-out solution.

Review Exercises

1.
(1) $3x + 5y = 5$,   (5,-2)
$3 \cdot 5 + 5(-2) = 15 - 10 = 5$

(2) $2x - y = 12$,   (5,-2)
$2 \cdot 5 - (-2) = 10 + 2 = 12$

(5,-2) is the solution to the system.

3.

$$x - 4y = -5 \ , \ \left(-1, \frac{3}{2}\right)$$

$$-1 - 4\left(\frac{3}{2}\right) = -1 - 6 = -7$$

#3 continued

$$3y + 2y = -6, \ \left(-1, \frac{3}{2}\right)$$
$$5y = -6$$
$$5\left(\frac{3}{2}\right) \neq -6$$

$\left(-1, \dfrac{3}{2}\right)$ is not the solution to either.

5.

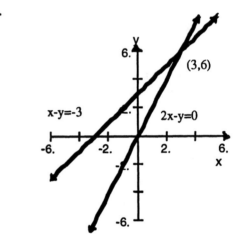

7.

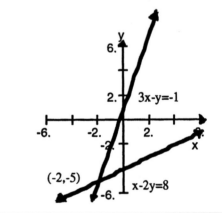

9.

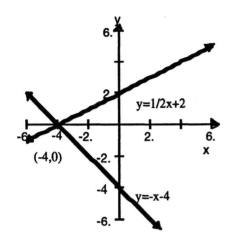

$y=1/2x+2$

(-4,0)

$y=-x-4$

11.

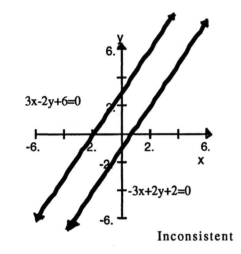

$3x-2y+6=0$

$-3x+2y+2=0$

Inconsistent

13.
(1) $3x - 7y = 8$
(2) $x = -2$

Substitute - 2 for x in (1).
$$3(-2) - 7y = 8$$
$$- 6 - 7y = 8$$
$$- 7y = 14$$
$$y = -2$$

Solution set: $\{(-2, -2)\}$

15.
(1) $x = 3y + 1$
(2) $2x - 5y = 3$

Substitute $3y + 1$ for x in (2).
$$2(3y + 1) - 5y = 3$$
$$6y + 2 - 5y = 3$$
$$y = 1$$

Substitute 1 for y in (1).
$$x = 3 \cdot 1 + 1$$
$$x = 4$$

Solution set: $\{(4, 1)\}$

17.
(1) $7x + 2y = 19$
(2) $y = x - 13$

Substitute $x - 13$ for y in (1).
$$7x + 2(x - 13) = 19$$
$$7x + 2x - 26 = 19$$
$$9x = 45$$
$$x = 5$$

Substitute 5 for x in (2).
$$y = 5 - 13$$
$$y = - 8$$

Solution set: $\{(5, -8)\}$

19.
(1) $3x + 8y = 1$
(2) $x + y = 2$

Solve (2) for x.

(3) $x = -y + 2$

Substitute $-y + 2$ for x in (1).
$3(-y + 2) + 8y = 1$
$-3y + 6 + 8y = 1$
$5y = -5$
$y = -1$

Substitute -1 for y in (3).
$x = -(-1) + 2$
$x = 3$

Solution set: $\{(3, -1)\}$

21.
(1) $2x - 6y = 4$
(2) $-x + 3y = -2$

Solve (2) for x.
(3) $x = 3y + 2$

Substitute $3y + 2$ for x in (1).
$2(3y + 2) - 6y = 4$
$6y + 4 - 6y = 4$
$4 = 4$

Dependent

23.
(1) $8x + 3y = 9$
(2) $6x + 5y = 26$

Solve (1) for y
$3y = -8x + 9$
(3) $y = -\dfrac{8}{3} x + 3$

Substitute $-\dfrac{8}{3} x + 3$ for y in (2).

$6x + 5\left(-\dfrac{8}{3} x + 3\right) = 26$

$6x - \dfrac{40}{3} x + 15 = 26$

#23   continued

$-\dfrac{22}{3} x = 11$

$x = -\dfrac{3}{2}$

Substitute $-\dfrac{3}{2}$ for x in (3).

$y = -\dfrac{8}{3}\left(-\dfrac{3}{2}\right) + 3$
$y = 4 + 3$
$y = 7$

Solution set: $\left\{\left(-\dfrac{3}{2}, 7\right)\right\}$

25.
(1) $\dfrac{2}{3} x - \dfrac{1}{5} y = 6$
(2) $\dfrac{3}{2} x + \dfrac{5}{4} y = -\dfrac{7}{2}$

Simplify by multiplying (1) by 15 and (2) by 4.
(3) $10x - 3y = 90$
(4) $6x + 5y = -14$

Solve (3) for y.
$-3y = -10x + 90$
(5) $y = \dfrac{10}{3} x - 30$

Substitute $\dfrac{10}{3} x - 30$ for y in (4).

$6x + 5\left(\dfrac{10}{3} x - 30\right) = -14$

$6x + \dfrac{50}{3} x - 150 = -14$

$\dfrac{68}{3} x = 136$

$x = 6$

Substitute 6 for x in (5).
$y = \dfrac{10}{3} \cdot 6 - 30$
$y = -10$

Solution set: $\{(6, -10)\}$

27.
(1) $x - 3y = 17$
(2) $\underline{- x - 4y = 18}$
$\qquad - 7y = 35$
$\qquad\quad y = -5$

Substitute -5 for y in (1).
$\quad x - 3(-5) = 17$
$\qquad x + 15 = 17$
$\qquad\qquad x = 2$

Solution set: $\{(2, -5)\}$

29.
(1) $6x - 5y = 14$
(2) $\underline{4x + 5y = -24}$
$\quad 10x = -10$
$\qquad\qquad x = -1$

Substitute -1 for x in (1).
$6(-1) - 5y = 14$
$\qquad -5y = 20$
$\qquad\quad y = -4$

Solution set: $\{(-1, -4)\}$

31.
(1) $-2x - y = 4$
(2) $6x + 3y = 5$

To eliminate the x's, multiply (1) by 3.
(3) $-6x - 3y = 12$
$\quad\underline{6x + 3y = 5}$
$\qquad\qquad 0 \neq 17$

Inconsistent. No solution.

33.
(1) $2x + 3y = 34$
(2) $5x - 4y = -7$

To eliminate the y's, multiply (1) by 4 and (2) by 3.
(3) $8x + 12y = 136$
$\quad\underline{15x - 12y = -21}$
$\quad 23x = 115$
$\qquad\qquad x = 5$

#33 continued

Substitute 5 for x in (1).
$\quad 2 \cdot 5 + 3y = 34$
$\qquad\quad 3y = 24$
$\qquad\quad\, y = 8$

Solution set: $\{(5, 8)\}$

35.
(1) $8x + 2y = 7$
(2) $7x + 3y = 3$

To eliminate the y's, multiply (1) by 3 and (2) by -2.
(3) $24x + 6y = 21$
$\quad\underline{- 14x - 6y = -6}$
$\quad10x = 15$
$\qquad\qquad x = \dfrac{3}{2}$

Substitute $\dfrac{3}{2}$ for x in (1).
$8 \cdot \dfrac{3}{2} + 2y = 7$
$\qquad 12 + 2y = 7$
$\qquad\qquad 2y = -5$
$\qquad\qquad\, y = -\dfrac{5}{2}$

Solution set: $\left\{\left(\dfrac{3}{2}, -\dfrac{5}{2}\right)\right\}$

37.
(1) $\dfrac{2}{3}x + \dfrac{7}{15}y = -1$
(2) $\dfrac{5}{4}x + \dfrac{9}{5}y = 12$

Simplify by multiplying (1) by 15 and (2) by 20.

(3) $10x + 7y = -15$
(4) $25x + 36y = 240$

To eliminate the x's, multiply (3) by 5 and (4) by -2.
(5) $50x + 35y = -75$
(6) $\underline{-50x - 72y = -480}$
$\qquad - 37y = -555$
$\qquad\qquad y = 15$

#37 continued

Substitute 15 for y in (3).

$10x + 7 \cdot 15 = -15$

$10x + 105 = -15$

$10x = -120$

$x = -12$

Solution set: {(-12, 15)}

39.

(1) $y = \frac{3}{4} x + 6$

(2) $x = \frac{4}{3} y - 8$

Place (1) and (2) in standard form and simplify by multiplying (1) by 4 and (2) by 3.

(3) $3x - 4y = -24$

(4) $\underline{3x - 4y = -24}$

$0 = 0$

Dependent

41.

Let x = first number

Let y = second number

The equatiions are

(1) $x + y = 12$

(2) $x = 5y + 30$

Place (2) in standard form as equation (3) below.

(1) $x + y = 12$

(3) $x - 5y = 30$

To eliminate the y's, multiply (1) by 5.

(4) $5x + 5y = 60$

(3) $\underline{x - 5y = 30}$

$6x = 90$

$x = 15$

Substitute 15 for x in (1).

$15 + y = 12$

$y = -3$

The numbers are 15 and -3.

43.

Let x = numerator

Let y = denominator

The equations are

(1) $x - y = 6$

(2) $\dfrac{x + 2}{y + 2} = \dfrac{5}{3}$

Place (2) in standard form as equation (3) below.

(1) $x - y = 6$

(3) $3x - 5y = 4$

To eliminate the x's, multiply (1) by -3.

(4) $-3x + 3y = -18$

(3) $\underline{3x - 5y = \quad 4}$

$-2y = -14$

$y = 7$

Substitute 7 for y in (1).

$x - 7 = 6$

$x = 13$

The original fraction is $\dfrac{13}{7}$.

45.

Let x = number of adult tickets

Let y = number of children's tickets

Let 24.75x = value of adult tickets sold

Let 18.50y = value of children's tickets sold

The equations are

(1) $x + y = 48$

(2) $24.75x + 18.50y = 925.50$

Simplify (2) by multiplying by 100.

(1) $x + y = 48$

(3) $2475x + 1850y = 92550$

To eliminate the y's, multiply (1) by - 1850.

(4) $-1850x - 1850y = -88,800$

(3) $\underline{2475x + 1850y = \quad 92,550}$

$625x = 3750$

$x = 6$

Substitute 6 for x in (1)

$6 + y = 48$

$y = 42$

There were 6 adult and 42 children's tickets.

47.
Let x = amount of Ann's
Let y = amount of Bennie's

	Amount ·	Percent =	Chili Powder
Ann's	x	.55	.55x
Bennie's	y	.07	.07y
Fred's	18	.23	.23(18)

The equations are
(1)   x + y = 18
(2)   .55x + .07y = .23(18)

Simplify (2) by multplying by 100.
(1)   x + y = 18
(3)   55x + 7y = 414

To eliminate the y's, multiply (1) by - 7.
(4)   -7x - 7y = - 126
(3)   <u>55x + 7y =   414</u>
            48x   = 288
               x = 6

Substitute 6 for x in (1).
         6 + y = 18
             y = 12

Use 6 oz of Ann's and 12 oz of Bennie's.

49. - 55.
The answers do not require worked-out solutions

Chapter 7 Test Solutions

1.

(1)   3x + 10y = 62
(2)   -4x + y = -11

#1   continued

Solve (2) for y.
(3)   y = 4x - 11
Substitute 4x - 11 for y in (1).
      3x + 10(4x - 11) = 62
        3x + 40x - 110 = 62
             43x - 110 = 62
                   43x = 172
                     x = 4
Substitute 4 for x in (3).
y = 4x - 11
y = 4 · 4 - 11   Since x = 4
y = 5
Solution set: {(4, 5)}

3.

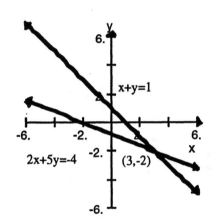

5.

(1)   3x + 7y   = 4
(2)   5x - 2y   = -7

To eliminate the y's, multiply (1) by 2 and (2) by 7.

(3)   6x + 14y =   8
(4)   <u>35x - 14y = -49</u>
           41x   = -41
                 x = -1

Substitute -1 for x in (1)

309

#5 continued

$$3(-1) + 7y = 4$$
$$-3 + 7y = 4$$
$$7y = 7$$
$$y = 1$$

Solution set: $\{(-1, 1)\}$

7.

The graph is in the answer section of the main text.

9.

Let x = amount invested at 3%.

Let y = amount invested at 2%.

Let $0.03x$ = amount of interest yield at 3%.

Let $0.02y$ = amount of interest yield at 2%.

The total amount of investment was $80,000 and the year's interest was $1,950.

The equations are:

$$(1) \qquad x + y = 80,000$$
$$(2) \ 0.03x + 0.02y = 1950$$

Simplify (2) by multiplying by 100.

$$(1) \qquad x + y = 80,000$$
$$(3) \qquad 3x + 2y = 195,000$$

To eliminate the x's, multiply (1) by -3.

$$(4) \ -3x - 3y = - 240,000$$
$$(3) \ \underline{\ 3x + 2y = \ 195,000\ }$$
$$-y = -45,000$$
$$y = 45,000$$

Substitute 45,000 for y in (1).

$$x + 45,000 = 80,000$$
$$x = 35,000$$

Answer: $45,000 at 2%, $35,000 at 3%.

Test Your Memory

1–7.

The graphs are in the answer section of the main text.

9.

$$(-5, 8), (7, -1)$$
$$m = \frac{-1 - 8}{7 + 5} = \frac{-9}{12} = -\frac{3}{4}$$

11.

$$m = \frac{-5}{2}, \ (-2, -4)$$

$$y + 4 = -\frac{5}{2}(x + 2)$$
$$2y + 8 = -5(x + 2) \qquad \text{Multiply by 2}$$
$$2y + 8 = -5x - 10$$
$$2y + 5x = -18 \qquad \text{Add 5x, subtract 8}$$

13.

$$(8, 7), (-4, 3)$$

$$m = \frac{3 - 7}{-4 - 8} = \frac{-4}{-12} = \frac{1}{3}$$

$$y - 7 = \frac{1}{3}(x - 8)$$

$$3y - 21 = x - 8 \qquad \text{Multiply by 3}$$
$$-13 = x - 3y, \text{ or} \qquad \text{Add 8, subtract 3y}$$
$$x - 3y = -13$$

15.

$$4x - 6y = 1 \qquad\qquad 2x + 3y = 12$$
$$-6y = -4x + 1 \qquad\qquad 3y = -2x + 12$$
$$y = \frac{-4x}{-6} - \frac{1}{6} \qquad\qquad y = -\frac{2}{3}x + 4$$
$$y = \frac{2}{3}x - \frac{1}{6} \qquad\qquad m = -\frac{2}{3}$$
$$m = \frac{2}{3}$$

The lines are neither parallel nor perpendicular.

17.

$$8 - 2(4x + 7) = 3(2x + 2)$$
$$8 - 8x - 14 = 6x + 6$$
$$-8x - 6 = 6x + 6$$
$$-6 = 14x + 6 \qquad \text{Add 8x}$$
$$-12 = 14x \qquad \text{Subtract -6}$$
$$-\frac{12}{14} = x \qquad \text{Divide by 14}$$
$$-\frac{6}{7} = x$$

Solution set: $\left\{-\frac{6}{7}\right\}$

19.

$$(3x + 3)(x - 5) = -15$$

#19 continued

$$3x^2 - 12x - 15 = -15$$
$$3x^2 - 12x = 0 \qquad \text{Add 15}$$
$$x^2 - 4x = 0 \qquad \text{Divide by 3}$$
$$x(x - 4) = 0$$
$$x = 0 \text{ or } x = 4$$

Solution set: $\{0, 4\}$

21.

$$\frac{2}{3x+1} = \frac{3}{5x-2}$$
$$2(5x - 2) = 3(3x + 1)$$
$$10x - 4 = 9x + 3$$
$$x - 4 = 3 \qquad \text{Subtract 9x}$$
$$x = 7 \qquad \text{Add 4}$$

Solution set: $\{7\}$

23.

$$\frac{2}{x - 2} - \frac{1}{x} = \frac{5}{4x} \qquad \text{LCD} = 4x(x - 2)$$
$$4x(x - 2)\left(\frac{2}{x - 2} - \frac{1}{x}\right) = \left(\frac{5}{4x}\right)4x(x - 2)$$
$$8x - 4(x - 2) = 5(x - 2)$$
$$8x - 4x + 8 = 5x - 10$$
$$4x + 8 = 5x - 10$$
$$8 = x - 10 \qquad \text{Subtract 4x}$$
$$18 = x \qquad \text{Add 10}$$

Solution set: $\{18\}$

25.

$$\frac{x}{x + 1} - \frac{1}{2x - 1} = \frac{1}{3} \qquad \text{LCD} = 3(x + 1)(2x - 1)$$
$$3(x + 1)(2x - 1)\left(\frac{x}{x + 1} - \frac{1}{2x - 1}\right)$$
$$= \left(\frac{1}{3}\right)3(x + 1)(2x-1)$$
$$3x(2x - 1) - 3(x + 1) = (x + 1)(2x - 1)$$
$$6x^2 - 3x - 3x - 3 = 2x^2 + x - 1$$
$$6x^2 - 6x - 3 = 2x^2 + x - 1$$
$$4x^2 - 7x - 2 = 0 \qquad \text{Subtract } 2x^2, \text{ add 1}$$
$$(x - 2)(4x + 1) = 0$$

#25 continued

$$x = 2 \qquad \text{or } x = -\frac{1}{4}$$

Solution set: $\left\{2, -\frac{1}{4}\right\}$

27.
$$\frac{(2x^3y^{-4})^{-2}}{(2x^{-1}y^{-7})-1} = \frac{2^{-2}x^{-6}y^8}{2^{-1}xy^7} = \frac{y^{8-7}}{2^{-1+2}x^{1+6}} = \frac{y}{2x^7}$$

29.

$$\frac{6x^2\left[\dfrac{1}{6} - \dfrac{1}{2x}\right]}{6x^2\left[\dfrac{1}{3x} - \dfrac{1}{x^2}\right]} \qquad \text{LCD} = 6x^2$$
$$= \frac{x^2 - 3x}{2x - 6} = \frac{x(x - 3)}{2(x - 3)} = \frac{x}{2}$$

31.

$$3x^2 + 7x - 4$$
$$3x - 2$$
$$9x^3 + 21x^2 - 12x$$
$$-6x^2 - 14x + 8$$
$$9x^3 + 15x^2 - 26x + 8$$

33.

$$\frac{4x^2 - 9y^2}{2y^2 + 11y + 12} \div \frac{10xy^2 - 15y^3}{3y^2 + 12y}$$
$$= \frac{(2x - 3y)(2x + 3y)}{(2y + 3)(y + 4)} \cdot \frac{3y(y + 4)}{5y^2(2x - 3y)} \qquad \text{Factor, invert}$$
$$= \frac{3x(2x + 3y)}{5y \cdot x(2y + 3)} = \frac{3(2x + 3y)}{5y(2y + 3)}$$

35.

$$\frac{3y}{3y} \cdot \frac{5}{2x} + \frac{4}{3y} \cdot \frac{2x}{2x} \qquad \text{LCD} = 6xy$$
$$= \frac{15y}{6xy} + \frac{8x}{6xy} = \frac{15y+8x}{6xy}$$

37.

$$\frac{2x}{x^2 - 5x + 6} - \frac{x - 18}{x^2 - 4}$$
$$= \frac{2x}{(x - 3)(x - 2)} - \frac{x - 18}{(x + 2)(x - 2)}$$
$$\text{LCD} = (x - 3)(x - 2)(x + 2)$$

#37 continued

$$= \frac{(x + 2)}{(x + 2)}\left(\frac{2x}{(x - 3)(x - 2)}\right)$$
$$- \left(\frac{x - 18}{(x + 2)(x-2)}\right)\frac{(x - 3)}{(x - 3)}$$

$$= \frac{2x^2 + 4x}{(x - 3)(x - 2)(x + 2)}$$
$$- \frac{(x^2 - 21x + 54)}{(x - 3)(x - 2)(x + 2)}$$

$$= \frac{2x^2 + 4x - x^2 + 21x - 54}{(x - 3)(x - 2)(x + 2)}$$

$$= \frac{x^2 + 25x - 54}{(x - 3)(x - 2)(x + 2)}$$

$$= \frac{(x - 2)(x + 27)}{(x - 3)(x - 2)(x + 2)} = \frac{x + 27}{(x - 3)(x + 2)}$$

39.

(1) $2x + 5y = 1$

(2) $3x - y = 10$

Solve (2) for y.

(3) $y = 3x - 10$

Substitute $3x - 10$ for y in (1).

$$2x + 5(3x - 10) = 1$$
$$2x + 15x - 50 = 1$$
$$17x = 51$$
$$x = 3$$

Substitute 3 for x in (3).

$$y = 3 \cdot 3 - 10$$
$$y = -1$$

Solution set: $\{(3, -1)\}$

41.

(1) $2x - 8y = 12$

(2) $3x - 12y = 18$

To eliminate the x's, multiply (1) by 3 and (2) by -2.

(3) $6x - 24y = 36$

(4) $\underline{-6x + 24y = -36}$

$$0 = 0$$

This indicates that the system is dependent.

43.

Let x = number of nickels.

Then, 3x - 2 = number of dimes.

Thus,

$$5x + 10(3x - 2) = 260$$
$$5x + 30x - 20 = 260$$
$$35x = 280$$
$$x = 8$$
$$3x - 2 = 22$$

Answer: 8 nickels, 22 dimes.

45.

Let x = amount invested at 4.5%.

Then 5000 - x = amount invested at 7.25%.

Thus,

$$0.045(x) + 0.0725(5000 - x) = 335$$
$$450x + 725(5000 - x) = 3,350,000 \quad \text{Multiply by 10,000}$$
$$450x + 3,625,000 - 725x = 3,350,000$$
$$-275x = -275,000$$
$$x = 1000$$
$$5000 - x = 4000$$

Answer: $1,000 in savings, $4000 in CD

47.

Let x = measurement of width.

Then $\frac{1}{2}x + 5$ = measurement of length.

Thus, $A = L \cdot W$

$$48 = \left(\frac{1}{2}x + 5\right)x$$

$$48 = \frac{1}{2}x^2 + 5x$$

$$96 = x^2 + 10x \qquad \text{Multiply by 2}$$
$$0 = x^2 + 10x - 96$$
$$0 = (x + 16)(x - 6)$$
$$x = -16 \text{ or } x = 6$$

Answer: 6m x 8m
Measurement is not negative.

49.

$$\frac{1}{8} + \frac{1}{x} = \frac{1}{3} \qquad \text{LCD} = 24x$$

$$24x\left(\frac{1}{8} + \frac{1}{x}\right) = \left(\frac{1}{3}\right)24x$$

$$3x + 24 = 8x$$

$$24 = 5x$$

$$\frac{24}{5} = x$$

Answer: $\frac{24}{5}$ hrs., or $4\frac{4}{5}$ hrs., or 4 hrs. 48 min.

CHAPTER 7 STUDY GUIDE

Self-Test Exercises

Determine the solutions of the following linear systems of equations using the indicated methods. If there are no solutions, write *inconsistent*. If there are infinitely many solutions, write *dependent*.

1. $2x - 3y = -11$ Substitution
   $x + 3y = 8$

2. $2x + 3y = 2$ Elimination
   $3x - 2y = 16$

3. $x + y = 3$ Graphing
   $x - 2y = 12$

4. $2x - 3y = 18$ Substitution
   $x - \frac{3}{2} y = 12$

5. $3x - 5y = 10$ Elimination
   $7x - 14 = 3y$

6. $x + 4y = 2$ Elimination
   $x - 8y = -7$

Graph the solution set of each of the following systems of linear inequalities.

7. $y < -2$
   $y \leq \frac{1}{2} x - 3$

8. $x + y \geq 1$
   $x - y \geq 1$

9. $3x + 2y \geq 7$
   $x - 5y < 5$

10. $3x - 5y \geq 15$
    $2x + 3y < 6$

11. $5x - 2y < 4$
    $3x + 4y \geq -8$

Use a linear system of equation to solve each of the following problems.

12. The difference of two numbers is 28. If the smaller is $\frac{1}{3}$ the larger, find the two numbers.

13. The sum of two numbers is 40. One number is 2 more than the other. Find the two numbers.

14. Margaret has $3.10 in nickels and quarters. If she has a total of 26 coins, how many of each kind of coin does she have?

15. Jamie bought 50 stamps. Some were 25¢ stamps, and some were 30¢ stamps. The cost was $14.15. How many of each type of stamp did he purchase?

The worked-out solutions begin on the next page.

Self-Test Solutions

1.

(1) $2x - 3y = -11$

(2) $x + 3y = 8$

Solve (2) for x.

(3) $x = -3y + 8$

Substitute $-3y + 8$ for x in (1).

$2(-3y + 8) - 3y = -11$

$-6y + 16 - 3y = -11$

$-9y = -27$

$y = 3$

Substitue 3 for y in (3).

$x = -3 \cdot 3 + 8$

$x = -1$

Solution set: $\{(-1, 3)\}$

2.

(1) $2x + 3y = 2$

(2) $3x - 2y = 16$

To eliminate the y's, multiply (1) by 2 and (2) by 3.

(3) $4x + 6y = 4$

(4) $\underline{9x - 6y = 48}$

$13x = 52$

$x = 4$

Substitute 4 for x in (1)

$2 \cdot 4 + 3y = 2$

$3y = -6$

$y = -2$

Solution set: $\{(4, -2)\}$.

3.

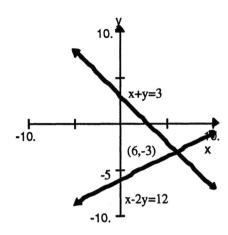

4.

(1) $2x - 3y = 18$

(2) $x - \dfrac{3}{2} y = 12$

Solve (2) for x.

(3) $x = \dfrac{3}{2} y + 12$

Substitute $\dfrac{3}{2} y + 12$ for x in (1).

$2\left(\dfrac{3}{2} y + 12\right) - 3y = 18$

$3y + 24 - 3y = 18$

$24 = 18$

The system is inconsistent.

5.

(1) $3x - 5y = 10$

(2) $7x - 14 = 3y$

Place (2) in standard form.

(1) $3x - 5y = 10$

(3) $7x - 3y = 14$

To eliminate the x's, multiply (1) by 7 and (3) by -3.

(4) $21x - 35y = 70$

(5) $\underline{-21x + 9y = -42}$

$-26y = 28$

$y = -\dfrac{28}{26} = -\dfrac{14}{13}$

#5 continued

Substitute $-\dfrac{14}{13}$ for y in (1).

$$3x - 5\left(-\dfrac{14}{13}\right) = 10$$

$$3x + \dfrac{70}{13} = 10$$

$$3x = 10 - \dfrac{70}{13}$$

$$3x = \dfrac{60}{13}$$

$$x = \dfrac{20}{13}$$

Solution set: $\left\{\left(\dfrac{20}{13},\ -\dfrac{14}{13}\right)\right\}$

6.
(1) $x + 4y = 2$
(2) $x - 8y = -7$
To eliminate the x's, multiply (2) by -1.
(1) $x + 4y = 2$
(3) $\underline{-x + 8y = 7}$
      $12y = 9$
      $y = \dfrac{9}{12} = \dfrac{3}{4}$

Substitute $\dfrac{3}{4}$ for y in (1).

$$x + 4\left(\dfrac{3}{4}\right) = 2$$
$$x + 3 = 2$$
$$x = -1$$

Solution set: $\left\{\left(-1,\ \dfrac{3}{4}\right)\right\}$

7.

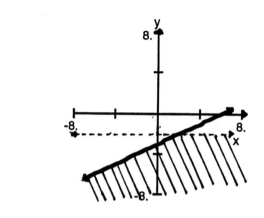

8.

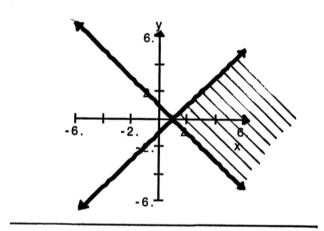

9.

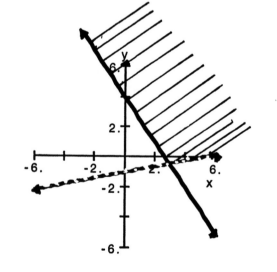

10.

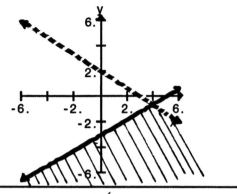

11.

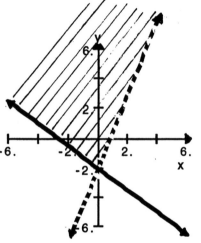

12.

Let x and y = two numbers.

Thus,

(1) x - y = 28

(2) $y = \frac{1}{3} x$

Multiply (2) by 3 and place in standard form.

(1)   x - y = 28

(3) $\underline{-x + 3y = 0}$

        2y = 28

        y = 14, x = 42

The two numbers are 14 and 42.

13.

Let x and y = two numbers

thus,

(1) x + y = 40

(2) x = y + 2

Place (2) in standard form.

#13 continued

(1) x + y = 40

(3) $\underline{x - y = \ \ 2}$

    2x   = 42

    x   = 21

Substitute 21 for x in (1)

  21 + y = 40

        y = 19

The two numbers are 21 and 19.

14.

Let x = number of nickels.

Let y = number of quarters.

(1)   x + y = 26

      5x + 25y = 310

To eliminate the x's, multiply (1) by -5.

(3)   -5x - 5y = -130

(2)   5x + 25y = 310

          20y = 180

          y = 9

Substitute 9 for y in (1).

        x + 9 = 26

        x = 17

Answer: 17 nickels and 9 quarters.

15.

Let x = number of 25¢ stamps

Let y = number of 30¢ stamps

Thus,

(1)   x + y = 50

(2) 25x + 30 y = 1415

Solve (1) for x and substitute 50 - y for x in (2)

(3) x = 50 - y

(2) 25(50 - y) + 30 y = 1415

  1250 - 25y + 30y = 1415

          5y = 165

          y = 33

Substitute 33 for y in (1)

  x + 33 = 50

      x = 17

Answer: 17-25¢ stamps, 33-30¢ stamps

CHAPTER 8   RADICAL EXPRESSIONS

Solutions   to   Text   Exercises

1. - 32.
The answers do not require worked-out solutions.

33.
$$\sqrt{\sqrt{81}} = \sqrt{9} = 3$$

First evaluate $\sqrt{81}$, then $\sqrt{9}$.

35. - 89.
The answers do not require worked-out solutions.

91.
$$\sqrt{\sqrt[3]{64}} = \sqrt{4} = 2$$

First evaluate $\sqrt[3]{64}$, then $\sqrt{4}$.

93. - 99.
The answers do not require worked-out solutions.

101.
$$\left(\sqrt[3]{(5x^2)^3}\right)^3 = (5x^2)^3 = 125x^6$$

103. - 119.
The answers do not require worked-out solutions.

121.
$$\sqrt{x^2 + 2x + 1} = \sqrt{(x + 1)^2} = x + 1$$

123.
$$\sqrt{4y^2 + 12y + 9} = \sqrt{(2y + 3)^2} = 2y + 3$$

125.
$$\sqrt{25x^2 + 20xy + 4y^2} = \sqrt{(5x + 2y)^2}$$
$$= 5x + 2y$$

1. - 5.
The answers do not require worked-out solutions.

7.
$$\sqrt{20} \cdot \sqrt{5} = \sqrt{100} = \sqrt{10^2} = 10$$

9.
$$\sqrt{3} \cdot \sqrt{12} = \sqrt{36} = \sqrt{6^2} = 6$$

11. - 15.
The answers do not require worked-out solutions.

17.
$$\sqrt{2x} \cdot \sqrt{8x} = \sqrt{16x^2} = \sqrt{(4x)^2} = 4x$$

19.
$$\sqrt{x} \cdot \sqrt{x^5} = \sqrt{x^6} = \sqrt{(x^3)^2} = x^3$$

21.
$$\sqrt{5x^3} \cdot \sqrt{5x^5} = \sqrt{25x^8} = \sqrt{(5x^4)^2} = 5x^4$$

23.
$$\sqrt{32x^7} \cdot \sqrt{2x^3} = \sqrt{64x^{10}} = \sqrt{(8x^5)^2} = 8x^5$$

25.
$$\sqrt{18} = \sqrt{9 \cdot 2} = \sqrt{9} \cdot \sqrt{2} = 3\sqrt{2}$$

27.
$$\sqrt{20} = \sqrt{4 \cdot 5} = \sqrt{4} \cdot \sqrt{5} = 2\sqrt{5}$$

29.
$\sqrt{65}$ Simplified

31.
$$\sqrt{150} = \sqrt{25 \cdot 6} = \sqrt{25} \cdot \sqrt{6} = 5\sqrt{6}$$

33.
$$\sqrt{40} = \sqrt{4 \cdot 10} = \sqrt{4} \cdot \sqrt{10} = 2\sqrt{10}$$

35.
$$\sqrt{54} = \sqrt{9 \cdot 6} = \sqrt{9} \cdot \sqrt{6} = 3\sqrt{6}$$

37.
$$\sqrt{7x^2} = \sqrt{x^2 \cdot 7} = \sqrt{x^2} \cdot \sqrt{7} = x\sqrt{7}$$

39. - 45.
The answers do not require worked-out solutions.

47.
$$\sqrt{121x} = \sqrt{121} \cdot \sqrt{x} = 11\sqrt{x}$$

49.
$$\sqrt{x^{11}} = \sqrt{x^{10} \cdot x} = \sqrt{x^{10}} \cdot \sqrt{x} = x^5\sqrt{x}$$

51.
$$\sqrt{64x^5} = \sqrt{64x^4 x} = \sqrt{64x^4} \cdot \sqrt{x} = 8x^2\sqrt{x}$$

53.
$$\sqrt{72x^2} = \sqrt{36x^2 \cdot 2} = \sqrt{36x^2} \cdot \sqrt{2} = 6x\sqrt{2}$$

55.
$$\sqrt{75x^3} = \sqrt{25x^2 \cdot 3x} = \sqrt{25x^2} \cdot \sqrt{3x} = 5x\sqrt{3x}$$

57.
$$\sqrt{48x^8y^6} = \sqrt{16x^8y^6 \cdot 3}$$
$$= \sqrt{16x^8y^6} \cdot \sqrt{3} = 4x^4y^3\sqrt{3}$$

59.
$$\sqrt{5x^6y^5} = \sqrt{x^6y^4 \cdot 5y}$$
$$= \sqrt{x^6y^4} \cdot \sqrt{5y} = x^3y^2\sqrt{5y}$$

61.
$$\sqrt{32x^9y} = \sqrt{16x^8 \cdot 2xy}$$
$$= \sqrt{16x^8} \cdot \sqrt{2xy} = 4x^4\sqrt{2xy}$$

63.
$$\sqrt{\frac{49}{4}} = \frac{\sqrt{49}}{\sqrt{4}} = \frac{7}{2}$$

65.
$$\sqrt{\frac{5}{64}} = \frac{\sqrt{5}}{\sqrt{64}} = \frac{\sqrt{5}}{8}$$

67.
$$\frac{\sqrt{6}}{\sqrt{96}} = \sqrt{\frac{6}{96}} = \sqrt{\frac{1}{16}} = \frac{\sqrt{1}}{\sqrt{16}} = \frac{1}{4}$$

69.
$$\frac{\sqrt{98}}{\sqrt{8}} = \sqrt{\frac{98}{8}} = \sqrt{\frac{49}{4}} = \frac{\sqrt{49}}{\sqrt{4}} = \frac{7}{2}$$

71.
$$\frac{\sqrt{200}}{\sqrt{8}} = \sqrt{\frac{200}{8}} = \sqrt{25} = 5$$

73.
$$\frac{\sqrt{48}}{\sqrt{50}} = \sqrt{\frac{48}{50}} = \sqrt{\frac{24}{25}} = \frac{\sqrt{24}}{5} = \frac{\sqrt{4 \cdot 6}}{5}$$
$$= \frac{\sqrt{4} \cdot \sqrt{6}}{5} = \frac{2\sqrt{6}}{5}$$

75.
$$\frac{\sqrt{160}}{\sqrt{45}} = \sqrt{\frac{160}{45}} = \sqrt{\frac{32}{9}} = \frac{\sqrt{32}}{\sqrt{9}} = \frac{\sqrt{16 \cdot 2}}{3}$$
$$= \frac{\sqrt{16} \cdot \sqrt{2}}{3} = \frac{4\sqrt{2}}{3}$$

77.
$$\frac{\sqrt{24}}{\sqrt{27}} = \sqrt{\frac{24}{27}} = \sqrt{\frac{8}{9}} = \frac{\sqrt{8}}{\sqrt{9}} = \frac{\sqrt{4 \cdot 2}}{3}$$
$$= \frac{\sqrt{4} \cdot \sqrt{2}}{3} = \frac{2\sqrt{2}}{3}$$

79. - 83.
The answers do not require worked-out solutions.

85.
$$\sqrt{\frac{24x^3}{49}} = \frac{\sqrt{24x^3}}{\sqrt{49}} = \frac{\sqrt{4x^2 \cdot 6x}}{7}$$
$$= \frac{\sqrt{4x^2} \cdot \sqrt{6x}}{7} = \frac{2x\sqrt{6x}}{7}$$

87.
$$\sqrt{\frac{112x}{81}} = \frac{\sqrt{112x}}{\sqrt{81}} = \frac{\sqrt{16 \cdot 7x}}{9}$$
$$= \frac{\sqrt{16} \cdot \sqrt{7x}}{9} = \frac{4\sqrt{7x}}{9}$$

89.

$$\sqrt{\frac{18x}{25y^2}} = \frac{\sqrt{18x}}{\sqrt{25y^2}} = \frac{\sqrt{9 \cdot 2x}}{5y}$$

$$= \frac{\sqrt{9} \cdot \sqrt{2x}}{5y} = \frac{3\sqrt{2x}}{5y}$$

91.

$$\sqrt{\frac{20x^4}{9y^4}} = \frac{\sqrt{20x^4}}{\sqrt{9y^4}} = \frac{\sqrt{4x^4 \cdot 5}}{3y^2}$$

$$= \frac{\sqrt{4x^4} \cdot \sqrt{5}}{3y^2} = \frac{2x^2\sqrt{5}}{3y^2}$$

93.

$$\sqrt{\frac{32x^3}{25y^8}} = \frac{\sqrt{32x^3}}{\sqrt{25y^8}} = \frac{\sqrt{16x^2 \cdot 2x}}{5y^4}$$

$$= \frac{\sqrt{16x^2} \cdot \sqrt{2x}}{5y^4} = \frac{4x\sqrt{2x}}{5y^4}$$

95.

$$\sqrt{\frac{75x^7}{16y^6}} = \frac{\sqrt{75x^7}}{\sqrt{16y^6}} = \frac{\sqrt{25x^6 \cdot 3x}}{4y^3}$$

$$= \frac{\sqrt{25x^6} \cdot \sqrt{3x}}{4y^3} = \frac{5x^3\sqrt{3x}}{4y^3}$$

97.

$$\sqrt[3]{81} = \sqrt[3]{27 \cdot 3} = \sqrt[3]{27} \cdot \sqrt[3]{3} = 3\sqrt[3]{3}$$

99.

$$\sqrt[3]{40} = \sqrt[3]{8 \cdot 5} = \sqrt[3]{8} \cdot \sqrt[3]{5} = 2\sqrt[3]{5}$$

101.

$$\sqrt[3]{250} = \sqrt[3]{125 \cdot 2} = \sqrt[3]{125} \cdot \sqrt[3]{2} = 5\sqrt[3]{2}$$

103.

$$\sqrt[3]{135x^3} = \sqrt[3]{27x^3 \cdot 5} = \sqrt[3]{27x^3} \cdot \sqrt[3]{5} = 3x\sqrt[3]{5}$$

105.

$$\sqrt[3]{27x^5} = \sqrt[3]{27x^3 \cdot x^2}$$

$$= \sqrt[3]{27x^3} \cdot \sqrt[3]{x^2} = 3x\sqrt[3]{x^2}$$

107.

$$\sqrt[3]{56x^4} = \sqrt[3]{8x^3 \cdot 7x} = \sqrt[3]{8x^3} \cdot \sqrt[3]{7x} = 2x\sqrt[3]{7x}$$

109. - 111.
The answers do not require worked-out solutions.

113.

$$\sqrt[3]{\frac{7x^4}{64}} = \frac{\sqrt[3]{7x^4}}{\sqrt[3]{64}} = \frac{\sqrt[3]{x^3 \cdot 7x}}{4}$$

$$= \frac{\sqrt[3]{x^3} \cdot \sqrt[3]{7x}}{4} = \frac{x\sqrt[3]{7x}}{4}$$

115.

$$\sqrt[3]{\frac{16x^4}{27y^3}} = \frac{\sqrt[3]{16x^4}}{\sqrt[3]{27y^3}} = \frac{\sqrt[3]{8x^3 \cdot 2x}}{3y}$$

$$= \frac{\sqrt[3]{8x^3} \cdot \sqrt[3]{2x}}{3y} = \frac{2x\sqrt[3]{2x}}{3y}$$

Exercises 8.3

1. - 23.
The answers do not need worked-out soluions.

25.

$$\sqrt{48} + \sqrt{12} = \sqrt{16 \cdot 3} + \sqrt{4 \cdot 3}$$
$$= \sqrt{16} \cdot \sqrt{3} + \sqrt{4} \cdot \sqrt{3}$$
$$= 4\sqrt{3} + 2\sqrt{3}$$
$$= 6\sqrt{3}$$

27.

$$2\sqrt{45} - 2\sqrt{20} = 2\sqrt{9 \cdot 5} - 2\sqrt{4 \cdot 5}$$
$$= 2\sqrt{9} \cdot \sqrt{5} - 2\sqrt{4} \cdot \sqrt{5}$$
$$= 2 \cdot 3\sqrt{5} - 2 \cdot 2\sqrt{5}$$
$$= 6\sqrt{5} - 4\sqrt{5}$$
$$= 2\sqrt{5}$$

29.

$$\sqrt{75} - \sqrt{24} = \sqrt{25 \cdot 3} - \sqrt{4 \cdot 6}$$
$$= \sqrt{25} \cdot \sqrt{3} - \sqrt{4} \cdot \sqrt{6}$$
$$= 5\sqrt{3} - 2\sqrt{6}$$

31.

$$3\sqrt{63} + 5\sqrt{7} = 3\sqrt{9 \cdot 7} + 5\sqrt{7} = 3\sqrt{9} \cdot \sqrt{7} + 5\sqrt{7}$$
$$= 3 \cdot 3\sqrt{7} + 5\sqrt{7}$$
$$= 9\sqrt{7} + 5\sqrt{7}$$
$$= 14\sqrt{7}$$

33.

$$\sqrt{8} - 2\sqrt{32} = \sqrt{4 \cdot 2} - 2\sqrt{16 \cdot 2}$$
$$= \sqrt{4} \cdot \sqrt{2} - 2\sqrt{16} \cdot \sqrt{2}$$
$$= 2\sqrt{2} - 2 \cdot 4\sqrt{2}$$
$$= 2\sqrt{2} - 8\sqrt{2}$$
$$= -6\sqrt{2}$$

35.

$$\sqrt{20} - \sqrt{5} - 3\sqrt{45} = \sqrt{4 \cdot 5} - \sqrt{5} - 3\sqrt{9 \cdot 5}$$
$$= \sqrt{4} \cdot \sqrt{5} - \sqrt{5} - 3\sqrt{9} \cdot \sqrt{5}$$
$$= 2\sqrt{5} - \sqrt{5} - 3 \cdot 3\sqrt{5}$$
$$= 2\sqrt{5} - \sqrt{5} - 9\sqrt{5}$$
$$= -8\sqrt{5}$$

37.

$$\sqrt{108} + 3\sqrt{27} + 2\sqrt{3}$$
$$= \sqrt{36 \cdot 3} + 3\sqrt{9 \cdot 3} + 2\sqrt{3}$$
$$= \sqrt{36} \cdot \sqrt{3} + 3\sqrt{9} \cdot \sqrt{3} + 2\sqrt{3}$$
$$= 6\sqrt{3} + 3 \cdot 3\sqrt{3} + 2\sqrt{3}$$
$$= 6\sqrt{3} + 9\sqrt{3} + 2\sqrt{3} = 17\sqrt{3}$$

39.

$$\sqrt{12} - 2\sqrt{48} + \sqrt{18}$$
$$= \sqrt{4 \cdot 3} - 2\sqrt{16 \cdot 3} + \sqrt{9 \cdot 2}$$
$$= \sqrt{4} \cdot \sqrt{3} - 2\sqrt{16} \cdot \sqrt{3} + \sqrt{9} \cdot \sqrt{2}$$
$$= 2\sqrt{3} - 2 \cdot 4\sqrt{3} + 3\sqrt{2}$$
$$= 2\sqrt{3} - 8\sqrt{3} + 3\sqrt{2} = -6\sqrt{3} + 3\sqrt{2}$$

41.

$$4\sqrt{3} + \sqrt{12} + 2\sqrt{18} + 2\sqrt{32}$$
$$= 4\sqrt{3} + \sqrt{4 \cdot 3} + 2\sqrt{9 \cdot 2} + 2\sqrt{16 \cdot 2}$$
$$= 4\sqrt{3} + \sqrt{4} \cdot \sqrt{3} + 2\sqrt{9} \cdot \sqrt{2} + 2\sqrt{16} \cdot \sqrt{2}$$
$$= 4\sqrt{3} + 2\sqrt{3} + 2 \cdot 3\sqrt{2} + 2 \cdot 4\sqrt{2}$$
$$= 6\sqrt{3} + 6\sqrt{2} + 8\sqrt{2} = 6\sqrt{3} + 14\sqrt{2}$$

43.

$$\sqrt{20} + \sqrt{54} - \sqrt{80} - \sqrt{150}$$
$$= \sqrt{4 \cdot 5} + \sqrt{9 \cdot 6} - \sqrt{16 \cdot 5} - \sqrt{25 \cdot 6}$$
$$= \sqrt{4} \cdot \sqrt{5} + \sqrt{9} \cdot \sqrt{6} - \sqrt{16} \cdot \sqrt{5} - \sqrt{25} \cdot \sqrt{6}$$

#43 continued

$$= 2\sqrt{5} + 3\sqrt{6} - 4\sqrt{5} - 5\sqrt{6} = -2\sqrt{5} - 2\sqrt{6}$$

45.

$$\sqrt{5} + \sqrt{45} - \sqrt{96} - 2\sqrt{6}$$
$$= \sqrt{5} + \sqrt{9 \cdot 5} - \sqrt{16 \cdot 6} - 2\sqrt{6}$$
$$= \sqrt{5} + \sqrt{9} \cdot \sqrt{5} - \sqrt{16} \cdot \sqrt{6} - 2\sqrt{6}$$
$$= \sqrt{5} + 3\sqrt{5} - 4\sqrt{6} - 2\sqrt{6} = 4\sqrt{5} - 6\sqrt{6}$$

47.

$$3\sqrt{12} + 2\sqrt{8} - \sqrt{75} - \sqrt{32}$$
$$= 3\sqrt{4 \cdot 3} + 2\sqrt{4 \cdot 2} - \sqrt{25 \cdot 3} - \sqrt{16 \cdot 2}$$
$$= 3\sqrt{4} \cdot \sqrt{3} + 2\sqrt{4} \cdot \sqrt{2} - \sqrt{25} \cdot \sqrt{3} - \sqrt{16} \cdot \sqrt{2}$$
$$= 3 \cdot 2\sqrt{3} + 2 \cdot 2\sqrt{2} - 5\sqrt{3} - 4\sqrt{2}$$
$$= 6\sqrt{3} + 4\sqrt{2} - 5\sqrt{3} - 4\sqrt{2} = \sqrt{3}$$

49.

$$2x\sqrt{45} + 4\sqrt{5x^2} = 2x\sqrt{9 \cdot 5} + 4\sqrt{x^2 \cdot 5}$$
$$= 2x\sqrt{9} \cdot \sqrt{5} + 4\sqrt{x^2} \cdot \sqrt{5}$$
$$= 2x \cdot 3\sqrt{5} + 4x\sqrt{5}$$
$$= 6x\sqrt{5} + 4x\sqrt{5}$$
$$= 10x\sqrt{5}$$

51.

$$\sqrt{24x^2} - 3\sqrt{54x^2} = \sqrt{4x^2 \cdot 6} - 3\sqrt{9x^2 \cdot 6}$$
$$= \sqrt{4x^2} \cdot \sqrt{6} - 3\sqrt{9x^2} \cdot \sqrt{6}$$
$$= 2x\sqrt{6} - 3 \cdot 3x\sqrt{6}$$
$$= 2x\sqrt{6} - 9x\sqrt{6}$$
$$= -7x\sqrt{6}$$

53.

$$x\sqrt{27} - x\sqrt{12} - \sqrt{3x^2}$$
$$= x\sqrt{9 \cdot 3} - x\sqrt{4 \cdot 3} - \sqrt{x^2 \cdot 3}$$
$$= x\sqrt{9} \cdot \sqrt{3} - x\sqrt{4} \cdot \sqrt{3} - \sqrt{x^2} \cdot \sqrt{3}$$
$$= x \cdot 3\sqrt{3} - x \cdot 2\sqrt{3} - x\sqrt{3}$$
$$= 3x\sqrt{3} - 2x\sqrt{3} - x\sqrt{3} = 0$$

55.

$$x\sqrt{10} + 2\sqrt{40x^2} - 2\sqrt{90x^2}$$
$$= x\sqrt{10} + 2\sqrt{4x^2 \cdot 10} - 2\sqrt{9x^2 \cdot 10}$$
$$= x\sqrt{10} + 2\sqrt{4x^2} \cdot \sqrt{10} - 2\sqrt{9x^2} \cdot \sqrt{10}$$
$$= x\sqrt{10} + 2 \cdot 2x\sqrt{10} - 2 \cdot 3x\sqrt{10}$$
$$= x\sqrt{10} + 4x\sqrt{10} - 6x\sqrt{10} = -x\sqrt{10}$$

57.

$$3\sqrt{20x} - \sqrt{45x} = 3\sqrt{4 \cdot 5x} - \sqrt{9 \cdot 5x}$$
$$= 3\sqrt{4} \cdot \sqrt{5x} - \sqrt{9} \cdot \sqrt{5x}$$
$$= 3 \cdot 2\sqrt{5x} - 3\sqrt{5x}$$
$$= 6\sqrt{5x} - 3\sqrt{5x} = 3\sqrt{5x}$$

59.

$$\sqrt{75x} + \sqrt{45x} = \sqrt{25 \cdot 3x} = \sqrt{9 \cdot 5x}$$
$$= \sqrt{25} \cdot \sqrt{3x} + \sqrt{9} \cdot \sqrt{5x}$$
$$= 5\sqrt{3x} + 3\sqrt{5x}$$

61.

$$\sqrt{10x} - 2\sqrt{40x} + 3\sqrt{160x}$$
$$= \sqrt{10x} - 2\sqrt{4 \cdot 10x} + 3\sqrt{16 \cdot 10x}$$
$$= \sqrt{10x} - 2\sqrt{4} \cdot \sqrt{10x} + 3\sqrt{16} \cdot \sqrt{10x}$$
$$= \sqrt{10x} - 2 \cdot 2\sqrt{10x} + 3 \cdot 4\sqrt{10x}$$
$$= \sqrt{10x} - 4\sqrt{10x} + 12\sqrt{10x} = 9\sqrt{10x}$$

63.

$$-\sqrt{54x} - \sqrt{96x} - \sqrt{24x}$$
$$= -\sqrt{9 \cdot 6x} - \sqrt{16 \cdot 6x} - \sqrt{4 \cdot 6x}$$
$$= -\sqrt{9} \cdot \sqrt{6x} - \sqrt{16} \cdot \sqrt{6x} - \sqrt{4} \cdot \sqrt{6x}$$
$$= -3\sqrt{6x} - 4\sqrt{6x} - 2\sqrt{6x} = -9\sqrt{6x}$$

65.

$$2x\sqrt{48x} + 2\sqrt{27x^3}$$
$$= 2x\sqrt{16 \cdot 3x} + 2\sqrt{9x^2 \cdot 3x}$$
$$= 2x\sqrt{16} \cdot \sqrt{3x} + 2\sqrt{9x^2} \cdot \sqrt{3x}$$
$$= 2x \cdot 4\sqrt{3x} + 2 \cdot 3x\sqrt{3x}$$
$$= 8x\sqrt{3x} + 6x\sqrt{3x} = 14x\sqrt{3x}$$

67.

$$-2\sqrt{50x^2y} - x\sqrt{72y}$$
$$= -2\sqrt{25x^2 \cdot 2y} - x\sqrt{36 \cdot 2y}$$
$$= -2\sqrt{25x^2} \cdot \sqrt{2y} - x\sqrt{36} \cdot \sqrt{2y}$$
$$= -2 \cdot 5x\sqrt{2y} - x \cdot 6\sqrt{2y}$$
$$= -10x\sqrt{2y} - 6x\sqrt{2y}$$
$$= -16x\sqrt{2y}$$

69.

$$\sqrt{96x^3} - 3\sqrt{6x^3} + 2x\sqrt{150x}$$
$$= \sqrt{16x^2 \cdot 6x} - 3\sqrt{x^2 \cdot 6x} + 2x\sqrt{25 \cdot 6x}$$
$$= \sqrt{16x^2} \cdot \sqrt{6x} - 3\sqrt{x^2} \cdot \sqrt{6x} + 2x\sqrt{25} \cdot \sqrt{6x}$$
$$= 4x\sqrt{6x} - 3x\sqrt{6x} + 2x \cdot 5\sqrt{6x}$$

#69 continued

$$= 4x\sqrt{6x} - 3x\sqrt{6x} + 10x\sqrt{6x}$$
$$= 11x\sqrt{6x}$$

71.

$$\sqrt{24xy^3} - 2y\sqrt{54xy} + \sqrt{12xy}$$
$$= \sqrt{4y^2 \cdot 6xy} - 2y\sqrt{9 \cdot 6xy} + \sqrt{4 \cdot 3xy}$$
$$= \sqrt{4y^2} \cdot \sqrt{6xy} - 2y\sqrt{9} \cdot \sqrt{6xy} + \sqrt{4} \cdot \sqrt{3xy}$$
$$= 2y\sqrt{6xy} - 2y \cdot 3\sqrt{6xy} + 2\sqrt{3xy}$$
$$= 2y\sqrt{6xy} - 6y\sqrt{6xy} + 2\sqrt{3xy}$$
$$= -4y\sqrt{6xy} + 2\sqrt{3xy}$$

73.

$$\sqrt{50x} + 2\sqrt{8x} + 2\sqrt{6x^3} + 2x\sqrt{96x}$$
$$= \sqrt{25 \cdot 2x} + 2\sqrt{4 \cdot 2x} + 2\sqrt{x^2 \cdot 6x}$$
$$\qquad\qquad\qquad + 2x\sqrt{16 \cdot 6x}$$
$$= \sqrt{25} \cdot \sqrt{2x} + 2\sqrt{4} \cdot \sqrt{2x} + 2\sqrt{x^2} \cdot \sqrt{6x}$$
$$\qquad\qquad\qquad + 2x\sqrt{16} \cdot \sqrt{6x}$$
$$= 5\sqrt{2x} + 4\sqrt{2x} + 2x\sqrt{6x} + 8x\sqrt{6x}$$
$$= 9\sqrt{2x} + 10x\sqrt{6x}$$

75.

$$2\sqrt{7x^2y} + 3x\sqrt{28y} - \sqrt{54xy^2} - \sqrt{6xy^2}$$
$$= 2\sqrt{x^2 \cdot 7y} + 3x\sqrt{4 \cdot 7y} - \sqrt{9y^2 \cdot 6x}$$
$$\qquad\qquad\qquad - \sqrt{y^2 \cdot 6x}$$
$$= 2\sqrt{x^2} \cdot \sqrt{7y} + 3x\sqrt{4} \cdot \sqrt{7y} - \sqrt{9y^2} \cdot \sqrt{6x}$$
$$\qquad\qquad\qquad - \sqrt{y^2} \cdot \sqrt{6x}$$
$$= 2x\sqrt{7y} + 3x \cdot 2\sqrt{7y} - 3y\sqrt{6x} - y\sqrt{6x}$$
$$= 8x\sqrt{7y} - 4y\sqrt{6x}$$

77.

$$x\sqrt{20} - 2x\sqrt{63y} + 3\sqrt{45x^2} + 5\sqrt{7x^2y}$$
$$= x\sqrt{4 \cdot 5} - 2x\sqrt{9 \cdot 7y} + 3\sqrt{9x^2 \cdot 5}$$
$$\qquad\qquad\qquad + 5\sqrt{x^2 \cdot 7y}$$
$$= x\sqrt{4} \cdot \sqrt{5} - 2x\sqrt{9} \cdot \sqrt{7y} + 3\sqrt{9x^2} \cdot \sqrt{5}$$
$$\qquad\qquad\qquad + 5\sqrt{x^2} \cdot \sqrt{7y}$$
$$= x \cdot 2\sqrt{5} - 2x \cdot 3\sqrt{7y} + 3 \cdot 3x\sqrt{5} + 5 \cdot x\sqrt{7y}$$
$$= 2x\sqrt{5} - 6x\sqrt{7y} + 9x\sqrt{5} + 5x\sqrt{7y}$$
$$= 11x\sqrt{5} - x\sqrt{7y}$$

79.

$$3\sqrt{48xy} - 4\sqrt{3xy} - \sqrt{36xy} - \sqrt{81xy}$$
$$= 3\sqrt{16 \cdot 3xy} - 4\sqrt{3xy} - \sqrt{36 \cdot xy} - \sqrt{81 \cdot xy}$$

#79 continued

$$= 3\sqrt{16} \cdot \sqrt{3xy} - 4\sqrt{3xy} - \sqrt{36} \cdot \sqrt{xy} \\ - \sqrt{81} \cdot \sqrt{xy}$$
$$= 3 \cdot 4\sqrt{3xy} - 4\sqrt{3xy} - 6\sqrt{xy} - 9\sqrt{xy}$$
$$= 8\sqrt{3xy} - 15\sqrt{xy}$$

81.

$$x\sqrt{40y^3} + 2y\sqrt{90x^2y} + 3xy\sqrt{160y} \\ + 4\sqrt{10x^2y^3}$$
$$= x\sqrt{4y^2 \cdot 10y} + 2y\sqrt{9x^2 \cdot 10y} \\ + 3xy\sqrt{16 \cdot 10y} + 4\sqrt{x^2y^2 \cdot 10y}$$
$$= x\sqrt{4y^2} \cdot \sqrt{10y} + 2y\sqrt{9x^2} \cdot \sqrt{10y} \\ + 3xy\sqrt{16} \cdot \sqrt{10y} + 4\sqrt{x^2y^2} \cdot \sqrt{10y}$$
$$= x \cdot 2y\sqrt{10y} + 2y \cdot 3x\sqrt{10y} + 3xy \cdot 4\sqrt{10y} \\ + 4 \cdot xy\sqrt{10y}$$
$$= 2xy\sqrt{10y} + 6xy\sqrt{10y} + 12xy\sqrt{10y} \\ + 4xy\sqrt{10y}$$
$$= 24xy\sqrt{10y}$$

83.

$$\sqrt[3]{54} + \sqrt[3]{250} = \sqrt[3]{27 \cdot 2} + \sqrt[3]{125 \cdot 2}$$
$$= \sqrt[3]{27} \cdot \sqrt[3]{2} + \sqrt[3]{125} \cdot \sqrt[3]{2}$$
$$= 3\sqrt[3]{2} + 5\sqrt[3]{2}$$
$$= 8\sqrt[3]{2}$$

85.

$$2\sqrt[3]{24} - 2\sqrt[3]{81} = 2\sqrt[3]{8 \cdot 3} - 2\sqrt[3]{27 \cdot 3}$$
$$= 2\sqrt[3]{8} \cdot \sqrt[3]{3} - 2\sqrt[3]{27} \cdot \sqrt[3]{3}$$
$$= 2 \cdot 2\sqrt[3]{3} - 2 \cdot 3\sqrt[3]{3}$$
$$= 4\sqrt[3]{3} - 6\sqrt[3]{3}$$
$$= -2\sqrt[3]{3}$$

87.

$$x\sqrt[3]{500} - 2\sqrt[3]{32x^3} = x\sqrt[3]{125 \cdot 4} - 2\sqrt[3]{8x^3 \cdot 4}$$
$$= x\sqrt[3]{125} \cdot \sqrt[3]{4} - 2\sqrt[3]{8x^3} \cdot \sqrt[3]{4}$$
$$= x \cdot 5\sqrt[3]{4} - 2 \cdot 2x\sqrt[3]{4}$$
$$= 5x\sqrt[3]{4} - 4x\sqrt[3]{4} = x\sqrt[3]{4}$$

89.

$$8\sqrt[3]{3x^4} + x\sqrt[3]{24x} = 8\sqrt[3]{x^3 \cdot 3x} + x\sqrt[3]{8 \cdot 3x}$$
$$= 8\sqrt[3]{x^3} \cdot \sqrt[3]{3x} + x\sqrt[3]{8} \cdot \sqrt[3]{3x}$$
$$= 8 \cdot x\sqrt[3]{3x} + x \cdot 2\sqrt[3]{3x}$$
$$= 8x\sqrt[3]{3x} + 2x\sqrt[3]{3x}$$
$$= 10x\sqrt[3]{3x}$$

Exercises 8.4

1.
$$2\sqrt{5} \cdot \sqrt{7} = 2\sqrt{5 \cdot 7} = 2\sqrt{35}$$

3.
$$9\sqrt{2} \cdot \sqrt{22} = 9\sqrt{44} = 9\sqrt{4 \cdot 11}$$
$$= 9\sqrt{4} \cdot \sqrt{11}$$
$$= 9 \cdot 2\sqrt{11}$$
$$= 18\sqrt{11}$$

5.
$$6\sqrt{15} \cdot \sqrt{6} = 6\sqrt{15 \cdot 6} = 6\sqrt{9 \cdot 10}$$
$$= 6\sqrt{9} \cdot \sqrt{10}$$
$$= 6 \cdot 3\sqrt{10}$$
$$= 18\sqrt{10}$$

7.
$$5\sqrt{3} \cdot 2\sqrt{7} = 10\sqrt{21}$$

9.
$$3\sqrt{10} \cdot 5\sqrt{14} = 3 \cdot 5\sqrt{10 \cdot 14}$$
$$= 15\sqrt{4 \cdot 35}$$
$$= 15\sqrt{4} \cdot \sqrt{35}$$
$$= 15 \cdot 2\sqrt{35}$$
$$= 30\sqrt{35}$$

11.
$$3\sqrt{7} \cdot 4\sqrt{6} = 12\sqrt{42}$$

13.
$$5\sqrt{3x} \cdot \sqrt{7y} = 5\sqrt{21xy}$$

15.
$$3\sqrt{11x} \cdot \sqrt{2y} = 3\sqrt{22xy}$$

323

17.

$$5\sqrt{2x} \cdot 7\sqrt{3y} = 35\sqrt{6xy}$$

19.

$$8\sqrt{2x} \cdot 2\sqrt{7xy} = 8 \cdot 2\sqrt{14x^2y}$$
$$= 16\sqrt{x^2} \cdot \sqrt{14y}$$
$$= 16x\sqrt{14y}$$

21.

$$3\sqrt{6x} \cdot 8\sqrt{10y} = 3 \cdot 8\sqrt{6x \cdot 10y}$$
$$= 24\sqrt{4 \cdot 15xy}$$
$$= 24\sqrt{4} \cdot \sqrt{15xy}$$
$$= 24 \cdot 2\sqrt{15xy}$$
$$= 48\sqrt{15xy}$$

23.

$$3\sqrt{50xy} \cdot 7\sqrt{2xy} = 3 \cdot 7\sqrt{50xy \cdot 2xy}$$
$$= 21\sqrt{100x^2y^2}$$
$$= 21 \cdot 10xy$$
$$= 210xy$$

25.

$$2\sqrt{3x} \cdot 4y\sqrt{10xy} = 2 \cdot 4y\sqrt{3x \cdot 10xy}$$
$$= 8y\sqrt{x^2 \cdot 30y}$$
$$= 8y\sqrt{x^2} \cdot \sqrt{30y}$$
$$= 8xy\sqrt{30y}$$

27.

$$9\sqrt{5xy} \cdot 4\sqrt{2xy} = 9 \cdot 4\sqrt{5xy \cdot 2xy}$$
$$= 36\sqrt{x^2y^2 \cdot 10}$$
$$= 36\sqrt{x^2y^2} \cdot \sqrt{10}$$
$$= 36xy\sqrt{10}$$

29.

$$xy\sqrt{2xy} \cdot 3x\sqrt{6y} = xy \cdot 3x\sqrt{2xy \cdot 6y}$$
$$= 3x^2y\sqrt{4y^2 \cdot 3x}$$
$$= 3x^2y\sqrt{4y^2} \cdot \sqrt{3x}$$
$$= 3x^2y \cdot 2y\sqrt{3x}$$
$$= 6x^2y^2\sqrt{3x}$$

31.

$$\sqrt{7}(\sqrt{3} + \sqrt{2}) = \sqrt{7} \cdot \sqrt{3} + \sqrt{7} \cdot \sqrt{2}$$
$$= \sqrt{21} + \sqrt{14}$$

33.

$$2\sqrt{5}(3\sqrt{6} + 4\sqrt{10}) = 2\sqrt{5} \cdot 3\sqrt{6}$$
$$+ 2\sqrt{5} \, 4\sqrt{10}$$
$$= 6\sqrt{30} + 8\sqrt{50}$$
$$= 6\sqrt{30} + 8\sqrt{25 \cdot 2}$$
$$= 6\sqrt{30} + 8\sqrt{25} \cdot \sqrt{2}$$
$$= 6\sqrt{30} + 8 \cdot 5\sqrt{2}$$
$$= 6\sqrt{30} + 40\sqrt{2}$$

35.

$$\sqrt{x}(\sqrt{5y} - \sqrt{3x}) = \sqrt{x} \cdot \sqrt{5y} - \sqrt{x} \cdot \sqrt{3x}$$
$$= \sqrt{5xy} - \sqrt{x^2 \cdot 3}$$
$$= \sqrt{5xy} - \sqrt{x^2} \cdot \sqrt{3}$$
$$= \sqrt{5xy} - x\sqrt{3}$$

37.

$$\sqrt{2}(\sqrt{3} + \sqrt{5} - \sqrt{7})$$
$$= \sqrt{2} \cdot \sqrt{3} + \sqrt{2} \cdot \sqrt{5} - \sqrt{2} \cdot \sqrt{7}$$
$$= \sqrt{6} + \sqrt{10} - \sqrt{14}$$

39.

$$5\sqrt{6}(2\sqrt{8} + 3\sqrt{5} - \sqrt{6})$$
$$= 5\sqrt{6} \cdot 2\sqrt{8} + 5\sqrt{6} \cdot 3\sqrt{5} - 5\sqrt{6} \cdot \sqrt{6}$$
$$= 10\sqrt{48} + 15\sqrt{30} - 5\sqrt{36}$$
$$= 10\sqrt{16 \cdot 3} + 15\sqrt{30} - 5\sqrt{36}$$
$$= 10\sqrt{16} \cdot \sqrt{3} + 15\sqrt{30} - 5\sqrt{36}$$
$$= 10 \cdot 4\sqrt{3} + 15\sqrt{30} - 5 \cdot 6$$
$$= 40\sqrt{3} + 15\sqrt{30} - 30$$

41.

$$2\sqrt{x}(3\sqrt{7} + \sqrt{2y} - 5\sqrt{x})$$
$$= 2\sqrt{x} \cdot 3\sqrt{7} + 2\sqrt{x} \cdot \sqrt{2y} - 2\sqrt{x} \cdot 5\sqrt{x}$$
$$= 6\sqrt{7x} + 2\sqrt{2xy} - 10\sqrt{x^2}$$
$$= 6\sqrt{7x} + 2\sqrt{2xy} - 10x$$

43.

$$(\sqrt{5} + \sqrt{2})(\sqrt{7} + \sqrt{3})$$

$$\quad F \qquad O \qquad I \qquad L$$
$$= \sqrt{5} \cdot \sqrt{7} + \sqrt{5} \cdot \sqrt{3} + \sqrt{2} \cdot \sqrt{7} + \sqrt{2} \cdot \sqrt{3}$$
$$= \sqrt{35} + \sqrt{15} + \sqrt{14} + \sqrt{6}$$

45.

$$\left(\sqrt{5} - \sqrt{3}\right)\left(\sqrt{6} - \sqrt{2}\right)$$

$$\begin{array}{cccc} \text{F} & \text{O} & \text{I} & \text{L} \end{array}$$
$$= \sqrt{5}\cdot\sqrt{6} - \sqrt{5}\cdot\sqrt{2} - \sqrt{3}\cdot\sqrt{6} + \sqrt{3}\cdot\sqrt{2}$$
$$= \sqrt{30} - \sqrt{10} - \sqrt{18} + \sqrt{6}$$
$$= \sqrt{30} - \sqrt{10} - \sqrt{9\cdot2} + \sqrt{6}$$
$$= \sqrt{30} - \sqrt{10} - 3\sqrt{2} + \sqrt{6}$$

47.

$$\left(\sqrt{x} + \sqrt{3}\right)\left(\sqrt{y} + \sqrt{2}\right)$$

$$\begin{array}{cccc} \text{F} & \text{O} & \text{I} & \text{L} \end{array}$$
$$= \sqrt{x}\cdot\sqrt{y} + \sqrt{x}\cdot\sqrt{2} + \sqrt{3}\cdot\sqrt{y} + \sqrt{3}\cdot\sqrt{2}$$
$$= \sqrt{xy} + \sqrt{2x} + \sqrt{3y} + \sqrt{6}$$

49.

$$\left(\sqrt{2x} + \sqrt{y}\right)\left(\sqrt{3x} - \sqrt{5y}\right)$$

$$\begin{array}{cccc} \text{F} & \text{O} & \text{I} & \text{L} \end{array}$$
$$= \sqrt{2x}\cdot\sqrt{3x} - \sqrt{2x}\cdot\sqrt{5y} + \sqrt{y}\cdot\sqrt{3x} - \sqrt{y}\cdot\sqrt{5y}$$
$$= \sqrt{6x^2} - \sqrt{10xy} + \sqrt{3xy} - \sqrt{5y^2}$$
$$= \sqrt{x^2}\cdot\sqrt{6} - \sqrt{10xy} + \sqrt{3xy} - \sqrt{y^2}\cdot\sqrt{5}$$
$$= x\sqrt{6} - \sqrt{10xy} + \sqrt{3xy} - y\sqrt{5}$$

51.

$$\left(\sqrt{5} + \sqrt{3}\right)\left(2\sqrt{5} - \sqrt{3}\right)$$

$$\begin{array}{cccc} \text{F} & \text{O} & \text{I} & \text{L} \end{array}$$
$$= \sqrt{5}\cdot2\sqrt{5} - \sqrt{5}\cdot\sqrt{3} + \sqrt{3}\cdot2\sqrt{5} - \sqrt{3}\cdot\sqrt{3}$$
$$= 2\sqrt{25} - \sqrt{15} + 2\sqrt{15} - \sqrt{9}$$
$$= 2\cdot5 - \sqrt{15} + 2\sqrt{15} - 3$$
$$= 10 - \sqrt{15} + 2\sqrt{15} - 3$$
$$= 7 + \sqrt{15}$$

53.

$$\left(2\sqrt{2} + 3\sqrt{3}\right)\left(\sqrt{2} + 4\sqrt{3}\right)$$

$$\begin{array}{cccc} \text{F} & \text{O} & \text{I} & \text{L} \end{array}$$
$$= 2\sqrt{2}\cdot\sqrt{2} + 2\sqrt{2}\cdot4\sqrt{3} + 3\sqrt{3}\cdot\sqrt{2} + 3\sqrt{3}\cdot4\sqrt{3}$$
$$= 2\sqrt{4} + 8\sqrt{6} + 3\sqrt{6} + 12\sqrt{9}$$
$$= 2\cdot2 + 8\sqrt{6} + 3\sqrt{6} + 12\cdot3$$
$$= 4 + 8\sqrt{6} + 3\sqrt{6} + 36$$
$$= 40 + 11\sqrt{6}$$

55.

$$\left(\sqrt{x} - \sqrt{3}\right)\left(2\sqrt{x} - 5\sqrt{3}\right)$$

#55 continued

$$\begin{array}{cccc} \text{F} & \text{O} & \text{I} & \text{L} \end{array}$$
$$= \sqrt{x}\cdot2\sqrt{x} - \sqrt{x}\cdot5\sqrt{3} - \sqrt{3}\cdot2\sqrt{x} + \sqrt{3}\cdot5\sqrt{3}$$
$$= 2\sqrt{x^2} - 5\sqrt{3x} - 2\sqrt{3x} + 5\sqrt{9}$$
$$= 2x - 7\sqrt{3x} + 15$$

57.

$$\left(\sqrt{x} + \sqrt{y}\right)\left(5\sqrt{x} - 9\sqrt{y}\right)$$

$$\begin{array}{cccc} \text{F} & \text{O} & \text{I} & \text{L} \end{array}$$
$$= \sqrt{x}\cdot5\sqrt{x} - \sqrt{x}\cdot9\sqrt{y} + \sqrt{y}\cdot5\sqrt{x} - \sqrt{y}\cdot9\sqrt{y}$$
$$= 5\sqrt{x^2} - 9\sqrt{xy} + 5\sqrt{xy} - 9\sqrt{y^2}$$
$$= 5x - 4\sqrt{xy} - 9y$$

59.

$$\left(\sqrt{5} + \sqrt{2}\right)\left(\sqrt{5} - \sqrt{2}\right)$$

The factors are a sum and difference of two terms. Thus,

$$\left(\sqrt{5} + \sqrt{2}\right)\left(\sqrt{5} - \sqrt{2}\right) = \left(\sqrt{5}\right)^2 - \left(\sqrt{2}\right)^2$$
$$= 5 - 2 = 3$$

61.

$$\left(\sqrt{7} + 2\right)\left(\sqrt{7} - 2\right) = \left(\sqrt{7}\right)^2 - 2^2$$
$$= 7 - 4 = 3$$

Note exercise 59 above.

63.

$$\left(\sqrt{x} + 3\right)\left(\sqrt{x} - 3\right) = \left(\sqrt{x}\right)^2 - 3^2 = x - 9$$

Note exercise 59 above.

65.

$$\left(\sqrt{x} + \sqrt{2y}\right)\left(\sqrt{x} - \sqrt{2y}\right)$$
$$= \left(\sqrt{x}\right)^2 - \left(\sqrt{2y}\right)^2 = x - 2y$$

Note exercise 59 above.

67.

$$\left(\sqrt{3} + \sqrt{2}\right)^2$$
$$= \left(\sqrt{3}\right)^2 + 2\left(\sqrt{3}\cdot\sqrt{2}\right) + \left(\sqrt{2}\right)^2$$
$$= 3 + 2\sqrt{6} + 2 = 5 + \sqrt{6}$$
Note the form: $(a + b)^2 = a^2 + 2(ab) + b^2$

69.

$$\left(3\sqrt{5} + 3\right)^2 = \left(3\sqrt{5}\right)^2 + 2\left(3\sqrt{5} \cdot 3\right) + 3^2$$
$$= 9 \cdot 5 + 18\sqrt{5} + 9$$
$$= 45 + 18\sqrt{5} + 9$$
$$= 54 + 18\sqrt{5}$$

Note exercise 67 above.

71.

$$\left(\sqrt{x} - \sqrt{5}\right)^2$$
$$= \left(\sqrt{x}\right)^2 - 2\left(\sqrt{x} \cdot \sqrt{5}\right) + \left(\sqrt{5}\right)^2$$
$$= x - 2\sqrt{5x} + 5$$

Note exercise 67 above.

73.

$$\left(4\sqrt{x} - 1\right)^2 = \left(\sqrt{4x}\right)^2 - 2\left(4\sqrt{x} \cdot 1\right) + (-1)^2$$
$$= 16x - 8\sqrt{x} + 1$$

Note exercise 67 above.

75.

$$\left(\sqrt{x} + 3\sqrt{y}\right)^2$$
$$= \left(\sqrt{x}\right)^2 + 2\left(\sqrt{x} \cdot 3\sqrt{y}\right) + \left(3\sqrt{y}\right)^2$$
$$= x + 6\sqrt{xy} + 9y$$

Note exercise 67 above.

77.

$$\left(\sqrt{x-4} - 2\right)^2$$
$$= \left(\sqrt{x-4}\right)^2 - 2\left(\sqrt{x-4} \cdot 2\right) + (-2)^2$$
$$= x - 4 - 4\sqrt{x-4} + 4$$
$$= x - 4\sqrt{x-4}$$

Note exercise 67 above.

Exercises 8.5

1.

$$\frac{1}{\sqrt{6}} = \frac{1}{\sqrt{6}} \cdot \frac{\sqrt{6}}{\sqrt{6}} = \frac{\sqrt{6}}{\sqrt{36}} = \frac{\sqrt{6}}{6}$$

3.

$$\frac{7}{\sqrt{2}} = \frac{7}{\sqrt{2}} \cdot \frac{\sqrt{2}}{\sqrt{2}} = \frac{7\sqrt{2}}{\sqrt{4}} = \frac{7\sqrt{2}}{2}$$

5.

$$\frac{15}{\sqrt{21}} = \frac{15}{\sqrt{21}} \cdot \frac{\sqrt{21}}{\sqrt{21}} = \frac{15\sqrt{21}}{\sqrt{441}} = \frac{15\sqrt{21}}{21}$$
$$= \frac{\cancel{3} \cdot 5\sqrt{21}}{\cancel{3} \cdot 7} = \frac{5\sqrt{21}}{7}$$

7.

$$\frac{\sqrt{6}}{\sqrt{5}} = \frac{\sqrt{6}}{\sqrt{5}} \cdot \frac{\sqrt{5}}{\sqrt{5}} = \frac{\sqrt{30}}{\sqrt{25}} = \frac{\sqrt{30}}{5}$$

9.

$$\sqrt{\frac{2}{3}} = \sqrt{\frac{2}{3} \cdot \frac{3}{3}} = \sqrt{\frac{6}{9}} = \frac{\sqrt{6}}{\sqrt{9}} = \frac{\sqrt{6}}{3}$$

11.

$$\sqrt{\frac{5}{2}} = \sqrt{\frac{5}{2} \cdot \frac{2}{2}} = \sqrt{\frac{10}{4}} = \frac{\sqrt{10}}{\sqrt{4}} = \frac{\sqrt{10}}{2}$$

13.

$$\frac{7}{2\sqrt{3}} = \frac{7}{2\sqrt{3}} \cdot \frac{\sqrt{3}}{\sqrt{3}} = \frac{7\sqrt{3}}{2\sqrt{9}} = \frac{7\sqrt{3}}{2 \cdot 3} = \frac{7\sqrt{3}}{6}$$

15.

$$\frac{8}{7\sqrt{2}} = \frac{8}{7\sqrt{2}} \cdot \frac{\sqrt{2}}{\sqrt{2}} = \frac{8\sqrt{2}}{7\sqrt{4}} = \frac{\cancel{2} \cdot 4\sqrt{2}}{7 \cdot \cancel{2}} = \frac{4\sqrt{2}}{7}$$

17.

$$\frac{6}{\sqrt{20}} = \frac{6}{\sqrt{4 \cdot 5}} = \frac{6}{2\sqrt{5}} = \frac{6}{2\sqrt{5}} \cdot \frac{\sqrt{5}}{\sqrt{5}} = \frac{6\sqrt{5}}{2\sqrt{25}} = \frac{6\sqrt{5}}{2 \cdot 5}$$
$$= \frac{6\sqrt{5}}{10} = \frac{\cancel{2} \cdot 3\sqrt{5}}{\cancel{2} \cdot 5} = \frac{3\sqrt{5}}{5}$$

19.

$$\frac{9}{\sqrt{12}} = \frac{9}{\sqrt{4 \cdot 3}} = \frac{9}{2\sqrt{3}} \cdot \frac{\sqrt{3}}{\sqrt{3}} = \frac{9\sqrt{3}}{2\sqrt{9}} = \frac{\cancel{3} \cdot 3\sqrt{3}}{2 \cdot \cancel{3}} = \frac{3\sqrt{3}}{2}$$

21.

$$\frac{\sqrt{63}}{\sqrt{72}} = \frac{\sqrt{9 \cdot 7}}{\sqrt{36 \cdot 2}} = \frac{\sqrt{9} \cdot \sqrt{7}}{\sqrt{36} \cdot \sqrt{2}} = \frac{3\sqrt{7}}{6\sqrt{2}} = \frac{\sqrt{7}}{2\sqrt{2}}$$
$$= \frac{\sqrt{7}}{2\sqrt{2}} \cdot \frac{\sqrt{2}}{\sqrt{2}} = \frac{\sqrt{14}}{2\sqrt{4}} = \frac{\sqrt{14}}{2 \cdot 2} = \frac{\sqrt{14}}{4}$$

23.

$$\frac{\sqrt{48}}{\sqrt{45}} = \sqrt{\frac{48}{45}} = \sqrt{\frac{16}{15}} = \frac{\sqrt{16}}{\sqrt{15}} = \frac{4}{\sqrt{15}} = \frac{4}{\sqrt{15}} \cdot \frac{\sqrt{15}}{\sqrt{15}}$$
$$= \frac{4\sqrt{15}}{\sqrt{15^2}} = \frac{4\sqrt{15}}{15}$$

25.

$$\frac{7\sqrt{18}}{5\sqrt{2}} = \frac{7 \cdot \sqrt{9 \cdot 2}}{5\sqrt{2}} = \frac{7 \cdot \sqrt{9} \cdot \sqrt{2}}{5 \cdot \sqrt{2}}$$
$$= \frac{7 \cdot 3 \cdot \sqrt{2}}{5 \cdot \sqrt{2}} = \frac{21}{5}$$

27.

$$\frac{3\sqrt{18}}{4\sqrt{45}} = \frac{3}{4} \cdot \sqrt{\frac{18}{45}} = \frac{3}{4} \cdot \sqrt{\frac{2}{5}} = \frac{3\sqrt{2}}{4\sqrt{5}} = \frac{3\sqrt{2}}{4\sqrt{5}} \cdot \frac{\sqrt{5}}{\sqrt{5}}$$
$$= \frac{3\sqrt{10}}{4\sqrt{25}} = \frac{3\sqrt{10}}{4 \cdot 5} = \frac{3\sqrt{10}}{20}$$

29.

$$\frac{2\sqrt{3} - 4\sqrt{2}}{\sqrt{5}} = \frac{2\sqrt{3} - 4\sqrt{2}}{\sqrt{5}} \cdot \frac{\sqrt{5}}{\sqrt{5}}$$
$$= \frac{2\sqrt{15} - 4\sqrt{10}}{5}$$

31.

$$\frac{5\sqrt{2} - 4\sqrt{3}}{\sqrt{3}} = \frac{5\sqrt{2} - 4\sqrt{3}}{\sqrt{3}} \cdot \frac{\sqrt{3}}{\sqrt{3}}$$
$$= \frac{5\sqrt{6} - 4\sqrt{9}}{3} = \frac{5\sqrt{6} - 4 \cdot 3}{3}$$
$$= \frac{5\sqrt{6} - 12}{3}$$

33.

$$\sqrt{\frac{2}{x}} = \sqrt{\frac{2}{x} \cdot \frac{x}{x}} = \sqrt{\frac{2x}{x^2}} = \frac{\sqrt{2x}}{\sqrt{x^2}} = \frac{\sqrt{2x}}{x}$$

35.

$$\sqrt{\frac{2}{7x}} = \sqrt{\frac{2}{7x} \cdot \frac{7x}{7x}} = \sqrt{\frac{14x}{49x^2}} = \frac{\sqrt{14x}}{\sqrt{49x^2}} = \frac{\sqrt{14x}}{7x}$$

37.

$$\sqrt{\frac{5x}{2y}} = \sqrt{\frac{5x}{2y} \cdot \frac{2y}{2y}} = \sqrt{\frac{10xy}{4y^2}} = \frac{\sqrt{10xy}}{\sqrt{4y^2}} = \frac{\sqrt{10xy}}{2y}$$

39.

$$\sqrt{\frac{11}{20x}} = \sqrt{\frac{11}{20x} \cdot \frac{5x}{5x}} = \sqrt{\frac{55x}{100x^2}} = \frac{\sqrt{55x}}{\sqrt{100x^2}}$$
$$= \frac{\sqrt{55x}}{10x}$$

41.

$$\sqrt{\frac{9x}{50y}} = \sqrt{\frac{9x}{50y} \cdot \frac{2y}{2y}} = \sqrt{\frac{9x \cdot 2y}{100y^2}} = \frac{\sqrt{9 \cdot 2xy}}{\sqrt{100y^2}}$$
$$= \frac{3\sqrt{2xy}}{10y}$$

43.

$$\sqrt{\frac{45x}{44y}} = \frac{\sqrt{45x}}{\sqrt{44y}} = \frac{\sqrt{9 \cdot 5x}}{\sqrt{4 \cdot 11y}} = \frac{\sqrt{9} \cdot \sqrt{5x}}{\sqrt{4} \cdot \sqrt{11y}}$$
$$= \frac{3\sqrt{5x}}{2\sqrt{11y}} = \frac{3\sqrt{5x}}{2\sqrt{11y}} \cdot \frac{\sqrt{11y}}{\sqrt{11y}} = \frac{3\sqrt{55xy}}{2\sqrt{(11y)^2}}$$
$$= \frac{3\sqrt{55xy}}{2 \cdot 11y} = \frac{3\sqrt{55xy}}{22y}$$

45.

$$\sqrt{\frac{9}{25x^6}} = \frac{\sqrt{9}}{\sqrt{25x^6}} = \frac{3}{5x^3}$$

47.

$$\sqrt{\frac{7}{2x^3}} = \sqrt{\frac{7}{2x^3} \cdot \frac{2x}{2x}} = \sqrt{\frac{14x}{4x^4}} = \frac{\sqrt{14x}}{\sqrt{4x^4}} = \frac{\sqrt{14x}}{2x^2}$$

49.

$$\sqrt{\frac{18}{7x^7}} = \frac{\sqrt{9 \cdot 2}}{\sqrt{x^6 \cdot 7x}} = \frac{\sqrt{9} \cdot \sqrt{2}}{\sqrt{x^6} \cdot \sqrt{7x}} = \frac{3\sqrt{2}}{x^3\sqrt{7x}}$$
$$= \frac{3\sqrt{2}}{x^3\sqrt{7x}} \cdot \frac{\sqrt{7x}}{\sqrt{7x}} = \frac{3\sqrt{14x}}{x^3\sqrt{(7x)^2}}$$
$$= \frac{3\sqrt{14x}}{x^3 \cdot 7x} = \frac{3\sqrt{14x}}{7x^4}$$

51.

$$\sqrt{\frac{81x^6}{16y^4}} = \frac{\sqrt{81x^6}}{\sqrt{16y^4}} = \frac{9x^3}{4y^2}$$

53.

$$\sqrt{\frac{11y}{2x^2}} = \frac{\sqrt{11y}}{\sqrt{2x^2}} = \frac{\sqrt{11y}}{\sqrt{x^2} \cdot \sqrt{2}} = \frac{\sqrt{11y}}{x\sqrt{2}}$$

#53 continued

$$= \frac{\sqrt{11y}}{x\sqrt{2}} \cdot \frac{\sqrt{2}}{\sqrt{2}} = \frac{\sqrt{22y}}{x\sqrt{4}} = \frac{\sqrt{22y}}{2x}$$

55.

$$\sqrt{\frac{72y^3}{5x^3}} = \sqrt{\frac{72y^3}{5x^3} \cdot \frac{5x}{5x}} = \sqrt{\frac{36y^2 \cdot 10xy}{25x^4}}$$

$$= \frac{\sqrt{36y^2 \cdot 10xy}}{\sqrt{25x^4}} = \frac{\sqrt{36y^2} \cdot \sqrt{10xy}}{\sqrt{25x^4}}$$

$$= \frac{6y\sqrt{10xy}}{5x^2}$$

57.

$$\frac{11}{\sqrt{10} - 2} = \frac{11}{\sqrt{10} - 2} \cdot \frac{\sqrt{10} + 2}{\sqrt{10} + 2}$$

$$= \frac{11(\sqrt{10} + 2)}{(\sqrt{10} - 2)(\sqrt{10} + 2)}$$

$$= \frac{11\sqrt{10} + 22}{(\sqrt{10})^2 - 2^2} = \frac{11\sqrt{10} + 22}{10 - 4}$$

$$= \frac{11\sqrt{10} + 22}{6} = \frac{11(\sqrt{10} + 2)}{6}$$

59.

$$\frac{15}{\sqrt{14} - 2} = \frac{15}{\sqrt{14} - 2} \cdot \frac{\sqrt{14} + 2}{\sqrt{14} + 2}$$

$$= \frac{15(\sqrt{14} + 2)}{(\sqrt{14} - 2)(\sqrt{14} + 2)}$$

$$= \frac{15\sqrt{14} + 30}{(\sqrt{14})^2 - (2)^2}$$

$$= \frac{15\sqrt{14} + 30}{14 - 4} = \frac{15\sqrt{14} + 30}{10}$$

$$= \frac{5(3\sqrt{14} + 6)}{5 \cdot 2} = \frac{3\sqrt{14} + 6}{2}$$

$$= \frac{3(\sqrt{14} + 2)}{2}$$

61.

$$\frac{1}{\sqrt{10} + \sqrt{3}} = \frac{1}{\sqrt{10} + \sqrt{3}} \cdot \frac{\sqrt{10} - \sqrt{3}}{\sqrt{10} - \sqrt{3}}$$

#61 continued

$$= \frac{\sqrt{10} - \sqrt{3}}{(\sqrt{10} + \sqrt{3})(\sqrt{10} - \sqrt{3})}$$

$$= \frac{\sqrt{10} - \sqrt{3}}{(\sqrt{10})^2 - (\sqrt{3})^2}$$

$$= \frac{\sqrt{10} - \sqrt{3}}{10 - 3} = \frac{\sqrt{10} - \sqrt{3}}{7}$$

63.

$$\frac{2}{\sqrt{5} - \sqrt{10}} = \frac{2}{\sqrt{5} - \sqrt{10}} \cdot \frac{\sqrt{5} + \sqrt{10}}{\sqrt{5} + \sqrt{10}}$$

$$= \frac{2(\sqrt{5} + \sqrt{10})}{(\sqrt{5} - \sqrt{10})(\sqrt{5} + \sqrt{10})}$$

$$= \frac{2\sqrt{5} + 2\sqrt{10}}{(\sqrt{5})^2 - (\sqrt{10})^2}$$

$$= \frac{2\sqrt{5} + 2\sqrt{10}}{5 - 10} = \frac{2\sqrt{5} + 2\sqrt{10}}{-5}$$

$$= -\frac{2(\sqrt{5} + \sqrt{10})}{5}$$

65.

$$\frac{5\sqrt{3}}{\sqrt{10} - 3} = \frac{5\sqrt{3}}{\sqrt{10} - 3} \cdot \frac{\sqrt{10} + 3}{\sqrt{10} + 3}$$

$$= \frac{5\sqrt{3}(\sqrt{10} + 3)}{(\sqrt{10} - 3)(\sqrt{10} + 3)}$$

$$= \frac{5\sqrt{30} + 15\sqrt{3}}{(\sqrt{10})^2 - (3)^2} = \frac{5\sqrt{30} + 15\sqrt{3}}{10 - 9}$$

$$= 5\sqrt{30} + 15\sqrt{3}$$

67.

$$\frac{5\sqrt{7}}{\sqrt{7} + 1} = \frac{5\sqrt{7}}{\sqrt{7} + 1} \cdot \frac{\sqrt{7} - 1}{\sqrt{7} - 1}$$

$$= \frac{5\sqrt{7}(\sqrt{7} - 1)}{(\sqrt{7} + 1)(\sqrt{7} - 1)}$$

$$= \frac{5\sqrt{49} - 5\sqrt{7}}{(\sqrt{7})^2 - (1)^2} = \frac{5 \cdot 7 - 5\sqrt{7}}{7 - 1}$$

$$= \frac{35 - 5\sqrt{7}}{6}$$

#75 continued

69.

$$\frac{\sqrt{10} + 1}{\sqrt{10} + 2} = \frac{\sqrt{10} + 1}{\sqrt{10} + 2} \cdot \frac{\sqrt{10} - 2}{\sqrt{10} - 2}$$

$$= \frac{\left(\sqrt{10} + 1\right)\left(\sqrt{10} - 2\right)}{\left(\sqrt{10} + 2\right)\left(\sqrt{10} - 2\right)}$$

$$= \frac{10 - \sqrt{10} - 2}{\left(\sqrt{10}\right)^2 - (2)^2} = \frac{8 - \sqrt{10}}{10 - 4}$$

$$= \frac{8 - \sqrt{10}}{6}$$

71.

$$\frac{\sqrt{10} + 2}{\sqrt{10} - 3} = \frac{\sqrt{10} + 2}{\sqrt{10} - 3} \cdot \frac{\sqrt{10} + 3}{\sqrt{10} + 3}$$

$$= \frac{\left(\sqrt{10} + 2\right)\left(\sqrt{10} + 3\right)}{\left(\sqrt{10} - 3\right)\left(\sqrt{10} + 3\right)}$$

$$= \frac{10 + 5\sqrt{10} + 6}{\left(\sqrt{10}\right)^2 - (3)^2} = \frac{16 + 5\sqrt{10}}{10 - 9}$$

$$= 16 + 5\sqrt{10}$$

73.

$$\frac{\sqrt{7} + \sqrt{2}}{\sqrt{7} - \sqrt{2}} = \frac{\sqrt{7} + \sqrt{2}}{\sqrt{7} - \sqrt{2}} \cdot \frac{\sqrt{7} + \sqrt{2}}{\sqrt{7} + \sqrt{2}}$$

$$= \frac{\left(\sqrt{7} + \sqrt{2}\right)^2}{\left(\sqrt{7} - \sqrt{2}\right)\left(\sqrt{7} + \sqrt{2}\right)}$$

$$= \frac{7 + 2\sqrt{14} + 2}{\left(\sqrt{7}\right)^2 - \left(\sqrt{2}\right)^2} = \frac{9 + 2\sqrt{14}}{7 - 2}$$

$$= \frac{9 + 2\sqrt{14}}{5}$$

75.

$$\frac{2\sqrt{10} - 3\sqrt{6}}{\sqrt{10} - \sqrt{6}} = \frac{2\sqrt{10} - 3\sqrt{6}}{\sqrt{10} - \sqrt{6}} \cdot \frac{\sqrt{10} + \sqrt{6}}{\sqrt{10} + \sqrt{6}}$$

$$= \frac{\left(2\sqrt{10} - 3\sqrt{6}\right)\left(\sqrt{10} + \sqrt{6}\right)}{\left(\sqrt{10}\right)^2 - \left(\sqrt{6}\right)^2}$$

$$= \frac{20 - 3\sqrt{60} + 2\sqrt{60} - 18}{10 - 6}$$

$$= \frac{2 - \sqrt{60}}{4}$$

$$= \frac{2 - \sqrt{4 \cdot 15}}{4}$$

$$= \frac{2 - 2\sqrt{15}}{4} = \frac{2(1 - \sqrt{15})}{4}$$

$$= \frac{\cancel{2}(1 - \sqrt{15})}{\cancel{2} \cdot 2} = \frac{1 - \sqrt{15}}{2}$$

77.

$$\frac{2\sqrt{3} + 5}{3\sqrt{3} - 2} = \frac{2\sqrt{3} + 5}{3\sqrt{3} - 2} \cdot \frac{3\sqrt{3} + 2}{3\sqrt{3} + 2}$$

$$= \frac{\left(2\sqrt{3} + 5\right)\left(3\sqrt{3} + 2\right)}{\left(3\sqrt{3}\right)^2 - (2)^2}$$

$$= \frac{18 + 15\sqrt{3} + 4\sqrt{3} + 10}{27 - 4}$$

$$= \frac{28 + 19\sqrt{3}}{23}$$

79.

$$\frac{5\sqrt{6} - 2}{3\sqrt{6} - 4} = \frac{5\sqrt{6} - 2}{3\sqrt{6} - 4} \cdot \frac{3\sqrt{6} + 4}{3\sqrt{6} + 4}$$

$$= \frac{\left(5\sqrt{6} - 2\right)\left(3\sqrt{6} + 4\right)}{\left(3\sqrt{6}\right)^2 - (4)^2}$$

$$= \frac{90 - 6\sqrt{6} + 20\sqrt{6} - 8}{54 - 16}$$

$$= \frac{82 + 14\sqrt{6}}{38} = \frac{2(41 + 7\sqrt{6})}{2 \cdot 19}$$

$$= \frac{41 + 7\sqrt{6}}{19}$$

81.

$$\frac{3\sqrt{2} + 1}{6\sqrt{7} - 4} = \frac{3\sqrt{2} + 1}{6\sqrt{7} - 4} \cdot \frac{6\sqrt{7} + 4}{6\sqrt{7} + 4}$$

$$= \frac{\left(3\sqrt{2} + 1\right)\left(6\sqrt{7} + 4\right)}{\left(6\sqrt{7}\right)^2 - (4)^2}$$

$$= \frac{18\sqrt{14} + 12\sqrt{2} + 6\sqrt{7} + 4}{252 - 16}$$

$$= \frac{\cancel{2}\left(9\sqrt{14} + 6\sqrt{2} + 3\sqrt{7} + 2\right)}{\cancel{2} \cdot 118}$$

$$= \frac{9\sqrt{14} + 6\sqrt{2} + 3\sqrt{7} + 2}{118}$$

83.

$$\frac{\sqrt{x}}{\sqrt{x} - 2} = \frac{\sqrt{x}}{\sqrt{x} - 2} \cdot \frac{\sqrt{x} + 2}{\sqrt{x} + 2} = \frac{\sqrt{x}\left(\sqrt{x} + 2\right)}{\left(\sqrt{x}\right)^2 - (2)^2}$$

$$= \frac{x + 2\sqrt{x}}{x - 4}$$

85.

$$\frac{\sqrt{2x}}{\sqrt{x} - 5} = \frac{\sqrt{2x}}{\sqrt{x} - 5} \cdot \frac{\sqrt{x} + 5}{\sqrt{x} + 5} = \frac{\sqrt{2x}\left(\sqrt{x} + 5\right)}{\left(\sqrt{x}\right)^2 - (5)^2}$$

$$= \frac{\sqrt{2x^2} + 5\sqrt{2x}}{x - 25} = \frac{x\sqrt{2} + 5\sqrt{2x}}{x - 25}$$

87.

$$\frac{\sqrt{x}}{\sqrt{x} + \sqrt{6}} = \frac{\sqrt{x}}{\sqrt{x} + \sqrt{6}} \cdot \frac{\sqrt{x} - \sqrt{6}}{\sqrt{x} - \sqrt{6}}$$

$$= \frac{\sqrt{x}\left(\sqrt{x} - \sqrt{6}\right)}{\left(\sqrt{x}\right)^2 - \left(\sqrt{6}\right)^2} = \frac{x - \sqrt{6x}}{x - 6}$$

89.

$$\frac{5\sqrt{x}}{\sqrt{2x} + \sqrt{7}} = \frac{5\sqrt{x}}{\sqrt{2x} + \sqrt{7}} \cdot \frac{\sqrt{2x} - \sqrt{7}}{\sqrt{2x} - \sqrt{7}}$$

$$= \frac{5\sqrt{x}\left(\sqrt{2x} - \sqrt{7}\right)}{\left(\sqrt{2x}\right)^2 - \left(\sqrt{7}\right)^2}$$

$$= \frac{5\sqrt{2x^2} - 5\sqrt{7x}}{2x - 7} = \frac{5x\sqrt{2} - 5\sqrt{7x}}{2x - 7}$$

91.

$$\frac{\sqrt{x} + 1}{\sqrt{x} + 5} = \frac{\sqrt{x} + 1}{\sqrt{x} + 5} \cdot \frac{\sqrt{x} - 5}{\sqrt{x} - 5}$$

$$= \frac{\left(\sqrt{x} + 1\right)\left(\sqrt{x} - 5\right)}{\left(\sqrt{x}\right)^2 - (5)^2} = \frac{x - 4\sqrt{x} - 5}{x - 25}$$

93.

$$\frac{2\sqrt{x} + 3}{\sqrt{x} - 1} = \frac{2\sqrt{x} + 3}{\sqrt{x} - 1} \cdot \frac{\sqrt{x} + 1}{\sqrt{x} + 1}$$

$$= \frac{\left(2\sqrt{x} + 3\right)\left(\sqrt{x} + 1\right)}{\left(\sqrt{x}\right)^2 - (1)^2}$$

$$= \frac{2x + 5\sqrt{x} + 3}{x - 1}$$

95.

$$\frac{\sqrt{x} - 3}{2\sqrt{x} - 3} = \frac{\sqrt{x} - 3}{2\sqrt{x} - 3} \cdot \frac{2\sqrt{x} + 3}{2\sqrt{x} + 3}$$

$$= \frac{\left(\sqrt{x} - 3\right)\left(2\sqrt{x} + 3\right)}{\left(2\sqrt{x}\right)^2 - (3)^2}$$

$$= \frac{2x - 3\sqrt{x} - 9}{4x - 9}$$

97.

$$\frac{\sqrt{x} + 2\sqrt{3}}{2\sqrt{x} - \sqrt{3}} = \frac{\sqrt{x} + 2\sqrt{3}}{2\sqrt{x} - \sqrt{3}} \cdot \frac{2\sqrt{x} + \sqrt{3}}{2\sqrt{x} + \sqrt{3}}$$

$$= \frac{\left(\sqrt{x} + 2\sqrt{3}\right)\left(2\sqrt{x} + \sqrt{3}\right)}{\left(2\sqrt{x}\right)^2 - \left(\sqrt{3}\right)^2}$$

$$= \frac{2x + 5\sqrt{3x} + 6}{4x - 3}$$

99.

$$\frac{\sqrt{x} - \sqrt{y}}{\sqrt{x} + \sqrt{y}} = \frac{\sqrt{x} - \sqrt{y}}{\sqrt{x} + \sqrt{y}} \cdot \frac{\sqrt{x} - \sqrt{y}}{\sqrt{x} - \sqrt{y}}$$

$$= \frac{\left(\sqrt{x} - \sqrt{y}\right)^2}{\left(\sqrt{x}\right)^2 - \left(\sqrt{y}\right)^2}$$

$$= \frac{x - 2\sqrt{xy} + y}{x - y}$$

101.

$$\frac{1}{\sqrt[3]{9}} = \frac{1}{\sqrt[3]{3^2}} \cdot \frac{\sqrt[3]{3}}{\sqrt[3]{3}} = \frac{\sqrt[3]{3}}{\sqrt[3]{3^3}} = \frac{\sqrt[3]{3}}{3}$$

103.

$$\sqrt[3]{\frac{10}{3}} = \sqrt[3]{\frac{10}{3} \cdot \frac{3^2}{3^2}} = \sqrt[3]{\frac{90}{3^3}} = \frac{\sqrt[3]{90}}{\sqrt[3]{3^3}} = \frac{\sqrt[3]{90}}{3}$$

105.

$$\sqrt[3]{\frac{5}{4x^3}} = \sqrt[3]{\frac{5}{2^2 x^3} \cdot \frac{2}{2}} = \sqrt[3]{\frac{10}{2^3 x^3}}$$

$$= \frac{\sqrt[3]{10}}{\sqrt[3]{2^3 x^3}} = \frac{\sqrt[3]{10}}{2x}$$

107.

$$\sqrt[3]{\frac{5x}{16y^3}} = \sqrt[3]{\frac{5x}{2^4 y^3} \cdot \frac{2^2}{2^2}} = \sqrt[3]{\frac{20x}{2^6 y^3}} = \frac{\sqrt[3]{20x}}{\sqrt[3]{2^6 y^3}}$$

$$= \frac{\sqrt[3]{20x}}{2^2 y} = \frac{\sqrt[3]{20x}}{4y}$$

109.

$$(\sqrt{a} + \sqrt{b})(\sqrt{a} - \sqrt{b}) = (\sqrt{a})^2 - (\sqrt{b})^2$$

$$= a - b$$

Exercises 8.6

1.

$$\sqrt{x} = 3$$
$$(\sqrt{x})^2 = (3)^2$$
$$x = 9 \qquad \text{Solution set: } \{9\}$$

3.

$$\sqrt{x} \neq -6 \qquad \text{Solution set: } \varnothing$$

5.

$$2\sqrt{x} = 7$$
$$(2\sqrt{x})^2 = (7)^2$$
$$4x = 49$$
$$x = \frac{49}{4} \qquad \text{Solution set: } \left\{\frac{49}{4}\right\}$$

7.

$$5\sqrt{2x} = 1$$
$$(5\sqrt{2x})^2 = (1)^2$$
$$50x = 1$$
$$x = \frac{1}{50} \qquad \text{Solution set: } \left\{\frac{1}{50}\right\}$$

9.

$$\sqrt{x - 2} = 3$$
$$(\sqrt{x - 2})^2 = (3)^2$$
$$x - 2 = 9$$
$$x = 11 \qquad \text{Solution set: } \{11\}$$

11.

$$\sqrt{2x + 1} = 0$$
$$(\sqrt{2x + 1})^2 = (0)^2$$
$$2x + 1 = 0$$
$$2x = -1$$
$$x = -\frac{1}{2} \qquad \text{Solution set: } \left\{-\frac{1}{2}\right\}$$

13.

$$\sqrt{8x - 1} \neq -4 \qquad \text{Solution set: } \varnothing$$

15.

$$\sqrt{3x - 1} = 4$$
$$(\sqrt{3x - 1})^2 = 4^2$$
$$3x - 1 = 16$$
$$3x = 17$$
$$x = \frac{17}{3} \qquad \text{Solution set: } \left\{\frac{17}{3}\right\}$$

17.

$$\sqrt{2x + 7} = 2$$
$$(\sqrt{2x + 7})^2 = 2^2$$
$$2x + 7 = 4$$
$$2x = -3$$
$$x = -\frac{3}{2} \qquad \text{Solution set: } \left\{-\frac{3}{2}\right\}$$

19.

$$\sqrt{5x - 2} = 6$$
$$(\sqrt{5x - 2})^2 = 6^2$$
$$5x - 2 = 36$$
$$5x = 38$$
$$x = \frac{38}{5} \qquad \text{Solution set: } \left\{\frac{38}{5}\right\}$$

21.

$$\sqrt{3x - 1} = \frac{1}{2}$$
$$(\sqrt{3x - 1})^2 = \left(\frac{1}{2}\right)^2$$
$$3x - 1 = \frac{1}{4}$$
$$3x = 1 + \frac{1}{4}$$
$$3x = \frac{5}{4}$$

#21 continued

$$x = \frac{5}{12} \qquad \text{Solution set: } \left\{ \frac{5}{12} \right\}$$

23.

$$\sqrt{2x + 3} = \frac{3}{4}$$

$$\left(\sqrt{2x + 3} \right)^2 = \left(\frac{3}{4} \right)^2$$

$$2x + 3 = \frac{9}{16}$$

$$2x = \frac{9}{16} - 3$$

$$2x = -\frac{39}{16}$$

$$x = -\frac{39}{32} \qquad \text{Solution set: } \left\{ -\frac{39}{32} \right\}$$

25.

$$3\sqrt{3x + 7} = 6$$

$$\left(3\sqrt{3x + 7} \right)^2 = 6^2$$

$$9(3x + 7) = 36$$

$$27x + 63 = 36$$

$$27x = -27$$

$$x = -1 \qquad \text{Solution set: } \{-1\}$$

27.

$$5\sqrt{2x - 1} = 2$$

$$\left(5\sqrt{2x - 1} \right)^2 = 2^2$$

$$25(2x - 1) = 4$$

$$50x - 25 = 4$$

$$50x = 29$$

$$x = \frac{29}{50} \qquad \text{Solution set: } \left\{ \frac{29}{50} \right\}$$

29.

$$\sqrt{x} + 3 = 9$$

$$\sqrt{x} = 6$$

$$\left(\sqrt{x} \right)^2 = 6^2$$

$$x = 36 \qquad \text{Solution set: } \{36\}$$

31.

$$\sqrt{x} + 8 = 3$$

$$\sqrt{x} \neq -5 \qquad \text{Solution set: } \emptyset$$

33.

$$2\sqrt{x} - 1 = 7$$

$$2\sqrt{x} = 8$$

$$\sqrt{x} = 4$$

$$\left(\sqrt{x} \right)^2 = 4^2$$

$$x = 16 \qquad \text{Solution set: } \{16\}$$

35.

$$3\sqrt{x} - 2 = 5$$

$$3\sqrt{x} = 7$$

$$\left(3\sqrt{x} \right)^2 = 7^2$$

$$9x = 49$$

$$x = \frac{49}{9} \qquad \text{Solution set: } \left\{ \frac{49}{9} \right\}$$

37.

$$\sqrt{x + 6} + 3 = 4$$

$$\sqrt{x + 6} = 1$$

$$\left(\sqrt{x + 6} \right)^2 = 1^2$$

$$x + 6 = 1$$

$$x = -5 \qquad \text{Solution set: } \{-5\}$$

39.

$$\sqrt{x - 6} - 2 = 1$$

$$\sqrt{x - 6} = 3$$

$$\left(\sqrt{x - 6} \right)^2 = 3^2$$

$$x - 6 = 9$$

$$x = 15 \qquad \text{Solution set: } \{15\}$$

41.

$$\sqrt{x - 5} + 6 = 3$$

$$\sqrt{x - 5} \neq -3 \qquad \text{Solution set: } \emptyset$$

43.

$$\sqrt{2x + 3} - 1 = 1$$

$$\sqrt{2x + 3} = 2$$

$$\left(\sqrt{2x + 3} \right)^2 = 2^2$$

$$2x + 3 = 4$$

$$2x = 1$$

$$x = \frac{1}{2} \qquad \text{Solution set: } \left\{ \frac{1}{2} \right\}$$

45.

$$\sqrt{4x + 9} + 1 = 4$$

$$\sqrt{4x + 9} = 3$$

#45 continued

$$\left(\sqrt{4x + 9}\right)^2 = 3^2$$
$$4x + 9 = 9$$
$$4x = 0$$
$$x = 0 \qquad \text{Solution set: } \{0\}$$

47.

$$\sqrt{3x + 10} + 2 = 4$$
$$\sqrt{3x + 10} = 2$$
$$\left(\sqrt{3x + 10}\right)^2 = 2^2$$
$$3x + 10 = 4$$
$$3x = -6$$
$$x = -2 \qquad \text{Solution set: } \{-2\}$$

49.

$$x = \sqrt{4x + 12}$$
$$x^2 = \left(\sqrt{4x + 12}\right)^2$$
$$x^2 = 4x + 12$$
$$x^2 - 4x - 12 = 0$$
$$(x + 2)(x - 6) = 0$$

$$x + 2 = 0 \text{ or } x - 6 = 0$$
$$x = -2 \text{ or } \quad x = 6$$

A check shows that the solution set is {6}.

51.

$$x = \sqrt{5x - 6}$$
$$x^2 = \left(\sqrt{5x - 6}\right)^2$$
$$x^2 = 5x - 6$$
$$x^2 - 5x + 6 = 0$$
$$(x - 2)(x - 3) = 0$$

$$x - 2 = 0 \text{ or } x - 3 = 0$$
$$x = 2 \text{ or } \quad x = 3$$

A check shows that the solution set is {2,3}.

53.

$$2x = \sqrt{8x - 3}$$
$$(2x)^2 = \left(\sqrt{8x - 3}\right)^2$$
$$4x^2 = 8x - 3$$
$$4x^2 - 8x + 3 = 0$$
$$(2x - 1)(2x - 3) = 0$$

$$2x - 1 = 0 \text{ or } 2x - 3 = 0$$
$$x = \frac{1}{2} \text{ or } \qquad x = \frac{3}{2}$$

#53 continued

A check shows that the solution set is $\left\{\frac{1}{2}, \frac{3}{2}\right\}$.

55.

$$3x = 2\sqrt{6x - 3}$$
$$(3x)^2 = \left(2\sqrt{6x - 3}\right)^2$$
$$9x^2 = 4(6x - 3)$$
$$9x^2 = 24x - 12$$
$$9x^2 - 24x + 12 = 0$$
$$3x^2 - 8x + 4 = 0 \qquad \text{Divide by 3.}$$
$$(3x - 2)(x - 2) = 0$$

$$3x - 2 = 0 \text{ or } x - 2 = 0$$
$$x = \frac{2}{3} \text{ or } \qquad x = 2$$

A check shows that the solution set is $\left\{\frac{2}{3}, 2\right\}$.

57.

$$\sqrt{5x - 3} = \sqrt{2x + 1}$$
$$\left(\sqrt{5x - 3}\right)^2 = \left(\sqrt{2x + 1}\right)^2$$
$$5x - 3 = 2x + 1$$
$$3x - 3 = 1$$
$$3x = 4$$
$$x = \frac{4}{3} \qquad \text{Solution set: } \left\{\frac{4}{3}\right\}$$

59.

$$\sqrt{2x - 1} = \sqrt{5x - 9}$$
$$\left(\sqrt{2x - 1}\right)^2 = \left(\sqrt{5x - 9}\right)^2$$
$$2x - 1 = 5x - 9$$
$$-1 = 3x - 9$$
$$8 = 3x$$
$$\frac{8}{3} = x \qquad \text{Solution set: } \left\{\frac{8}{3}\right\}$$

61.

$$\sqrt{2x^2 + 3x - 2} = \sqrt{x^2 + 3x + 2}$$
$$\left(\sqrt{2x^2 + 3x - 2}\right)^2 = \left(\sqrt{x^2 + 3x + 2}\right)^2$$
$$2x^2 + 3x - 2 = x^2 + 3x + 2$$
$$x^2 - 4 = 0$$
$$(x - 2)(x + 2) = 0$$

#61 continued

$x - 2 = 0$ or $x + 2 = 0$
     $x = 2$ or $x = -2$
Solution set: $\{2, -2\}$

63.

$$\sqrt{4x^2 + 4x - 1} = \sqrt{2x^2 - x + 2}$$
$$\left(\sqrt{4x^2 + 4x - 1}\right)^2 = \left(\sqrt{2x^2 - x + 2}\right)^2$$
$$4x^2 + 4x - 1 = 2x^2 - x + 2$$
$$2x^2 + 5x - 3 = 0$$
$$(2x - 1)(x + 3) = 0$$

$2x - 1 = 0$ or $x + 3 = 0$

$x = \dfrac{1}{2}$ or $x = -3$

Solution set: $\left\{\dfrac{1}{2}, -3\right\}$

65.

$$2\sqrt{x + 2} = 3\sqrt{x - 3}$$
$$\left(2\sqrt{x + 2}\right)^2 = \left(3\sqrt{x - 3}\right)^2$$
$$4(x + 2) = 9(x - 3)$$
$$4x + 8 = 9x - 27$$
$$8 = 5x - 27$$
$$35 = 5x$$
$$7 = x \qquad \text{Solution set:} \ \{7\}$$

67.

$$3\sqrt{2x + 12} = 5\sqrt{2x - 4}$$
$$\left(3\sqrt{2x + 12}\right)^2 = \left(5\sqrt{2x - 4}\right)^2$$
$$9(2x + 12) = 25(2x - 4)$$
$$18x + 108 = 50x - 100$$
$$108 = 32x - 100$$
$$208 = 32x$$
$$\dfrac{208}{32} = x$$
$$\dfrac{13}{2} = x \qquad \text{Solution set:} \ \left\{\dfrac{13}{2}\right\}$$

69.

$$\sqrt{x - 2} + 2 = x$$
$$\sqrt{x - 2} = x - 2$$
$$\left(\sqrt{x - 2}\right)^2 = (x - 2)^2$$
$$x - 2 = x^2 - 4x + 4$$
$$0 = x^2 - 5x + 6$$
$$0 = (x - 2)(x - 3)$$

#69 continued

$x - 2 = 0$ or $x - 3 = 0$
     $x = 2$ or $x = 3$ Solution set: $\{2, 3\}$

71.

$$\sqrt{5x + 19} - 1 = x$$
$$\sqrt{5x + 19} = x + 1$$
$$\left(\sqrt{5x + 19}\right)^2 = (x + 1)^2$$
$$5x + 19 = x^2 + 2x + 1$$
$$0 = x^2 - 3x - 18$$
$$0 = (x + 3)(x - 6)$$

$x + 3 = 0$ or $x - 6 = 0$
   $x = -3$ or $x = 6$

A check shows that the solution set is $\{6\}$.

73.

$$\sqrt{4x - 1} = 2x$$
$$\left(\sqrt{4x - 1}\right)^2 = (2x)^2$$
$$4x - 1 = 4x^2$$
$$0 = 4x^2 - 4x + 1$$
$$0 = (2x - 1)^2$$
$$2x - 1 = 0$$
$$x = \dfrac{1}{2}$$

A check shows that the solution set is $\left\{\dfrac{1}{2}\right\}$.

75.

$$\sqrt{2x + 3} + 2x = 3$$
$$\sqrt{2x + 3} = 3 - 2x$$
$$\left(\sqrt{2x + 3}\right)^2 = (3 - 2x)^2$$
$$2x + 3 = 9 - 12x + 4x^2$$
$$0 = 4x^2 - 14x + 6$$
$$0 = 2x^2 - 7x + 3$$
$$0 = (2x - 1)(x - 3)$$

$2x - 1 = 0$ or $x - 3 = 0$

$x = \dfrac{1}{2}$ or $x = 3$

A check shows that the solution set is $\left\{\dfrac{1}{2}\right\}$

77.

$$\sqrt{3x + 4} - 2x = 2$$
$$\sqrt{3x + 4} = 2x + 2$$
$$\left(\sqrt{3x + 4}\right)^2 = (2x + 2)^2$$
$$3x + 4 = 4x^2 + 8x + 4$$
$$0 = 4x^2 + 5x$$
$$0 = x(4x + 5)$$

$x = 0$ or $4x + 5 = 0$

$x = 0$ or $\quad x = -\dfrac{5}{4}$

A check shows that the solution set is $\{0\}$.

79.

$$2\sqrt{2x + 6} + 7 = 3x$$
$$2\sqrt{2x + 6} = 3x - 7$$
$$\left(2\sqrt{2x + 6}\right)^2 = (3x - 7)^2$$
$$4(2x + 6) = 9x^2 - 42x + 49$$
$$8x + 24 = 9x^2 - 42x + 49$$
$$0 = 9x^2 - 50x + 25$$
$$0 = (9x - 5)(x - 5)$$

$9x - 5 = 0$ or $x - 5 = 0$

$\quad x = \dfrac{5}{9}$ or $\quad x = 5$

A check shows that the solution set is $\{5\}$.

81.

$$\sqrt{x + 8} = \sqrt{x} + 2$$
$$\left(\sqrt{x + 8}\right)^2 = \left(\sqrt{x} + 2\right)^2$$
$$x + 8 = x + 4\sqrt{x} + 4$$
$$4 = 4\sqrt{x} \qquad \text{Isolate the square root.}$$
$$1 = \sqrt{x} \qquad \text{Divide by 4.}$$
$$(1)^2 = (\sqrt{x})^2$$
$$1 = x \qquad \text{Solution set: } \{1\}.$$

83.

$$\sqrt{2x + 7} = \sqrt{x} + 2$$
$$\left(\sqrt{2x + 7}\right)^2 = \left(\sqrt{x} + 2\right)^2$$
$$2x + 7 = x + 4\sqrt{x} + 4$$
$$x + 3 = 4\sqrt{x} \qquad \text{Isolate the square root.}$$
$$(x + 3)^2 = (4\sqrt{x})^2$$
$$x^2 + 6x + 9 = 16x$$
$$x^2 - 10x + 9 = 0$$

#83 continued

$$(x - 1)(x - 9) = 0$$
$$x = 1 \quad \text{or} \quad x = 9 \qquad \text{Solution set: } \{1, 9\}$$

85.

$$\sqrt{4x + 3} = 2\sqrt{x} + 1$$
$$\left(\sqrt{4x + 3}\right)^2 = \left(2\sqrt{x} + 1\right)^2$$
$$4x + 3 = 4x + 4\sqrt{x} + 1$$
$$2 = 4\sqrt{x} \qquad \text{Isolate the square root}$$
$$(2)^2 = (4\sqrt{x})^2$$
$$4 = 16x$$
$$\frac{1}{4} = x \qquad \text{Solution set: } \left\{\frac{1}{4}\right\}.$$

87.

$$2\sqrt{x - 5} = 3\sqrt{x} - 5$$
$$\left(2\sqrt{x - 5}\right)^2 = \left(3\sqrt{x} - 5\right)^2$$
$$4(x - 5) = 9x - 30\sqrt{x} + 25$$
$$4x - 20 = 9x - 30\sqrt{x} + 25$$
$$30\sqrt{x} = 5x + 45$$
$$\qquad\qquad \text{Isolate the square root}$$
$$6\sqrt{x} = x + 9 \quad \text{Divide by 5}$$
$$(6\sqrt{x})^2 = (x + 9)^2$$
$$36x = x^2 + 18x + 81$$
$$0 = x^2 - 18x + 81$$
$$0 = (x - 9)^2$$
$$x - 9 = 0$$
$$x = 9 \qquad \text{Solution set: } \{9\}.$$

Review Exercises

1. - 19.
The answers do not require worked-out solutions.

21.
$$\sqrt{7} \cdot \sqrt{28} = \sqrt{7 \cdot 28} = \sqrt{196} = 14$$

23.
$$\sqrt{3y} \cdot \sqrt{3y} = \sqrt{9y^2} = 3y$$

25.
$$\sqrt{63} = \sqrt{9 \cdot 7} = \sqrt{9} \cdot \sqrt{7} = 3\sqrt{7}$$

27.

$$\sqrt{245} = \sqrt{49 \cdot 5} = \sqrt{49} \cdot \sqrt{5} = 7\sqrt{5}$$

29.

$$\sqrt{13x^{10}} = \sqrt{x^{10} \cdot 13} = \sqrt{x^{10}} \cdot \sqrt{13} = x^5\sqrt{13}$$

31.

$$\sqrt{x^{15}} = \sqrt{x^{14} \cdot x} = \sqrt{x^{14}} \cdot \sqrt{x} = x^7\sqrt{x}$$

33.

$$\sqrt{80x^4y^7} = \sqrt{16x^4y^6 \cdot 5y} = \sqrt{16x^4y^6} \cdot \sqrt{5y}$$
$$= 4x^2y^3\sqrt{5y}$$

35.

$$\sqrt{\frac{11}{144}} = \frac{\sqrt{11}}{\sqrt{144}} = \frac{\sqrt{11}}{12}$$

37.

$$\frac{\sqrt{18}}{\sqrt{98}} = \sqrt{\frac{18}{98}} = \sqrt{\frac{9}{49}} = \frac{\sqrt{9}}{\sqrt{49}} = \frac{3}{7}$$

39.

$$\sqrt{\frac{121x^4}{y^{10}}} = \frac{\sqrt{121x^4}}{\sqrt{y^{10}}} = \frac{11x^2}{y^5}$$

41.

$$\sqrt{\frac{90x}{169y^2}} = \frac{\sqrt{90x}}{\sqrt{169y^2}} = \frac{\sqrt{9 \cdot 10x}}{13y}$$
$$= \frac{\sqrt{9} \cdot \sqrt{10x}}{13y} = \frac{3\sqrt{10x}}{13y}$$

43.

$$\sqrt[3]{200} = \sqrt[3]{(8 \cdot 25)} = \sqrt[3]{8} \cdot \sqrt[3]{25} = 2\sqrt[3]{25}$$

45.

$$\sqrt[3]{\frac{x^3}{64y^3}} = \frac{\sqrt[3]{x^3}}{\sqrt[3]{64y^3}} = \frac{x}{4y}$$

47. - 53.
The answers do not require worked-out solutions.

55.

$$\sqrt{48} - \sqrt{75} = \sqrt{16 \cdot 3} - \sqrt{25 \cdot 3}$$
$$= \sqrt{16} \cdot \sqrt{3} - \sqrt{25} \cdot \sqrt{3}$$
$$= 4\sqrt{3} - 5\sqrt{3} = -\sqrt{3}$$

57.

$$7\sqrt{28} - 2\sqrt{7} + \sqrt{175}$$
$$= 7\sqrt{4 \cdot 7} - 2\sqrt{7} + \sqrt{25 \cdot 7}$$
$$= 7\sqrt{4} \cdot \sqrt{7} - 2\sqrt{7} + \sqrt{25} \cdot \sqrt{7}$$
$$= 7 \cdot 2\sqrt{7} - 2\sqrt{7} + 5\sqrt{7}$$
$$= 14\sqrt{7} - 2\sqrt{7} + 5\sqrt{7} = 17\sqrt{7}$$

59.

$$15x\sqrt{24} + 3\sqrt{54x^2} = 15x\sqrt{4 \cdot 6} + 3\sqrt{9x^2 \cdot 6}$$
$$= 15x\sqrt{4} \cdot \sqrt{6} + 3\sqrt{9x^2} \cdot \sqrt{6}$$
$$= 15x \cdot 2\sqrt{6} + 3 \cdot 3x\sqrt{6}$$
$$= 30x\sqrt{6} + 9x\sqrt{6} = 39x\sqrt{6}$$

61.

$$\sqrt{63x} + \sqrt{147x} = \sqrt{9 \cdot 7x} + \sqrt{49 \cdot 3x}$$
$$= \sqrt{9} \cdot \sqrt{7x} + \sqrt{49} \cdot \sqrt{3x}$$
$$= 3\sqrt{7x} + 7\sqrt{3x}$$

63.

$$\sqrt{72x^3y} - x\sqrt{288xy} + \sqrt{24xy}$$
$$= \sqrt{36x^2 \cdot 2xy} - x\sqrt{144 \cdot 2xy} + \sqrt{4 \cdot 6xy}$$
$$= \sqrt{36x^2} \cdot \sqrt{2xy} - x\sqrt{144} \cdot \sqrt{2xy} + \sqrt{4} \cdot \sqrt{6xy}$$
$$= 6x\sqrt{2xy} - 12x\sqrt{2xy} + 2\sqrt{6xy}$$
$$= -6x\sqrt{2xy} + 2\sqrt{6xy}$$

65.

$$\sqrt[3]{108} + \sqrt[3]{32} = \sqrt[3]{(27 \cdot 4)} + \sqrt[3]{(8 \cdot 4)}$$
$$= \sqrt[3]{27} \cdot \sqrt[3]{4} + \sqrt[3]{8} \cdot \sqrt[3]{4}$$
$$= 3\sqrt[3]{4} + 2\sqrt[3]{4} = 5\sqrt[3]{4}$$

67.

$$5\sqrt{14} \cdot \sqrt{21} = 5\sqrt{14 \cdot 21} = 5\sqrt{2 \cdot 7 \cdot 3 \cdot 7}$$
$$= 5\sqrt{49 \cdot 6} = 5\sqrt{49} \cdot \sqrt{6}$$
$$= 5 \cdot 7\sqrt{6} = 35\sqrt{6}$$

69.

$6\sqrt{6} \cdot 7\sqrt{60} = 42\sqrt{6 \cdot 60}$

$= 42\sqrt{360} = 42\sqrt{36 \cdot 10}$

$= 42\sqrt{36} \cdot \sqrt{10} = 42 \cdot 6\sqrt{10} = 252\sqrt{10}$

71.

$12\sqrt{5x} \cdot \sqrt{3x} = 12\sqrt{15x^2} = 12\sqrt{x^2 \cdot 15}$

$\qquad = 12\sqrt{x^2} \cdot \sqrt{15} = 12x\sqrt{15}$

73.

$2\sqrt{7x} \cdot 6\sqrt{28xy} = 12\sqrt{7 \cdot 28x^2 y} = 12\sqrt{196x^2 y}$

$\qquad = 12\sqrt{196x^2} \cdot \sqrt{y} = 12 \cdot 14x\sqrt{y}$

$\qquad = 168x\sqrt{y}$

75.

$3\sqrt{10xy} \cdot 5\sqrt{3xy} = 15\sqrt{30x^2 y^2}$

$\qquad = 15\sqrt{x^2 y^2} \cdot \sqrt{30}$

$\qquad = 15xy\sqrt{30}$

77.

$\sqrt{14}(\sqrt{6} + \sqrt{5}) = \sqrt{14} \cdot \sqrt{6} + \sqrt{14} \cdot \sqrt{5}$

$\qquad = \sqrt{84} + \sqrt{70}$

$\qquad = \sqrt{4 \cdot 21} + \sqrt{70}$

$\qquad = \sqrt{4} \cdot \sqrt{21} + \sqrt{70}$

$\qquad = 2\sqrt{21} + \sqrt{70}$

79.

$2\sqrt{6}\left(7\sqrt{10} - 3\sqrt{5} + \sqrt{21}\right)$

$= 2\sqrt{6} \cdot 7\sqrt{10} - 2\sqrt{6} \cdot 3\sqrt{5} + 2\sqrt{6} \cdot \sqrt{21}$

$= 14\sqrt{60} - 6\sqrt{30} + 2\sqrt{126}$

$= 14\sqrt{4 \cdot 15} - 6\sqrt{30} + 2\sqrt{9 \cdot 14}$

$= 14\sqrt{4} \cdot \sqrt{15} - 6\sqrt{30} + 2\sqrt{9} \cdot \sqrt{14}$

$= 14 \cdot 2\sqrt{15} - 6\sqrt{30} + 2 \cdot 3\sqrt{14}$

$= 28\sqrt{15} - 6\sqrt{30} + 6\sqrt{14}$

81.

$\left(\sqrt{5} + 5\right)\left(\sqrt{6} - 7\right)$

$\qquad$ F $\qquad$ O $\qquad$ I $\qquad$ L

$= \sqrt{5} \cdot \sqrt{6} - \sqrt{5} \cdot 7 + 5\sqrt{6} - 5 \cdot 7$

$= \sqrt{30} - 7\sqrt{5} + 5\sqrt{6} - 35$

83.

$\left(\sqrt{5x} + \sqrt{y}\right)\left(\sqrt{2x} - \sqrt{5}\right)$

$\qquad$ F $\qquad$ O $\qquad$ I $\qquad$ L

$= \sqrt{5x} \cdot \sqrt{2x} - \sqrt{5x} \cdot \sqrt{5} + \sqrt{y} \cdot \sqrt{2x} - \sqrt{y} \cdot \sqrt{5}$

$= \sqrt{10x^2} - \sqrt{25x} + \sqrt{2xy} - \sqrt{5y}$

$= x\sqrt{10} - 5\sqrt{x} + \sqrt{2xy} - \sqrt{5y}$

85.

$\left(4\sqrt{x} + \sqrt{3}\right)\left(3\sqrt{x} + 2\sqrt{3}\right)$

$\qquad$ F $\qquad$ O $\qquad$ I $\qquad$ L

$= 4\sqrt{x} \cdot 3\sqrt{x} + 4\sqrt{x} \cdot 2\sqrt{3} + \sqrt{3} \cdot 3\sqrt{x} + \sqrt{3} \cdot 2\sqrt{3}$

$= 12\sqrt{x^2} + 8\sqrt{3x} + 3\sqrt{3x} + 2\sqrt{9}$

$= 12x + 11\sqrt{3x} + 6$

87.

$\left(3\sqrt{11} + 8\right)\left(3\sqrt{11} - 8\right) = \left(3\sqrt{11}\right)^2 - (8)^2$

$\qquad\qquad\qquad\qquad = 9 \cdot 11 - 64$

$\qquad\qquad\qquad\qquad = 99 - 64 = 35$

89.

$\left(\sqrt{7} - 4\right)^2 = \left(\sqrt{7}\right)^2 - 2\left(4\sqrt{7}\right) + 4^2$

$\qquad\qquad = 7 - 8\sqrt{7} + 16$

$\qquad\qquad = 23 - 8\sqrt{7}$

91.

$\left(\sqrt{x} + 5\sqrt{y}\right)^2$

$= \left(\sqrt{x}\right)^2 + 2\left(\sqrt{x} \cdot 5\sqrt{y}\right) + \left(5\sqrt{y}\right)^2$

$= x + 10\sqrt{xy} + 25y$

93.

$\dfrac{2}{\sqrt{7}} = \dfrac{2}{\sqrt{7}} \cdot \dfrac{\sqrt{7}}{\sqrt{7}} = \dfrac{2\sqrt{7}}{\sqrt{49}} = \dfrac{2\sqrt{7}}{7}$

95.

$\sqrt{\dfrac{13}{6}} = \sqrt{\dfrac{13}{6} \cdot \dfrac{6}{6}} = \sqrt{\dfrac{78}{36}} = \dfrac{\sqrt{78}}{\sqrt{36}} = \dfrac{\sqrt{78}}{6}$

97.

$$\frac{12}{\sqrt{18}} = \frac{12}{\sqrt{9 \cdot 2}} = \frac{12}{3\sqrt{2}} = \frac{4}{\sqrt{2}} = \frac{4}{\sqrt{2}} \cdot \frac{\sqrt{2}}{\sqrt{2}} = \frac{4\sqrt{2}}{2}$$

$$= \frac{2 \cdot \not{2}\sqrt{2}}{\not{2}} = 2\sqrt{2}$$

99.

$$\frac{3\sqrt{60}}{7\sqrt{50}} = \frac{3}{7} \cdot \frac{\sqrt{60}}{\sqrt{50}} = \frac{3}{7}\sqrt{\frac{60}{50}} = \frac{3}{7}\sqrt{\frac{6}{5}} = \frac{3}{7}\sqrt{\frac{6}{5} \cdot \frac{5}{5}}$$

$$= \frac{3}{7}\sqrt{\frac{30}{25}} = \frac{3\sqrt{30}}{7\sqrt{25}} = \frac{3\sqrt{30}}{7 \cdot 5} = \frac{3\sqrt{30}}{35}$$

101.

$$\sqrt{\frac{11}{3x}} = \sqrt{\frac{11}{3x} \cdot \frac{3x}{3x}} = \sqrt{\frac{33x}{9x^2}} = \frac{\sqrt{33x}}{\sqrt{9x^2}} = \frac{\sqrt{33x}}{3x}$$

103.

$$\sqrt{\frac{16x}{54y}} = \sqrt{\frac{8x}{27y}} = \sqrt{\frac{8x}{3^3y} \cdot \frac{3y}{3y}} = \sqrt{\frac{24xy}{3^4y^2}}$$

$$= \frac{\sqrt{4 \cdot 6xy}}{\sqrt{3^4y^2}} = \frac{\sqrt{4} \cdot \sqrt{6xy}}{3^2y} = \frac{2\sqrt{6xy}}{9y}$$

105.

$$\sqrt{\frac{20}{13x^3}} = \sqrt{\frac{20}{13x^3} \cdot \frac{13x}{13x}} = \sqrt{\frac{20 \cdot 13x}{13^2x^4}}$$

$$= \frac{\sqrt{4 \cdot 5 \cdot 13x}}{\sqrt{13^2x^4}} = \frac{\sqrt{4} \cdot \sqrt{65x}}{13x^2} = \frac{2\sqrt{65x}}{13x^2}$$

107.

$$\frac{5}{\sqrt{11} - 3} = \frac{5}{\sqrt{11} - 3} \cdot \frac{\sqrt{11} + 3}{\sqrt{11} + 3}$$

$$= \frac{5(\sqrt{11} + 3)}{(\sqrt{11} - 3)(\sqrt{11} + 3)}$$

$$= \frac{5\sqrt{11} + 15}{(\sqrt{11})^2 - (3)^2} = \frac{5\sqrt{11} + 15}{11 - 9}$$

$$= \frac{5\sqrt{11} + 15}{2} = \frac{5(\sqrt{11} + 3)}{2}$$

109.

$$\frac{7\sqrt{2}}{\sqrt{2} + 3} = \frac{7\sqrt{2}}{\sqrt{2} + 3} \cdot \frac{\sqrt{2} - 3}{\sqrt{2} - 3}$$

#109 continued

$$= \frac{7\sqrt{2}(\sqrt{2} - 3)}{(\sqrt{2} + 3)(\sqrt{2} - 3)}$$

$$= \frac{14 - 21\sqrt{2}}{(\sqrt{2})^2 - (3)^2}$$

$$= \frac{14 - 21\sqrt{2}}{2 - 9} = \frac{\not{7}(2 - 3\sqrt{2})}{-1 \cdot \not{7}}$$

$$= -(2 - 3\sqrt{2}) = 3\sqrt{2} - 2$$

111.

$$\frac{3\sqrt{7} - \sqrt{3}}{\sqrt{7} - 2\sqrt{3}} = \frac{3\sqrt{7} - \sqrt{3}}{\sqrt{7} - 2\sqrt{3}} \cdot \frac{\sqrt{7} + 2\sqrt{3}}{\sqrt{7} + 2\sqrt{3}}$$

$$= \frac{(3\sqrt{7} - \sqrt{3})(\sqrt{7} + 2\sqrt{3})}{(\sqrt{7} - 2\sqrt{3})(\sqrt{7} + 2\sqrt{3})}$$

$$= \frac{21 + 6\sqrt{21} - \sqrt{21} - 6}{(\sqrt{7})^2 - (2\sqrt{3})^2}$$

$$= \frac{15 + 5\sqrt{21}}{7 - 12} = \frac{15 + 5\sqrt{21}}{-5}$$

$$= \frac{\not{5}(3 + \sqrt{21})}{-1 \cdot \not{5}} = -(3 + \sqrt{21})$$

$$= -3 - \sqrt{21}$$

113.

$$\frac{\sqrt{x}}{3\sqrt{x} + 2} = \frac{\sqrt{x}}{3\sqrt{x} + 2} \cdot \frac{3\sqrt{x} - 2}{3\sqrt{x} - 2}$$

$$= \frac{\sqrt{x}(3\sqrt{x} - 2)}{(3\sqrt{x} + 2)(3\sqrt{x} - 2)}$$

$$= \frac{3x - 2\sqrt{x}}{(3\sqrt{x})^2 - (2)^2} = \frac{3x - 2\sqrt{x}}{9x - 4}$$

115.

$$\frac{\sqrt{x} + 3}{\sqrt{x} + 5} = \frac{\sqrt{x} + 3}{\sqrt{x} + 5} \cdot \frac{\sqrt{x} - 5}{\sqrt{x} - 5}$$

$$= \frac{(\sqrt{x} + 3)(\sqrt{x} - 5)}{(\sqrt{x} + 5)(\sqrt{x} - 5)}$$

$$= \frac{x - 5\sqrt{x} + 3\sqrt{x} - 15}{(\sqrt{x})^2 - (5)^2}$$

#115 continued

$$= \frac{x - 2\sqrt{x} - 15}{x - 25}$$

117.

$$\frac{\sqrt{x} - 2\sqrt{5y}}{\sqrt{x} - \sqrt{5y}} = \frac{\sqrt{x} - 2\sqrt{5y}}{\sqrt{x} - \sqrt{5y}} \cdot \frac{\sqrt{x} + \sqrt{5y}}{\sqrt{x} + \sqrt{5y}}$$

$$= \frac{\left(\sqrt{x} - 2\sqrt{5y}\right)\left(\sqrt{x} + \sqrt{5y}\right)}{\left(\sqrt{x} - \sqrt{5y}\right)\left(\sqrt{x} + \sqrt{5y}\right)}$$

$$= \frac{x + \sqrt{5xy} - 2\sqrt{5xy} - 10y}{\left(\sqrt{x}\right)^2 - \left(\sqrt{5y}\right)^2}$$

$$= \frac{x - \sqrt{5xy} - 10y}{x - 5y}$$

119.

$$\sqrt[3]{\frac{5}{4}} = \sqrt[3]{\frac{5}{4} \cdot \frac{2}{2}} = \sqrt[3]{\frac{10}{8}} = \frac{\sqrt[3]{10}}{\sqrt[3]{8}} = \frac{\sqrt[3]{10}}{2}$$

121.

$$\sqrt[3]{\frac{2x}{81y^3}} = \sqrt[3]{\frac{2x}{3^4y^3} \cdot \frac{3^2}{3^2}} = \sqrt[3]{\frac{18x}{3^6y^3}} = \frac{\sqrt[3]{18x}}{\sqrt[3]{3^6y^3}}$$

$$= \frac{\sqrt[3]{18x}}{3^2y} = \frac{\sqrt[3]{18x}}{9y}$$

123.

$$\sqrt{x - 7} = 4$$

$$\left(\sqrt{x - 7}\right)^2 = 4^2$$

$$x - 7 = 16$$

$$x = 23 \qquad \text{Solution Set:} \quad \{23\}$$

125.

$$\sqrt{3x - 1} = \frac{1}{3}$$

$$\left(\sqrt{3x - 1}\right)^2 = \left(\frac{1}{3}\right)^2$$

$$3x - 1 = \frac{1}{9}$$

$$3x = \frac{1}{9} + 1$$

#125 continued

$$3x = \frac{10}{9}$$

$$x = \frac{10}{27} \qquad \text{Solution Set:} \quad \left\{\frac{10}{27}\right\}$$

127.

$$3\sqrt{x} + 8 = 5$$

$$3\sqrt{x} \ne -3 \qquad \text{Solution Set:} \quad \emptyset$$

129.

$$\sqrt{5x + 1} + 1 = 5$$

$$\sqrt{5x + 1} = 4$$

$$\left(\sqrt{5x + 1}\right)^2 = 4^2$$

$$5x + 1 = 16$$

$$5x = 15$$

$$x = 3 \qquad \text{Solution Set:} \quad \{3\}$$

131.

$$\sqrt{3x + 5} = \sqrt{7x - 3}$$

$$\left(\sqrt{3x + 5}\right)^2 = \left(\sqrt{7x - 3}\right)^2$$

$$3x + 5 = 7x - 3$$

$$5 = 4x - 3$$

$$8 = 4x$$

$$2 = x \qquad \text{Solution Set:} \quad \{2\}$$

133.

$$5\sqrt{x + 2} = 3\sqrt{4x - 3}$$

$$\left(5\sqrt{x + 2}\right)^2 = \left(3\sqrt{4x - 3}\right)^2$$

$$25(x + 2) = 9(4x - 3)$$

$$25x + 50 = 36x - 27$$

$$50 = 11x - 27$$

$$77 = 11x$$

$$7 = x \qquad \text{Solution Set} \quad \{7\}$$

135.

$$\sqrt{3x + 33} - 5 = x$$

$$\sqrt{3x + 33} = x + 5$$

$$\left(\sqrt{3x + 33}\right)^2 = (x + 5)^2$$

$$3x + 33 = x^2 + 10x + 25$$

$$0 = x^2 + 7x - 8$$

$$0 = (x + 8)(x - 1)$$

#135 continued

$x + 8 = 0$ or $x - 1 = 0$
        $x = -8$ or $x = 1$

A check shows that the solution set is {1}.

137.

$3\sqrt{4x - 3} - 1 = 2x$
$3\sqrt{4x - 3} = 2x + 1$
$\left(3\sqrt{4x - 3}\right)^2 = (2x + 1)^2$
$9(4x - 3) = 4x^2 + 4x + 1$
$36x - 27 = 4x^2 + 4x + 1$
$0 = 4x^2 - 32x + 28$
$0 = x^2 - 8x + 7$
$0 = (x - 7)(x - 1)$

$x - 7 = 0$ or $x - 1 = 0$
     $x = 7$ or $x = 1$ Solution Set: {1,7}

139.

$\sqrt{7x - 12} = \sqrt{x} + 2$
$\left(\sqrt{7x - 12}\right)^2 = \left(\sqrt{x} + 2\right)^2$
$7x - 12 = x + 4\sqrt{x} + 4$
$6x - 16 = 4\sqrt{x}$
$3x - 8 = 2\sqrt{x}$ Divide by 2
$(3x - 8)^2 = \left(2\sqrt{x}\right)^2$
$9x^2 - 48x + 64 = 4x$
$9x^2 - 52x + 64 = 0$
$(9x - 16)(x - 4) = 0$

$9x - 16 = 0$ or $x - 4 = 0$
     $x = \dfrac{16}{9}$ or $x = 4$

A check shows that the solution set is {4}.

Chapter 8 Test Solutions

1–3.

The answers do not require worked-out solutions.

5.

$\sqrt{90} = \sqrt{9 \cdot 10} = \sqrt{9} \cdot \sqrt{10} = 3\sqrt{10}$

7.

$\sqrt{48x^6y^{11}} = \sqrt{3 \cdot 16 \cdot x^6 \cdot y^{10} \cdot y}$
$= \sqrt{16x^6y^{10} \cdot 3 \cdot y}$
$= \sqrt{16x^6y^{10}} \cdot \sqrt{3y}$
$= 4x^3y^5\sqrt{3y}$

9.

$\sqrt[3]{\dfrac{54}{x^6}} = \sqrt[3]{\dfrac{27 \cdot 2}{x^6}} = \dfrac{\sqrt[3]{27 \cdot 2}}{\sqrt[3]{x^6}} = \dfrac{3\sqrt[3]{2}}{x^2}$

11.

$\sqrt{5x} \cdot \sqrt{10y} = \sqrt{50 \cdot xy} = \sqrt{25 \cdot 2xy}$
$= \sqrt{25} \cdot \sqrt{2xy}$
$= 5\sqrt{2xy}$

13.

$5\sqrt{3} - 3\sqrt{5} - \sqrt{3} - 4\sqrt{5}$
$= 5\sqrt{3} - \sqrt{3} - 3\sqrt{5} - 4\sqrt{5}$
$= 4\sqrt{3} - 7\sqrt{5}$

15.

$2\sqrt{xy}\left(3\sqrt{2} + \sqrt{5x} - \sqrt{9y}\right)$
$= 6\sqrt{2xy} + 2\sqrt{5x^2y} - 2\sqrt{9xy^2}$
$= 6\sqrt{2xy} + 2x\sqrt{5y} - 2 \cdot 3y\sqrt{x}$
$= 6\sqrt{2xy} + 2x\sqrt{5y} - 6y\sqrt{x}$

17.

$\left(\sqrt{3x} + \sqrt{y}\right)\left(\sqrt{x} - \sqrt{3}\right)$

$= \overset{F}{\sqrt{3x^2}} - \overset{O}{\sqrt{9x}} + \overset{I}{\sqrt{xy}} - \overset{L}{\sqrt{3y}}$
$= x\sqrt{3} - 3\sqrt{x} + \sqrt{xy} - \sqrt{3y}$

19.

$\sqrt[3]{81} - \sqrt[3]{375} = \sqrt[3]{27 \cdot 3} - \sqrt[3]{125 \cdot 3}$
$= 3\sqrt[3]{3} - 5\sqrt[3]{3} = -2\sqrt[3]{3}$

21.

$$8x\sqrt{20y} - \sqrt{45xy^2} - 3\sqrt{5x^2y} + 4y\sqrt{80x}$$

$$=8x\sqrt{4 \cdot 5y} - \sqrt{9 \cdot 5xy^2}$$
$$- 3\sqrt{5x^2y} + 4y\sqrt{16 \cdot 5x}$$

$$=8x\sqrt{4 \cdot 5y} - \sqrt{9y^2 \cdot 5x}$$
$$- 3\sqrt{x^2 \cdot 5y} + 4y\sqrt{16 \cdot 5x}$$

$$=8x \cdot 2\sqrt{5y} - 3y\sqrt{5x} - 3x\sqrt{5y} + 4y \cdot 4\sqrt{5x}$$

$$=16x\sqrt{5y} - 3y\sqrt{5x} - 3x\sqrt{5y} + 16y\sqrt{5x}$$

$$=16x\sqrt{5y} - 3x\sqrt{5y} - 3y\sqrt{5x} + 16y\sqrt{5x}$$

$$=13x\sqrt{5y} + 13y\sqrt{5y}$$

23.

$$\frac{3}{\sqrt{5}} \cdot \frac{\sqrt{5}}{\sqrt{5}} = \frac{3\sqrt{5}}{5}$$

25.

$$\sqrt{\frac{50}{12}} = \sqrt{\frac{25}{6}} = \frac{\sqrt{25}}{\sqrt{6}} \qquad \text{Reduce}$$

$$= \frac{5}{\sqrt{6}} = \frac{5}{\sqrt{6}} \cdot \frac{\sqrt{6}}{\sqrt{6}} = \frac{5\sqrt{6}}{6}$$

27.

$$\frac{\sqrt{x}}{5\sqrt{x} + 3} \cdot \frac{5\sqrt{x} - 3}{5\sqrt{x} - 3}$$

$$= \frac{\sqrt{x}(5\sqrt{x} - 3)}{25x - 9} = \frac{5x - 3\sqrt{x}}{25x - 9}$$

29.

$\sqrt{3x + 10} - 4 = 3$	Isolate radical
$\sqrt{3x + 10} = 7$	Add 4
$3x + 10 = 49$	Square
$3x = 39$	Subtract 10
$x = 13$	Divide by 3

Solution set: {13}

31.

$\sqrt{5x + 4} = \sqrt{x} + 4$	
$5x + 4 = x + 8\sqrt{x} + 16$	Square
$4x - 12 = 8\sqrt{x}$	Subtract x, 16
$x - 3 = 2\sqrt{x}$	Divide by 4

#31 continued

$x^2 - 6x + 9 = 4x$	Square
$x^2 - 10x + 9 = 0$	Subtract 4x
$(x - 1)(x - 9) = 0$	
$x = 1$ or $x = 9$	

Solution set: {9}
A check shows that 1 is not a solution.

Test Your Memory

1–3.

The graph is in the answer section of the main text.

5.

$m = -\frac{1}{4}$, (-3, 0)

$y - 0 = -\frac{1}{4}(x + 3)$	
$4y = -(x + 3)$	Multiply by 4
$4y = -x - 3$	
$x + 4y = -3$	Add x

7.

(7, 9), (-5, -1)

$$m = \frac{-1 - 9}{-5 - 7} = \frac{-10}{-12} = \frac{5}{6}$$

$$y - 9 = \frac{5}{6}(x - 7)$$

$6y - 54 = 5(x - 7)$	Multiply by 6
$6y - 54 = 5x - 35$	
$-54 = 5x - 6y - 35$	Subtract 6y, add 35
$-19 = 5x - 6y$ or	
$5x - 6y = -19$	

9.

$5 + 3(x - 7) = 10 - 2(3x + 1)$	
$5 + 3x - 21 = 10 - 6x - 2$	
$3x - 16 = -6x + 8$	
$9x = 24$	Add 6x, 16
$x = \frac{24}{9}$ or $\frac{8}{3}$	Reduce

#9 continued

Solution set: $\left\{\dfrac{8}{3}\right\}$

11.

$$(4x - 4)(x + 4) = -25$$

$$4x^2 + 12x - 16 = -25$$

$$4x^2 + 12x + 9 = 0$$

$$(2x + 3)(2x + 3) = 0$$

$$x = -\dfrac{3}{2}$$

Solution set: $\left\{-\dfrac{3}{2}\right\}$

13.

$$\dfrac{3}{4x - 3} = \dfrac{2}{2x + 1}$$

$$3(2x + 1) = 2(4x - 3)$$

$$6x + 3 = 8x - 6$$

$$3 = 2x - 6 \qquad \text{Subtract } 6x$$

$$9 = 2x \qquad \text{Add } 6$$

$$\dfrac{9}{2} = x$$

Solution set: $\left\{\dfrac{9}{2}\right\}$.

15.

$$\dfrac{3}{x + 1} - \dfrac{2}{x} = \dfrac{1}{3x} \qquad \text{LCD} = 3x(x + 1)$$

$$3x(x + 1)\left(\dfrac{3}{x + 1} - \dfrac{2}{x}\right) = \left(\dfrac{1}{3x}\right) 3x(x + 1)$$

$$9x - 6(x + 1) = x + 1$$

$$9x - 6x - 6 = x + 1$$

$$3x - 6 = x + 1$$

$$2x = 7 \qquad \text{Subtract x, add 6}$$

$$x = \dfrac{7}{2}$$

Solution set: $\left\{\dfrac{7}{2}\right\}$

17.

$$\sqrt{2x - 1} = 3$$

$$2x - 1 = 9 \qquad \text{Square}$$

$$2x = 10$$

$$x = 5$$

Solution set: {5}

19.

$$\sqrt{5x+1} - x = 1$$

$$\sqrt{5x+1} = 1 + x \qquad \text{Add x}$$

$$5x + 1 = 1 + 2x + x^2 \qquad \text{Square}$$

$$0 = x^2 - 3x \qquad \text{Subtract 1, 5x}$$

$$0 = x(x - 3)$$

$$x = 0 \text{ or } x = 3$$

Solution set: {0, 3}

21.

$$\dfrac{(5x^3y^{-2})^{-1}}{(5x^{-4}y^{-5})^{-2}} = \dfrac{5^{-1}x^{-3}y^2}{5^{-2}x^8y^{10}} = \dfrac{5^{-1+2}}{x^{8+3}y^{10-2}} = \dfrac{5}{x^{11}y^8}$$

23–25.

The exercises do not require worked-out solutions.

27.

$$\sqrt{72x^8y^9} = \sqrt{36 \cdot 2x^8y^8 \cdot y}$$

$$= \sqrt{36x^8y^8 2y} = 6x^4y^4\sqrt{2y}$$

29.

$$\dfrac{2x^2 + 6x}{12x^3 - 24x^2} \cdot \dfrac{12x^3 - 30x^2 + 12x}{2x^2 + 5x - 3}$$

$$= \dfrac{2x(x + 3)}{12x^2(x - 2)} \cdot \dfrac{6x(2x - 1)(x - 2)}{(2x - 1)(x + 3)} = 1$$

31.

$$\dfrac{5x}{x - 3} + \dfrac{2}{x - 1}$$

$$= \dfrac{(x - 1)}{(x - 1)} \cdot \left(\dfrac{5x}{x - 3}\right) + \left(\dfrac{2}{x - 1}\right) \cdot \dfrac{(x - 3)}{(x - 3)}$$

$$\text{LCD} = (x - 3)(x - 1)$$

$$= \dfrac{5x^2 - 5x}{(x - 1)(x - 3)} + \dfrac{2x - 6}{(x - 3)(x - 1)}$$

#31 continued

$$= \frac{5x^2 - 3x - 6}{(x-1)(x-3)}$$

33.

$$\sqrt{6x} \cdot \sqrt{15y} = \sqrt{90xy} = \sqrt{9 \cdot 10xy} = 3\sqrt{10xy}$$

35.

$$\left(\sqrt{7x} + \sqrt{5y}\right)\left(\sqrt{7x} - \sqrt{7y}\right)$$

$$= \left(\sqrt{7x}\right)^2 - \left(\sqrt{5y}\right)^2$$

Note: $(a + b)(a - b) = a^2 - b^2$

$$= 7x - 5y$$

37.

$$\sqrt{48} + 2\sqrt{27} - 4\sqrt{12}$$

$$= \sqrt{16 \cdot 3} + 2\sqrt{9 \cdot 3} - 4\sqrt{4 \cdot 3}$$

$$4\sqrt{3} + 6\sqrt{3} - 8\sqrt{3} = 2\sqrt{3}$$

39.

$$\sqrt{\frac{2}{3} \cdot \frac{3}{3}} = \sqrt{\frac{6}{9}} = \frac{\sqrt{6}}{3}$$

41.

$$\frac{3}{\sqrt{5} + 1} \cdot \frac{\sqrt{5} - 1}{\sqrt{5} - 1} = \frac{3\left(\sqrt{5} - 1\right)}{5 - 1} = \frac{3\sqrt{5} - 3}{4}$$

43.

(1) $3x - 2y = -4$

(2) $9x + 4y = 3$

To eliminate the y's, multiply (1) by 2.

(3) $6x - 4y = -8$

(2) $\underline{9x + 4y = 3}$

$\quad\quad 15x = -5$

$$x = \frac{-5}{15} \text{ or } -\frac{1}{3}$$

Substitute $-\frac{1}{3}$ for x in (2).

$$9\left(-\frac{1}{3}\right) + 4y = 3$$

$$-3 + 4y = 3$$

#43 continued

$$4y = 6 \quad\quad\quad \text{Add 3}$$

$$y = \frac{6}{4} \text{ or } \frac{3}{2}$$

Solution set: $\left\{\left(-\frac{1}{3}, \frac{3}{2}\right)\right\}$

45.

Let x = number of dimes.

Then 23 - x = number of quarters.

Thus,

$$10x + 25(23 - x) = 380$$

$$10x + 575 - 25x = 380$$

$$-15x = -195$$

$$x = 13$$

$$23 - x = 10$$

Answer: 10 quarters, 13 dimes

47.

Let x = measurement of height.

Then 4x - 1 = measurement of base.

Thus, $A = \frac{1}{2} bh$.

$$30 = \frac{1}{2}(4x - 1)(x)$$

$$60 = (4x - 1)x \quad\quad \text{Multiply by 2}$$

$$60 = 4x^2 - x$$

$$0 = 4x^2 - x - 60 \quad\quad \text{Subtract -60}$$

$$0 = (x - 4)(4x + 15)$$

$$x = 4 \text{ or } x = -\frac{15}{4}$$

Answer: height, 4ft.; base 15 ft.
Measurement is not negative.

49.

$$60x\left(\frac{1}{30} + \frac{1}{20}\right) = \left(\frac{1}{x}\right)60x \quad\quad \text{LCD} = 60x$$

$$2x + 3x = 60$$

$$5x = 60$$

$$x = 12$$

Answer: 12 minutes.

CHAPTER 8 STUDY GUIDE

Self-Test Exercises

Simplify the following radicals. Assume that all variables represent positive real numbers.

1. $\sqrt{\dfrac{49}{4}}$

2. $\sqrt[3]{-27}$

3. $\sqrt{144}$

4. $\sqrt{64y^2}$

5. $\sqrt{40}$

6. $\sqrt[3]{54}$

7. $\sqrt{75x^9y^{12}}$

8. $\sqrt{\dfrac{9x^7}{y^4}}$

9. $\sqrt[3]{\dfrac{72}{x^9}}$

Perform the indicated operations and simplify your answers. Assume that all variables represent positive real numbers.

10. $\left(3\sqrt{7}\right)^2$

11. $\sqrt{2x} \cdot \sqrt{12y}$

12. $\dfrac{\sqrt{5}}{\sqrt{75}}$

13. $7\sqrt{2} - 5\sqrt{5} - \sqrt{2} - 2\sqrt{5}$

14. $2\sqrt{8x} \cdot 3\sqrt{6xy}$

15. $\left(\sqrt{2x} + \sqrt{5}\right)\left(\sqrt{x} - \sqrt{2y}\right)$

16. $\left(2\sqrt{3} - 1\right)\left(3\sqrt{3} + 3\right)$

17. $\sqrt[3]{81} - \sqrt[3]{192}$

18. $3x\sqrt{75y} - 14\sqrt{12xy^2} - 2y\sqrt{48x}$

19. $\left(\sqrt{x} - 3\sqrt{y}\right)^2$

Rationalize the denominators of the following radical expressions. Assume that all variables represent positive real numbers.

20. $\dfrac{5}{\sqrt{7}}$

21. $\dfrac{3}{\sqrt[3]{4}}$

22. $\sqrt{\dfrac{9y^5}{28x^6}}$

23. $\dfrac{3\sqrt{3} + \sqrt{5}}{2\sqrt{3} + 4\sqrt{5}}$

Find the solution set of each radical equation.

24. $\sqrt{x+7} + 5 = x$

25. $\sqrt{x-5} + \sqrt{x} = 5$

The worked-out solutions begin on the next page.

Self-Test Solutions

1. $\sqrt{\dfrac{49}{9}} = \dfrac{7}{2}$ 2. $\sqrt[3]{-27} = -3$

3. $\sqrt{144} = 12$ 4. $\sqrt{64y^2} = 8y$

5.

$\sqrt{40} = \sqrt{4 \cdot 10} = 2\sqrt{10}$

6.

$\sqrt[3]{54} = \sqrt[3]{27 \cdot 2} = 3\sqrt[3]{2}$

7.

$\sqrt{75x^9y^{12}} = \sqrt{25 \cdot x^8 y^{12} \cdot 3x}$
$\qquad = 5x^4 y^6 \sqrt{3x}$

8.

$\sqrt{\dfrac{9x^7}{y^4}} = \dfrac{\sqrt{9x^7}}{y^2} = \dfrac{\sqrt{9x^6 \cdot x}}{y^2} = \dfrac{3x^3\sqrt{x}}{y^2}$

9.

$\sqrt[3]{\dfrac{72}{x^9}} = \dfrac{\sqrt[3]{72}}{\sqrt[3]{x^9}} = \dfrac{\sqrt[3]{8 \cdot 9}}{x^3} = \dfrac{2\sqrt[3]{9}}{x^3}$

10.

$\left(3\sqrt{7}\right)^2 = 9 \cdot 7 = 63$

11.

$\sqrt{2x} \cdot \sqrt{12y} = \sqrt{24xy} = \sqrt{4 \cdot 6xy} = 2\sqrt{6xy}$

12.

$\dfrac{\sqrt{5}}{\sqrt{75}} = \sqrt{\dfrac{5}{75}} = \sqrt{\dfrac{1}{15} \cdot \dfrac{15}{15}} = \dfrac{\sqrt{15}}{\sqrt{15^2}} = \dfrac{\sqrt{15}}{15}$

13.

$7\sqrt{2} - \sqrt{2} - 5\sqrt{5} - 2\sqrt{5} = 6\sqrt{2} - 7\sqrt{5}$

14.

$2\sqrt{8x} \cdot 3\sqrt{6xy} = 6\sqrt{48x^2y} = 6\sqrt{16x^2 \cdot 3y}$
$\qquad = 6 \cdot 4x\sqrt{3y} = 24x\sqrt{3y}$

15.

$\left(\sqrt{2x} + \sqrt{5}\right)\left(\sqrt{x} - \sqrt{2y}\right)$

$\qquad\quad$ F $\qquad$ O $\qquad$ I $\qquad$ L
$= \sqrt{2x^2} - \sqrt{4xy} + \sqrt{5x} - \sqrt{10y}$
$= x\sqrt{2} - 2\sqrt{xy} + \sqrt{5x} - \sqrt{10y}$

16.

$\qquad\qquad\qquad\qquad$ F $\qquad$ O $\qquad$ I $\qquad$ L
$\left(2\sqrt{3} - 1\right)\left(3\sqrt{3} + 3\right) = 6 \cdot 3 + 6\sqrt{3} - 3\sqrt{3} - 3$
$\qquad\qquad\qquad\qquad = 18 + 6\sqrt{3} - 3\sqrt{3} - 3$
$\qquad\qquad\qquad\qquad = 15 + 3\sqrt{3}$

17.

$\sqrt[3]{81} - \sqrt[3]{192} = \sqrt[3]{27 \cdot 3} - \sqrt[3]{64 \cdot 3}$
$\qquad\qquad\qquad = 3\sqrt[3]{3} - 4\sqrt[3]{3} = -\sqrt[3]{3}$

18.

$3x\sqrt{75y} - 14\sqrt{12xy^2} - 2y\sqrt{48x}$
$= 3x\sqrt{25 \cdot 3y} - 14\sqrt{4y^2 \cdot 3x} - 2y\sqrt{16 \cdot 3x}$
$= 15x\sqrt{3y} - 28y\sqrt{3x} - 8y\sqrt{3x}$
$= 15x\sqrt{3y} - 36y\sqrt{3x}$

19.

$\left(\sqrt{x} - 3\sqrt{y}\right)^2$
$= \left(\sqrt{x}\right)^2 - 2\left(3\sqrt{y} \cdot \sqrt{x}\right) + \left(3\sqrt{y}\right)^2$
$= x - 6\sqrt{xy} + 9y$

20.

$\dfrac{5}{\sqrt{7}} \cdot \dfrac{\sqrt{7}}{\sqrt{7}} = \dfrac{5\sqrt{7}}{7}$

21.

$\dfrac{3}{\sqrt[3]{4}} \cdot \dfrac{\sqrt[3]{2}}{\sqrt[3]{2}} = \dfrac{3\sqrt[3]{2}}{\sqrt[3]{8}} = \dfrac{3\sqrt[3]{2}}{2}$

22.

$$\sqrt{\frac{9y^5}{28x^6}} = \sqrt{\frac{9y^5}{4x^6 \cdot 7} \cdot \frac{7}{7}}$$

$$= \sqrt{\frac{3^2 y^4 \cdot 7y}{2^2 \cdot 7^2 \cdot x^6}}$$

$$= \frac{3y^2\sqrt{7y}}{2 \cdot 7 \cdot x^3} = \frac{3y^2\sqrt{7y}}{14x^3}$$

23.

$$\frac{3\sqrt{3} + \sqrt{5}}{2\sqrt{3} + 4\sqrt{5}} \cdot \frac{2\sqrt{3} - 4\sqrt{5}}{2\sqrt{3} - 4\sqrt{5}}$$

$$= \frac{6 \cdot 3 - 12\sqrt{15} + 2\sqrt{15} - 4 \cdot 5}{4 \cdot 3 - 16 \cdot 5}$$

$$= \frac{18 - 10\sqrt{15} - 20}{12 - 80} = \frac{-2 - 10\sqrt{15}}{-68}$$

$$= \frac{\cancel{-2}\left(1 + 5\sqrt{15}\right)}{\cancel{-2} \cdot 34} = \frac{1 + 5\sqrt{15}}{34}$$

24.

$$\sqrt{x + 7} + 5 = x$$

$$\left(\sqrt{x + 7}\right)^2 = (x - 5)^2$$

$$x + 7 = x^2 - 10x + 25$$

$$0 = x^2 - 11x + 18$$

$$0 = (x - 9)(x - 2)$$

$$x - 9 = 0 \text{ or } \quad x - 2 = 0$$

$$x = 9 \quad x = 2$$

Solution set: {9}
2 does not check.

25.

$$\sqrt{x - 5} = 5 - \sqrt{x}$$

$$\left(\sqrt{x - 5}\right)^2 = \left(5 - \sqrt{x}\right)^2$$

$$x - 5 = 25 - 10\sqrt{x} + x$$

$$-30 = -10\sqrt{x}$$

$$(3)^2 = \left(\sqrt{x}\right)^2 \qquad \text{Multiply by } -\frac{1}{10},$$
$$\text{square}$$

$$9 = x$$

Solution set: {9}

CHAPTER 9   QUADRATIC EQUATIONS

Solutions to Text Exercises

1. - 27.
The answers do not require worked-out solutions.

29.
$(5+2i)+(4-3i)=5+4+2i-3i=9-i$

31.
$$\left(\frac{1}{2}-\frac{1}{3}i\right)+\left(\frac{1}{4}+\frac{2}{9}i\right)=\frac{1}{2}+\frac{1}{4}-\frac{1}{3}i+\frac{2}{9}i$$
$$=\left(\frac{2}{4}+\frac{1}{4}\right)+\left(-\frac{3}{9}+\frac{2}{9}\right)i$$
$$=\frac{3}{4}-\frac{1}{9}i$$

33.
$$(2+4i)-(5-i)=2+4i-5+i=2-5+4i+i$$
$$=-3+5i$$

35.
$3+(2-5i)=3+2-5i=5-5i$

37.
$8i-(2-3i)=8i-2+3i=-2+8i+3i=-2+11i$

39.
$$\left(-1+\frac{2}{5}i\right)-\left(-\frac{1}{3}+\frac{3}{4}i\right)=-1+\frac{2}{5}i+\frac{1}{3}-\frac{3}{4}i$$
$$=\left(-1+\frac{1}{3}\right)+\left(\frac{2}{5}-\frac{3}{4}\right)i$$
$$=\left(-\frac{3}{3}+\frac{1}{3}\right)+\left(\frac{8}{20}-\frac{15}{20}\right)i$$
$$=-\frac{2}{3}-\frac{7}{20}i$$

41.
$-6(-2-3i)=-6(-2)-6(-3i)=12+18i$

43.
$$-2i(3-i)=-2i\cdot 3-2i(-i)=-6i+2i^2$$
$$=-6i+2(-1)$$
$$=-6i-2$$
$$=-2-6i$$

45.
$$(3-2i)(5-i)=15-3i-10i+2i^2$$
$$=15-13i+2(-1)$$
$$=15-13i-2$$
$$=13-13i$$

47.
$$(-1-3i)(-2-5i)=2+5i+6i+15i^2$$
$$=2+11i+15(-1)$$
$$=2+11i-15$$
$$=-13+11i$$

49.
$(-2+7i)(-2-7i)$

These factors represent the product of the sum and difference of two terms.
Thus,
$$(-2+7i)(-2-7i)=(-2)^2-(7i)^2=4-49i^2$$
$$=4-49(-1)$$
$$=4+49$$
$$=53$$

51.
$$(1-3i)^2=1^2-2(1\cdot 3i)+(-3i)^2=1-6i+9i^2$$
$$=1-6i+9(-1)$$
$$=1-6i-9$$
$$=-8-6i$$

53.

$$\frac{13+11i}{2-5i} = \frac{13+11i}{2-5i} \cdot \frac{2+5i}{2+5i} = \frac{(13+11i)(2+5i)}{(2-5i)(2+5i)}$$

$$= \frac{26+65i+22i+55i^2}{4-25i^2}$$

$$= \frac{26+87i+55(-1)}{4-25(-1)}$$

$$= \frac{26+87i-55}{4+25}$$

$$= \frac{-29+87i}{29}$$

$$= \frac{\cancel{29}(-1+3i)}{\cancel{29}}$$

$$= -1+3i$$

55.

$$\frac{5-i}{2+i} = \frac{5-i}{2+i} \cdot \frac{2-i}{2-i} = \frac{(5-i)(2-i)}{(2+i)(2-i)} = \frac{10-7i+i^2}{4-i^2}$$

$$= \frac{10-7i+(-1)}{4-(-1)}$$

$$= \frac{9-7i}{4+1}$$

$$= \frac{9-7i}{5}$$

$$= \frac{9}{5} - \frac{7}{5}i$$

57.

$$\frac{-1+6i}{5+2i} = \frac{-1+6i}{5+2i} \cdot \frac{5-2i}{5-2i} = \frac{(-1+6i)(5-2i)}{(5+2i)(5-2i)}$$

$$= \frac{-5+32i-12i^2}{25-4i^2}$$

$$= \frac{-5+32i-12(-1)}{25-4(-1)}$$

$$= \frac{-5+32i+12}{25+4}$$

$$= \frac{7+32i}{29}$$

$$= \frac{7}{29} + \frac{32}{29}i$$

59.

$$\frac{6+3i}{i} = \frac{6+3i}{i} \cdot \frac{i}{i} = \frac{(6+3i)i}{i^2} = \frac{6i+3i^2}{i^2} = \frac{6i+3(-1)}{-1}$$

$$= \frac{-3+6i}{-1} = 3-6i$$

61.

$$\frac{2-5i}{-i} = \frac{2-5i}{-i} \cdot \frac{i}{i} = \frac{(2-5i)i}{-i^2} = \frac{2i-5i^2}{-i^2}$$

$$= \frac{2i-5(-1)}{-(-1)} = \frac{2i+5}{1} = 5+2i$$

63.

$$\frac{12+\sqrt{-64}}{4} + \frac{12+8i}{4} = \frac{\cancel{4}(3+2i)}{\cancel{4}} = 3+2i$$

65.

$$\frac{-18+\sqrt{-36}}{6} = \frac{-18+6i}{6} = \frac{\cancel{6}(-3+i)}{\cancel{6}} = -3+i$$

67.

$$\frac{-38-\sqrt{-36}}{2} = \frac{-38-6i}{2} = \frac{\cancel{2}(-19-3i)}{\cancel{2}} = -19-3i$$

69.

$$\frac{8-\sqrt{-64}}{8} = \frac{8-8i}{8} = \frac{\cancel{8}(1-i)}{\cancel{8}} = 1-i$$

71.

$$\frac{12+\sqrt{-100}}{6} = \frac{12+10i}{6} = \frac{\cancel{2}(6+5i)}{\cancel{2}\cdot 3} = \frac{6+5i}{3} = 2+\frac{5}{3}i$$

73.

$$\frac{8-\sqrt{-4}}{10} = \frac{8-2i}{10} = \frac{\cancel{2}(4-i)}{\cancel{2}\cdot 5} = \frac{4-i}{5} = \frac{4}{5} - \frac{1}{5}i$$

75.

$$\frac{-15+\sqrt{-25}}{10} = \frac{-15+5i}{10} = \frac{\cancel{5}(-3+i)}{\cancel{5}\cdot 2}$$

$$= \frac{-3+i}{2} = -\frac{3}{2} + \frac{1}{2}i$$

77.

$$\frac{-6-\sqrt{-36}}{4} = \frac{-6-6i}{4} = \frac{\cancel{2}(-3-3i)}{\cancel{2}\cdot 2} = \frac{-3-3i}{2}$$
$$= -\frac{3}{2} - \frac{3}{2}i$$

Exercises 9.2

1. - 11.
The answers do not require worked out solutions.

13.
$x^2 = -32$

$\quad x = \pm\sqrt{-32}$ Extraction of
                         Roots Property

$\quad x = \pm\sqrt{16(-1)\cdot 2}$

$\quad x = \pm\sqrt{16}\cdot\sqrt{-1}\cdot\sqrt{2}$

$\quad x = \pm 4i\sqrt{2}$

    Solution set: $\left\{\pm 4i\sqrt{2}\right\}$

15.

$$4x^2 - 25 = 0$$

$(2x-5)(2x+5) = 0$ Factor

$\quad\quad 2x - 5 = 0 \quad\text{or}\quad 2x + 5 = 0$

$\quad\quad\quad\quad x = \frac{5}{2} \quad\text{or}\quad\quad x = -\frac{5}{2}$

          Solution set: $\left\{\pm\frac{5}{2}\right\}$

17.
$9x^2 - 7 = 0$

$\quad 9x^2 = 7$

$\quad x^2 = \frac{7}{9}$

$\quad x = \pm\sqrt{\frac{7}{9}}$ Extraction of Roots Property

---

#17 continued

$\quad x = \pm\frac{\sqrt{7}}{3}$ Simplify

Solution set: $\left\{\pm\frac{\sqrt{7}}{3}\right\}$

19.
$9x^2 - 20 = 0$

$\quad 9x^2 = 20$ Add 20

$\quad x^2 = \frac{20}{9}$ Divide by 9

$\quad x = \pm\sqrt{\frac{20}{9}}$ Extraction of
                   Roots Property

$\quad x = \pm\frac{\sqrt{20}}{3}$

$\quad x = \pm\frac{2\sqrt{5}}{3}$ Simplify

    Solution set: $\left\{\pm\frac{2\sqrt{5}}{3}\right\}$

21.
$3x^2 - 1 = 0$

$\quad 3x^2 = 1$ Add 1

$\quad x^2 = \frac{1}{3}$ Divide by 3

$\quad x = \pm\sqrt{\frac{1}{3}}$

$\quad x = \pm\frac{\sqrt{3}}{3}$ Rationalize the
                denominator

    Solution set: $\left\{\pm\frac{\sqrt{3}}{3}\right\}$

23.
$3x^2 - 4 = 0$

$\quad 3x^2 = 4$

$\quad x^2 = \frac{4}{3}$

#23 continued

$x = \pm\sqrt{\dfrac{4}{3}}$ Extraction of Roots Property

$x = \pm\dfrac{2}{\sqrt{3}}$ Simplify

$x = \pm\dfrac{2\sqrt{3}}{3}$ Rationalize the denominator

Solution set: $\left\{\pm\dfrac{2\sqrt{3}}{3}\right\}$

25.

$25x^2 + 4 = 0$

$25x^2 = -4$ Subtract 4

$x^2 = \dfrac{-4}{25}$ Divide by 25

$x = \pm\sqrt{-\dfrac{4}{25}}$ Extraction of Roots Property

$x = \pm\dfrac{\sqrt{4}}{\sqrt{25}}\cdot\sqrt{-1} = \pm\dfrac{2}{5}i$

Solution set: $\left\{\pm\dfrac{2}{5}i\right\}$

27.

$4x^2 + 81 = 0$

$4x^2 = -81$ Add -81

$x^2 = -\dfrac{81}{4}$ Divide by 4

$x = \pm\sqrt{-\dfrac{81}{4}}$ Extraction of Roots Property

$x = \pm\dfrac{\sqrt{81}}{\sqrt{4}}\cdot\sqrt{-1}$

$x = \pm\dfrac{9}{2}i$ Simplify

Solution set: $\left\{\pm\dfrac{9}{2}i\right\}$

29.

$(x+5)^2 = 1$

$x + 5 = \pm\sqrt{1}$ Extraction of Roots Property

$x + 5 = \pm 1$

$x = \pm 1 - 5$

$x = 1 - 5 = -4$ or $x = -1 - 5 = -6$

Solution set: $\{-6, -4\}$

31.

$\left(x - \dfrac{1}{3}\right)^2 = \dfrac{25}{9}$

$x - \dfrac{1}{3} = \pm\sqrt{\dfrac{25}{9}}$ Extraction of Roots Property

$x = \pm\dfrac{5}{3} + \dfrac{1}{3}$ Add $\dfrac{1}{3}$

$x = \dfrac{5}{3} + \dfrac{1}{3} = \dfrac{6}{3} = 2$ or $x = -\dfrac{5}{3} + \dfrac{1}{3} = -\dfrac{4}{3}$

Solution set: $\left\{-\dfrac{4}{3},\ 2\right\}$

33.

$\left(x - \dfrac{1}{3}\right)^2 = \dfrac{1}{25}$

$x - \dfrac{1}{3} = \pm\sqrt{\dfrac{1}{25}}$ Extraction of Roots Property

$x = \pm\dfrac{1}{5} + \dfrac{1}{3}$ Add $\dfrac{1}{3}$

$x = \dfrac{1}{5} + \dfrac{1}{3} = \dfrac{8}{15}$ or $x = -\dfrac{1}{5} + \dfrac{1}{3} = \dfrac{2}{15}$

Solution set: $\left\{\dfrac{8}{15},\ \dfrac{2}{15}\right\}$

35.

$(x+3)^2 = 38$

$x + 3 = \pm\sqrt{38}$ Extraction of Roots Property

$x = -3 \pm\sqrt{38}$ Add -3

Solution set: $\left\{-3 \pm\sqrt{38}\right\}$

37.

$(x+2)^2 = 32$

$x + 2 = \pm\sqrt{32}$ Extraction of Roots Property

$x = -2 \pm\sqrt{32}$ Add -2

$x = -2 \pm 4\sqrt{2}$ Simplify

Solution set: $\left\{-2 \pm 4\sqrt{2}\right\}$

39.

$(x-9)^2 = -49$

$x - 9 = \pm\sqrt{-49}$ Extraction of Roots Property

$x = 9 \pm \sqrt{-49}$ Add 9

$x = 9 \pm 7i$ Simplify

Solution set: $\{9 \pm 7i\}$

41.

$(3x-1)^2 = 16$

$3x - 1 = \pm\sqrt{16}$ Extraction of Roots Property

$3x - 1 = \pm 4$ Simplify

$3x = 1 \pm 4$ Add 1

$x = \dfrac{1 \pm 4}{3}$ Divide by 3

$x = \dfrac{1+4}{3} = \dfrac{5}{3}$ or $x = \dfrac{1-4}{3} = -1$

Solution set: $\left\{-1, \ \dfrac{5}{3}\right\}$

43.

$(5x+1)^2 = 100$

$5x + 1 = \pm\sqrt{100}$ Extraction of Roots Property

$5x + 1 = \pm 10$ Simplify

$5x = -1 \pm 10$ Add -1

$x = \dfrac{-1 \pm 10}{5}$ Divide by 5

$x = \dfrac{-1+10}{5} = \dfrac{9}{5}$ or $x = \dfrac{-1-10}{5} = -\dfrac{11}{5}$

Solution set: $\left\{-\dfrac{11}{5}, \ \dfrac{9}{5}\right\}$

45.

$(2x-5)^2 = 26$

$2x - 5 = \pm\sqrt{26}$ Extraction of Roots Property

$2x = 5 \pm \sqrt{26}$ Add 5

$x = \dfrac{5 \pm \sqrt{26}}{2}$ Divide by 2

Solution set: $\left\{\dfrac{5 \pm \sqrt{26}}{2}\right\}$

47.

$(3x-2)^2 = 12$

$3x - 2 = \pm\sqrt{12}$ Extraction of Roots Property

$3x - 2 = \pm 2\sqrt{3}$ Simplify

$3x = 2 \pm 2\sqrt{3}$ Add 2

$x = \dfrac{2 \pm 2\sqrt{3}}{3}$ Divide by 2

Solution set: $\left\{\dfrac{2 \pm 2\sqrt{3}}{3}\right\}$

49.

$(6x-2)^2 = 48$

$6x - 2 = \pm\sqrt{48}$ Extraction of Roots Property

$6x - 2 = \pm 4\sqrt{3}$ Simplify

$6x = 2 \pm 4\sqrt{3}$ Add 2

$x = \dfrac{2 \pm 4\sqrt{3}}{6}$ Divide by 6

$x = \dfrac{1 \pm 2\sqrt{3}}{3}$ Simplify

Solution set: $\left\{\dfrac{1 \pm 2\sqrt{3}}{3}\right\}$

51.

$(3x+6)^2 = 63$

$3x+6 = \pm\sqrt{63}$	Extraction of Roots Property
$3x+6 = \pm3\sqrt{7}$	Simplify
$3x = -6 \pm 3\sqrt{7}$	Add -6
$x = \dfrac{-6 \pm 3\sqrt{7}}{3}$	Divide by 3
$x = -2 \pm \sqrt{7}$	Simplify

Solution set: $\left\{-2 \pm \sqrt{7}\right\}$

53.

$(7x-3)^2 = -81$

$7x-3 = \pm\sqrt{-81}$	Extraction of Roots Property
$7x-3 = \pm9i$	Simplify
$7x = 3 \pm 9i$	Add 3
$x = \dfrac{3}{7} \pm \dfrac{9}{7}i$	Divide by 7

Solution set: $\left\{\dfrac{3}{7} \pm \dfrac{9}{7}i\right\}$

55.

$(2x-4)^2 = -100$

$2x-4 = \pm\sqrt{-100}$	Extraction of Roots Property
$2x-4 = \pm10i$	Simplify
$2x = 4 \pm 10i$	Add 4
$x = \dfrac{4}{2} \pm \dfrac{10}{2}i$	Divide by 2
$x = 2 \pm 5i$	Simplify

Solution set: $\{2 \pm 5i\}$

57.

$(4x+10)^2 = -4$

$4x+10 = \pm\sqrt{-4}$ Extraction of Roots Property

#57 continued

$4x+10 = \pm2i$	Simplify
$4x = -10 \pm 2i$	Add -10
$x = \dfrac{-10}{4} \pm \dfrac{2}{4}i$	Divide by 4
$x = -\dfrac{5}{2} \pm \dfrac{1}{2}i$	Simplify

Solution set: $\left\{-\dfrac{5}{2} \pm \dfrac{1}{2}i\right\}$

59.

$(5x+2)^2 - 1 = 48$

$(5x+2)^2 = 49$	Add 1
$5x+2 = \pm\sqrt{49}$	Extraction of Roots Property
$5x+2 = \pm7$	Simplify
$5x = -2 \pm 7$	Add -2
$x = \dfrac{-2 \pm 7}{5}$	Divide by 5

$x = \dfrac{-2+7}{5} = 1$ or $x = \dfrac{-2-7}{5} = -\dfrac{9}{5}$

Solution set: $\left\{-\dfrac{9}{5},\ 1\right\}$

61.

$(4x-6)^2 - 5 = 23$

$(4x-6)^2 = 28$	
$4x-6 = \pm\sqrt{28}$	Extraction of Roots Property
$4x-6 = \pm2\sqrt{7}$	Simplify
$4x = 6 \pm 2\sqrt{7}$	Add 6
$x = \dfrac{6}{4} \pm \dfrac{2\sqrt{7}}{4}$	Divide by 4
$x = \dfrac{3}{2} \pm \dfrac{\sqrt{7}}{2}$	Simplify.

Solution set: $\left\{\dfrac{3}{2} \pm \dfrac{\sqrt{7}}{2}\right\}$

63.

$$(6x-9)^2 - 4 = -13$$

$(6x-9)^2 = -9$	Add 4
$6x-9 = \pm\sqrt{-9}$	Extraction of Roots Property
$6x - 9 = \pm 3i$	Simplify
$6x = 9 \pm 3i$	Add 9
$x = \dfrac{9}{6} \pm \dfrac{3}{6}i$	Divide by 6
$x = \dfrac{3}{2} \pm \dfrac{1}{2}i$	Simplify

Solution set: $\left\{\dfrac{3}{2} \pm \dfrac{1}{2}i\right\}$

65.

The equation is $(x-3)^2 = 30$.

$$x - 3 = \pm\sqrt{30}$$
$$x = 3 \pm\sqrt{30}$$

The number is $3 + \sqrt{30}$ or $3 - \sqrt{30}$.

67.

The equation is $(2x+5)^2 = 36$

$2x + 5 = \pm\sqrt{36}$	Extraction of Roots Property
$2x + 5 = \pm 6$	Simplify
$2x = -5 \pm 6$	Add -5
$x = \dfrac{-5 \pm 6}{2}$	Divide by 2

$$x = \frac{-5+6}{2} = \frac{1}{2} \quad\text{or}\quad x = \frac{-5-6}{2} = -\frac{11}{2}$$

The number is $\dfrac{1}{2}$ or $-\dfrac{11}{2}$.

69.

Let x = side length of the smaller square.
Let 2x = side length of the larger square.
Let x^2 = area of smaller square.
Let $4x^2$ = area of larger square.

The equation is $x^2 + 4x^2 = 80$.

$$5x^2 = 80$$

#69 continued

$x^2 = 16$	Divide by 5
$x = \pm 4$	Extraction of Roots Property

The negative value is rejected.
The shorter side is 4 ft. and the longer side is 8 ft.

71.

$$d = 16t^2$$
$$256 = 16t^2$$

$16 = t^2$	Divide by 16
$\pm\sqrt{16} = t$	
$\pm 4 = t$	Extraction of Roots Property

The negative value is rejected.

It takes 4 seconds to travel 256 ft.

$$d = 16t^2$$
$$784 = 16t^2$$

$49 = t^2$	Divide by 16
$\pm\sqrt{49} = t$	Extraction of Roots Property
$\pm 7 = t$	

The negative value is rejected.

It takes 7 seconds to hit the ground.

Exercises 9.3

1.

$x^2 + 14x$

$b = 14$

$$\left(\frac{1}{2}b\right)^2 = \left(\frac{1}{2} \cdot 14\right)^2 = (7)^2 = 49$$

So, $x^2 + 14x + 49 = (x+7)^2$

3.

$x^2 + 9x$

$b = 9$

$$\left(\frac{1}{2}b\right)^2 = \left(\frac{1}{2}\cdot 9\right)^2 = \left(\frac{9}{2}\right)^2 = \frac{81}{4}$$

So, $x^2 + 9x + \frac{81}{4} = \left(x + \frac{9}{2}\right)^2$

5.

$x^2 - 6x$

$b = -6$

$$\left(\frac{1}{2}b\right)^2 = \left[\frac{1}{2}(-6)\right]^2 = (-3)^2 = 9$$

So, $x^2 - 6x + 9 = (x - 3)^2$

7.

$x^2 - x$

$b = -1$

$$\left(\frac{1}{2}b\right)^2 = \left[\frac{1}{2}(-1)\right]^2 = \left(-\frac{1}{2}\right)^2 = \frac{1}{4}$$

So, $x^2 - x + \frac{1}{4} = \left(x - \frac{1}{2}\right)^2$

9.

$x^2 + \frac{4}{3}x$

$b = \frac{4}{3}$

$$\left(\frac{1}{2}b\right)^2 = \left(\frac{1}{2}\cdot\frac{4}{3}\right)^2 = \left(\frac{2}{3}\right)^2 = \frac{4}{9}$$

So, $x^2 + \frac{4}{3}x + \frac{4}{9} = \left(x + \frac{2}{3}\right)^2$

11.

$x^2 + \frac{8}{5}x$

$b = \frac{8}{5}$

$$\left(\frac{1}{2}b\right)^2 = \left(\frac{1}{2}\cdot\frac{8}{5}\right)^2 = \left(\frac{4}{5}\right)^2 = \frac{16}{25}$$

So, $x^2 + \frac{8}{5}x + \frac{16}{25} = \left(x + \frac{4}{5}\right)^2$

13.

$x^2 + \frac{1}{4}x$

$b = \frac{1}{4}$

$$\left(\frac{1}{2}b\right)^2 = \left(\frac{1}{2}\cdot\frac{1}{4}\right)^2 = \left(\frac{1}{8}\right)^2 = \frac{1}{64}$$

So, $x^2 + \frac{1}{4}x + \frac{1}{64} = \left(x + \frac{1}{8}\right)^2$

15.

$x^2 + \frac{3}{7}x$

$b = \frac{3}{7}$

$$\left(\frac{1}{2}b\right)^2 = \left(\frac{1}{2}\cdot\frac{3}{7}\right)^2 = \left(\frac{3}{14}\right)^2 = \frac{9}{196}$$

So, $x^2 + \frac{3}{7}x + \frac{9}{196} = \left(x + \frac{3}{14}\right)^2$

17.

$x^2 - \frac{2}{7}x$

$b = -\frac{2}{7}$

$$\left(\frac{1}{2}b\right)^2 = \left[\frac{1}{2}\left(-\frac{2}{7}\right)\right]^2 = \left(-\frac{1}{7}\right)^2 = \frac{1}{49}$$

So, $x^2 - \frac{2}{7}x + \frac{1}{49} = \left(x - \frac{1}{7}\right)^2$

19.

$$x^2 - \frac{6}{5}x$$

$$b = -\frac{6}{5}$$

$$\left(\frac{1}{2}b\right)^2 = \left[\frac{1}{2}\left(-\frac{6}{5}\right)\right]^2 = \left(-\frac{3}{5}\right)^2 = \frac{9}{25}$$

So, $x^2 - \frac{6}{5}x + \frac{9}{25} = \left(x - \frac{3}{5}\right)^2$

21.

$$x^2 - \frac{1}{3}x$$

$$b = -\frac{1}{3}$$

$$\left(\frac{1}{2}b\right)^2 = \left[\frac{1}{2}\left(-\frac{1}{3}\right)\right]^2 = \left(-\frac{1}{6}\right)^2 = \frac{1}{36}$$

So, $x^2 - \frac{1}{3}x + \frac{1}{36} = \left(x - \frac{1}{6}\right)^2$

23.

$$x^2 - \frac{7}{2}x$$

$$b = -\frac{7}{2}$$

$$\left(\frac{1}{2}b\right)^2 = \left[\frac{1}{2}\left(-\frac{7}{2}\right)\right]^2 = \left(-\frac{7}{4}\right)^2 = \frac{49}{16}$$

So, $x^2 - \frac{7}{2}x + \frac{49}{16} = \left(x - \frac{7}{4}\right)^2$

25.

$$x^2 - 5x + 6 = 0$$

$x^2 - 5x = -6$	Add -6
$x^2 - 5x + \frac{25}{4} = -6 + \frac{25}{4}$	Add $\left(\frac{-5}{2}\right)^2$
$x^2 - 5x + \frac{25}{4} = \frac{1}{4}$	Simplify
$\left(x - \frac{5}{2}\right)^2 = \frac{1}{4}$	Factor
$x - \frac{5}{2} = \pm\sqrt{\frac{1}{4}}$	Extraction

#25 continued

$x - \frac{5}{2} = \pm\frac{1}{2}$	Simplify.
$x = \frac{5}{2} \pm \frac{1}{2}$	Add $\frac{5}{2}$.
$x = \frac{5}{2} + \frac{1}{2} = 3$ or $x = \frac{5}{2} - \frac{1}{2} = 2$	

Solution set: $\{3, 2\}$

27.

$$x^2 + 4x - 5 = 0$$

$x^2 + 4x = 5$	Add 5
$x^2 + 4x + 4 = 5 + 4$	Add $\left(\frac{4}{2}\right)^2$
$(x + 2)^2 = 9$	Factor
$x + 2 = \pm\sqrt{9}$	Extraction of Roots Property
$x + 2 = \pm 3$	Simplify
$x = -2 \pm 3$	Add -2
$x = -2 + 3 = 1$ or $x = -2 - 3 = -5$	

Solution set: $\{-5, 1\}$

29.

$$x^2 - 16 = 0$$
$$(x + 4)(x - 4) = 0$$
$$x = -4 \quad \text{or} \quad x = 4$$

Solution set: $\{-4, 4\}$

31.

$$x^2 - 2x - 1 = 0$$

$x^2 - 2x = 1$	Add 1
$x^2 - 2x + 1 = 1 + 1$	Add $\left(\frac{-2}{2}\right)^2$
$(x - 1)^2 = 2$	Factor
$x - 1 = \pm\sqrt{2}$	Extraction of Roots Property
$x = 1 \pm \sqrt{2}$	Add 1

Solution set: $\left\{1 \pm \sqrt{2}\right\}$

33.
$$x^2 + 4x + 1 = 0$$

$$x^2 + 4x = -1 \qquad \text{Add } -1$$

$$x^2 + 4x + 4 = -1 + 4 \qquad \text{Add } \left(\frac{4}{2}\right)^2$$

$$(x+2)^2 = 3 \qquad \text{Factor}$$

$$x + 2 = \pm\sqrt{3} \qquad \begin{array}{l}\text{Extraction of} \\ \text{Roots Property}\end{array}$$

$$x = -2 \pm \sqrt{3} \qquad \text{Add } -2$$

$$\text{Solution set: } \left\{-2 \pm \sqrt{3}\right\}$$

35.
$$x^2 + 8x + 14 = 0$$

$$x^2 + 8x = -14 \qquad \text{Add } -14$$

$$x^2 + 8x + 16 = -14 + 16 \qquad \text{Add } \left(\frac{8}{2}\right)^2$$

$$(x+4)^2 = 2 \qquad \text{Factor}$$

$$x + 4 = \pm\sqrt{2} \qquad \begin{array}{l}\text{Extraction of} \\ \text{Roots Property}\end{array}$$

$$x = -4 \pm \sqrt{2} \qquad \text{Add } -4$$

$$\text{Solution set: } \left\{-4 \pm \sqrt{2}\right\}$$

37.
$$x^2 - 10x + 18 = 0$$

$$x^2 - 10x = -18 \qquad \text{Add } -18$$

$$x^2 - 10x + 25 = -18 + 25 \qquad \text{Add } \left(\frac{-10}{2}\right)^2$$

$$(x-5)^2 = 7 \qquad \text{Factor}$$

$$x - 5 = \pm\sqrt{7} \qquad \begin{array}{l}\text{Extraction of} \\ \text{Roots Property}\end{array}$$

$$x = 5 \pm \sqrt{7} \qquad \text{Add } 5$$

$$\text{Solution set: } \left\{5 \pm \sqrt{7}\right\}$$

39.
$$x^2 - 4x + 5 = 0$$

$$x^2 - 4x = -5 \qquad \text{Add } -5$$

$$x^2 - 4x + 4 = -5 + 4 \qquad \text{Add } \left(\frac{-4}{2}\right)^2$$

$$(x-2)^2 = -1 \qquad \text{Factor}$$

$$x - 2 = \pm\sqrt{-1} \qquad \begin{array}{l}\text{Extraction of} \\ \text{Roots Property}\end{array}$$

$$x - 2 = \pm i \qquad \text{Simplify}$$

$$x = 2 \pm i \qquad \text{Add } 2$$

$$\text{Solution set: } \left\{2 \pm i\right\}$$

41.
$$x^2 + 8x + 25 = 0$$

$$x^2 + 8x = -25 \qquad \text{Add } -25$$

$$x^2 + 8x + 16 = -25 + 16 \qquad \text{Add } \left(\frac{8}{2}\right)^2$$

$$(x+4)^2 = -9 \qquad \text{Factor}$$

$$x + 4 = \pm\sqrt{-9} \qquad \begin{array}{l}\text{Extraction of} \\ \text{Roots Property}\end{array}$$

$$x + 4 = \pm 3i \qquad \text{Simplify}$$

$$x = -4 \pm 3i \qquad \text{Add } -4$$

$$\text{Solution set: } \left\{-4 \pm 3i\right\}$$

43.
$$-x^2 + 6x - 3 = 0$$

$$x^2 - 6x + 3 = 0 \qquad \text{Multiply by } -1$$

$$x^2 - 6x = -3 \qquad \text{Add } -3$$

$$x^2 - 6x + 9 = -3 + 9 \qquad \text{Add } \left(\frac{-6}{2}\right)^2$$

$$(x-3)^2 = 6 \qquad \text{Factor}$$

$$x - 3 = \pm\sqrt{6} \qquad \begin{array}{l}\text{Extraction of} \\ \text{Roots Property}\end{array}$$

$$x = 3 \pm \sqrt{6} \qquad \text{Add } 3$$

$$\text{Solution set: } \left\{3 \pm \sqrt{6}\right\}$$

45.

$$-x^2 - 8x - 13 = 0$$

$$x^2 + 8x + 13 = 0 \qquad \text{Multiply by -1}$$

$$x^2 + 8x = -13 \qquad \text{Add -13}$$

$$x^2 + 8x + 16 = -13 + 16 \qquad \text{Add } \left(\frac{8}{2}\right)^2$$

$$(x + 4)^2 = 3 \qquad \text{Factor}$$

$$x + 4 = \pm\sqrt{3} \qquad \text{Extraction of}$$
$$\text{Roots Property}$$

$$x = -4 \pm \sqrt{3} \qquad \text{Add -4}$$

$$\text{Solution set:} \quad \left\{-4 \pm \sqrt{3}\right\}$$

47.

$$2x^2 + 3x - 9 = 0$$

$$x^2 + \frac{3}{2}x - \frac{9}{2} = 0 \qquad \text{Divide by } 2$$

$$x^2 + \frac{3}{2}x = \frac{9}{2} \qquad \text{Add } \frac{9}{2}$$

$$x^2 + \frac{3}{2}x + \frac{9}{16} = \frac{9}{2} + \frac{9}{16} \qquad \text{Add } \left(\frac{1}{2} \cdot \frac{3}{2}\right)^2$$

$$\left(x + \frac{3}{4}\right)^2 = \frac{81}{16} \qquad \text{Factor}$$

$$x + \frac{3}{4} = \pm\sqrt{\frac{81}{16}} \qquad \text{Extraction \quad of}$$
$$\text{Roots Property}$$

$$x + \frac{3}{4} = \pm\frac{9}{4} \qquad \text{Simplify}$$

$$x = -\frac{3}{4} \pm \frac{9}{4} \qquad \text{Add } -\frac{3}{4}$$

$$x = -\frac{3}{4} + \frac{9}{4} = \frac{3}{2} \quad \text{or} \quad x = \frac{-3}{4} - \frac{9}{4} = -3$$

$$\text{Solution set:} \qquad \left\{-3, \; \frac{3}{2}\right\}$$

49.

$$3x^2 - 14x + 8 = 0$$

$$x^2 - \frac{14}{3}x + \frac{8}{3} = 0 \qquad \text{Divide by 3}$$

$$x^2 - \frac{14}{3}x = -\frac{8}{3} \qquad \text{Add } -\frac{8}{3}$$

$$x^2 - \frac{14}{3}x + \frac{49}{9} = -\frac{8}{3} + \frac{49}{9} \qquad \text{Add } \left(\frac{-14}{3} \cdot \frac{1}{2}\right)^2$$

$$\left(x - \frac{7}{3}\right)^2 = \frac{25}{9} \qquad \text{Factor}$$

$$x - \frac{7}{3} = \pm\sqrt{\frac{25}{9}} \qquad \text{Extraction of}$$
$$\text{Roots Property}$$

$$x - \frac{7}{3} = \pm\frac{5}{3} \qquad \text{Simplify}$$

$$x = \frac{7}{3} \pm \frac{5}{3}$$

$$x = \frac{7}{3} + \frac{5}{3} = 4 \quad \text{or} \quad x = \frac{7}{3} - \frac{5}{3} = \frac{2}{3}$$

$$\text{Solution set:} \quad \left\{\frac{2}{3}, \; 4\right\}$$

51.

$$4x^2 - 4x - 3 = 0$$

$$x^2 - x - \frac{3}{4} = 0 \qquad \text{Divide by 4}$$

$$x^2 - x = \frac{3}{4} \qquad \text{Add } \frac{3}{4}$$

$$x^2 - x + \frac{1}{4} = \frac{3}{4} + \frac{1}{4} \qquad \text{Add } \left(-1 \cdot \frac{1}{2}\right)^2$$

$$\left(x - \frac{1}{2}\right)^2 = 1 \qquad \text{Factor}$$

$$x - \frac{1}{2} = \pm\sqrt{1} \qquad \text{Extraction}$$

$$x - \frac{1}{2} = \pm 1 \qquad \text{Simplify}$$

$$x = \frac{1}{2} \pm 1 \qquad \text{Add } \frac{1}{2}$$

$$x = \frac{1}{2} + 1 = \frac{3}{2} \quad \text{or} \quad x = \frac{1}{2} - 1 = -\frac{1}{2}$$

$$\text{Solution set:} \quad \left\{-\frac{1}{2}, \; \frac{3}{2}\right\}$$

53.

$9x^2 - 6x - 1 = 0$

$x^2 - \dfrac{2}{3}x - \dfrac{1}{9} = 0$ Divide by 9

$x^2 - \dfrac{2}{3}x = \dfrac{1}{9}$ Add $\dfrac{1}{9}$

$x^2 - \dfrac{2}{3}x + \dfrac{1}{9} = \dfrac{1}{9} + \dfrac{1}{9}$ Add $\left(-\dfrac{2}{3} \cdot \dfrac{1}{2}\right)^2$

$\left(x - \dfrac{1}{3}\right)^2 = \dfrac{2}{9}$ Factor

$x - \dfrac{1}{3} = \pm\sqrt{\dfrac{2}{9}}$ Extraction

$x - \dfrac{1}{3} = \pm\dfrac{\sqrt{2}}{3}$ Simplify

$x = \dfrac{1}{3} \pm \dfrac{\sqrt{2}}{3}$ Add $\dfrac{1}{3}$

Solution set: $\left\{\dfrac{1 \pm \sqrt{2}}{3}\right\}$

55.

$4x^2 - 8x - 1 = 0$

$x^2 - 2x - \dfrac{1}{4} = 0$ Divide by 4

$x^2 - 2x = \dfrac{1}{4}$ Add $\dfrac{1}{4}$

$x^2 - 2x + 1 = \dfrac{1}{4} + 1$ Add $\left(\dfrac{-2}{2}\right)^2$

$(x - 1)^2 = \dfrac{5}{4}$ Factor

$x - 1 = \pm\sqrt{\dfrac{5}{4}}$ Extraction

$x - 1 = \pm\dfrac{\sqrt{5}}{2}$ Simplify

$x = 1 \pm \dfrac{\sqrt{5}}{2}$ Add 1

Solution set: $\left\{\dfrac{2 \pm \sqrt{5}}{2}\right\}$

57.

$4x^2 + 12x + 3 = 0$

$x^2 + 3x + \dfrac{3}{4} = 0$ Divide by 4

#57. continued.

$x^2 + 3x = -\dfrac{3}{4}$ Add $-\dfrac{3}{4}$

$x^2 + 3x + \dfrac{9}{4} = -\dfrac{3}{4} + \dfrac{9}{4}$ Add $\left(\dfrac{3}{2}\right)^2$

$\left(x + \dfrac{3}{2}\right)^2 = \dfrac{6}{4}$ Factor

$x + \dfrac{3}{2} = \pm\sqrt{\dfrac{6}{4}}$ Extraction

$x + \dfrac{3}{2} = \pm\dfrac{\sqrt{6}}{2}$ Simplify

$x = -\dfrac{3}{2} \pm \dfrac{\sqrt{6}}{2}$ Add $-\dfrac{3}{2}$

Solution set: $\left\{\dfrac{-3 \pm \sqrt{6}}{2}\right\}$

59.

$4x^2 + 4x - 7 = 0$

$x^2 + x - \dfrac{7}{4} = 0$ Divide by 4

$x^2 + x = \dfrac{7}{4}$ Add $\dfrac{7}{4}$

$x^2 + x + \dfrac{1}{4} = \dfrac{7}{4} + \dfrac{1}{4}$ Add $\left(\dfrac{1}{2}\right)^2$

$\left(x + \dfrac{1}{2}\right)^2 = \dfrac{8}{4}$ Factor

$x + \dfrac{1}{2} = \pm\sqrt{2}$ Extraction

$x = -\dfrac{1}{2} \pm \sqrt{2}$ Add $-\dfrac{1}{2}$

Solution set: $\left\{\dfrac{-1 \pm 2\sqrt{2}}{2}\right\}$

61.

$9x^2 - 24x + 4 = 0$

$x^2 - \dfrac{8}{3}x + \dfrac{4}{9} = 0$ Divide by 9

$x^2 - \dfrac{8}{3}x = -\dfrac{4}{9}$ Add $-\dfrac{4}{9}$

#61 continued

$$x^2 - \frac{8}{3}x + \frac{16}{9} = -\frac{4}{9} + \frac{16}{9} \quad \text{Add} \left(-\frac{8}{3} \cdot \frac{1}{2}\right)^2$$

$$\left(x - \frac{4}{3}\right)^2 = \frac{12}{9} \qquad \text{Factor}$$

$$x - \frac{4}{3} = \pm\sqrt{\frac{12}{9}} \qquad \text{Extraction}$$

$$x - \frac{4}{3} = \pm\frac{2\sqrt{3}}{3} \qquad \text{Simplify}$$

$$x = \frac{4}{3} \pm \frac{2\sqrt{3}}{3}$$

Solution set: $\left\{\dfrac{4 \pm 2\sqrt{3}}{3}\right\}$

63.

$$4x^2 + 24x + 31 = 0$$

$$x^2 + 6x + \frac{31}{4} = 0 \qquad \text{Divide by 4}$$

$$x^2 + 6x = -\frac{31}{4} \qquad \text{Add } -\frac{31}{4}$$

$$x^2 + 6x + 9 = -\frac{31}{4} + 9 \quad \text{Add} \left(\frac{6}{2}\right)^2$$

$$(x + 3)^2 = \frac{5}{4} \qquad \text{Factor}$$

$$x + 3 = \pm\sqrt{\frac{5}{4}} \qquad \text{Extraction}$$

$$x + 3 = \pm\frac{\sqrt{5}}{2} \qquad \text{Simplify}$$

$$x = -3 \pm \frac{\sqrt{5}}{2} \qquad \text{Add } -3$$

Solution set: $\left\{\dfrac{-6 \pm \sqrt{5}}{2}\right\}$

65.

$$9x^2 - 18x + 10 = 0$$

$$x^2 - 2x + \frac{10}{9} = 0 \qquad \text{Divide by 9}$$

$$x^2 - 2x = -\frac{10}{9} \qquad \text{Add } -\frac{10}{9}$$

#65 continued

$$x^2 - 2x + 1 = -\frac{10}{9} + 1 \quad \text{Add} \left(\frac{-2}{2}\right)^2$$

$$(x - 1)^2 = -\frac{1}{9} \qquad \text{Factor}$$

$$x - 1 = \pm\sqrt{-\frac{1}{9}} \qquad \text{Extraction}$$

$$x - 1 = \pm\frac{1}{3}i \qquad \text{Simplify}$$

$$x = 1 \pm \frac{1}{3}i \qquad \text{Add 1}$$

Solution set: $\left\{1 \pm \dfrac{1}{3}i\right\}$

67.

$$2x^2 + 2x + 13 = 0$$

$$x^2 + x + \frac{13}{2} = 0 \qquad \text{Divide by 2}$$

$$x^2 + x = -\frac{13}{2} \qquad \text{Add } -\frac{13}{2}$$

$$x^2 + x + \frac{1}{4} = -\frac{13}{2} + \frac{1}{4} \quad \text{Add} \left(\frac{1}{2}\right)^2$$

$$\left(x + \frac{1}{2}\right)^2 = -\frac{25}{4} \qquad \text{Factor}$$

$$x + \frac{1}{2} = \pm\sqrt{-\frac{25}{4}} \qquad \text{Extraction}$$

$$x + \frac{1}{2} = \pm\frac{5}{2}i \qquad \text{Simplify}$$

$$x = -\frac{1}{2} \pm \frac{5}{2}i$$

Solution set: $\left\{-\dfrac{1}{2} \pm \dfrac{5}{2}i\right\}$

69.

Let x = the number.

The equation is $x^2 + 4x = 7$.

$$x^2 + 4x + 4 = 7 + 4 \qquad \text{Add} \left(\frac{4}{2}\right)^2$$

$$(x + 2)^2 = 11 \qquad \text{Factor}$$

$$x + 2 = \pm\sqrt{11} \qquad \text{Extraction}$$

$$x = -2 \pm \sqrt{11} \qquad \text{Add } -2$$

The number is $-2 + \sqrt{11}$ or $-2 - \sqrt{11}$.

71.

Let x = the number.

The equation is $2x^2 - 2x = 1$.

$$x^2 - x = \frac{1}{2} \qquad \text{Divide by 2}$$

$$x^2 - x + \frac{1}{4} = \frac{1}{2} + \frac{1}{4} \qquad \text{Add } \left(-\frac{1}{2}\right)^2$$

$$\left(x - \frac{1}{2}\right)^2 = \frac{3}{4} \qquad \text{Factor}$$

$$x - \frac{1}{2} = \pm\sqrt{\frac{3}{4}} \qquad \text{Extraction of}$$
$$\text{Roots Property}$$

$$x - \frac{1}{2} = \pm\frac{\sqrt{3}}{2} \qquad \text{Simplify}$$

$$x = \frac{1}{2} \pm \frac{\sqrt{3}}{2}$$

The number is $\dfrac{1+\sqrt{3}}{2}$ or $\dfrac{1-\sqrt{3}}{2}$.

73.

Let x = the width.

Let 4x + 2 = the length.

The equation is $x(4x + 2) = 10$.

$$4x^2 + 2x = 10$$

$$x^2 + \frac{1}{2}x = \frac{10}{4} \qquad \text{Divide by 4.}$$

$$x^2 + \frac{1}{2}x + \frac{1}{16} = \frac{10}{4} + \frac{1}{16} \quad \text{Add } \left(\frac{1}{2} \cdot \frac{1}{2}\right)^2$$

$$\left(x + \frac{1}{4}\right)^2 = \frac{41}{16} \qquad \text{Factor.}$$

$$x + \frac{1}{4} = \pm\sqrt{\frac{41}{16}} \qquad \text{Extraction of}$$
$$\text{Roots Property}$$

$$x + \frac{1}{4} = \pm\frac{\sqrt{41}}{4} \qquad \text{Simplify.}$$

$$x = -\frac{1}{4} \pm \frac{\sqrt{41}}{4} \qquad \text{Add } -\frac{1}{4}.$$

The width is $\dfrac{-1+\sqrt{41}}{4}$ ft.

Note: The negative solution is rejected.

75.

$$d = -16t^2 + 80t$$

$$\text{Let } d = 0$$

$$0 = -16t^2 + 80t$$

$$0 = t^2 - 5t \qquad \text{Divide by -16.}$$

$$\frac{25}{4} = t^2 - 5t + \frac{25}{4} \qquad \text{Add } \left(\frac{-5}{2}\right)^2.$$

$$\frac{25}{4} = \left(t - \frac{5}{2}\right)^2 \qquad \text{Factor.}$$

$$\pm\sqrt{\frac{25}{4}} = t - \frac{5}{2} \qquad \text{Extraction of}$$
$$\text{Roots Property}$$

$$\pm\frac{5}{2} = t - \frac{5}{2} \qquad \text{Simplify}$$

$$\frac{5}{2} \pm \frac{5}{2} = t \qquad \text{Add } \frac{5}{2}$$

$$t = \frac{5}{2} + \frac{5}{2} = 5 \quad \text{or} \quad t = \frac{5}{2} - \frac{5}{2} = 0$$

The ball hits the ground after 5 seconds.

Exercises 9.4

1. - 3.

The answers do not require worked-out solutions.

5.

$$3x^2 + 2x - 11 = x - 4$$

$$3x^2 + x - 11 = -4 \qquad \text{Add - x.}$$

$$3x^2 + x - 7 = 0 \qquad \text{Add 4.}$$

$$a = 3, \ b = 1, c = -7$$

7.

$$8x^2 + 2x - 1 = 5x^2 + x + 8$$

$$3x^2 + 2x - 1 = x + 8 \qquad \text{Add } -5x^2$$

$$3x^2 + x - 1 = 8 \qquad \text{Add - x.}$$

$$3x^2 + x - 9 = 0 \qquad \text{Add - 8.}$$

$$a = 3, b = 1, c = -9$$

9.

$$-x^2 + 4x + 8 = 2(2x + 5)$$

$$-x^2 + 4x + 8 = 4x + 10$$

$$-x^2 - 2 = 0 \qquad \text{Add } -4x \text{ and } -10$$

$$a = -1, \ b = 0, c = -2$$

11.

$$5x^2 + 3x + 9 = 3(x + 3)$$

$$5x^2 + 3x + 9 = 3x + 9$$

$$5x^2 = 0 \qquad \text{Add } -3x \text{ and } -9$$

$$a = 5, \ b = 0, \ c = 0$$

13.

$$(4x + 1)(2x + 3) = 2$$

$$8x^2 + 14x + 3 = 2 \qquad \text{Multiply.}$$

$$8x^2 + 14x + 1 = 0 \qquad \text{Add } -2.$$

$$a = 8, \ b = 14, \ c = 1$$

15.

$$(2x + 1)(x - 5) = (x + 1)(x - 2)$$

$$2x^2 - 9x - 5 = x^2 - x - 2$$

$$x^2 - 8x - 3 = 0 \qquad \text{Add } -x^2, \ x, \text{ and } 2.$$

$$a = 1, \ b = -8, \ c = -3$$

17.

$$3x^2 - 8x - 3 = 0$$

$$a = 3, \ b = -8, \ c = -3$$

$$x = \frac{-(-8) \pm \sqrt{(-8)^2 - 4(3)(-3)}}{2(3)}$$

$$x = \frac{8 \pm \sqrt{64 + 36}}{6}$$

$$x = \frac{8 \pm \sqrt{100}}{6} = \frac{8 \pm 10}{6}$$

$$x = \frac{8 + 10}{6} = 3 \quad \text{or} \quad x = \frac{8 - 10}{6} = -\frac{1}{3}$$

#17 continued

Solution set: $\left\{ -\frac{1}{3}, \ 3 \right\}$

19.

$$x^2 - 16 = 0$$

$$a = 1, \ b = 0, \ c = -16$$

$$x = \frac{0 \pm \sqrt{0^2 - 4(1)(-16)}}{2 \cdot 1}$$

$$x = \frac{\pm \sqrt{64}}{2}$$

$$x = \frac{\pm 8}{2}$$

$$x = \frac{8}{2} = 4 \quad \text{or} \quad x = \frac{-8}{2} = -4$$

Solution set: $\{-4, 4\}$

21.

$$2x^2 - 3x + 7 = 2x + 7$$

$$2x^2 - 5x = 0$$

$$a = 2, \ b = -5, \ c = 0$$

$$x = \frac{-(-5) \pm \sqrt{(-5)^2 - 4(2)(0)}}{2 \cdot 2}$$

$$x = \frac{5 \pm \sqrt{25 - 0}}{4}$$

$$x = \frac{5 \pm 5}{4}$$

$$x = \frac{5 + 5}{4} = \frac{5}{2} \quad \text{or} \quad x = \frac{5 - 5}{4} = 0$$

Solution set: $\left\{ 0, \ \frac{5}{2} \right\}$

23.

$$3x^2 - 2x - 13 = 2x^2 + x - 9$$

$$x^2 - 3x - 4 = 0$$

$$a = -1, \ b = -3, \ c = 4$$

$$x = \frac{-(-3) \pm \sqrt{(-3)^2 - 4(1)(-4)}}{2 \cdot 1}$$

#23 continued

$$x = \frac{3 \pm \sqrt{9+16}}{2}$$

$$x = \frac{3 \pm \sqrt{25}}{2} = \frac{3 \pm 5}{2}$$

$$x = \frac{3+5}{2} = 4 \quad \text{or} \quad x = \frac{3-5}{2} = -1$$

Solution set: $\{-1, \ 4\}$

25.

$$(5x+1)(x-2) = (x-1)(3x-1)$$

$$5x^2 - 9x - 2 = 3x^2 - 4x + 1$$

$$2x^2 - 5x - 3 = 0$$

$a = 2, \ b = -5, \ c = -3$

$$x = \frac{-(-5) \pm \sqrt{(-5)^2 - 4(2)(-3)}}{2 \cdot 2}$$

$$x = \frac{5 \pm \sqrt{25+24}}{4}$$

$$x = \frac{5 \pm \sqrt{49}}{4}$$

$$x = \frac{5 \pm 7}{4}$$

$$x = \frac{5+7}{4} = 3 \quad \text{or} \quad x = \frac{5-7}{4} = -\frac{1}{2}$$

Solution set: $\left\{ -\frac{1}{2}, \ 3 \right\}$

27.

$$7x^2 - 3 = 0$$

$a = 7, \ b = 0, \ c = -3$

$$x = \frac{-0 \pm \sqrt{0^2 - 4(7)(-3)}}{2 \cdot 7}$$

$$x = \frac{\pm \sqrt{84}}{14}$$

$$x = \frac{\pm 2\sqrt{21}}{14} \qquad \text{Simplify}$$

$$x = \frac{\pm \sqrt{21}}{7} \qquad \text{Reduce}$$

Solution set: $\left\{ \frac{\pm \sqrt{21}}{7} \right\}$

29.

$$2x^2 + 7x + 2 = 0$$

$a = 2, \ b = 7, \ c = 2$

$$x = \frac{-7 \pm \sqrt{7^2 - 4(2)(2)}}{2 \cdot 2}$$

$$x = \frac{-7 \pm \sqrt{49-16}}{4}$$

$$x = \frac{-7 \pm \sqrt{33}}{4}$$

Solution set: $\left\{ \frac{-7 \pm \sqrt{33}}{4} \right\}$

31.

$$x^2 + 3x - 9 = 0$$

$a = 1, \ b = 3, \ c = -9$

$$x = \frac{-3 \pm \sqrt{3^2 - 4(1)(-9)}}{2 \cdot 1}$$

$$x = \frac{-3 \pm \sqrt{9+36}}{2}$$

$$x = \frac{-3 \pm \sqrt{45}}{2}$$

$$x = \frac{-3 \pm 3\sqrt{5}}{2} \qquad \text{Simplify}$$

Solution set: $\left\{ \frac{-3 \pm 3\sqrt{5}}{2} \right\}$

33.

$$2x^2 - 7x + 4 = 4x + 5$$

$$2x^2 - 11x - 1 = 0$$

$a = 2, \ b = -11, \ c = -1$

$$x = \frac{-(-11) \pm \sqrt{(-11)^2 - 4(2)(-1)}}{2 \cdot 2}$$

$$x = \frac{11 \pm \sqrt{121+8}}{4}$$

$$x = \frac{11 \pm \sqrt{129}}{4}$$

Solution set: $\left\{ \frac{11 \pm \sqrt{129}}{4} \right\}$

35.

$(3x+1)(x-4) = 2x(x-2)$

$3x^2 - 11x - 4 = 2x^2 - 4x$

$x^2 - 7x - 4 = 0$

$a = 1, b = -7, c = -4$

$$x = \frac{-(-7) \pm \sqrt{(-7)^2 - 4(1)(-4)}}{2 \cdot 1}$$

$$x = \frac{7 \pm \sqrt{49 + 16}}{2}$$

$$x = \frac{7 \pm \sqrt{65}}{2}$$

Solution set: $\left\{ \dfrac{7 \pm \sqrt{65}}{2} \right\}$

37.

$x^2 - 6x + 6 = 0$

$a = 1, b = -6, c = 6$

$$x = \frac{-(-6) \pm \sqrt{(-6)^2 - 4(1)(6)}}{2 \cdot 1}$$

$$x = \frac{6 \pm \sqrt{36 - 24}}{2}$$

$$x = \frac{6 \pm \sqrt{12}}{2} = 3 \pm \sqrt{3} \qquad \text{Simplify}$$

Solution set: $\left\{ 3 \pm \sqrt{3} \right\}$

39.

$x^2 - 4x - 4 = 0$

$a = 1, b = -4, c = -4$

$$x = \frac{-(-4) \pm \sqrt{(-4)^2 - 4(1)(-4)}}{2 \cdot 1}$$

$$x = \frac{4 \pm \sqrt{16 + 16}}{2}$$

$$x = \frac{4 \pm \sqrt{32}}{2} = \frac{4 \pm 4\sqrt{2}}{2}$$

$x = 2 \pm 2\sqrt{2} \qquad \text{Simplify}$

Solution set: $\left\{ 2 \pm 2\sqrt{2} \right\}$

41.

$4x^2 + 24x + 31 = 0$

$a = 4, b = 24, c = 31$

#41 continued

$$x = \frac{-24 \pm \sqrt{24^2 - 4(4)(31)}}{2 \cdot 4}$$

$$x = \frac{-24 \pm \sqrt{576 - 496}}{8}$$

$$x = \frac{-24 \pm \sqrt{80}}{8} = \frac{-24 \pm 4\sqrt{5}}{8} = \frac{-6 \pm \sqrt{5}}{2}$$

Solution set: $\left\{ \dfrac{-6 \pm \sqrt{5}}{2} \right\}$

43.

$9x^2 - 12x - 71 = 0$

$a = 9, b = -12, c = -71$

$$x = \frac{-(-12) \pm \sqrt{(-12)^2 - 4(9)(-71)}}{2 \cdot 9}$$

$$x = \frac{12 \pm \sqrt{144 + 2556}}{18}$$

$$x = \frac{12 \pm \sqrt{2700}}{18}$$

$$x = \frac{12 \pm 30\sqrt{3}}{18} \qquad \text{Simplify}$$

$$x = \frac{2 \pm 5\sqrt{3}}{3} \qquad \text{Reduce}$$

Solution set: $\left\{ \dfrac{2 \pm 5\sqrt{3}}{3} \right\}$

45.

$4x^2 + 10x + 11 = 3x^2 - 8$

$x^2 + 10x + 19 = 0$

$a = 1, b = 10, c = 19$

$$x = \frac{-10 \pm \sqrt{10^2 - 4(1)(19)}}{2 \cdot 1}$$

$$x = \frac{-10 \pm \sqrt{100 - 76}}{2} = \frac{-10 \pm \sqrt{24}}{2}$$

$$x = \frac{-10 \pm 2\sqrt{6}}{2} \qquad \text{Simplify}$$

$x = -5 \pm \sqrt{6} \qquad \text{Reduce}$

Solution set: $\left\{ -5 \pm \sqrt{6} \right\}$

47.

$17x^2 - x - 1 = x^2 + 7x + 10$

$16x^2 - 8x - 11 = 0$

$a = 16, b = -8, c = -11$

$$x = \frac{-(-8) \pm \sqrt{(-8)^2 - 4(16)(-11)}}{2 \cdot 16}$$

$$x = \frac{8 \pm \sqrt{64 + 704}}{32} = \frac{8 \pm \sqrt{768}}{32}$$

$$x = \frac{8 \pm 16\sqrt{3}}{32} \qquad \text{Simplify}$$

$$x = \frac{1 \pm 2\sqrt{3}}{4} \qquad \text{Reduce}$$

Solution set: $\left\{ \dfrac{1 \pm 2\sqrt{3}}{4} \right\}$

49.

$(2x - 3)(x + 4) = (x + 3)(x + 2)$

$2x^2 + 5x - 12 = x^2 + 5x + 6$

$x^2 - 18 = 0$

$a = 1, b = 0, c = -18$

$$x = \frac{0 \pm \sqrt{0 - 4(1)(-18)}}{2 \cdot 1} = \frac{\pm\sqrt{72}}{2}$$

$$x = \frac{\pm 6\sqrt{2}}{2} \qquad \text{Simplify}$$

$$x = \pm 3\sqrt{2} \qquad \text{Reduce}$$

Solution set: $\left\{ \pm 3\sqrt{2} \right\}$

51.

$(3x - 19)(2x + 1) = x(2x - 47)$

$6x^2 - 35x - 19 = 2x^2 - 47x$

$4x^2 + 12x - 19 = 0$

$a = 4, b = 12, c = -19$

$$x = \frac{-12 \pm \sqrt{12^2 - 4(4)(-19)}}{2 \cdot 4}$$

$$x = \frac{-12 \pm \sqrt{144 + 304}}{8} = \frac{-12 \pm \sqrt{448}}{8}$$

$$x = \frac{-12 \pm 8\sqrt{7}}{8} \qquad \text{Simplify}$$

$$x = \frac{-3 \pm 2\sqrt{7}}{2} \qquad \text{Reduce}$$

Solution set: $\left\{ \dfrac{-3 \pm 2\sqrt{7}}{2} \right\}$

53.

$9x^2 + 25 = 0$

$a = 9, b = 0, c = 25$

$$x = \frac{0 \pm \sqrt{0^2 - 4(9)(25)}}{2 \cdot 9}$$

$$x = \frac{0 \pm \sqrt{-900}}{18} = \pm \frac{30i}{18} \qquad \text{Simplify}$$

$$x = \pm \frac{5}{3}i \qquad \text{Reduce}$$

Solution set: $\left\{ \pm \dfrac{5}{3}i \right\}$

55.

$x^2 - 2x + 10 = 0$

$a = 1, b = -2, c = 10$

$$x = \frac{-(-2) \pm \sqrt{(-2)^2 - 4(1)(10)}}{2 \cdot 1}$$

$$x = \frac{2 \pm \sqrt{4 - 40}}{2} = \frac{2 \pm \sqrt{-36}}{2}$$

$$x = \frac{2 \pm 6i}{2} \qquad \text{Simplify}$$

$$x = 1 \pm 3i \qquad \text{Reduce}$$

Solution set: $\{1 \pm 3i\}$

57.

$2x^2 + 5x + 28 = x^2 - 5x - 1$

$x^2 + 10x + 29 = 0$

$a = 1, b = 10, c = 29$

$$x = \frac{-10 \pm \sqrt{10^2 - 4(1)(29)}}{2 \cdot 1}$$

$$x = \frac{-10 \pm \sqrt{100 - 116}}{2} = \frac{-10 \pm \sqrt{-16}}{2}$$

$$x = \frac{-10 \pm 4i}{2} \qquad \text{Simplify}$$

$$x = -5 \pm 2i \qquad \text{Reduce}$$

Solution set: $\{-5 \pm 2i\}$

59.

$8x^2 - 12x + 5 = 0$

$a = 8, b = -12, c = 5$

$$x = \frac{-(-12) \pm \sqrt{(-12)^2 - 4(8)(5)}}{2 \cdot 8}$$

#59 continued

$$x = \frac{12 \pm \sqrt{144 - 160}}{16} = \frac{12 \pm \sqrt{-16}}{16}$$

$$x = \frac{12 \pm 4i}{16} \qquad \text{Simplify}$$

$$x = \frac{3 \pm i}{4} \text{ or } \frac{3}{4} \pm \frac{1}{4}i \qquad \text{Reduce}$$

Solution set: $\left\{ \frac{3}{4} \pm \frac{1}{4}i \right\}$

61.

$$(16x + 15)(x + 1) = 15x + 2$$
$$16x^2 + 31x + 15 = 15x + 2$$
$$16x^2 + 16x + 13 = 0$$
$$a = 16, \ b = 16, \ c = 13$$

$$x = \frac{-16 \pm \sqrt{16^2 - 4(16)(13)}}{2 \cdot 16}$$

$$x = \frac{-16 \pm \sqrt{256 - 832}}{32} = \frac{-16 \pm \sqrt{-576}}{32}$$

$$x = \frac{-16 \pm 24i}{32} = \frac{-2 \pm 3i}{4} = -\frac{1}{2} \pm \frac{3}{4}i$$

Solution set: $\left\{ -\frac{1}{2} \pm \frac{3}{4}i \right\}$

63.

Let x = the number.
The equation is $x^2 - 4x = 3$.
$$x^2 - 4x - 3 = 0$$
$$a = 1, \ b = -4, \ c = -3$$

$$x = \frac{-(-4) \pm \sqrt{(-4)^2 - 4(1)(-3)}}{2 \cdot 1}$$

$$x = \frac{4 \pm \sqrt{16 + 12}}{2} = \frac{4 \pm \sqrt{28}}{2}$$

$$x = \frac{4 \pm 2\sqrt{7}}{2} \qquad \text{Simplify}$$

$$x = 2 \pm \sqrt{7}$$

The number is $2 + \sqrt{7}$ or $2 - \sqrt{7}$.

65.

Let x = the number.
The equation is $2x^2 + 2x = -5$.
$$2x^2 + 2x + 5 = 0$$

#65 continued

$$a = 2, \ b = 2, \ c = 5$$

$$x = \frac{-2 \pm \sqrt{2^2 - 4(2)(5)}}{2 \cdot 2}$$

$$x = \frac{-2 \pm \sqrt{4 - 40}}{4} = \frac{-2 \pm \sqrt{-36}}{4}$$

$$x = \frac{-2 \pm 6i}{4} = \frac{-1 \pm 3i}{2} = -\frac{1}{2} \pm \frac{3}{2}i$$

The number is $-\frac{1}{2} + \frac{3}{2}i$ or $-\frac{1}{2} - \frac{3}{2}i$.

67.

Let x = the width.
Let $3x + 1$ = the length.
The equation is $x(3x + 1) = 3$.
$$3x^2 + x - 3 = 0$$
$$a = 3, \ b = 1, \ c = -3$$

$$x = \frac{-1 \pm \sqrt{1^2 - 4(3)(-3)}}{2 \cdot 3}$$

$$x = \frac{-1 \pm \sqrt{1 + 36}}{6} = \frac{-1 \pm \sqrt{37}}{6}$$

The width is $\frac{-1 + \sqrt{37}}{6}$ ft.

Measurement is not negative, so $\frac{-1 - \sqrt{37}}{6}$ is rejected.

69.

$$d = -16t^2 + 96t + 32$$
$$176 = -16t^2 + 96t + 32$$
$$0 = -16t^2 + 96t - 144$$
$$0 = -t^2 + 6t - 9 \qquad \text{Multiply by } \frac{1}{16}$$
$$a = -1, \ b = 6, \ c = -9$$

$$t = \frac{-6 \pm \sqrt{6^2 - 4(-1)(-9)}}{2(-1)}$$

$$t = \frac{-6 \pm \sqrt{36 - 36}}{-2}$$

$$t = \frac{-6 \pm 0}{-2} \qquad \text{Simplify}$$

$$t = 3$$

The ball is 176 ft. high at 3 seconds.

Exercises 9.5

1.
$$(2x+1)(2x-9) = -16x$$
$$4x^2 - 16x - 9 = -16x$$
$$4x^2 - 9 = 0 \qquad \text{Add } 16x$$
$$(2x-3)(2x+3) = 0 \qquad \text{Factor}$$
$$2x - 3 = 0 \quad \text{or} \quad 2x + 3 = 0$$
$$x = \frac{3}{2} \quad \text{or} \quad x = -\frac{3}{2}$$

Solution set: $\left\{\pm\frac{3}{2}\right\}$

3.
$$x^2 - 7x + 2 = 0$$
$$a = 1, \ b = -7, \ c = 2$$
$$x = \frac{-(-7) \pm \sqrt{(-7)^2 - 4(1)(2)}}{2 \cdot 1}$$
$$x = \frac{7 \pm \sqrt{49 - 8}}{2} = \frac{7 \pm \sqrt{41}}{2}$$

Solution set: $\left\{\frac{7 \pm \sqrt{41}}{2}\right\}$

5.
$$5x^2 - 2 = 0$$
$$5x^2 = 2 \qquad \text{Add } 2$$
$$x^2 = \frac{2}{5} \qquad \text{Divide by 5}$$
$$x = \pm\sqrt{\frac{2}{5}} \qquad \text{Extraction}$$
$$x = \pm\frac{\sqrt{10}}{5} \qquad \text{Simplify}$$

Solution set: $\left\{\pm\frac{\sqrt{10}}{5}\right\}$

7.
$$(2x-3)^2 - 11 = 14$$
$$(2x-3)^2 = 25 \qquad \text{Add 11}$$
$$2x - 3 = \pm\sqrt{25} \qquad \text{Extraction}$$

#7 continued
$$2x - 3 = \pm 5 \qquad \text{Simplify}$$
$$2x = 3 \pm 5 \qquad \text{Add 3}$$
$$= \frac{3 \pm 5}{2} \qquad \text{Divide by 2}$$
$$x = \frac{3 + 5}{2} = 4 \quad \text{or} \quad x = \frac{3 - 5}{2} = -1$$

Solution set: $\{-1, \ 4\}$

9.
$$(4x+1)(x-7) = -7x - 36$$
$$4x^2 - 27x - 7 = -7x - 36$$
$$4x^2 - 20x + 29 = 0 \qquad \text{Add 7x and 36.}$$
$$a = 4, b = -20, c = 29$$
$$x = \frac{-(-20) \pm \sqrt{(-20)^2 - 4(4)(29)}}{2 \cdot 4}$$
$$x = \frac{20 \pm \sqrt{400 - 464}}{8}$$
$$x = \frac{20 \pm \sqrt{-64}}{8}$$
$$x = \frac{20 \pm 8i}{8} \qquad \text{Simplify}$$
$$x = \frac{5 \pm 2i}{2} \qquad \text{Reduce}$$

Solution set: $\left\{\frac{5}{2} \pm i\right\}$

11.
$$(5x+3)(x-2) = (2x-1)(x-3)$$
$$5x^2 - 7x - 6 = 2x^2 - 7x + 3$$
$$3x^2 - 9 = 0$$
$$x^2 - 3 = 0 \qquad \text{Divide by 3}$$
$$x^2 = 3$$
$$x = \pm\sqrt{3} \qquad \text{Extraction}$$

Solution set: $\left\{\pm\sqrt{3}\right\}$

13.
$$6x^2 - 10x = 0$$
$$3x^2 - 5x = 0 \qquad \text{Divide by 2}$$

#13 continued

$$x(3x - 5) = 0 \qquad \text{Factor}$$
$$x = 0 \quad \text{or} \quad 3x - 5 = 0$$
$$x = 0 \quad \text{or} \quad x = \frac{5}{3}$$

Solution set: $\left\{ 0, \frac{5}{3} \right\}$

15.

$$(2x + 5)(x - 3) = 0$$
$$2x + 5 = 0 \quad \text{or} \quad x - 3 = 0$$
$$x = -\frac{5}{2} \quad \text{or} \quad x = 3$$

Solution set: $\left\{ -\frac{5}{2}, \ 3 \right\}$

17.

$$(3x + 2)(6x + 1) = 9$$
$$18x^2 + 15x + 2 = 9$$
$$18x^2 + 15x - 7 = 0 \qquad \text{Add } -9$$
$$(6x + 7)(3x - 1) = 0 \qquad \text{Factor}$$
$$6x + 7 = 0 \quad \text{or} \quad 3x - 1 = 0$$
$$x = -\frac{7}{6} \quad \text{or} \quad x = \frac{1}{3}$$

Solution set: $\left\{ -\frac{7}{6}, \ \frac{1}{3} \right\}$

19.

$$(2x + 1)(2x + 5) = 4(3x + 1)$$
$$4x^2 + 12x + 5 = 12x + 4$$
$$4x^2 + 1 = 0 \qquad \text{Add } -12x \text{ and } -4$$
$$4x^2 = -1 \qquad \text{Add } -4$$
$$x^2 = -\frac{1}{4} \qquad \text{Divide by 4}$$
$$x = \pm\sqrt{-\frac{1}{4}} \qquad \text{Extraction}$$
$$x = \pm\frac{1}{2}i \qquad \text{Simplify}$$

Solution set: $\left\{ \pm\frac{1}{2}i \right\}$

21.

$$(4x + 1)^2 = 13$$
$$4x + 1 = \pm\sqrt{13} \qquad \text{Extraction}$$
$$4x = -1 \pm \sqrt{13} \qquad \text{Add } -1$$
$$x = \frac{-1 \pm \sqrt{13}}{4}$$

Solution set: $\left\{ \frac{-1 \pm \sqrt{13}}{4} \right\}$

23.

$$2x^2 + 3x - 3 = 0$$
$$a = 2, \ b = 3, \ c = -3$$
$$x = \frac{-3 \pm \sqrt{3^2 - 4(2)(-3)}}{2 \cdot 2}$$
$$x = \frac{-3 \pm \sqrt{9 + 24}}{4}$$
$$x = \frac{-3 \pm \sqrt{33}}{4}$$

Solution set: $\left\{ \frac{-3 \pm \sqrt{33}}{4} \right\}$

25.

$$2x^2 - 6x + 3 = 0$$
$$a = 2, \ b = -6, \ c = 3$$
$$x = \frac{-(-6) \pm \sqrt{(-6)^2 - 4(2)(3)}}{2 \cdot 2}$$
$$x = \frac{6 \pm \sqrt{36 - 24}}{4}$$
$$x = \frac{6 \pm \sqrt{12}}{4}$$
$$x = \frac{6 \pm 2\sqrt{3}}{4} \qquad \text{Simplify}$$
$$x = \frac{3 \pm \sqrt{3}}{2}$$

Solution set: $\left\{ \frac{3 \pm \sqrt{3}}{2} \right\}$

27.

$$(3x - 1)(3x - 2) = 3x$$
$$9x^2 - 9x + 2 = 3x$$
$$9x^2 - 12x + 2 = 0$$

#27   continued

$a = 9,\ b = -12,\ c = 2$

$$x = \frac{-(-12) \pm \sqrt{(-12)^2 - 4(9)(2)}}{2 \cdot 9}$$

$$x = \frac{12 \pm \sqrt{144 - 72}}{18}$$

$$x = \frac{12 \pm \sqrt{72}}{18}$$

$$x = \frac{12 \pm 6\sqrt{2}}{18} \qquad \text{Simplify}$$

$$x = \frac{2 \pm \sqrt{2}}{3}$$

Solution set: $\left\{ \dfrac{2 \pm \sqrt{2}}{3} \right\}$

29.

$(2x + 4)(x - 3) = (x + 3)(x - 1)$

$2x^2 - 2x - 12 = x^2 + 2x - 3$

$x^2 - 4x - 9 = 0$

$a = 1,\ b = -4,\ c = -9$

$$x = \frac{-(-4) \pm \sqrt{(-4)^2 - 4(1)(-9)}}{2 \cdot 1}$$

$$x = \frac{4 \pm \sqrt{16 + 36}}{2} = \frac{4 \pm \sqrt{52}}{2} = \frac{4 \pm 2\sqrt{13}}{2}$$

$x = 2 \pm \sqrt{13}$

Solution set: $\left\{ 2 \pm \sqrt{13} \right\}$

31.

$6x^2 + 19x + 8 = 0$

$(3x + 8)(2x + 1) = 0$

$\qquad 3x + 8 = 0 \qquad$ or $\qquad 2x + 1 = 0$

$\qquad\qquad x = -\dfrac{8}{3} \qquad$ or $\qquad\qquad x = -\dfrac{1}{2}$

Solution set: $\left\{ -\dfrac{8}{3},\ -\dfrac{1}{2} \right\}$

33.

$(x + 3)^2 + 5 = 1$

$\qquad (x + 3)^2 = -4$

#33   continued

$x + 3 = \pm\sqrt{-4} \qquad$ Extraction

$x + 3 = \pm 2i \qquad\quad$ Simplify

$x = -3 \pm 2i \qquad\quad$ Add -3

Solution set: $\{-3 \pm 2i\}$

35.

$(6x - 9)(x + 2) = -18$

$6x^2 + 3x - 18 = -18$

$\qquad 6x^2 + 3x = 0 \qquad$ Add 18

$\qquad 2x^2 + x = 0 \qquad$ Divide by 3

$\qquad x(2x + 1) = 0 \qquad$ Factor

$\qquad\qquad x = 0 \quad$ or $\quad 2x + 1 = 0$

$\qquad\qquad x = 0 \quad$ or $\qquad\qquad x = -\dfrac{1}{2}$

Solution set: $\left\{ 0,\ -\dfrac{1}{2} \right\}$

37.

$(2x + 1)(x - 1) = (x + 7)(x + 8)$

$\qquad 2x^2 - x - 1 = x^2 + 15x + 56$

$x^2 - 16x - 57 = 0$

$(x - 19)(x + 3) = 0 \qquad$ Factor

$\qquad\qquad x - 19 = 0 \quad$ or $\quad x + 3 = 0$

$\qquad\qquad x = 19 \quad$ or $\qquad x = -3$

Solution set: $\{-3,\ 19\}$

39.

$16x^2 + 8x - 5 = 0$

$a = 16,\ b = 8,\ c = -5$

$$x = \frac{-8 \pm \sqrt{8^2 - 4(16)(-5)}}{2 \cdot 16}$$

$$x = \frac{-8 \pm \sqrt{64 + 320}}{32} = \frac{-8 \pm \sqrt{384}}{32}$$

$$x = \frac{-8 \pm 8\sqrt{6}}{32} \qquad \text{Simplify}$$

$$x = \frac{-1 \pm \sqrt{6}}{4}$$

Solution set: $\left\{ \dfrac{-1 \pm \sqrt{6}}{4} \right\}$

41.

$(4x-1)(x+1) = 3x+4$

$4x^2+3x-1 = 3x+4$

$4x^2-5 = 0$ Add $-3x$ and -4

$4x^2 = 5$ Add 5

$x^2 = \dfrac{5}{4}$ Divide by 4

$x = \pm\sqrt{\dfrac{5}{4}}$ Extraction

$x = \pm\dfrac{\sqrt{5}}{2}$ Simplify

Solution set: $\left\{\pm\dfrac{\sqrt{5}}{2}\right\}$

43.

$x^2-8x+17 = 0$

$a=1,\ b=-8,\ c=17$

$x = \dfrac{-(-8)\pm\sqrt{(-8)^2-4(1)(17)}}{2\cdot1}$

$x = \dfrac{8\pm\sqrt{64-68}}{2} = \dfrac{8\pm\sqrt{-4}}{2}$

$x = \dfrac{8\pm2i}{2}$ Simplify

$x = 4\pm i$ Reduce

Solution set: $\{4\pm i\}$

Exercises 9.6

1. - 23.
The answers do not require worked-out solutions.

25.

$y = x^2-4x+8$

$y = x^2-4x+(4-4)+8$

$y = x^2-4x+4+(-4+8) = (x-2)^2+4$

Vertex: $(2,4)$

Note: $\left(\dfrac{-4}{2}\right)^2 = 4$, so add $4+(-4)$ to complete the square.

27.

$y = x^2-2x-2$

$y = x^2-2x+(1-1)-2$

$y = x^2-2x+1+(-1-2)$

$y = (x-1)^2-3$

Vertex: $(1,-3)$

Note: $\left(\dfrac{-2}{2}\right)^2 = 1$, so add $1+(-1)$ to complete the square.

29.

$y = x^2-6x+9$

$y = x^2-6x+(9-9)+9$

$y = x^2-6x+9+(-9+9)$

$y = (x-3)^2+0$

Vertex: $(3,0)$

Note: $\left(\dfrac{-6}{2}\right)^2 = 9$, so add $9-9$ to complete the square.

31.

$y = x^2+6x+10$

$y = x^2+6x+(9-9)+10$

$y = x^2+6x+9+(-9+10)$

$y = (x+3)^2+1$

Vertex: $(-3,1)$

Note: $\left(\dfrac{6}{2}\right)^2 = 9$, so add $9-9$ to complete the square.

33.

$y = x^2+2x-5$

$y = x^2+2x+(1-1)-5$

$y = x^2+2x+1+(-1-5)$

$y = (x+1)^2-6$

Vertex: $(-1,-6)$

Note: $\left(\dfrac{2}{2}\right)^2 = 1$, so add $1-1$ to complete the square.

Review Exercises

1.
$$\sqrt{-81} = \sqrt{81} \cdot \sqrt{-1} = 9i$$

3.
$$\sqrt{\frac{-4}{25}} = \frac{\sqrt{-4}}{\sqrt{25}} = \frac{\sqrt{4} \cdot \sqrt{-1}}{\sqrt{25}} = \frac{2i}{5} = \frac{2}{5}i$$

5.
$$(-8 + 3i) + (5 + 9i) = (-8 + 5) + (3i + 9i) = -3 + 12i$$

7.
$$4i - (-11 - 7i) = 4i + 11 + 7i = 11 + (4i + 7i) = 11 + 11i$$

9.
$$-4(5 - 9i) = -4 \cdot 5 - (-4) \cdot 9i \quad \text{Distributive property}$$
$$= -20 + 36i$$

11.
$$(5 + 6i)(1 + 4i)$$
$$\quad \text{F} \quad \text{O} \quad \text{I} \quad \text{L}$$
$$= 5 \cdot 1 + 5 \cdot 4i + 6i \cdot 1 + 6i \cdot 4i$$
$$= 5 + 20i + 6i + 24i^2$$
$$= 5 + 26i + 24(-1)$$
$$= 5 + 26i - 24$$
$$= -19 + 26i$$

13.
$$(9 + 2i)(9 - 2i) = 9^2 - (2i)^2$$
$$= 81 - 4i^2$$
$$= 81 - 4(-1)$$
$$= 81 + 4 = 85$$
The solution is the product of two conjugates.
$$(a + bi)(a - bi) = a^2 + b^2$$

15.
$$\frac{21 + i}{4 + i} = \frac{21 + i}{4 + i} \cdot \frac{4 - i}{4 - i} = \frac{(21 + i)(4 - i)}{(4 + i)(4 - i)}$$
$$= \frac{84 - 17i - i^2}{16 - i^2}$$
$$= \frac{84 - 17i - (-1)}{16 - (-1)}$$
$$= \frac{84 - 17i + 1}{16 + 1}$$
$$= \frac{85 - 17i}{17}$$
$$= \frac{17(5 - i)}{17} = 5 - i$$

17.
$$\frac{3 - 11i}{i} = \frac{3 - 11i}{i} \cdot \frac{i}{i} = \frac{i(3 - 11i)}{i^2} = \frac{3i - 11i^2}{i^2}$$
$$= \frac{3i - 11(-1)}{-1} = \frac{3i + 11}{-1} = -11 - 3i$$

19.
$$\frac{-12 + \sqrt{-36}}{6} = \frac{-12 + \sqrt{36} \cdot \sqrt{-1}}{6} = \frac{-12 + 6i}{6}$$
$$= \frac{\cancel{6}(-2 + i)}{\cancel{6}} = -2 + i$$

21.
$$x^2 = 16$$
$$x = \pm\sqrt{16} \qquad \text{Extraction}$$
$$x = \pm 4 \qquad \text{Simplify}$$
Solution set: $\{\pm 4\}$

23.
$$x^2 = -50$$
$$x = \pm\sqrt{-50} \qquad\qquad \text{Extraction}$$
$$x = \pm\sqrt{25(-1) \cdot 2}$$
$$x = \pm\sqrt{25}\sqrt{-1}\sqrt{2}$$
$$x = \pm 5i\sqrt{2} \qquad\qquad \text{Simplify}$$
Solution set: $\left\{\pm 5i\sqrt{2}\right\}$

25.

$49x^2 - 20 = 0$

$49x^2 = 20$

$x^2 = \dfrac{20}{49}$ Divide by 49.

$x = \pm\sqrt{\dfrac{20}{49}}$ Extraction of Roots Property

$x = \pm\dfrac{\sqrt{20}}{7}$

$x = \pm\dfrac{2\sqrt{5}}{7}$ Simplify.

Solution set: $\left\{\pm\dfrac{2\sqrt{5}}{7}\right\}$

27.

$9x^2 + 25 = 0$

$9x^2 = -25$ Add -25.

$x^2 = -\dfrac{25}{9}$ Divide by 9.

$x = \pm\sqrt{-\dfrac{25}{9}}$ Extraction of Roots Property

$x = \sqrt{\dfrac{25}{9}} \cdot \sqrt{-1}$

$x = \pm\dfrac{5}{3}i$ Simplify.

Solution set: $\left\{\pm\dfrac{5}{3}i\right\}$

29.

$\left(x - \dfrac{3}{4}\right)^2 = \dfrac{25}{16}$

$x - \dfrac{3}{4} = \pm\sqrt{\dfrac{25}{16}}$ Extraction of Roots Property

$x - \dfrac{3}{4} = \pm\dfrac{5}{4}$ Simplify.

$x = \dfrac{3}{4} \pm \dfrac{5}{4}$ Add $\dfrac{3}{4}$.

$x = \dfrac{3}{4} + \dfrac{5}{4} = 2$ or $x = \dfrac{3}{4} - \dfrac{5}{4} = -\dfrac{1}{2}$

Solution set: $\left\{-\dfrac{1}{2}, \ 2\right\}$

31.

$(x + 1)^2 = 80$

$x + 1 = \pm\sqrt{80}$ Extraction

$x + 1 = \pm 4\sqrt{5}$ Simplify

$x = -1 \pm 4\sqrt{5}$ Add -1

Solution set: $\left\{-1 \pm 4\sqrt{5}\right\}$

33.

$(5x + 11)^2 = 36$

$5x + 11 = \pm\sqrt{36}$ Extraction

$5x + 11 = \pm 6$ Simplify

$5x = -11 \pm 6$ Add -11.

$x = \dfrac{-11 \pm 6}{5}$ Divide by 5.

$x = \dfrac{-11 + 6}{5} = -1$ or $x = \dfrac{-11 - 6}{5} = \dfrac{-17}{5}$

Solution set: $\left\{-1, \ \dfrac{-17}{5}\right\}$

35.

$(4x - 1)^2 = -9$

$4x - 1 = \pm\sqrt{-9}$ Extraction

$4x - 1 = \pm 3i$ Simplify

$4x = 1 \pm 3i$ Add 1

$x = \dfrac{1 \pm 3i}{4}$ Divide by 4

Solution set: $\left\{\dfrac{1}{4} \pm \dfrac{3}{4}i\right\}$

37.

$(6x - 5)^2 + 11 = 27$

$(6x - 5)^2 = 16$

$6x - 5 = \pm\sqrt{16}$ Extraction

$6x - 5 = \pm 4$ Simplify

$6x = 5 \pm 4$ Add 5

$x = \dfrac{5 \pm 4}{6}$ Divide by 6

$x = \dfrac{5 + 4}{6} = \dfrac{3}{2}$ or $x = \dfrac{5 - 4}{6} = \dfrac{1}{6}$

Solution set: $\left\{\dfrac{1}{6}, \ \dfrac{3}{2}\right\}$

39.

Let x = the number.

The equation is

$(x-5)^2 = 12$

$x - 5 = \pm\sqrt{12}$ Extraction of Roots Property

$x - 5 = \pm 2\sqrt{3}$ Simplify.

$x = 5 \pm 2\sqrt{3}$ Add 5.

The number is $5 + 2\sqrt{3}$ or $5 - 2\sqrt{3}$.

41.

Let x = length of the side of the smaller square.

Let $2x$ = length of the side of the larger square.

The equation is $x^2 + (2x)^2 = 180$.

$$x^2 + 4x^2 = 180$$

$$5x^2 = 180$$

$$x^2 = 36 \qquad \text{Divide by 5}$$

$$x = \pm\sqrt{36} \quad \text{Extraction}$$

$$x = \pm 6$$

The shorter side is 6 ft. and the longer side is 12 ft.

Note: The negative number is rejected since measurement is positive.

43.

$$x^2 - 12x$$

$$b = -12$$

$$\left(\frac{1}{2}b\right)^2 = \left[\frac{1}{2}(-12)\right]^2 = (-6)^2 = 36$$

So, $x^2 - 12x + 36 = (x-6)^2$

45.

$$x^2 + \frac{6}{7}x$$

$$b = \frac{6}{7}$$

$$\left(\frac{1}{2}b\right)^2 = \left(\frac{1}{2} \cdot \frac{6}{7}\right)^2 = \left(\frac{3}{7}\right)^2 = \frac{9}{49}$$

So, $x^2 + \frac{6}{7}x + \frac{9}{49} = \left(x + \frac{3}{7}\right)^2$

47.

$$x^2 - 9x + 14 = 0$$

$$x^2 - 9x = -14 \qquad \text{Add -14.}$$

$$x^2 - 9x + \frac{81}{4} = -14 + \frac{81}{4} \qquad \text{Add } \left(-\frac{9}{2}\right)^2.$$

$$\left(x - \frac{9}{2}\right)^2 = \frac{25}{4} \qquad \text{Factor.}$$

$$x - \frac{9}{2} = \pm\sqrt{\frac{25}{4}} \qquad \text{Extraction of Roots Property}$$

$$x - \frac{9}{2} = \pm\frac{5}{2} \qquad \text{Simplify}$$

$$x = \frac{9}{2} \pm \frac{5}{2} \qquad \text{Add } \frac{9}{2}$$

$$x = \frac{9}{2} + \frac{5}{2} = 7 \quad \text{or} \quad x = \frac{9}{2} - \frac{5}{2} = 2$$

Solution set: $\{2, 7\}$

49.

$$x^2 + 2x - 6 = 0$$

$$x^2 + 2x = 6 \qquad \text{Add 6}$$

$$x^2 + 2x + 1 = 6 + 1 \qquad \text{Add}\left(\frac{2}{2}\right)^2$$

$$(x + 1)^2 = 7 \qquad \text{Factor}$$

$$x + 1 = \pm\sqrt{7} \qquad \text{Extraction}$$

$$x = -1 \pm \sqrt{7} \qquad \text{Add } -1$$

Solution set: $\left\{-1 \pm \sqrt{7}\right\}$

51.

$$-x^2 + 10x - 1 = 0$$

$$x^2 - 10x + 1 = 0 \qquad \text{Divide by } -1$$

$$x^2 - 10x = -1 \qquad \text{Add } -1$$

$$x^2 - 10x + 25 = -1 + 25 \qquad \text{Add } \left(\frac{-10}{2}\right)^2$$

$$(x - 5)^2 = 24 \qquad \text{Factor}$$

$$x - 5 = \pm\sqrt{24} \qquad \text{Extraction}$$

$$x - 5 = \pm 2\sqrt{6} \qquad \text{Simplify}$$

$$x = 5 \pm 2\sqrt{6} \qquad \text{Add 5}$$

#51 continued

Solution set: $\left\{5 \pm 2\sqrt{6}\right\}$

53.

$4x^2 - 12x + 1 = 0$

$x^2 - 3x + \dfrac{1}{4} = 0$	Divide by 4
$x^2 - 3x = -\dfrac{1}{4}$	Add $-\dfrac{1}{4}$
$x^2 - 3x + \dfrac{9}{4} = -\dfrac{1}{4} + \dfrac{9}{4}$	Add $\dfrac{9}{4}$
$\left(x - \dfrac{3}{2}\right)^2 = 2$	Factor
$x - \dfrac{3}{2} = \pm\sqrt{2}$	Extraction
$x = \dfrac{3}{2} \pm \sqrt{2}$ or $\dfrac{3 \pm 2\sqrt{2}}{2}$	

Solution set: $\left\{\dfrac{3 \pm 2\sqrt{2}}{2}\right\}$

55.

Let $x =$ the number.

The equation is $x^2 + 8x = 5$

$x^2 + 8x + 16 = 5 + 16$	Add$\left(\dfrac{8}{2}\right)^2$
$(x + 4)^2 = 21$	Factor
$x + 4 = \pm\sqrt{21}$	Extraction
$x = -4 \pm \sqrt{21}$	Add -4

The number is $-4 + \sqrt{21}$ or $-4 - \sqrt{21}$.

57.

Let $x =$ measurement of the width.
Let $5x + 2 =$ measurement of the length.
The equation is $x(5x + 2) = 8$.

$5x^2 + 2x = 8$	
$x^2 + \dfrac{2}{5}x = \dfrac{8}{5}$	Divide by 5
$x^2 + \dfrac{2}{5}x + \dfrac{1}{25} = \dfrac{8}{5} + \dfrac{1}{25}$	Add$\left(\dfrac{2}{5} \cdot \dfrac{1}{2}\right)^2$
$\left(x + \dfrac{1}{5}\right)^2 = \dfrac{41}{25}$	Factor

#57 continued

$x + \dfrac{1}{5} = \pm\dfrac{\sqrt{41}}{5}$

$x = -\dfrac{1}{5} \pm \dfrac{\sqrt{41}}{5}$ or $\dfrac{-1 \pm \sqrt{41}}{5}$

The width is $\dfrac{-1 + \sqrt{41}}{5}$ feet.

Measurement is not negative, so $\dfrac{-1 - \sqrt{41}}{5}$ is rejected.

59.

$4x^2 - 7x = 0$
$4x^2 - 7x + 0 = 0$
$a = 4, b = -7, c = 0$

61.

$(2x - 7)(x + 2) = 5$

$2x^2 - 3x - 14 = 5$	Multiply
$2x^2 - 3x - 19 = 0$	Add 5

$a = 2, b = -3, c = -19$

63

$3x(3x - 4) = (x - 3)(x + 1)$
$9x^2 - 12x = x^2 - 2x - 3$
$8x^2 - 10x + 3 = 0$
$a = 8, b = -10, c = 3$

$x = \dfrac{-(-10) \pm \sqrt{(-10)^2 - 4(8)(3)}}{2 \cdot 8}$

$x = \dfrac{10 \pm \sqrt{100 - 96}}{16}$

$x = \dfrac{10 \pm \sqrt{4}}{16}$

$x = \dfrac{10 \pm 2}{16}$

$x = \dfrac{10 + 2}{16} = \dfrac{3}{4}$ or $x = \dfrac{10 - 2}{16} = \dfrac{1}{2}$

Solution set: $\left\{\dfrac{1}{2}, \dfrac{3}{4}\right\}$

65.

$3x^2 - 11 = 0$
$a = 3, b = 0, c = -11$

$x = \dfrac{0 \pm \sqrt{0^2 - 4(3)(-11)}}{2 \cdot 3} = \pm\dfrac{\sqrt{132}}{6}$

$x = \pm\dfrac{2\sqrt{33}}{6}$	Simplify

#65 continued

$$x = \pm\frac{\sqrt{33}}{3} \qquad \text{Reduce}$$

Solution set: $\left\{\pm\frac{\sqrt{33}}{3}\right\}$

67.

$(x + 4)(2x - 5) = (3x - 1)(x - 2)$

$2x^2 + 3x - 20 = 3x^2 - 7x + 2$

$\qquad\qquad 0 = x^2 - 10x + 22$

$a = 1, b = -10, c = 22$

$$x = \frac{-(-10) \pm \sqrt{(-10)^2 - 4(1)(22)}}{2 \cdot 1}$$

$$x = \frac{10 \pm \sqrt{100 - 88}}{2}$$

$$x = \frac{10 \pm \sqrt{12}}{2}$$

$$x = \frac{10 \pm 2\sqrt{3}}{2} \qquad \text{Simplify}$$

$$x = 5 \pm \sqrt{3} \qquad \text{Reduce}$$

Solution set: $\left\{5 \pm \sqrt{3}\right\}$

69.

$4x^2 - 3x + 8 = 3x^2 + 2x - 5$

$x^2 - 5x + 13 = 0$

$a = 1, b = -5, c = 13$

$$x = \frac{-(-5) \pm \sqrt{(-5)^2 - 4(1)(13)}}{2 \cdot 1}$$

$$x = \frac{5 \pm \sqrt{25 - 52}}{2} = \frac{5 \pm \sqrt{-27}}{2}$$

$$x = \frac{5 \pm 3i\sqrt{3}}{2} \quad \text{or} \quad \frac{5}{2} \pm \frac{3\sqrt{3}}{2}i \qquad \text{Simplify}$$

Solution set: $\left\{\frac{5}{2} \pm \frac{3\sqrt{3}}{2}i\right\}$

71.

Let x = the number.

The equation is $2x^2 + 2x = -1$

$2x^2 + 2x + 1 = 0$

$a = 2, b = 2, c = 1$

#71 continued

$$x = \frac{-2 \pm \sqrt{2^2 - 4(2)(1)}}{2 \cdot 2} = \frac{-2 \pm \sqrt{4 - 8}}{4}$$

$$x = \frac{-2 \pm \sqrt{-4}}{4} = \frac{-2 \pm 2i}{4} \qquad \text{Simplify}$$

$$x = \frac{-1 \pm i}{2} \quad \text{or} \quad -\frac{1}{2} \pm \frac{1}{2}i \qquad \text{Reduce}$$

The number is $-\frac{1}{2} + \frac{1}{2}i$ or $-\frac{1}{2} - \frac{1}{2}i$.

73.

$\qquad d = -16t^2 + 128t + 44$

$300 = -16t^2 + 128t + 44$

$\qquad 0 = -16t^2 + 128t - 256$

$\qquad 0 = -t^2 + 8t - 16$

$\qquad a = -1, b = 8, c = -16$

$$t = \frac{-8 \pm \sqrt{8^2 - 4(-1)(-16)}}{2(-1)}$$

$$t = \frac{-8 \pm \sqrt{64 - 64}}{-2} = \frac{-8 \pm \sqrt{0}}{-2}$$

$t = 4$

The brick is 300 ft. high at 4 seconds.

75.

$(3x + 2)(x + 2) = (x + 3)(x - 3)$

$3x^2 + 8x + 4 = x^2 - 9$

$2x^2 + 8x + 13 = 0$

$a = 2, b = 8, c = 13$

$$x = \frac{-8 \pm \sqrt{8^2 - 4(2)(13)}}{2 \cdot 2}$$

$$x = \frac{-8 \pm \sqrt{64 - 104}}{4} = \frac{-8 \pm \sqrt{-40}}{4}$$

$$x = \frac{-8 \pm 2i\sqrt{10}}{4} \qquad \text{Simplify}$$

$$x = \frac{-4 \pm i\sqrt{10}}{2} \quad \text{or} \quad -2 \pm \frac{1}{2}i\sqrt{10} \qquad \text{Reduce}$$

Solution set: $\left\{-2 \pm \frac{1}{2}i\sqrt{10}\right\}$

77.

$\qquad 2x^2 + 7x - 9 = 0$

$(2x + 9)(x - 1) = 0$

$2x + 9 = 0 \quad \text{or} \quad x - 1 = 0$

#77 continued

$$x = -\frac{9}{2} \quad \text{or} \quad x = 1$$

Solution set: $\left\{-\frac{9}{2}, 1\right\}$

79.

$$3x^2 - 24 = 0$$
$$3x^2 = 24$$
$$x^2 = 8$$
$$x = \pm\sqrt{8}$$
$$x = \pm 2\sqrt{2}$$

Solution set: $\left\{\pm 2\sqrt{2}\right\}$

81.

$$5(4x - 1) = (2x + 1)(x + 5)$$
$$20x - 5 = 2x^2 + 11x + 5$$
$$0 = 2x^2 - 9x + 10$$
$$0 = (x - 2)(2x - 5)$$
$$x - 2 = 0 \quad \text{or} \quad 2x - 5 = 0$$
$$x = 2 \qquad\qquad x = \frac{5}{2}$$

Solution set: $\left\{2, \frac{5}{2}\right\}$

83.-85.

The answers do not require worked-out solutions.

87.

$$y = x^2 - 6x + 8$$
$$y = x^2 - 6x + (9 - 9) + 8$$
$$y = x^2 - 6x + 9 + (-9 + 8)$$
$$y = (x - 3)^2 - 1$$

Vertex: $(3, -1)$

Note: $\left(\frac{-6}{2}\right)^2 = 9$, so add $9 - 9$ to complete the square.

Chapter 9 Test Solutions

1.

$$(5 - 7i) + (-6 - 3i) = 5 - 6 - 7i - 3i$$
$$= -1 - 10i$$

3.

$$\qquad\qquad\qquad \text{F} \quad \text{O} \quad \text{I} \quad \text{L}$$
$$(2 - 5i)(3 + i) = 6 + 2i - 15i - 5i^2$$
$$= 6 - 13i + 5 \qquad i^2 = -1$$
$$= 11 - 13i$$

5.

$$\frac{4 - 9i}{i} \cdot \frac{i}{i} = \frac{4i - 9i^2}{i^2} = \frac{4i + 9}{-1} = -9 - 4i$$

Note: $i^2 = -1$

7.

$$x^2 - 12x + 28 = 0$$
$$x^2 - 12x = -28 \qquad\qquad \text{Subtract } 28$$
$$x^2 - 12x + 36 = -28 + 36 \qquad \text{Add } \left(\frac{-12}{2}\right)^2$$
$$(x - 6)^2 = 8 \qquad\qquad \text{Factor}$$
$$x - 6 = \pm\sqrt{8} \qquad\qquad \text{Extraction of roots property}$$
$$x - 6 = \pm 2\sqrt{2} \qquad\qquad \text{Simplify}$$
$$x = 6 \pm 2\sqrt{2} \qquad\qquad \text{Add } 6$$

Solution set: $\left\{6 \pm 2\sqrt{2}\right\}$

9.

$$(x - 5)^2 = 12$$
$$x - 5 = \pm\sqrt{12} \qquad \text{Extraction of roots}$$
$$x - 5 = \pm 2\sqrt{3}$$
$$x = 5 \pm 2\sqrt{3}$$

Solution set: $\left\{5 \pm 2\sqrt{3}\right\}$

11.

One method is by the quadratic formula.

$$x^2 + 8x + 5 = 0$$
$$a = 1, b = 8, c = 5$$

#11 continued

$$x = \frac{-8 \pm \sqrt{64-4(1)(5)}}{2}$$

$$x = \frac{-8 \pm \sqrt{64-20}}{2} = \frac{-8 \pm \sqrt{44}}{2}$$

$$x = \frac{-8 \pm 2\sqrt{11}}{2} = -4 \pm \sqrt{11}$$

Solution set: $\left\{-4 \pm \sqrt{11}\right\}$

13.

$$(x + 5)(2x - 1) = (x - 2)(x - 1)$$
$$2x^2 + 9x - 5 = x^2 - 3x + 2$$
$$x^2 + 12x - 7 = 0$$

Use completing the square.

$$x^2 + 12x = 7$$
$$x^2 + 12x + 36 = 7 + 36$$
$$(x + 6)^2 = 43$$
$$x + 6 = \pm\sqrt{43}$$
$$x = -6 \pm \sqrt{43}$$

Solution set: $\left\{-6 \pm \sqrt{43}\right\}$

15.

$$8x^2 - 14x + 5 = 0$$
$$(2x - 1)(4x - 5) = 0$$
$$x = \frac{1}{2} \text{ or } x = \frac{5}{4}$$

Solution set: $\left\{\frac{1}{2}, \frac{5}{4}\right\}$

17.

$$x^2 + (4x)^2 = 153$$
$$x^2 + 16x^2 = 153$$
$$17x^2 = 153$$
$$x^2 = 9 \qquad \text{Divide by 17}$$
$$x = \pm 3 \qquad \text{Extraction of roots}$$

Since measurement is not negative, x = 3 and 4x = 12. Answer: 3 ft. and 12 ft.

19.

The graph is in the answer section of the main text.

1–3.

The graphs are in the answer section of the main text.

5.

(1, 4), (-2, 4)

The line is horizontal passing through y = 4.

7.

$$3x - 4(2x - 1) = 5 - (x + 4)$$
$$3x - 8x + 4 = 5 - x - 4$$
$$-5x + 4 = -x + 1$$
$$4 = 4x + 1 \qquad \text{Add 5x}$$
$$3 = 4x \qquad \text{Subtract 1}$$
$$\frac{3}{4} = x$$

Solution set: $\left\{\frac{3}{4}\right\}$

9.

$$(2x + 3)(x + 6) = 18$$
$$2x^2 + 15x + 18 = 18$$
$$2x^2 + 15x = 0 \qquad \text{Subtract 18}$$
$$x(2x + 15) = 0$$
$$x = 0 \text{ or } x = -\frac{15}{2}$$

Solution set: $\left\{0, -\frac{15}{2}\right\}$

11.

$$\frac{2x+3}{x+5} = \frac{7}{5}$$
$$5(2x + 3) = 7(x + 5) \qquad \text{Cross multiply}$$
$$10x + 15 = 7x + 35$$
$$3x = 20 \qquad \text{Subtract 7x, 15}$$
$$x = \frac{20}{3}$$

Solution set: $\left\{\frac{20}{3}\right\}$

13.

$$6(x+1)(x+2)\left(\frac{1}{x+1} - \frac{2}{x+2}\right) = \left(-\frac{1}{6}\right)6(x+1)(x+2)$$
$$\text{LCD} = 6(x + 1)(x + 2)$$

#13 continued

$$6(x + 2) - 12(x + 1) = -(x + 1)(x + 2)$$
$$6x + 12 - 12x - 12 = -x^2 - 3x - 2$$
$$-6x = -x^2 - 3x - 2$$
$$x^2 - 3x + 2 = 0 \quad \text{Add } x^2, 3x, 2$$
$$(x - 2)(x - 1) = 0$$
$$x = 2 \text{ or } x = 1$$

Solution set: {1, 2}

15.

$$\sqrt{5x+4} = 7$$
$$5x + 4 = 49 \quad \text{Square}$$
$$5x = 45$$
$$x = 9$$

Solution set: {9}

17.

$$(x + 3)^2 = 4$$
$$x + 3 = \pm\sqrt{4}$$
$$x + 3 = \pm 2$$
$$x = -3 \pm 2$$
$$x = -1 \text{ or } x = -5$$

Solution set: {-1, -5}

19.

$$3x^2 + 8x - 3 = 0$$
$$a = 3, b = 8, c = -3$$
$$x = \frac{-8 \pm \sqrt{64-4(3)(-3)}}{2 \cdot 3} = \frac{-8 \pm \sqrt{64 + 36}}{6}$$
$$x = \frac{-8 \pm \sqrt{100}}{6} = \frac{-8 \pm 10}{6}$$
$$x = \frac{2}{6} \text{ or } \frac{1}{3}$$
$$x = \frac{-18}{6} \text{ or } -3$$

Solution set: $\left\{\frac{1}{3}, -3\right\}$

21.

$$\frac{(4^{-1}x^{-4}y^{-3})^{-2}}{(4x^3y^{-4})^{-1}} = \frac{4^2x^8y^6}{4^{-1}x^{-3}y^4} = 4^{2+1}x^{8+3}y^{6-4}$$
$$= 4^3x^{11}y^2 \text{ or } 64x^{11}y^2$$

23.

The exercise does not require a worked-out solution.

25.

$$\sqrt{20x^5y^9} = \sqrt{4 \cdot 5 \cdot x^4 \cdot y^8 \cdot y}$$
$$= \sqrt{4x^4y^8 \cdot 5xy}$$
$$= 2x^2y^4\sqrt{5xy}$$

27.

Factor, invert the divisor, and multiply.

$$\frac{(2x+1)(2x+1)}{(2x+1)(x-2)} \cdot \frac{8x^2(x-2)}{4x(2x+1)(x+3)}$$
$$= \frac{8x^2}{4x(x + 3)} = \frac{2 \cdot 4 \cdot x \cdot x}{4 \cdot x(x + 3)} = \frac{2x}{x + 3}$$

29.

$$\frac{3x + 2}{3x + 2} \cdot \frac{3}{2x} - \frac{4}{3x + 2} \cdot \frac{2x}{2x} \quad \text{LCD} = 2x(3x + 2)$$
$$= \frac{9x + 6}{2x(3x + 2)} - \frac{8x}{2x(3x + 2)}$$
$$= \frac{9x + 6 - 8x}{2x(3x + 2)} = \frac{x + 6}{2x(3x + 2)}$$

31.

$$5\sqrt{2x}\left(3\sqrt{6x} + 4\sqrt{y} - 5\sqrt{2y}\right)$$
$$= 15\sqrt{12x^2} + 20\sqrt{2xy} - 25\sqrt{4xy}$$
$$= 15\sqrt{4x^2 \cdot 3} + 20\sqrt{2xy} - 25\sqrt{4xy}$$
$$= 15 \cdot 2x\sqrt{3} + 20\sqrt{2xy} - 25 \cdot 2\sqrt{xy}$$
$$= 30x\sqrt{3} + 20\sqrt{2xy} - 50\sqrt{xy}$$

33.

$$3\sqrt{80} - \sqrt{45} - 2\sqrt{20}$$
$$= 3\sqrt{16 \cdot 5} - \sqrt{9 \cdot 5} - 2\sqrt{4 \cdot 5}$$
$$= 3 \cdot 4\sqrt{5} - 3\sqrt{5} - 2 \cdot 2\sqrt{5}$$
$$= 12\sqrt{5} - 3\sqrt{5} - 4\sqrt{5} = 5\sqrt{5}$$

35.

$$(3 + 2i) - (4 - 7i)$$
$$= 3 + 2i - 4 + 7i$$
$$= -1 + 9i$$

37.

$$\frac{3-8i}{i} \cdot \frac{i}{i} = \frac{3i - 8i^2}{i^2}$$

$$= \frac{3i + 8}{-1} \qquad i^2 = -1$$

$$= -8 - 3i$$

39.

$$\sqrt{\frac{7y}{3x} \cdot \frac{3x}{3x}} = \sqrt{\frac{21xy}{9x^2}} = \frac{\sqrt{21xy}}{3x}$$

41.

(1) $\qquad 6x - 3y = 21$

(2) $\qquad 4x - 2y = 14$

To eliminate the x's, multiply (1) by 4 and (2) by -6.

(3) $\qquad 24x - 12y = 84$

(4) $\qquad \underline{-24x + 12y = -84}$

$$\qquad\qquad 0 = 0$$

The system is dependent.

43.

Let x be the number.

Thus,

$$x^2 - 2x = 6$$

$$x^2 - 2x - 6 = 0$$

$$a = 1, b = -2, c = -6$$

$$x = \frac{2 \pm \sqrt{(-2)^2 - 4(1)(-6)}}{2 \cdot 1} = \frac{2 \pm \sqrt{4 + 24}}{2}$$

$$x = \frac{2 \pm \sqrt{28}}{2} = \frac{2 \pm \sqrt{4 \cdot 7}}{2}$$

$$x = \frac{2 \pm 2\sqrt{7}}{2} \quad \text{or} \quad 1 \pm \sqrt{7}$$

The numbers are $1 + \sqrt{7}$ and $1 - \sqrt{7}$.

45.

Let x = ounces 12% juice.

Let 40 - x = ounces 28% juice.

Thus,

$$0.12x + 0.28(40 - x) = 0.18(40)$$

$$12x + 28(40 - x) = 18(40) \text{ Multiply by } 100$$

#45 continued

$$12x + 1120 - 28x = 720$$

$$-16x = -400$$

$$x = 25$$

$$40 - x = 15$$

Andy's: 25 oz.; Alan's: 15 oz.

47.

$$\frac{cups}{servings} = \frac{\frac{1}{3}}{6} = \frac{x}{45}$$

$$\frac{1}{3}(45) = 6x$$

$$15 = 6x$$

$$\frac{15}{6} = x$$

$$\frac{5}{2} = x$$

Answer: $\frac{5}{2}$ cups.

49.

Let x = speed in still water.

	R ·	D	= T
Down	x + 3	$\frac{8}{x + 3}$	8
Up	x - 3	$\frac{8}{x - 3}$	8

The total trip took 2 hours.

Thus,

$$\frac{8}{x + 3} + \frac{8}{x - 3} = 2$$

$$8(x - 3) + 8(x + 3) = 2(x + 3)(x - 3)$$

$$8x - 24 + 8x + 24 = 2x^2 - 18$$

$$16x = 2x^2 - 18$$

$$8x = x^2 - 9$$

$$0 = x^2 - 8x - 9$$

$$0 = (x - 9)(x + 1)$$

$$x = 9 \text{ or } x = -1$$

Answer: 9 m.p.h.

CHAPTER 9 STUDY GUIDE

Self-Test Exercises

Perform the indicated operations. Express all numbers in the form *a + bi*.

1. $(6 - 5i) + (-5 - 3i)$

2. $(-5 - 2i) - (7 - 5i)$

3. $(3 - 3i)(5 + i)$

4. $(2 - 5i)^2$

5. $\dfrac{9 - 4i}{i}$

6. $\dfrac{11 - i}{3 + 7i}$

7. $(-5 - 7i)(-5 + 7i)$

8. $(7 - i) - (-1 + 3i) + (4 - 2i)$

Find the solution set of each quadratic equation by completing the square.

9. $x^2 - 12x + 35 = 0$

10. $7x^2 + 34x - 5 = 0$

11. $x^2 - 4x - 1 = 0$

12. $5x^2 - 16x + 3 = 0$

Using the most efficient method, find the solution set of each quadratic equation.

13. $(x - 4)^2 = 25$

14. $y - 2 = (y - 3)(y + 2)$

15. $2x^2 - 3x - 6 = 0$

16. $4x^2 + x + 2 = x^2 + 9x$

17. $(x - 2)(3x + 1) = (x + 3)(x - 1)$

18. $(x - 3)^2 = 2x + 10$

19. $36x^2 + 84x + 49 = 0$

20. $5x^2 - 2 = 0$

21. $\dfrac{x^2}{2} + 2x + \dfrac{2}{3} = 0$

22. $6x^2 = -x + 1$

Use quadratic equations to find the solutions of the following problems.

23. The product of two consecutive positive even integers is 48. Find the two even integers.

24. A stone is shot with a sling shot vertically upward with an initial velocity of 128 feet per second. The equation that gives the height of the stone is $h = -16t^2 + 128t$. Find the time when it will hit the ground.

25. Sketch the graph and determine the vertex of the quadratic equation: $y = (x - 1)^2 + 3$.

26. Sketch the graph of the equation: $y = x^2 + 4x + 7$. Find the vertex by completing the square.

The worked-out solutions begin on the next page.

Self-Test Solutions

1.

$(6 - 5i) + (-5 - 3i) = 6 - 5 - 5i - 3i = 1 - 8i$

2.

$(-5 - 2i) - (7 - 5i) = -5 - 2i - 7 + 5i$
$$= -12 + 3i$$

3.

$$\text{F \quad O \quad I \quad L}$$

$(3 - 3i)(5 + i) = 15 + 3i - 15i - 3i^2$
$$= 15 + 3i - 15i + 3$$
$$= 18 - 12i$$

4.

$(2 - 5i)^2 = 2^2 - 2(2 \cdot 5i) + (5i)^2$
$= 4 - 20i + 25i^2 = 4 - 20i - 25 = -21 - 20i$

5.

$$\frac{9 - 4i}{i} \cdot \frac{i}{i} \cdot \frac{9i - 4i^2}{i^2} = \frac{9i + 4}{-1} = -4 - 9i$$

6.

$$\frac{11 - i}{3 + 7i} \cdot \frac{3 - 7i}{3 - 7i} = \frac{33 - 77i - 3i + 7i^2}{9 - 49i^2}$$

$$= \frac{33 - 80i - 7}{9 + 49}$$

$$= \frac{26 - 80i}{58} = \frac{13 - 40i}{29}$$

7.

$(-5 - 7i)(-5 + 7i) = (-5)^2 - (7i)^2$
$$= 25 - 49i^2 = 25 + 49 = 74$$

8.

$(7 - i) - (-1 + 3i) + (4 - 2i)$
$= 7 - i + 1 - 3i + 4 - 2i = 12 - 6i$

9.

$$x^2 - 12x + 35 = 0$$
$$x^2 - 12x + 36 = -35 + 36$$
$$(x - 6)^2 = 1$$
$$x - 6 = \pm 1$$

#9 continued

$$x = 6 \pm 1; \quad x = 6 + 1 = 7$$
$$x = 6 - 1 = 5$$

Solution set: $\{5, 7\}$

10.

$$7x^2 + 34x - 5 = 0$$
$$x^2 + \frac{34}{7}x + \frac{289}{49} = \frac{5}{7} + \frac{289}{49} \quad \text{Multiply by } \frac{1}{7}$$
$$\left(x + \frac{17}{7}\right)^2 = \frac{324}{49}$$
$$x + \frac{17}{7} = \pm\sqrt{\frac{324}{49}} = \pm\frac{18}{7}$$
$$x = \frac{-17 \pm 18}{7}; \quad x = \frac{-17 + 18}{7} = \frac{1}{7}$$
$$x = \frac{-17 - 18}{7} = -5$$

Solution set: $\left\{\frac{1}{7}, -5\right\}$

11.

$$x^2 - 4x - 1 = 0$$
$$x^2 - 4x + 4 = 1 + 4$$
$$(x - 2)^2 = 5$$
$$x - 2 = \pm\sqrt{5}$$
$$x = 2 \pm \sqrt{5}$$

Solution set: $\left\{2 \pm \sqrt{5}\right\}$

12.

$$5x^2 - 16x + 3 = 0$$
$$x^2 - \frac{16}{5}x = -\frac{3}{5} \quad \text{Multiply by } \frac{1}{5}$$
$$x^2 - \frac{16}{5}x + \frac{64}{25} = -\frac{3}{5} + \frac{64}{25}$$
$$\left(x - \frac{8}{5}\right)^2 = \frac{49}{25}$$
$$x - \frac{8}{5} = \pm\frac{7}{5}$$
$$x = \frac{8}{5} \pm \frac{7}{5}; \quad x = \frac{8}{5} + \frac{7}{5} = \frac{15}{5} = 3$$
$$x = \frac{8}{5} - \frac{7}{5} = \frac{1}{5}$$

#12 continued

Solution set: $\left\{3, \dfrac{1}{5}\right\}$

13.

$$(x - 4)^2 = 25$$
$$x - 4 = \pm\sqrt{25}$$
$$x = 4 \pm 5; \quad x = 4 + 5 = 9$$
$$x = 4 - 5 = -1$$

Solution set: $\{9, -1\}$

14.

$$y - 2 = (y - 3)(y + 2)$$
$$y - 2 = y^2 - y - 6$$
$$0 = y^2 - 2y - 4$$
$$a = 1, b = -2, c = -4$$
$$y = \frac{2 \pm \sqrt{4+16}}{2} = \frac{2 \pm \sqrt{20}}{2} = \frac{2 \pm 2\sqrt{5}}{2} = 1 \pm \sqrt{5}$$

Solution set: $\left\{1 \pm \sqrt{5}\right\}$

15.

$$2x^2 - 3x - 6 = 0$$
$$a = 2, b = -3, c = -6$$
$$x = \frac{3 \pm \sqrt{9 + 48}}{2 \cdot 2}$$
$$x = \frac{3 \pm \sqrt{57}}{4}$$

Solution set: $\left\{\dfrac{3 \pm \sqrt{57}}{4}\right\}$

16.

$$4x^2 + x - 3 = x^2 + 9x$$
$$3x^2 - 8x - 3 = 0$$
$$(3x + 1)(x - 3) = 0$$
$$3x + 1 = 0 \text{ or } x - 3 = 0$$
$$x = -\frac{1}{3} \quad x = 3$$

Solution set: $\left\{-\dfrac{1}{3}, 3\right\}$

17.

$$(x - 2)(3x + 1) = (x + 3)(x - 1)$$
$$3x^2 - 5x - 2 = x^2 + 2x - 3$$
$$2x^2 - 7x + 1 = 0$$
$$a = 2, b = -7, c = 1$$
$$x = \frac{7 \pm \sqrt{49 - 8}}{2 \cdot 2}$$
$$x = \frac{7 \pm \sqrt{41}}{4}$$

Solution set: $\left\{\dfrac{7 \pm \sqrt{41}}{4}\right\}$

18.

$$(x - 3)^2 = 2x + 10$$
$$x^2 - 6x + 9 = 2x + 10$$
$$x^2 - 8x - 1 = 0$$
$$a = 1, b = -8, c = -1$$
$$x = \frac{8 \pm \sqrt{64 + 4}}{2}$$
$$x = \frac{8 \pm \sqrt{68}}{2} = \frac{8 \pm 2\sqrt{17}}{2} = 4 \pm \sqrt{17}$$

Solution set: $\left\{4 \pm \sqrt{17}\right\}$

19.

$$36x^2 + 84x + 49 = 0$$
$$(6x + 7)(6x + 7) = 0$$
$$(6x + 7)^2 = 0$$
$$6x + 7 = 0$$
$$x = -\frac{7}{6}$$

Solution set: $\left\{-\dfrac{7}{6}\right\}$

20.

$$5x^2 - 2 = 0$$
$$5x^2 = 2$$
$$x^2 = \frac{2}{5}$$
$$x = \pm\sqrt{\frac{2}{5}} = \pm\frac{\sqrt{10}}{5}$$

#20 continued

Solution set: $\left\{\pm\dfrac{\sqrt{10}}{5}\right\}$

21

$\dfrac{x^2}{2} + 2x + \dfrac{2}{3} = 0$

$3x^2 + 12x + 4 = 0$ Multiply by 6

$a = 3, \ b = 12, \ c = 4$

$x = \dfrac{-12 \pm \sqrt{144 - 48}}{2 \cdot 3}$

$x = \dfrac{-12 \pm \sqrt{96}}{6} = \dfrac{-12 \pm 4\sqrt{6}}{6}$

$x = \dfrac{-6 \pm 2\sqrt{6}}{3}$ Reduce

Solution set: $\left\{\dfrac{-6 \pm 2\sqrt{6}}{3}\right\}$

22.

$$6x^2 = -x + 1$$
$$6x^2 + x - 1 = 0$$
$$(2x + 1)(3x - 1) = 0$$
$$2x + 1 = 0 \ \text{ or } \ 3x - 1 = 0$$
$$x = -\dfrac{1}{2} \qquad x = \dfrac{1}{3}$$

Solution set: $\left\{-\dfrac{1}{2}, \ \dfrac{1}{3}\right\}$

23.

Let x and x + 2 = two consecutive positive even integers.

Thus, $x(x + 2) = 48$

$x^2 + 2x - 48 = 0$

$(x - 6)(x + 8) = 0$

$x - 6 = 0$ or $x + 8 = 0$

     $x = 6$ $\cancel{x = -8}$

     $x + 2 = 8$

Answer: 6 and 8

24.

$h = -16t^2 + 128t$

Let $h = 0$

#24 continued

$0 = -16t(t - 8)$

     $-16t = 0$ or $t - 8 = 0$

        $t = 0$ $t = 8$

Answer: after 8 seconds

25.

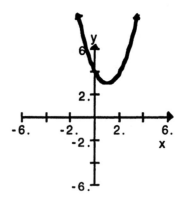

Vertex: (1, 3).

26.

$y = x^2 + 4x + 7$

$y = x^2 + 4x + 4 + 7 - 4$

$y = (x + 2)^2 + 3$

Vertex: (-2, 3).

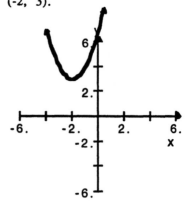

APPENDIX: SELECTED TOPICS FROM GEOMETRY

Solutions to Text Exercises

#41 continued

Exercises A.1

1–41.

The answers do not require worked-out solutions.

Exercises A.2

1–45.

The answers do not require worked-out solutions.

Exercises A.3

1–19.

The answers do not require worked-out solutions.

21.
$94° + 94° + 140° + 165° + 116° = 609°$
m ($\angle$ A) = $720° - 609° = 111°$

23.
$130° + 124° + 88° + 97° = 439°$
m ($\angle$ 0) = $540° - 439° = 101°$

25.
$4(152°) + 4(135°) = 608° + 540° = 1148°$
m ($\angle$ A) = $1260° - 1148° = 112°$

27–37.
The answers do not require worked-out solutions.

39.
m $\angle$ C is equal in measurement to $\angle$ Y.
m $\angle$ Y = m $\angle$ C = $360° - 2(53°)$
$\qquad = 254° + 2 = 127°$

41.
$\dfrac{15}{35} = \dfrac{18}{RS}$
15 RS = 35 · 18 $\qquad$ Cross multiply
15 RS = 630
$\quad$ RS = 42

_____

#41 continued

$\dfrac{15}{35} = \dfrac{21}{TU}$
15TU = 735 $\qquad$ Cross multiply
$\quad$ TU = 49

43.
$\dfrac{64}{28} = \dfrac{40}{IJ}$
64IJ = 28(40) $\qquad$ Cross multiply
64IJ = 1120
$\quad$ IJ = 17.5 = KL

_____

$\dfrac{56}{64} = \dfrac{JK}{40}$
64JK = 56(40) $\qquad$ Cross multiply
64JK = 2240
$\quad$ JK = 35

45.
$\dfrac{92}{70} = \dfrac{34.5}{ZY}$
92YZ = 34.5(70) $\qquad$ Cross multiply
92YZ = 2415
$\quad$ YZ = 26.25 = AB
$\quad$ YB = 34.5

Exercises A.4

1–23.
The answers do not require worked-out solutions.

25.

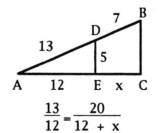

$\triangle$ ABC ~ $\triangle$ ADE

$\dfrac{13}{12} = \dfrac{20}{12 + x}$

13(12 + x) = 12(20)

156 + 13x = 240

$\qquad$ 13x = 84

#25 continued

$$x \approx 6.46$$

Answer: About 6.5 feet

27.

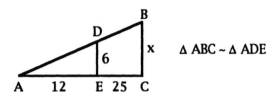

$\Delta ABC \sim \Delta ADE$

$$\frac{6}{12} = \frac{x}{37}$$

$$12x = 6(37)$$

$$12x = 222$$

$$x = 18.5$$

Answer: 18.5 feet

29.

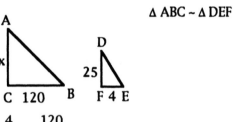

$\Delta ABC \sim \Delta DEF$

$$\frac{4}{25} = \frac{120}{x}$$

$$4x = 3000$$

$$x = 750$$

Answer: 750 feet

31.

$\Delta QRS \sim \Delta QTR \sim \Delta RTS$

$$\frac{16}{RT} = \frac{RT}{4}$$

$$4(16) = RT^2$$

$$64 = RT^2$$

$$8 = RT$$

33.

$$\frac{14}{YB} = \frac{8}{14}$$

$$14^2 = 8YB$$

$$196 = 8YB$$

$$24.5 = YB$$

35.

$$\frac{5}{7} = \frac{13}{7 + KJ}$$

$$5(7 + KJ) = 7(13)$$

$$35 + 5KJ = 91$$

$$5KJ = 56$$

$$KJ = 11.2$$

Exercises A.5

1-5.
The solutions involve single formulas.

7.
Perimeter = $\frac{1}{2}$ circumference of top half circle plus three sides of the bottom rectangle.

$$P = \frac{1}{2} \cdot \Pi \cdot 8 + 8 + 2 \cdot 6$$

$$P = 4\Pi + 8 + 12 = 4\Pi + 20 \text{ ft} \approx 32.56 \text{ ft}$$

$$\text{Area} = 6 \cdot 8 + \frac{1}{2} \cdot \Pi \cdot 4^2 = 48 + 8\Pi \text{ ft}^2$$

$$\approx 73.12 \text{ ft}^2$$

9.
For perimeter:
There are four 5 yds, or 20 yards. There are eight 2 yds or 16 yds.

So, P = 20 + 16 = 36 yds.

For area:
There are four 2 yd x 5 yd rectangles.
Total area = 40 yd^2. There is one large middle 5 yd by 5 yd rectangle:

Area = 25 yd^2
So, A = 40 + 25 = 65 yd^2

11.
P = perimeter of two sides of the rectangle plus circumference of $\frac{1}{2}$ a circle.

$$P = 2 \cdot 12 + 8 + \frac{1}{2} \Pi \cdot 8$$

$$P = 24 + 8 + 4\Pi = 32 + 4\Pi \text{ in.} \approx 44.56 \text{ in.}$$

A = area of the total rectangle less area of the half circle.

$$A = 8 \cdot 12 - \frac{1}{2} \pi \cdot 4^2 = 96 - 8\pi \text{ in}^2 \approx 70.88 \text{ in}^2$$

13.

$P = 16 + 16 + \Pi \cdot 4 = 32 + 4\Pi \text{ ft} \approx 44.56 \text{ ft}$

$A = 4 \cdot 16 + \Pi \cdot 2^2 = 64 + 4\Pi \text{ ft}^2 \approx 76.56 \text{ ft}^2$,
where r = 2.

15.

$P = 20 + \frac{1}{2}\Pi \cdot 20 = 20 + 10\Pi \text{ yd} \approx 51.4 \text{ yd}$

$A = \frac{1}{2}\Pi \cdot 10^2 = 50\Pi \text{ yds}^2 \approx 157 \text{ yd}^2$

17.

$A = 12^2 - 3.14 \cdot 6^2 = 144 - 113.04 \approx 30.96 \text{ m}^2$

19.

$A = 8 \cdot 10 - \frac{1}{2} \, 3 \cdot 5 = 80 - 7.5 = 72.5 \text{ yd}^2$

21.

$A = 10 \cdot 3 - \frac{1}{2} \cdot 10 \cdot 3 = 30 - 15 = 15 \text{ mm}^2$

23.

$A = 5 \cdot 6 - 2 \cdot 4 = 30 - 8 = 22 \text{ in}^2$

25.

$A = \Pi \cdot 8^2 - \Pi \cdot 4^2 = 64\Pi - 16\Pi$

$\qquad = 48\Pi \text{ in}^2$

$\qquad \approx 150.72 \text{ in}^2$

27.

$\quad b = 6, h = 2$

$A = \frac{1}{2} \cdot 6 \cdot 2 = 6$

29.

Step 1: Find area of pool hall.

$\quad A = 40 \cdot 75 = 3,000 \text{ ft}^2$

Step 2: Find the cost.

$\quad \text{Cost} = 3,000 \cdot 1.20 = \$3600$

31.

The less expensive fencing will be 140 ft. The more expensive fencing will be 80 ft. The less expensive fencing will cost 140 · 2 or $280.

#31 continued

The more expensive fencing will cost 80 · 5 = $400.

Total cost: 280 + 400 = $680.

Exercises A.6

1.

$SA = 6s^2 = 6 \cdot 4^2 = 96 \text{ m}^2$

$V = s^3 = 4^3 = 64 \text{ m}^3$

3.

$SA = 4\Pi r^2 = 4\Pi \cdot 12^2 = 576\Pi \text{ yd}^2$

$\qquad\qquad\qquad \approx 1808.64 \text{ yd}^2$

$V = \frac{4}{3}\Pi r^3 = \frac{4}{3} \cdot \Pi \cdot 12^3 = 2304 \; \Pi \cdot \text{yd}^3$

$\qquad\qquad\qquad \approx 7234.56 \text{ yd}^3$

5.

$SA = 2\ell w + 2\ell h + 2hw$

$SA = 2 \cdot 2 \cdot 7 + 2 \cdot 2 \cdot 1 + 2 \cdot 1 \cdot 7$

$SA = 28 + 4 + 14 = 46 \text{ mm}^2$

$V = \ell wh = 2 \cdot 7 \cdot 1 = 14 \text{ m}^3$

7.

$SA = 2\Pi r^2 + 2\Pi rh$

$SA = 2\Pi \cdot 6^2 + 2\Pi \cdot 6 \cdot 12 = 72\Pi + 144\Pi$

$\qquad\qquad\qquad \approx 678.24 \text{ m}^2$

$V = \Pi r^2 h = \Pi \cdot 6^2 \cdot 12 = 432\Pi \text{ m}^3 \approx 1356.48 \text{ m}^3$

9.

$SA = \frac{4\Pi r^2}{2} + \Pi r^2 = \frac{1}{2} \cdot 4\Pi \cdot 12^2 + \Pi \cdot 12^2$

$\quad = 288\Pi + 144\Pi$

$\quad = 432\Pi \text{ ft}^2 \approx 1356.48 \text{ ft}^2$

$V = \dfrac{\frac{4}{3}\Pi r^3}{2} = \frac{1}{2} \cdot \frac{4}{3} \cdot \Pi \cdot 12^3 = \frac{2304}{2}\Pi$

$\qquad\qquad\qquad = 1152\Pi \text{ ft}^2$

$\qquad\qquad\qquad \approx 3617.28 \text{ ft}^3$

11.

$SA = 6s^2 = 6 \cdot 16^2 = 1536$ mm

$V = s^3 = 16^3 = 4096$ mm^3

13.

$SA = 2\ell w + 2\ell h + 2hw$

$= 2 \cdot 12 \cdot 2 + 2 \cdot 12 \cdot 3 + 2 \cdot 3 \cdot 2$

$= 48 + 72 + 12 = 132$ in^2

$V = \ell \cdot w \cdot h$

$= 12 \cdot 2 \cdot 3 = 72$ in^3

15.

$SA = \Pi r^2 + 2\Pi rh + \dfrac{4\Pi r^2}{2}$

$= \Pi 2^2 + 2\Pi \cdot 2 \cdot 8 + \dfrac{4\Pi \cdot 2^2}{2}$

$= 4\Pi + 32\Pi + 8\Pi$

$= 44\Pi$ ft$^2 = 138.16$ ft^2

$V = \Pi r^2 h + \dfrac{\frac{4}{3}\Pi r^3}{2}$

$= \Pi \cdot 2^2 \cdot 8 + \dfrac{\frac{4}{3}\Pi \cdot 2^3}{2}$

$= 32\Pi + \dfrac{16\Pi^2}{3}$

$= \dfrac{96\Pi}{3} + \dfrac{16\Pi}{3} = \dfrac{112\Pi}{3}$ ft^3

$= 117.23$ ft^3

17.

$SA = 2\Pi rh + 4\Pi r^2$

$= 2\Pi \cdot 9 \cdot 12 + 4\Pi \cdot 9^2$

$= 216\Pi + 324\Pi = 540\Pi$ mm^2

≈ 1695.6 mm^2

$V = \Pi r^2 h + \dfrac{4}{3}\Pi r^3$

$= \Pi \cdot 9^2 \cdot 12 + \dfrac{4}{3}\Pi + 9^3$

$= 972\Pi + 972\Pi = 1944\Pi$ mm^3

≈ 6104.16 mm^3

19.

$SA = 2\ell w + 2\ell h + 2hw$

$= 2 \cdot 16 \cdot 7 + 2 \cdot 16 \cdot 6 + 2 \cdot 6 \cdot 7$

$= 224 + 192 + 84 = 500$ in^2

$V = \ell \cdot w \cdot h$

$V = 16 \cdot 7 \cdot 6 = 672$ in^3

21.

Step 1: Find surface area of all sides except the top and bottom.

$SA = 2\ell h + 2hw$

$= 2 \cdot 4 \cdot 3 + 2 \cdot 3 \cdot 4$

$= 24 + 24 = 48$ ft^2

Step 2 Find surface area of the top and bottom.

SA $2\ell w = 2 \cdot 4 \cdot 4 = 32$ ft^2

Step 3 Find the cost.

Cost $= 48 \cdot 2 + 32 \cdot 3 \cdot 50$

$= 96 + 112 = \$208$

23.

The box has measurements $\ell = 26$ cm, $w = 12$ cm, $h = 3$ cm.

$V = \ell \cdot w \cdot h$

$= 26 \cdot 12 \cdot 3 = 936$ cm^3

$SA = \ell w + 2\ell h + 2hw$

$= 26 \cdot 12 + 2 \cdot 26 \cdot 3 + 2 \cdot 3 \cdot 12$

$= 312 + 156 + 72 = 540$ cm^2

25.

$SA = 2\Pi rh + 4\Pi r^2$

$= 2\Pi \cdot 9 \cdot 40 + 4\Pi \cdot 9^2$

$= 720\Pi + 324\Pi = 1044\Pi$ yd^2

≈ 3278.16 yd^2

$V = \Pi r^2 h + \dfrac{4}{3}\Pi r^3$

$= \Pi \cdot 9^2 \cdot 40 + \dfrac{4}{3}\Pi \cdot 9^3$

$= 3240\Pi + 972\Pi = 4212\Pi$ yd^3

$\approx 13{,}225.68$ yd^3

Exercises A.7

1.

$c^2 = a^2 + b^2$, a = 3 in, b = 4 in
$c^2 = 3^2 + 4^2$
$c^2 = 9 + 16 = 25$
  c = 5 in.

3.

$a^2 + b^2 = c^2$, b = 8 ft., c = 17 ft
$a^2 + 8^2 = 17^2$
    $a^2 = 289 - 64$
    $a^2 = 225$
     a = 15 ft

5.

$c^2 = a^2 + b^2$, a = 10 yd, b = 24 yd
$c^2 = 10^2 + 24^2$
$c^2 = 100 + 576 = 676$
  c = 26 yd

7.

$a^2 + b^2 = c^2$, b = 8 m, c = 17 m
$a^2 + 64 = 289$
    $a^2 = 225$
     a = 15 m

9.

$a^2 + b^2 = c^2$, a = 12 cm, c = 13 cm
$144 + b^2 = 169$
     $b^2 = 25$
      b = 5 cm

11.

$c^2 = a^2 + b^2$, a = 9 mm, b = 12 mm
       $c^2 = 81 + 144$
       $c^2 = 125$
        c = 15 mm

13.

$c^2 = a^2 + cb^2$, a = 3 m, b = 3 m
     $c^2 = 9 + 9$
     $c^2 = 18$
      c = $3\sqrt{2}$ m ≈ 4.243 m

15.

$a^2 + b^2 = c^2$, a = $\sqrt{6}$ cm, c = $\sqrt{13}$ cm
  $6 + b^2 = 13$
     $b^2 = 7$
      b = $\sqrt{7}$ cm ≈ 2.646 cm

17.

$c^2 = a^2 + b^2$, a = $\sqrt{7}$ mm, b = $2\sqrt{5}$
$c^2 = 7 + 4 \cdot 5$
$c^2 = 7 + 20 = 27$
 c = $3\sqrt{3}$ mm ≈ 5.196 mm

19.

$a^2 + b^2 = c^2$, b = 2 in, c = $\sqrt{53}$ in
$a^2 + 4 = 53$
   $a^2 = 49$
    a = 7 in

21.

$a^2 + b^2 = c^2$, a = 6 ft., c = $6\sqrt{2}$ ft
$36 + b^2 = 36 \cdot 2$
     $b^2 = 72 - 36$
     $b^2 = 36$
      b = 6 ft

23.

$c^2 = a^2 + b^2$, a = $2\sqrt{3}$ yd., b = 2 yd
$c^2 = 4 \cdot 3 + 4$
$c^2 = 16$
 c = 4 yd

25.

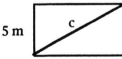

Let a = 5 m, 6 = 12 m.
 $c^2 = a^2 + b^2$
 $c^2 = 25 + 144$
 $c^2 = 169$
  c = 13 m

27.

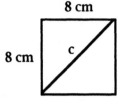

Let $a = b = 8$ cm
$$c^2 = a^2 + b^2$$
$$c^2 = 64 + 64$$
$$c^2 = 128$$
$$c = 8\sqrt{2} \text{ cm} \approx 11.314$$

29.

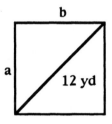

$a = b$, $c = 12$ yd, Area $= a \cdot a = a^2$
$$a^2 + b^2 = c^2$$
$$a^2 + a^2 = 144 \qquad \text{Since } b = a$$
$$2a^2 = 144$$
$$a^2 = 72 \text{ yd}^2$$

31.

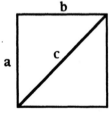

$a = b$, so $a^2 = 36$cm^2 and $a = 6$ cm
Thus, $\quad c^2 = a^2 + b^2$
$$c^2 = a^2 + a^2 \quad \text{since } b = a$$
$$c^2 = 36 + 36$$
$$c^2 = 72$$
$$c = 6\sqrt{2} \text{ cm} \approx 8.485 \text{ cm}$$

33.

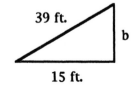

#33 continued

$a^2 + b^2 = c^2$, $a = 15$ ft, $c = 39$ ft, find b.
$$225 + b^2 = 1521$$
$$b^2 = 1296$$
$$b = 36 \text{ ft}$$

35.

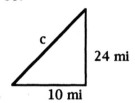

$c^2 = a^2 + b^2$, $a = 10$ mi. $6 = 24$ mi, find c.
$$c^2 = 100 + 576$$
$$c^2 = 676$$
$$c = 26 \text{ mi}$$

Exercises A.8

1–23.
Answers do not require worked-out solutions.

25.

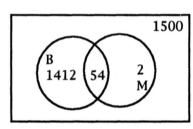

$1500 - (1412 + 54 + 2) = 32$

27.

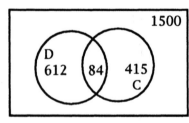

$1500 - (612 + 84 + 475) = 389$

29.
Beth, Ivy, Sue, Karen

Exercises A.9

1–49.
The exercises do not require worked-out solutions.
